Power Systems

Electrical power has been the technological foundation of industrial societies for many years. Although the systems designed to provide and apply electrical energy have reached a high degree of maturity, unforeseen problems are constantly encountered, necessitating the design of more efficient and reliable systems based on novel technologies. The book series Power Systems is aimed at providing detailed, accurate and sound technical information about these new developments in electrical power engineering. It includes topics on power generation, storage and transmission as well as electrical machines. The monographs and advanced textbooks in this series address researchers, lecturers, industrial engineers and senior students in electrical engineering.

Power Systems is indexed in Scopus

More information about this series at http://www.springer.com/series/4622

José Carlos Goulart de Siqueira ·
Benedito Donizeti Bonatto

Introduction to Transients in Electrical Circuits

Analytical and Digital Solution Using
an EMTP-based Software

 Springer

José Carlos Goulart de Siqueira
Institute of Electrical Systems and Energy
Federal University of Itajubá
Itajubá, Brazil

Benedito Donizeti Bonatto ⓘ
Institute of Electrical Systems and Energy
Federal University of Itajubá
Itajubá, Brazil

ISSN 1612-1287   ISSN 1860-4676 (electronic)
Power Systems
ISBN 978-3-030-68251-4   ISBN 978-3-030-68249-1 (eBook)
https://doi.org/10.1007/978-3-030-68249-1

This Springer imprint is published by the registered company Springer Nature Switzerland AG
The registered company address is: Gewerbestrasse 11, 6330 Cham, Switzerland

Preface

The book *Introduction to Transients in Electrical Circuits: Analytical and Digital Solution Using an EMTP-based Software* integrates the analytical solution with the digital solution through the ATP—Alternative Transients Program software (recognized for its use all over the world in academia and in the electric power industry) in a didactic approach and mathematical and scientific rigour suitable for undergraduate and graduate programs, as well as for industry professionals.

It is clear that many other classical and advanced books indeed cover more extensive subjects in electromagnetic transients. This book fills a gap between classic works in the field of electrical circuits and more advanced works in the field of transients in electrical power systems, facilitating a full understanding of analytical and digital modelling and solution of transients in basic circuits. Therefore, it fulfils its role as an introductory book, but completes in its learning proposal in undergraduate and graduate courses. Finally, the book presents a significant number of examples and proposed problems, all with answers, without exception. The book fills an important gap between the analytical solution of traditional electrical circuits with the digital solution using a well-established tool (the ATP program) in the power engineering community. It may be particularly important when teaching some basic power system transients that can be simulated with linear networks and lumped parameters.

This book presents an approach to solving singular function differential equations representing the transient and steady-state dynamics of a circuit in a structured manner and without the need for physical reasoning to set initial conditions to zero plus (0+). The method indeed facilitates the mathematical solution. That does not mean that physical reasoning and proper modelling are unimportant in engineering education. Real applications may come after a strong background in these concepts. It also presents, for each presented problem, the exact analytical solution as well as the corresponding digital solution through a computer program based on the EMTP —Electromagnetics Transients Program. The andragogical approach (science and art in adult education) integrates existing knowledge and starts from the simple to the complex, with rich exposure to a variety of problems solved (all with tested and validated solutions).

The teacher and the apprentice are encouraged to use modern computational tools used in the electrical power industry for the calculation of electromagnetic transients, but are warned about their limitations and modelling constraints. This integrates the "know-how" and "know why" necessary for the safe analysis of simulated results. The teacher can direct students in solving specific problems in the textbook or even propose that students present real day-to-day problems for solution, as the introductory methodology for modelling and calculation of digital and analytical solution of circuit transients in electrical systems are fully presented in the textbook. Technical–scientific articles should be consulted, as well as textbooks for more advanced studies on transients in electrical power systems can be consulted, depending on the interest and need of the problem to be solved.

It is important to clarify that the motivation and intent primacy of the main author, Prof. José Carlos Goulart de Siqueira, are to leave a legacy to the teachers and students of UNIFEI, where he magistrate with excellence for decades, having been the first rector of UNIFEI-Federal University of Itajubá, which has been completed, on November 23rd 2020, 107 years of history of contributions to higher education in Brazil.

We think that the book is intended for engineering students who already have been successfully approved in introductory circuit analysis, calculus and ordinary differential equations courses. Our experience as educators is that, when it comes to truly understand and apply the mathematical knowledge in electrical circuits transients, the majority of students have much more difficulties. Therefore, this book supplies, in an integrated way, all the necessary knowledge and practice through extensive examples.

This book is organized as follows: Chap. 1 presents an introduction to transients in electrical circuits with a discussion about fundamentals of circuit analysis, physical phenomena and the need for mathematical modelling and simulation. The student is encouraged from the beginning to become familiar with Appendix A—processing at the ATP. Chapter 2 presents singular functions for the analytical solution, and Appendix B shows the main relations involving singular functions. Chapter 3 presents the solution of differential equations. It emphasizes the solution using the classical method in the time domain. For the operational method in the complex frequency domain using the Laplace transform, Appendix C—Laplace transform properties, Appendix D—Laplace transform pairs and Appendix E—Heaviside expansion theorem can be helpful.

Chapter 4 presents the digital solution of transients in basic electrical circuits. The fundamental algorithm of EMTP-based program is introduced, and the problem of numerical oscillations due to the trapezoidal method is briefly discussed. Chapter 5 presents transients in first-order circuits. Extensive number of examples are provided with their analytical and digital solution using the ATPDraw. Chapter 6 presents transients in circuits of any order, exploring the solution methods consolidated so far. Finally, Chap. 7 introduces switching transients using the injection of sources method. All chapters provide useful references to enhance the learning process, as well as to instigate further investigation about advanced topics. Enjoy it!

Itajubá, Brazil

José Carlos Goulart de Siqueira
Benedito Donizeti Bonatto

Acknowledgments We would like to express our gratitude to many people who have helped to bring this book project to its conclusion. We deeply thank all members of our families, friends (Antonio Eduardo Hermeto…), colleagues who by listening and encouraging have provided the fundamental support and personal care. We give a special thanks to Alexa Bonelli Bonatto and Aline Bonelli Bonatto for their work in the initial translation and editing. Professional editing was provided by Springer's team to whom we are deeply gratefull. Finally, we thank many students who were challenged in their studies during their courses at UNIFEI by the many proposed problems.

Contents

List of Figures

List of Tables

Chapter 1
Introduction to Fundamental Concepts in Electric Circuit Analysis

1.1 Introduction

This opening chapter has two main objectives. The first of them, which has a strictly revisionary character, is to approach fundamental concepts of physics, as well as laws, equations and formulas of electricity and electromagnetism. The second is to anticipate, through an example, the real raison d'être of this book, which is the solution, both analytical and digital, of transient phenomena in linear, invariant and concentrated circuits. This chapter presents then the main elements of an electrical circuit with concentrated parameters that do not vary with the frequency, as well as the dynamic response of basic circuits when disturbances occur, such as energizing or de-energizing switches, enabling a physical understanding of the phenomena and the need for mathematical modelling and computer simulations. However, it is assumed that the reader already has fundamental knowledge of analysis of electrical circuits, laws, theorems and methods of solution, as well as of solving ordinary differential equations. Thus, this chapter aims to create the motivations for the study of mathematical and computational techniques and solutions presented in the following chapters.

The challenges associated with overvoltages and overcurrents caused by external and internal disturbances to the electrical networks require an understanding of the phenomena, their adequate modelling, simulation and interpretation of the results in the search for solutions that minimize the damage to equipment and that in certain more critical cases, are causing major interruptions (blackouts), with catastrophic social and economic impacts. Fundamental knowledge on circuit analysis [1–9] is essential for a better understanding of the growing complexity of modern power systems.

In the emerging scenario of greater penetration of renewable and distributed energy sources (DER—Distributed Energy Resources) in current electrical systems, which tend to become more intelligent electrical networks (Smart Grids), the challenges are even greater and more complex, as they also require considering the

© The Author(s), under exclusive license to Springer Nature Switzerland AG 2021
J. C. Goulart de Siqueira and B. D. Bonatto, *Introduction to Transients in Electrical Circuits*, Power Systems, https://doi.org/10.1007/978-3-030-68249-1_1

phenomena and effects of integrated power electronics, telecommunications and control systems on the performance of the interconnected power grid. The complete understanding of the modelling and simulation of transients in basic electrical circuits is fundamental to the approach aiming at understanding phenomena in more complex circuits, such as, for example, power electronics converter circuits (DC/DC, DC/AC, AC/AC) used at the interfaces between renewable sources and electrical systems, or transient in HVDC or AC transmission lines in high or extra-high voltage, or transients in electrical machines (transformers, generators and motors) including effects of couplings as well as effects of non-linearity (due to lightning arresters, surge arresters, saturation in transformers, etc.), and also the consideration of the variation of parameters (resistance, inductance) with frequency. For these and several complex studies, there are very advanced technical-scientific books and articles available in the world literature.

There are indeed classical books and new ones covering more detailed and extensive subjects about electromagnetic transients in electric power systems. This book fills a gap between classic works in the field of electrical circuits and more advanced works in the field of transients in electrical power systems, facilitating a full understanding of digital and analytical modelling and solution of transients in basic circuits. Therefore, it fulfils its role as an introductory book, but complete in its learning proposal in undergraduate and graduate courses.

In Sect. 1.2, the concepts of linear, analogous, dual, time-invariant and concentrated systems are reviewed. In Sect. 1.3, under the generic title of electrical quantities, fundamental ideas associated with the electric charge are reviewed—Coulomb's law and the concept of electrostatic field are discussed. The important concepts of work and electrostatic potential are remembered, defining, next, a difference in potential or voltage. Then, this section ends dealing with the electric current in all its aspects. In Sect. 1.4, the important concepts of power and energy are reviewed, emphasizing the conventions used to distinguish the supply and absorption of power and energy.

In Sect. 1.5, under the title of circuit elements, the electrical resistance is initially addressed—in addition to Ohm's law, the issue of increasing resistance with increasing temperature in conductive materials is addressed, as well as decreasing it in semiconductor materials. Then a detailed survey is made on capacitors and capacitance, including by discussing, throughout an example, the issue of the between voltage and current for sinusoidal signals, in addition to the issue of polarization of the capacitor dielectric. Before focusing on inductance, a wide approach is made to permanent magnets, taking advantage of the opportunity to introduce the concept of magnetic field and define the magnetic induction vector B and the magnetic flux Φ. Then, Oersted's experience and Faraday's law of induction are discussed, as well as Lenz's law, used to determine the polarity of the induced electromotive force. Only then is the inductance defined, initially on a coil and then generalizing. Closing the section, comments are made on the voltage and current sources, both independent, controlled or linked.

Section 1.6 is dedicated to Kirchhoff's laws, currents and voltages, discussing their limitations on the applicability to concentrated for elements, as well as on the issue of obtaining linearly independent equations, taking advantage of the definition of branch, node, loop and mesh. In Sect. 1.7, a complete example of energizing a capacitance, linear and invariant over time, it is presented by a constant voltage source, that is, by a battery. The analytical solution is initially presented in great detail using the nodal method to obtain voltage, current, power and energy in all elements of the circuit. Then, the corresponding digital solution of the same circuit is made using the software ATPDraw, for the given numerical values. Finally, Sect. 1.8 presents the conclusions with the main motivations for electromagnetic transients' studies in modern power systems analysis.

1.2 Preliminary Concepts

System

A system is a collection of different interconnected components subject to one or more inputs, also called stimuli or excitations, and which produce one or more outputs, also called responses. In Fig. 1.1a is represented, symbolically, a system subject to an input x(t) and that produces an output y(t). In Fig. 1.1b, as an example, a mechanical system composed of mass, spring and friction is shown, in which the only input is the applied force, and whose output may be the speed resulting from the mass or the potential energy stored in the spring. Note, therefore, that the term output refers to the magnitude to be observed in the dynamics of the system.

Model

In the development of mathematical methods of analysis of physical systems two steps are fundamental. The first is the mathematical description of the individual components or elements of the system. The second is the establishment of a mathematical description of the effect of the interconnection of the different components, formulating the so-called interconnection laws—in the case of electrical circuits, these are Kirchhoff's laws. Thus, a given physical system can then be described, approximately, by an idealized model or system. After that, the input–output relationships can then be established, using known mathematical methods.

Linear System

Systems can be classified by placing restrictions on the input–output relationship. A system is called linear when it satisfies two principles. The first is

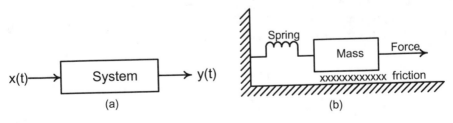

Fig. 1.1 **a** System input and output; **b** Mechanical system composed of mass, spring and friction

that of homogeneity, and the second is that of superposition. The homogeneity principle states that:

> When the input $x(t)$ of a system is multiplied by a factor k, then the output $y(t)$ is multiplied by the same factor k.

And the principle of superposition establishes the following:

> If two entries are applied, simultaneously or not, to the same system, then the total response will be the sum of the individual responses to each of the entries.

If $y_1(t)$ and $y_2(t)$ denote the responses corresponding to the independent entries $x_1(t)$ and $x_2(t)$, then a system will be linear if, and only if, the response to

$$x(t) = k_1 x_1(t) + k_2 x_2(t) \tag{1.1}$$

is

$$y(t) = k_1 y_1(t) + k_2 y_2(t), \tag{1.2}$$

for all entries and for all values of the constants k_1 and k_2.

Note, for example, that for $k_2 = 0$ Eqs. 1.1 and 1.2 speak of homogeneity. And that when $k_1 = k_2 = 1$ they speak of superposition. Note, too, that linearity does not necessarily imply that the output time function must have the same waveform as the input. In another way, it can be said that a linear system is one whose behavior is described by linear equations, whether algebraic equations, or differential equations. Therefore, a non-linear system will always be described by non-linear equations, which are much more difficult to solve. The system described by $y(t) = x(t).[dx(t)/dt]$, for example, is non-linear, because when doubling the input $x(t)$ the output $y(t)$ is quadrupled, contrary to the property of homogencity.

Analog and Dual Systems

Sometimes the same equation describes a number of different physical situations. In such cases, the different physical systems are called analogues. Therefore, two or more systems are analogous when they are described by the same equation. When, however, the analogy occurs between physical systems of the same nature, such as two electrical circuits for example, although they have different physical functioning, they are called dual. Therefore, duality means analogy between two physical systems of the same nature. The systems described by Eqs. 1.3 and 1.4 are analogous, because although the letters are different, mathematically they are saying the same thing, namely: that one quantity is proportional to the derivative of the other. But the systems described by Eqs. 1.4 and 1.5 are dual, since in both the quantities inserted are electric.

$$f(t) = M\frac{d}{dt}v(t) \tag{1.3}$$

$$e(t) = L\frac{d}{dt}i(t) \tag{1.4}$$

$$i(t) = C \frac{d}{dt} e(t) \tag{1.5}$$

Time Invariant or Fixed System

Delaying a function $f(t)$ of a value t_0 corresponds to shifting its waveform (curve) to the right of a value equal to t_0 and replacing the independent variable t, in the expression that describes it, by $t - t_0$. An input system $x(t)$ and output $y(t)$ is called a time-invariant or fixed if its input–output relationship is independent of time. That is, if your input $x(t)$ is delayed by t_0, your output $y(t)$ will simply be delayed by the same t_0, but maintaining its waveform and amplitude. Thus, for a time-invariant system, the response to input $x(t - t_0)$ must be $y(t - t_0)$, as long as the response to $x(t)$ is $y(t)$. The system described by the input–output ratio $y(t) = t \frac{d}{dt} x(t)$ is not invariant, since the output corresponding to input $x(t - t_0)$ is given by

$$
\begin{aligned}
y_1(t) = t \frac{d}{dt} x(t - t_0) &= t \frac{dx(t - t_0)}{d(t - t_0)} \cdot \frac{d(t - t_0)}{dt} = (t - t_0 + t_0) \frac{dx(t - t_0)}{d(t - t_0)} \cdot (1 - 0) \\
&= (t - t_0) \frac{dx(t - t_0)}{d(t - t_0)} + t_0 \cdot \frac{dx(t - t_0)}{d(t - t_0)} = y(t - t_0) + t_0 \cdot \frac{dx(t - t_0)}{d(t - t_0)} \neq y(t - t_0).
\end{aligned}
$$

In time-varying systems, the form and amplitude of the response depends on the instant in which the input is applied. Thus, the response to $x(t - t_0)$ is not $y(t - t_0)$.

Concentrated Systems and Distributed Systems

Obviously, systems with concentrated parameters are those consisting of concentrated elements; the same can be said of systems with distributed parameters. The fundamental property associated with the concentrated elements is their small size, compared to the wavelength λ of their natural frequency of operation $f-$ with $v = \lambda f$, where v is the wave propagation speed. As $T = 1/f$, it can also be written that $T = \lambda/v$, with the period T corresponding to the time necessary for the wave to propagate a distance exactly equal to the wavelength λ. Thus, for a circuit element of length l the propagation time will be τ, so that $\tau = l/v$. **Therefore, the condition for a circuit element of length l to be considered as concentrated is that $l \ll \lambda$, or that $\tau \ll T$.**

In other words, a circuit element is considered to be concentrated when the propagation time of the electrical signal through it can be considered negligible. This means that the actions are instantaneous; there is simultaneity. At any moment, whatever is happening at the beginning of a concentrated circuit element will be happening at the end. In a circuit with concentrated parameters, therefore, the electrical quantities are functions exclusively of time t, and the equations obtained from it are **ordinary differential equations**. In the case of systems with distributed parameters, as in electrical transmission lines, when propagation time cannot be neglected, and unidirectional propagation is assumed, electrical quantities are functions of distance and time, respectively x and t. Therefore, the equations obtained are **partial differential equations**.

In summary, **what defines whether a circuit element is concentrated or distributed is its size and the frequency of the signal to which it is subjected, for the same propagation speed**. A small cable $l = 1$ [m] in length transmitting a sinusoidal signal at frequency $f = 500$ [MHz], with a propagation speed close to that of light, $v = 300,000$ [km/s] $= 3 \times 10^8$ [m/s], corresponding to a wavelength $\lambda = v/f = 3 \times 10^8/500 \times 10^6 = 0.6$ [m], cannot be treated as a concentrated element, since $l > \lambda$. But, itself, with a signal at frequency $f = 60$ [Hz] and wavelength $\lambda = v/f = 3 \times 10^8/60 = 5,000$ [km] must be treated as a concentrated component, because $l \ll \lambda$.

1.3 Electrical Quantities

Electric Charge

The existence of the electric charge dates from approximately 600 years B.C., when the Greeks observed that amber rubbed with wool could attract small light objects, such as pieces of dry straw, for example.

Matter, which is something that has volume and mass, is made up of very small particles called atoms. Every atom is made up of a nucleus and its surroundings, called the electrosphere. Protons and neutrons are located in the nucleus. In the electrosphere, electrons. **The electron is the fundamental negative charge of electricity**, since it is indivisible. **The proton is the fundamental positive charge of electricity**. And the neutron, as the name implies, is the fundamental neutral charge of electricity. In their natural state, the atom of each chemical element contains an equal number of electrons and protons, called an atomic number. Since the negative charge of each electron has the same absolute value as the positive charge of each proton, the totals of positive and negative charges are canceled. And an atom in this condition is called electrically neutral or in equilibrium.

Electrons orbit the nucleus of the atom in what are called quantized layers of energy, for a total of seven. They are the layers K, L, M, N, O, P, Q. Each energy layer of an atom can contain only a certain maximum number of electrons. In a copper atom, for example, with atomic number 29, 29 protons are electrically neutralized by 29 electrons, which fill the K layer with 2 electrons and the L layer with 8 electrons. The remaining 19 electrons fill the M layer with 18, leaving 1 electron in the N layer. In an equilibrium or stable atom, the amount of energy is equal to the sum of the energies of its electrons. And **the energy level of an electron is proportional to its distance from the nucleus**. Therefore, the energy levels of electrons in layers farther from the atom's nucleus are higher than those of electrons in layers closer to it. The electrons located in the outermost energy layer of an atom are called valence electrons. And the outermost energy layer of an atom can contain a maximum of 8 electrons.

When external energy is applied to certain materials in the form of heat, light or electrical energy, electrons acquire energy, and this can cause them to move to a higher energy layer. It is said, then, that an atom in which this happened is in an excited state. An atom in this condition is unstable. When the electron is displaced to the outermost energy layer of an atom, it is minimally attracted by the positive charges of the protons in its nucleus. Therefore, if sufficient energy is applied to the atom, some valence electrons will abandon the atom. These electrons are called **free electrons. It is the movement of these free electrons that produces the electric current in the metallic conductors.**

Atomic electrons are attached to their nuclei by forces that, while reasonably intense, are not insuperable. Therefore, they can be transferred from one body to another when their substances come into contact. Thus, in the process of friction of two bodies, many electrons can be transferred from one to another. When this happens, one of the bodies is left with an excess of electrons, while the other is left with a deficiency of them. The one with excess electrons is negatively charged, and the one with lack of electrons is positively charged. Therefore, **the amount of electrical charge that a body has is determined by the difference between the number of protons and the number of electrons that the body contains.** It is worth remembering that the **electric charge of a body is quantized**, that is, that it appears as a whole number of electronic charges.

In his experiments, Benjamin Franklin (1706–1790) showed that:

1. The rubbing of cat skin with hard rubber causes the cat skin to transfer electrons to the rubber. Thus, the cat's skin is left with excess of protons and the hard rubber with excess of electrons. It is said, then, that the cat's fur is positively charged ($+$) and that the rubber is negatively charged ($-$).
2. Rubbing silk with glass causes the glass to transfer electrons to the silk. Thus, silk has an excess of negative charges and the glass has an excess of positive charges. It is said, then, that the glass is positively charged ($+$) and that the silk is negatively charged ($-$).

As described above, Benjamin Franklin defined, **in a completely arbitrary way**, the charge of the electrified hard rubber as negative and the charge of the electrified glass as positive. These conventions persist to this day, although perhaps it would have been more appropriate to have adopted the inverse convention, as this would avoid the need to work with the **conventional direction of electric current, assuming that it is the positive charges that move in the metallic conductors.**

When a pair of bodies contains the same type of charge, that is, both positive ($+$) or both negative ($-$), the two bodies are said to have **equal charges.** When a pair of bodies contain different charges, that is, one is positively charged ($+$) and the other is negatively charged ($-$), they are said to have **opposite or unequal charges.**

A fundamental law of electrical charge, called the **law of charges**, is as follows:

Equal charges repel, opposite charges attract.

In addition, in the electrical circuits there is the **principle of charge conservation**, which means that **the total electrical charge remains constant**—the electrical charge cannot be generated or destroyed.

The charge of an electron or proton, while fundamental to electricity, is too small to be the basic unit of charge. Thus, the charge unit in SI (acronym SI, from the French Système International d'unités), the International System of Units, is Coulomb, with the symbol of unit C. Thus, the combined charge of 6.242×10^{18} electrons is equal to -1 [C]. And that of 6.242×10^{18} protons is equal to $+1$ [C]. Conversely, the charge of an electron is -1.602×10^{-19} [C] and that of a proton is $+1.602 \times 10^{-19}$ [C].

Coulomb's Law

The French physicist Charles Augustin de Coulomb (1736–1806), in 1784, established the law that quantitatively expresses the force between two point charges, that is, charges of infinitesimally small bodies, which contain mass, but with negligible volumes. He found that the force between the charges q_1 and q_2 is directly proportional to the modulus of each of the charges and inversely proportional to the square of the distance between them. He also noted that the force is directed along the straight line that passes through the point charges. Logically, the meaning is given by the law of charges. Mathematically speaking, Coulomb's law is expressed by Eq. 1.6.

$$F = k\frac{q_1 q_2}{d^2}$$

(1.6)

If $q_1 = q_2 = q = 1$ [C] and $d = 1$ [m], then $F = 8.98742$ [N]. Therefore, the proportionality constant is $k = 8.98742$ [N m^2/C^2].

Thus, one can define, indirectly, 1 [C] with that total punctual charge that when separated by 1 [m] of a similar charge generates a force of 8.98742 [N].

It must be emphasized that Coulomb's law applies strictly only to point charges. They are, however, necessary to deduce the expressions of forces between large distributions of charges or in large bodies. Furthermore, Coulomb's law is universal and also applies to the force between two electrons or between an electron and a proton.

Electrostatic Field

The electric charge has a very important property, which is the ability to create a vector field of forces in the space that surrounds it, and that field, in turn, transmits forces to other bodies also electrically charged immersed in it and, therefore, affect their movements. Therefore, in the case of electric fields, electric charges constitute the sources of these fields. Thus, the utility of the electric field comes from its action as an intermediate agent in the transmission of forces from one or more charged bodies to others. **The electrostatic field refers to the electric field created by charges at rest. Moving electric charges can also establish magnetic fields in space and these, in turn, can exert forces on charged objects in motion.**

At any spatial point, characterized by its orthogonal Cartesian coordinates (x, y, z), the electric field vector E is defined as the quotient of the force F acting on a **positive test charge** q and itself. Like this,

$$E(x, y, z) = \frac{F(x, y, z)}{q} \quad (q \to 0) \tag{1.7}$$

The observation that the test charge q is infinitesimally small $(q \to 0)$ is intended to alert that, in a measurement process, it cannot influence the charge (s) that created the force field. In other words, you need to ensure that E is always independent of q. An alternative way of expressing this requirement is to define the electrostatic field as

$$E = \lim_{q \to 0} \frac{F}{q} \tag{1.8}$$

Evidently, in SI, the electric field unit is [N/C].

Example 1.1 A point charge $+Q$ is located at the origin of an orthogonal Cartesian coordinate system, as shown in Fig. 1.2. Find the electrostatic field it produces.

Fig. 1.2 Force F acting on q according to Coulomb's law

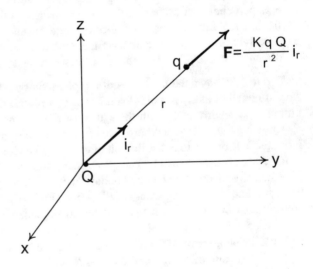

Solution
First, it is necessary to obtain the force that a test charge q located at a distance r from the origin experiences due to the charge $+Q$. According to Coulomb's law, Eq. 1.6, the force F acting on q is given by

$$F = k\frac{qQ}{r^2}i_r = k\frac{qQ}{r^2} \cdot \frac{r}{r} \tag{1.9}$$

In Eq. 1.9, $i_r = \frac{r}{r}$ is a unit vector pointing outwards, from the origin, where the charge $+Q$ is located, towards the observation point, where the test charge q is located. So, according to Eq. 1.7,

$$E = \frac{F}{q} = k\frac{Q}{r^2}i_r \qquad (1.10)$$

Therefore, the electrostatic field E also points radially outward, and is spherically symmetrical, since its modulus depends only on the distance r from the origin, for a given generator charge $+Q$. And that its modulus varies in inverse proportion to the square of the distance r.

It was the British physicist Michael Faraday (1791–1867) who proposed the adoption of the so-called field lines or lines of force, as a geometric means of representing the electric field, and which allow to illustrate how the electric fields behave. The construction of these field lines follows the rules:

1. The lines of force are drawn and oriented in the direction of the force that a positive test charge would experience at each spatial point in the field.
2. The density of the field lines, expressed by the number of lines crossing a unit of area perpendicular to their direction, is a measure of the intensity of the electric field. Thus, higher density of lines of force means more intense electric field.
3. **The field lines can never intercept**. As the force at any point can have only one direction, it is evident that the field lines can never intersect.

Figure 1.3 shows the field lines of the point charge $+Q$ in Example 1.1, which extend radially out of the point load. It is easy to intuit that the number of lines per unit area varies inversely with the square of the distance from the charge (origin).

Knowing the electric field produced in the whole space by a single charge, as in Example 1.1, or due to a distribution of charges, the Coulomb force exerted on any point charge q can always be calculated using the expression $F = qE$.

When two electrically charged bodies, with opposite polarities, are placed close to each other, constituting what is called an **electric dipole**, the electrostatic field is concentrated in the region between them, as indicated by the lines of force in the illustration shown in Fig. 1.4.

It can be demonstrated that the electric field produced outside a hollow spherical shell, of very thin thickness, and of radius R, with a uniform charge density, totaling a charge $+Q$, presents an electric field outside the sphere, at a distant point r from its center, given exactly by Eq. 1.10. On the other hand, the electric field at any point within the spherical shell charged with uniform charge density is identically zero. It is also possible to demonstrate that any spherically symmetric charge distribution produces an electric field outside the charge distribution identical to the field that would be produced if the entire charge were placed in the center of the charge distribution.

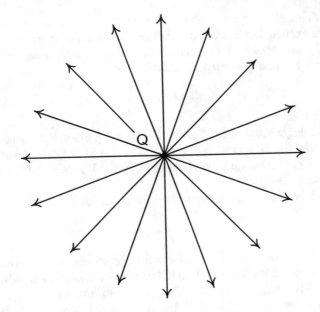

Fig. 1.3 Force lines of the point charge $+Q$

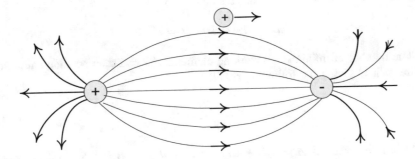

Fig. 1.4 Electrostatic field in an electric dipole

Regarding the electrostatic field, it can be added that:

(a) The electric field within any **conductor** in equilibrium is zero.
(b) Any total charge produced by a conductor **resides on its surface**.
(c) The electric field exactly outside a conductor is **normal to its surface** at each point.

Voltage or Potential Difference

Work

Mechanical work can be performed by any moving force. Therefore, **if there is force but there is no movement caused by it, there is no work**. In this way, a person standing with his arms on a concrete wall is not doing work. In other words, when there is force exerted, but it does not produce displacement, there is no work done.

The total work done by a constant force F that moves through a straight displacement r is a **scalar quantity** W defined by

$$W = F \cdot r \cdot \cos \theta, \tag{1.11}$$

where θ is the angle between the directions of F and r. So, if F and r have the same vector direction, $\theta = 0°$ and $W = F.r$. Obviously, in SI the unit of work is [N.m], also called joule, [J]. **As the term energy, in general, refers to the ability to do work, its unit is also joule**.

Equation 1.11 can also be written in vector language as

$$W = F \cdot r \tag{1.12}$$

In the most general case, when the force is not constant during the displacement and the displacement itself is not straight but rather a trajectory C, which extends from a point $P_1(x_1, y_1, z_1)$ to a point $P_2(x_2, y_2, z_2)$, as illustrated in Fig. 1.5, the total displacement r can be considered to be the vectorial sum of infinitesimal displacements dr along the path C, in each of which the force F can be considered as constant. At any point P, for example, the infinitesimal work dW associated with displacement dr, according to Eqs. 1.11 and 1.12, is

$$dW = F \cdot \cos \theta \cdot dr = F \cdot dr$$

The number of infinitesimal working elements dW can then be added along the entire path C giving the following result:

$$W = \int_C F \cdot dr \tag{1.13}$$

Given the vectors $A = A_x i_x + A_y i_y + A_z i_z$ and $B = B_x i_x + B_y i_y + B_z i_z$, where i_x, i_y, i_z are the **unit vectors** of the coordinate axes of the orthogonal Cartesian system, their scalar product is given by

Fig. 1.5 Infinitesimal work dW associated with displacement dr

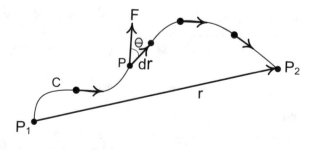

$$A \cdot B = A_x B_x + A_y B_y + A_z B_z$$

So, if the scalar product $F \cdot dr$ is expressed in terms of the components of vectors F and dr, the integral of Eq. 1.13 can be separated into three, as follows:

$$W = \int_C F \cdot dr = \int_{P_1}^{P_2} \left(F_x d_x + F_y d_y + F_z d_z \right)$$

$$= \int_{x_1}^{x_2} F_x dx + \int_{y_1}^{y_2} F_y dy + \int_{z_1}^{z_2} F_z dz$$

(1.14)

Example 1.2 A body moves along the ABC path in the xy plane shown in Fig. 1.6. The force $F = k(x i_x + 2y i_y + 0 \cdot i_z)$ [N], where $k = 2$ [N/m], acts on the body during its displacement. Find the work done on it by moving it from A to C.

Fig. 1.6 Body moviment along the ABC path in the xy plane

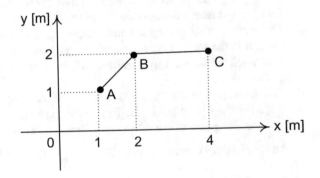

Solution
Applying Eq. 1.14 to segment AB, it follows that:

$$W_{AB} = \int_1^2 F_x dx + \int_1^2 F_y dy = \int_1^2 2x dx + \int_1^2 4y dy = x^2 \big|_1^2 + 2y^2 \big|_1^2 = 3 + 6 = 9 \ [\text{J}].$$

For the BC segment:

$$W_{BC} = \int_2^4 F_x dx + \int_2^2 F_y dy = \int_2^4 2x = x^2 \big|_2^4 = 12 \ [\text{J}].$$

The total work done on the body when it is moved by the force F along the path from A to C is the sum of the two contributions:

$$W = W_{AB} + W_{BC} = 9 + 12 = 21 \, [\text{J}].$$

The potential energy of a system is the energy that it can have due to its position or displacement. In other words, the total work that a body can do on the environment by virtue of its position is called potential energy. Thus, the potential energy of a mass m that is raised to a height h above the earth's surface, considered as a **reference**, is $m.g.h$, since it is capable of carrying out this work when it descends to it, where $g = 9.81 \, [\text{m/s}^2]$ is the acceleration of Earth's gravity. Forces such as frictional forces, for which it is not possible to specify a defined potential energy, are called **non-conservative forces**. Therefore, there are forces that do not accumulate potential energy and from which work cannot be recovered by releasing the accumulated potential energy. So, it is impossible to associate a potential energy with a non-conservative force. **Conservative systems** are those in which only conservative forces operate.

Conservative forces have the following properties:

1. The work performed by a conservative force when moving from point A to point B along a given trajectory is the negative of the work performed by the same force on the return through the same trajectory, that is, from B to A.
2. The work carried out by a conservative force when moving from point A to point B is the same, **regardless of the path taken** to go from A to B.
3. **The work performed by a conservative force when moving along a closed path is zero**.
4. The work performed by a conservative force when moving from A to B is $U_P(B) - U_P(A)$, where $U_P(x, y, z)$ is the potential energy associated with the force at any point of the orthogonal Cartesian coordinates x, y, z.

An infinitesimal element of potential energy associated with a conservative force

$\boldsymbol{F} = F_x \boldsymbol{i}_x + F_y \boldsymbol{i}_y + F_z \boldsymbol{i}_z$ is given by
$$dU_P(x, y, z) = -F_x dx - F_y dy - F_z dz$$

So,

$$U_P(x, y, z) = -\int_0^x F_x dx - \int_0^y F_y dy - \int_0^z F_z dz, \tag{1.15}$$

where the lower limit of the integrals was taken equal to zero due to the origin of the coordinate system being taken as a reference, that is, that $U_P(0, 0, 0) = 0$.

Electrostatic Potential

Suppose that a fixed flat plate, of small thickness and large horizontal dimensions, with a uniform distribution of positive electrical charges—such as a glass plate rubbed with a silk scarf—presents, on one of its faces, an electrostatic field $\boldsymbol{E}$,

Fig. 1.7 The electrostatic
potential of point P

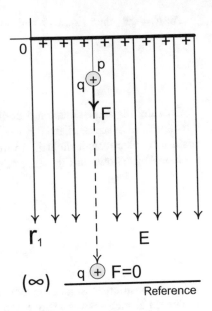

represented by their lines of force parallel and directed downwards, as illustrated in
Fig. 1.7. Obviously, a test load $+q$ positioned at any point P within this field will
experience a Coulomb force that will move it vertically downwards to a point where
the force acting on it is so weak that it can be considered to be zero. This point—
actually, plane—where the force no longer acts on the charge $+q$ and, therefore, no
longer performs work on it, is called a **reference**. Mathematically, it is customary to
consider that this reference is located at infinity. Therefore, the work done by the
field to move the charge $+q$ from the initial point P to the reference, according to
Eq. 1.13, is given by

$$W_P = \int_C F \cdot dr_1 = \int_P^\infty F \cdot dr_1 \tag{1.16}$$

   This Eq. 1.16 **also represents the energy that was expended by the electric
field to carry out the work of moving the charge** $+q$ **from point** P **to the
reference**. As the electrostatic field is a conservative field, it is therefore **a potential
energy**.
   The **electrostatic potential** of point P, in Fig. 1.7, is defined as the quotient of
the potential energy of point P by the charge q, and called e_P. Then,

$$e_P = \frac{W_P}{q} = \frac{1}{q}\int_C F \cdot dr_1 = \frac{1}{q}\int_P^\infty F \cdot dr_1 \tag{1.17}$$

As $F = qE$, then Eq. 1.17 can be rewritten as

$$e_P = \int_P^\infty E \cdot dr_1 \tag{1.18}$$

Consider the same situation described earlier, but now with the inversion of the direction of displacement, from r_1 to r, and imagine two charges $+q$ being displaced against the electrostatic field E, from the reference, one to position A and one to position B, as shown in Fig. 1.8. So, the potential energy of each of these points is

$$e_A = -\int_{-\infty}^A E \cdot dr \tag{1.19}$$

$$e_B = -\int_{-\infty}^B E \cdot dr \tag{1.20}$$

By the positions assumed for points A and B, $e_A > e_B$ and $e_A - e_B > 0$.
The difference between the electrostatic potentials at points A and B is what is called the voltage between A and B, designed as e_{AB}. Therefore,

$$e_{AB} = e_A - e_B = -\left[\int_{-\infty}^A E \cdot dr - \int_{-\infty}^B E \cdot dr \right]$$

$$= -\left[\int_{-\infty}^B E \cdot dr + \int_B^A E \cdot dr - \int_{-\infty}^B E \cdot dr \right]$$

Fig. 1.8 Two charges $+q$ being displaced against the electrostatic field E, from the reference, one to position A and the other one to position B

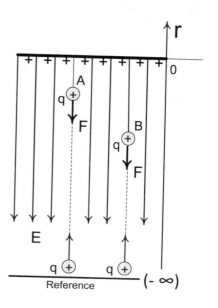

So,

$$e_{AB} = e_A - e_B = -\int_B^A \mathbf{E} \cdot d\mathbf{r} = \frac{W_A}{q} - \frac{W_B}{q} \qquad (1.21)$$

As the potential energy is expressed in joules and the electric charge in cou-lombs, then the voltage, like the electrostatic potential, has the unit of **joules per Coulomb**, also called **volts**. Therefore, e_{AB} [J/C] $= e_{AB}$ [V]. It is evident, then, that **$+1$ [C] of charge produces 1 [J] of work (energy) in its environment, when suffering a potential decrease of 1 [V].**

Swapping A with B in Eq. 1.21, it follows that

$$e_{BA} = e_B - e_A = -\int_A^B \mathbf{E} \cdot d\mathbf{r} = \int_B^A \mathbf{E} \cdot d\mathbf{r} = -e_{AB}. \quad \text{Therefore,} \quad \text{if} \quad e_{AB} = 6 \,[\text{V}],$$

$e_{BA} = -6$ [V]. This is a consequence of the electrostatic field being a field of conservative forces.

Once the voltage unit in volts is established, it is possible to establish an equivalent unit for the electric field, since

$$\left[\frac{V}{m}\right] = \left[\frac{J/C}{m}\right] = \left[\frac{N\,m}{C\,m}\right] = \left[\frac{N}{C}\right]$$

It is more common to use the unit **volts per meter** than newtons per coulomb for the electric field.

Since the voltage is defined as the **difference** between electrostatic potentials, according to Eq. 1.21, and **not in absolute value**, one can choose the reference, where it is assumed that the electrostatic potential is zero, anywhere, and not necessarily at infinity. It is for this reason that in the solution of electrical circuits by the nodal method it is common to adopt a node of the circuit itself as a reference, to which the grounding symbol is assigned, with zero potential, and to measure the potentials of the other nodes in relation to it, calling them as **nodal voltages**. In this book, the voltage will always be represented by an open arrow at the tip $(\leftarrow)$, where the tip refers to the highest potential $(+)$ and the tail refers to the lowest potential $(-)$.

An alternative and indirect way to define voltage or potential difference in any circuit element is:

The elementary energy dW **absorbed** by a two-terminal circuit element, when a differential amount of charge $(+)$ dq moves through it, from the tip to the tail of the voltage reference arrow e, is given by

$$dW = e \cdot dq \qquad (1.22)$$

or

$$e = \frac{dW}{dq} \qquad (1.23)$$

Therefore, **a negative dW value indicates that power is being supplied** by the circuit element. Then, **the sign of $e(t)$ is positive if, and only if, the circuit element is absorbing energy when positive charges move from the tip to the tail of the voltage reference arrow.**

Example 1.3 Work that consumed 110 [J] was performed to move 4.575×10^{18} electrons from one point to another in an element of an electrical circuit. What is the magnitude of the voltage that was established between these two points?

Solution
For this case, Eq. 1.23 can be written as $e = W/Q$. So,

$$|e| = \frac{110\,[\mathrm{J}]}{4.575 \times 10^{18}\,[\mathrm{e}] \times 1.602 \times 10^{-19}\,\left[\frac{\mathrm{C}}{\mathrm{e}}\right]} = 150.09\,\left[\frac{\mathrm{J}}{\mathrm{C}}\right] \cong 150\,[\mathrm{V}].$$

Electric Current

It was the invention of voltaic batteries by Alessandro Volta (1745–1827), now called batteries, that made it possible to chemically generate differences in electrical potentials, capable of producing greater fluxes of charges continuously through metallic conductors. As a consequence, the discoveries that were made since the beginning of the nineteenth century were notable.

When a charge flow occurs within conductors, the conditions within the conductive material are no longer those of electrostatic equilibrium. Therefore, it is no longer valid to consider the electric field to be zero everywhere in the conductor. On the contrary, **the electric field must be different from zero to establish and maintain the flow of charges.**

The term electrical current means charges in motion. In a metallic conductor, as already mentioned, the current is formed by the movement of so-called **free electrons**. Whenever an **external force** acts on these free electrons of a conductive substance, as occurs when there is an **imposition of an electric field**, these negative free charges will move, producing, then, the electric (electronic) current. The positive charges of the conductor are trapped in the nuclei of the conductor atoms and will never move freely. In a copper conductor, for example, approximately 8.5×10^{28} electrons per cubic meter are free to move. This is equivalent to -1.36×10^{10} [C] of charge that can be moved in each cubic meter.

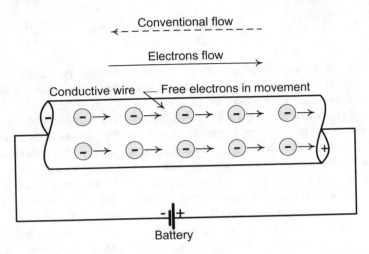

Fig. 1.9 Electric current produced by the potential difference of the battery applied to both ends of the conductor

Figure 1.9 shows the situation where a lead wire is connected to a battery. The potential difference of the battery applied to both ends of the conductor produces the electric current.

As shown in Fig. 1.9, the direction of movement of the electrons is from the negative battery terminal, passing through the conductive wire, and back to the positive battery terminal. In other words, electrons are repelled by the battery's negative terminal and attracted to its positive terminal—and thus the battery is discharged. The direction of electron flow is from a negative potential point to a positive potential point. The continuous arrow in Fig. 1.9 indicates the direction of the current as a function of the flow of electrons. The direction of the **supposed** movement of the positive charges, opposite to that of the electron flow, is considered as the **conventional current flow** and is indicated by the dashed arrow in Fig. 1.9. Any circuit can be analyzed either through the flow of electrons or the conventional flow in the opposite direction. **In this book, the current will always be considered in the conventional direction**. Had the electron charge been considered positive and the proton charge negative, there would be no need to define a conventional current direction.

Let Δq be the amount of positive moving charge passing through a given point P of a conductor wire over a period of time Δt. So, the electric current I is defined by

$$I = \frac{\Delta q}{\Delta t} \tag{1.24}$$

For currents that vary with time, the **instantaneous current** $i(t)$ is defined as the rate (speed) at which the charge moves passing through point P, and whose unit in SI is expressed in coulombs per second, called Ampère. Then, mathematically,

$$i(t) = \lim_{\Delta t \to 0} \frac{\Delta q}{\Delta t} = \frac{dq}{dt} \tag{1.25}$$

Equations 1.24 and 1.25, alone, are not sufficient to completely define an electric current. This is because it is necessary to clarify whether positive electrical charges are moving, in the conductor, from end A to end B, or from end B to end A. To add this information, then a **reference arrow** is used together with the current designation $i(t)$. The reference arrow then represents the **direction assumed** to be positive for current $i(t)$. In this book, the reference will always be represented by an arrow with a small triangle at the end, as shown in Fig. 1.10. Therefore, the current $i(t)$ will be positive if the positive charges are moving in the direction of the arrow, from A to B. If, however, the positive charges are moving in the opposite direction to the reference arrow, that is, from B to A, then the current i(t) is negative. Thus, a current of 10 [A], from A to B, is equivalent to a current of -10 [A], from B to A.

A current is generally positive during some time intervals and negative during others. And the outline of a current versus time, as shown in Fig. 1.11, is called the **current waveform**. The current reference arrow can be fixed in any direction, as long as its waveform is consistent with the direction chosen for the reference arrow. Therefore, if the waveform of current $i(t)$ in Fig. 1.11 refers to the current in Fig. 1.10, then the positive charges move **to the right** at a constant rate of 2 [C/s] = 2 [A], during the time interval $0 < t < 1$ [s], and **to the left**, at a constant rate of 1 [C/s] = 1 [A], during the time interval 1 [s] $< t < 3$ [s].

According to Eq. 1.25, if the charge is $q(t) = Q = $ constant, then the current $i(t) = 0$[A]. This means that there is no electric charge movement in the conductor, but only a body charged with Q charge. A permanently constant current is called **direct current**. And a current whose module varies continuously with time, and whose direction is periodically inverted is called **alternating current**.

If the current is the derivative of the charge, as expressed by Eq. 1.25, then, inversely, the charge is the integral of the current, i.e.

$$q(t) = \int i(t)dt \tag{1.26}$$

But an indefinite integral requires the determination of an integration constant, which is somewhat abstract and is not very practical in determining the waveform of the charge from the current waveform. Fortunately, an indefinite integral can be replaced by a well-defined definite integral, according to the following rule:

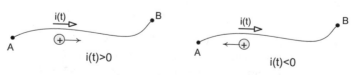

Fig. 1.10 Reference arrow representing the direction assumed to be positive for current $i(t)$

Fig. 1.11 Graphic of current versus time, called the current waveform

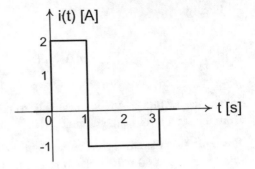

$$\int f(x)dx = \int\limits_{-\infty}^{x} f(\lambda)d\lambda \qquad (1.27)$$

Note that in the transition from one to another, the lower limit of the defined integral is always $-\infty$, the upper limit is the primitive integration variable x, and the integration variable must be given a new name—in the present case, λ. The introduction of the variable λ is necessary to distinguish the integration variable from the upper limit of the defined integral. It is sometimes called the **dumb variable**, because after the integration process, and the replacement of the upper and lower limits, it disappears, and the result is expressed, correctly, as a function of variable x.

Using the rule of Eq. 1.27, then Eq. 1.26 can also be written as

$$q(t) = \int\limits_{-\infty}^{t} i(\lambda)d\lambda \qquad (1.28)$$

The advantage of changing integrals is that a definite integral allows you to interpret it as an area. Thus, according to Eq. 1.28, the charge value q at any instant t is equal to the net area under the current waveform from $-\infty$ to instant t. Furthermore, the integral in Eq. 1.28 can be separated, for example, into two, as follows:

$$q(t) = \int\limits_{-\infty}^{t_0} i(\lambda)d\lambda + \int\limits_{t_0}^{t} i(\lambda)d\lambda$$

But according to Eq. 1.28, when the upper limit t is replaced by t_0,

$$\int\limits_{-\infty}^{t_0} i(\lambda)d\lambda = q(t_0)$$

Finally,

$$q(t) = q(t_0) + \int_{t_0}^{t} i(\lambda)d\lambda \tag{1.29}$$

The use of Eq. 1.29 is the best way to obtain the waveform of the charge $q(t)$ from the waveform of the current $i(t)$.

Example 1.4 Electrons pass to the right through the cross section of a conductive wire, at a rate of 60×10^{23} electrons per hour. What is the conventional current in the conductor?

Solution

The electron charge is $\bar{e} = -1.602 \times 10^{-19}$ [C]. So, the electronic current is

$$I[A] = \frac{\Delta q\,[C]}{\Delta t\,[s]} = \frac{60 \times 10^{23}\,[\bar{e}]}{1\,[h]} \times \frac{-1.602 \times 10^{-19}\,[C]}{1\,[\bar{e}]} \times \frac{1\,[h]}{3.600\,[s]} = -267\,[A].$$

Therefore, the conventional current is 267 [A] to the left.

Example 1.5 Lightning produced a current of 6 [kA] for 5×10^{-5} [min]. How many coulombs of charge were contained in the lightning?

Solution

$$\Delta q\,[C] = I\,[A] \cdot \Delta t\,[s] = 6 \times 10^3\,[A] \times 5 \times 10^{-5}\,[min] \times \frac{60\,[s]}{1\,[min]} = 18\,[C].$$

Example 1.6 Find the analytical expression of the charge $q(t)$ corresponding to the current waveform $i(t)$ in Fig. 1.11, knowing that $q(0) = 0$ [C].

Solution

Considering that $q(0) = 0$ [C] and that the charge is the net area under the current curve, it is already possible to state that $q(1) = 2$ [A] $\cdot 1$ [s] $= 2$ [C], which is the current first pulse area; and that $q(3) = 2.1 + (-1) \cdot 2 = 0$ [C], which is the total area under the two current pulses.

Taking $t_0 = 0$ in Eq. 1.29, it follows that

$$q(t) = q(0) + \int_{0}^{t} i(\lambda)d\lambda$$

For the interval $0 \le t \le 1$ [s],

$$q(t) = 0 + \int_{0}^{t} 2d\lambda = 2\lambda|_{0}^{t} = 2t\ [C]$$

So, $q(1) = 2$ [C].

Fig. 1.12 Waveform of the charge $q(t)$

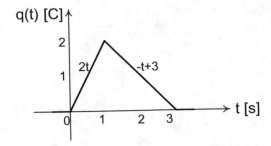

For the interval $1\,[\mathrm{s}] \le t \le 3\,[\mathrm{s}]$, with $t_0 = 1\,[\mathrm{s}]$ in Eq. 1.29,

$$q(t) = q(1) + \int_1^t i(\lambda)d\lambda = 2 + \int_1^t (-1)d\lambda = 2 - \lambda\big|_1^t = 2 - t + 1 = -t + 3\,[\mathrm{C}].$$

Therefore, $q(3) = 0\,[\mathrm{C}]$.

For the interval $3\,[\mathrm{s}] \le t < \infty$, with $t_0 = 3\,[\mathrm{s}]$ in Eq. 1.29,

$$q(t) = q(3) + \int_3^t i(\lambda)d\lambda = 0 + \int_3^t 0 \cdot d\lambda = 0\,[\mathrm{C}].$$

Finally,

$$q(t) = \begin{cases} 2t\,[\mathrm{C}] & for\,0\,[\mathrm{s}] \le t \le 1\,[\mathrm{s}] \\ -t+3\,[\mathrm{C}] & for\,1\,[\mathrm{s}] \le t \le 3\,[\mathrm{s}] \\ 0\,[\mathrm{C}] & for\,3\,[\mathrm{s}] \le t < \infty. \end{cases} \tag{1.30}$$

With the expressions contained in Eq. 1.30, the waveform of the charge $q(t)$ shown in Fig. 1.12 is constructed.

It is interesting to note that the integration of a waveform having finite discontinuities always results in a continuous waveform.

1.4 Power and Energy

The circuit elements in this book have two accessible terminals, as shown in Fig. 1.13a. As it is not possible to accumulate electrical moving charge within these elements, then, as a consequence, any current that enters through one terminal must necessarily exit through the other. According to Eq. 1.22, **the elementary energy dW absorbed** by the circuit element of Fig. 1.13b, when a differential amount of charge $(+)$ dq moves through the element, from terminal A, positive $(+)$, to the terminal B, negative $(-)$, that is, from the tip to the tail of the voltage reference arrow $e(t) = e_A(t) - e_B(t)$, is

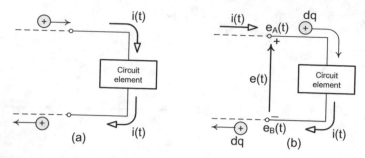

Fig. 1.13 Power absorbed by the circuit element

$$dW = e \cdot dq$$

(As previously mentioned, a negative dW value means that power is being **supplied** by the circuit element).

But, according to Eq. 1.25,

$$dq = i \cdot dt$$

Therefore, the energy absorbed in dt seconds by the circuit element in Fig. 1.13b is

$$dW = e \cdot i \cdot dt, \tag{1.31}$$

and the power $p(t)$ absorbed by it, that is, the rate (speed) of energy absorption is

$$p(t) = \frac{dW}{dt} = e(t) \cdot i(t) \tag{1.32}$$

where p is expressed in SI in joules per second, called watts. Therefore, $1\,[\mathrm{W}] = 1\left[\frac{\mathrm{J}}{\mathrm{s}}\right]$.

Since time t does not recede, always $dt > 0$. Then the sign of $p(t)$, according to Eq. 1.32, depends exclusively on the sign of dW. Since $dW > 0$ means energy absorption, $p(t) > 0$ also means that the circuit element is **absorbing** energy. On the other hand, as $dW < 0$ means power supply, $p(t) < 0$ also means that the circuit element is **supplying** power.

An alternative interpretation, obtained from the situation described in relation to Fig. 1.13b, is that the power is positive when the voltage and current **reference arrows** are in **opposition** in the circuit element. And then, he is a **consumer**. Therefore, $p(t) = +e(t) \cdot i(t)$. So, when the reference arrows agree, $p(t) = -e(t) \cdot i(t)$ and the circuit element is a **supplier or source** of energy. These conclusions are summarized in Fig. 1.14a, b.

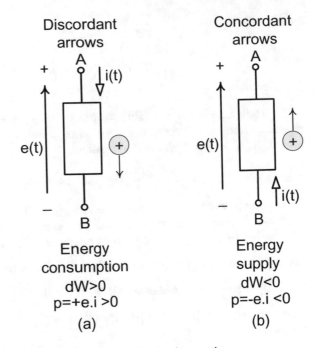

Fig. 1.14 Voltage and current reference arrows and power and energy being **a** consumed or **b** supplied by the circuit element

From Eq. 1.32, the net energy absorbed by a circuit element is

$$W(t) = \int p(t)dt = \int e(t)i(t)dt \tag{1.33}$$

Using the rule of Eq. 1.27, Eq. 1.33 can be written as

$$W(t) = \int_{-\infty}^{t} p(\lambda)d\lambda = \int_{-\infty}^{t} e(\lambda)i(\lambda)d\lambda \tag{1.34}$$

And then, Eq. 1.34 is showing that the energy absorbed by a circuit element at any time t is the liquid area under the power curve that flowed into it, from $-\infty$ to time t. In addition, integral of Eq. 1.34 can be separated into two as follows:

$$W(t) = \int_{-\infty}^{t_0} p(\lambda)d\lambda + \int_{t_0}^{t} p(\lambda)d\lambda \tag{1.35}$$

But, according to Eq. 1.34 itself,

$$\int_{-\infty}^{t_0} p(\lambda)d\lambda = W(t_0)$$

So, finally,

$$W(t) = W(t_0) + \int_{t_0}^{t} p(\lambda)d\lambda \qquad (1.36)$$

Equation 1.36 is more suitable for determining the energy waveform from the power waveform, especially when the latter is defined by sections. Obviously, if $t_0 = 0$, $W(t_0) = 0$ and $p(t) = P = $ constant,

$$W(t) = P.t \quad for\, t \geq 0 \qquad (1.37)$$

Another power unit, used mainly to specify electric motors, is [CV]. A [CV] is the power needed to raise the mass from 75 [kg] to a height of 1 [m] in 1 [s]. And the relationship between [CV] and watts is obtained as follows:

Force [N] = mass [kg] × acceleration [m/s²] $\therefore f = m \times g$ [kg · m/s²].

Work [J] = force [N] × distance [m] $\therefore W = f \times d = m \times g \times d$ [kg · m²/s²].

Power [W] = work [J] ÷ time [s] $\therefore P = W/t = \frac{m \times g \times d}{t} = mgv$ [kg · m²/s³].
   Therefore,

$$P = m[\text{kg}] \times g\left[\tfrac{\text{m}}{\text{s}^2}\right] \times v\left[\tfrac{\text{m}}{\text{s}}\right] = 75 \times 9.81 \times 1 = 735,75\,[\text{W}] \cong 736\,[\text{W}]. \text{ So,}$$
$$1\,[\text{CV}] = 736\,[\text{W}]. \qquad (1.38)$$

In the British system there is also a unit called *horsepower* [HP], equivalent to 746 [W].

The kilowatt-hour [kWh] is a unit of energy commonly used to designate large amounts of electrical energy. And it is calculated simply by making the product of the power in kilowatts [kW] by the time in hours [h] during which the power is supplied.

Example 1.7 If in Fig. 1.13b, $e(t) = 2\cos 10t$ [V] and $i(t) = 4\cos 10t$ [A], determine the expression of the power absorbed by the circuit element. How much energy will be absorbed for $0 \leq t \leq \frac{2\pi}{10}$ [s]?

Solution
The instantaneous power is
   $p(t) = +e(t) \cdot i(t) = 8\cos^2 10t$ [W].

Since the absorbed energy is cumulative, then the absorbed energy for $0 \leq t \leq \frac{2\pi}{10}$ [s] is

$$W\left(\frac{2\pi}{10}\right) = \int_0^{2\pi/10} 8\cos^2 10t\, dt = \int_0^{2\pi/10} 4(1 + \cos 20t)dt = 4t\Big|_0^{2\pi/10} + 4\frac{sen\, 20t}{20}\Big|_0^{2\pi/10}$$

$$= \frac{8\pi}{10} = 2.513 \,[\text{J}]$$

Example 1.8 How much energy in joules does a 5 [CV] engine consume in two hours? And in kilowatt-hours?

Solution
By Eq. 1.37,

$$W = \text{P.t} = 5\,[\text{CV}] \cdot \frac{736\,[\text{W}]}{1\,[\text{CV}]} \cdot 2[\text{h}] \cdot \frac{3,600\,[\text{s}]}{1\,[\text{h}]} = 26,496,000\,[\text{J}] = 26.496\,[\text{MJ}].$$

$$W = 5\,[\text{CV}] \cdot \frac{0.736\,[\text{kW}]}{1\,[\text{CV}]} \cdot 2\,[\text{h}] = 7.36\,[\text{kWh}].$$

1.5   Circuit Elements

The circuit elements of two terminals can be divided into two large groups: the **active** ones, which are the independent sources of voltage and current, which are capable of supplying energy to the **network**, that is, the rest of the circuit, and the **passive circuits**, which can absorb or store energy.

More precisely, a circuit element is called passive if the total net energy supplied to it by the rest of the circuit, calculated by Eq. 1.36, is always non-negative, that is: $W \geq 0$. The power delivered to a passive circuit element can be **dissipated** in the form of heat, in a component called **resistor**, when it will be irretrievably **lost**, that is, no longer recoverable by the circuit, or it can be **stored**, and recovered, in the components called **capacitor** and **inductor**. The elements properties used to approximate the behavior of these components are resistance R, capacitance C and inductance L.

One must also consider, as a component of two electrical circuit terminals, an **ideal switch**, which closes or opens instantly, thus approximating the behavior of a real switch. When closed, it can be connecting an independent voltage source to a network, or attaching more load to a circuit already in operation, or, still, simulating a short circuit between any two points of a circuit. When it opens, it can be disconnecting an independent voltage source, or connecting an independent current source to a network, or even disconnecting a segmented part of a circuit that is in permanent operation.

Resistance

As mentioned, the portion of an electrical circuit in which electrical energy is converted to heat is called a resistor.

When passing through a conductor, the free electrons collide with each other and with the atoms of the conductive material and lose part of their kinetic energy, which is transformed into heat. These collisions are responsible for opposing the current flow and, therefore, for the **resistance** of the circuit. Thus, the resistance of a conductor is defined as the opposition it presents to the movement of free electrons, which constitute the electric current, which, in order to exist, requires the application of a voltage between the conductor terminals. Note, however, that opposing does not mean preventing!

A conductor's resistance depends on its dimensions, the material from which it is made—which determines the distribution and nature of its atoms, and the temperature, which determines the vibration of the atoms. Generally, in metallic conductors, an increase in temperature causes an increase in resistance, since atoms more often collide with the displacement of electrons. However, in a certain class of materials, called **semiconductors**, the resistance decreases with increasing temperature. In semiconductors, resistance also tends to increase with increasing temperature. However, semiconductors have the peculiar property that additional charge carriers appear with increasing temperature, which establish an increase in current flow, which implies less resistance. This property of semiconductors overcomes the effect of increased resistance caused by atomic vibration and thus the resulting effect is a decrease in resistance with increasing temperature. Examples of semiconductors are Carbon, Silicon and Germanium.

There are also materials that are neither **conductors** nor **semiconductors**. These are called **dielectric** or **insulating materials**. They can provide physical support without significant leakage of current, or even cover the conductive wires preventing current leakage when they touch or are touched by us.

Mathematically, the definition of resistance is expressed by the so-called Ohm's Law, in honor of the German physicist and mathematician Georg Simon Ohm (1789–1854). It is as follows:

$$e(t) = R \cdot i(t) \tag{1.39}$$

It indicates that the voltage and current are directly proportional. Thus, the curves of $e(t)$ and $i(t)$, plotted as a function of the variable t, always have the same waveform, differing only in amplitude. The unit of resistance in the SI is volts per amp or ohm, with the symbol $[\Omega]$. The **conductance**, defined by $G = 1/R$, has the Siemens unit, with the symbol $[S]$. Therefore, $i(t) = G \cdot e(t)$. Equation 1.39 implies arrows voltage and current in opposition. When they agree, a minus sign is required, as shown in Fig. 1.15.

In the case of the opposing arrows, positive charges enter resistance R with a higher level of energy (tip of the voltage arrow) and exit at a lower level (voltage arrow tail). Therefore, when passing through the resistance, electrical charges lose energy, which is absorbed by the resistance and, by joule effect, transformed into heat.

From Eq. 1.32, the power absorbed by resistance R in Fig. 1.15 is

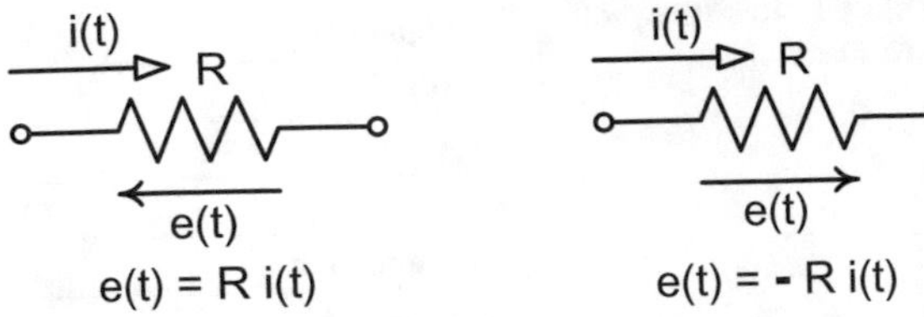

Fig. 1.15 Reference arrows for voltage between the terminal nodes and current across a resistor

$$e(t) = R\,i(t)$$

$$e(t) = -\,R\,i(t)$$

$$p(t) = e(t) \cdot i(t) = R[i(t)]^2 = G[e(t)]^2 = [e(t)]^2/R \qquad (1.40)$$

The energy absorbed by a resistance up to an instant t can be calculated by Eq. 1.36.

The zero value resistance (zero voltage) is called a **short-circuit** and the zero value conductance, that is, infinite resistance (zero current) is called an **open circuit**. Although all metallic conductors have some resistance, resistors are components built to make resistance the dominant effect. In circuit analysis it is assumed that the resistance of a wire connecting two components is either negligible or is represented by some component concentrated in the circuit. Thus, all the **interconnection wires** shown in the circuits must be interpreted as short-circuits, with no voltage between their two ends.

The heat dissipated power is distributed over the entire volume of a resistor, and this distribution is a function of several factors. A change in this distribution can cause a change in the resistance value. The phenomenon known as the **skin effect** or **pelicular effect**, which is a combination of high frequency currents in the vicinity of a conductor's surface, is an example. The higher the frequency of the sinusoidal current, the lower current density results in the center of the conductor. This phenomenon can cause the resistance of a conductor to be significantly higher for alternating currents of very high frequency than for direct currents. It can therefore be said that $R = R(T,f)$, where T is the temperature and f, the frequency of the current signal.

At a fixed temperature, the resistance of a cylindrical conductor of length l [m] and cross-sectional area A [m^2] is given by

$$R = \rho \frac{l}{A} \qquad (1.41)$$

where ρ is the factor that depends on the type of material the resistor is made of, called **resistivity**, whose unit in the *SI* is the ôhmetro, with symbol [$\Omega\,m$]. Table 1.1 shows the resistivity of some materials at a temperature of 20 [°C].

In the upper part of Table 1.1, some of the best conductors are listed, that is, those with less resistivity, and greater capacity to conduct electric current. You can also see why copper is a better conductor than aluminum. Below are the

Table 1.1 Resistivity of some materials

Material	$\rho[\Omega\,m]$ at 20 [°C]
Silver	1.64×10^{-8}
Annealed copper	1.72×10^{-8}
Gold	2.44×10^{-8}
Aluminium	2.83×10^{-8}
Tungsten	5.52×10^{-8}
Platinum	10.00×10^{-8}
Iron	12.30×10^{-8}
Bronze	18.00×10^{-8}
Lead	22.00×10^{-8}
Manganite	44.00×10^{-8}
Constantan	49.00×10^{-8}
Steel	60.00×10^{-8}
Nichrome	100.00×10^{-8}
Carbon	4×10^{-5}
Germanium	47×10^{-2}
Silicon	6.4×10^{2}
Paper	1×10^{10}
Transformer oil	1×10^{11}
Mica	5×10^{11}
Glass	1×10^{12}
Teflon	3×10^{12}
Porcelain	1×10^{16}
Quartz	1×10^{17}

Manganite: alloy with 80–85% copper, 12–15% manganese, 2–4% nickel

Constantan: alloy with 55% copper, 45% nickel

semiconductors Carbon, Germanium and silicon. And in the final part of the table, the insulators, with resistivity from 10^{10} [Ω m].

The resistance of most good conductors increases approximately as a straight line when the temperature increases beyond the normal operating range, as shown in Fig. 1.16, with the exception of semiconductors.

If the straight line in Fig. 1.16 extends to the left, it will touch the temperature axis at $T = T_0$, corresponding to $R = 0$ [Ω]. However, any resistance can only be equal to zero at the **zero absolute temperature** of the thermodynamic scale or Kelvin scale, corresponding to -273.15 [°C], when the movement of the molecules ceases. For this reason, T_0 is called the **inferred absolute temperature for zero resistance**.

If T_0 is known and if the resistance R_1 at a temperature T_1 is also known, then the resistance R_2 at temperature T_2 can be obtained using the proportionality between the sides of the similar rectangular triangles:

Fig. 1.16 Resistance of most good conductors as a function of temperature

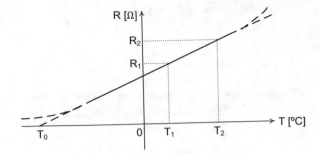

$$\frac{R_2}{R_1} = \frac{T_2 - T_0}{T_1 - T_0}$$

Therefore,

$$R_2 = \frac{T_2 - T_0}{T_1 - T_0} \cdot R_1 \tag{1.42}$$

Table 1.2 shows the inferred absolute temperature for zero resistance for some common conductive materials.

An equivalent way of determining the resistance R_2 is through the formula

$$R_2 = R_1[1 + \alpha_1(T_2 - T_1)], \tag{1.43}$$

where α_1 is the **temperature coefficient of the resistance at temperature** T_1. Generally,

Table 1.2 Inferred absolute temperature for zero resistance for some common conductive materials

Material	T_0 [°C]
Iron	−162
Tungsten	−202
Copper	−234.5
Aluminium	−236
Silver	−243
Constantan	−125.000
Carbon	2.020

$T_1 = 20[°C]$. For this condition,

$$\alpha_1 = \frac{1}{T_1 - T_0} \tag{1.44}$$

Example 1.9 A semiconductor material has $\alpha_1 = -3 \times 10^{-2} [°C]^{-1}$ for $T_1 = 20[°C]$. How much is T_0 worth? What is the resistance R_2 when $T_2 = 40[°C]$?

Solution

From Eq. 1.44, we have

$$T_0 = T_1 - \frac{1}{\alpha_1} = 20 - \frac{1}{-3 \times 10^{-2}} = 53.3 \, [°C].$$

From Eq. 1.43,

$$R_2 = R_1 \left[1 - 3 \times 10^{-2}(40 - 20) \right] = 0.4 \, R_1.$$

Example 1.10 The resistance of an aluminum power transmission line is $100 [\Omega]$ at $20[°C]$. What is its resistance when the sun raises its temperature to $40[°C]$?

Solution

According to Table 1.2, $T_0 = -236[°C]$. From the statement of the problem, $T_2 = 40[°C]$, $R_1 = 100[\Omega]$ and $T_1 = 20[°C]$. Therefore, according to Eq. 1.42,

$$R_2 = \frac{40 - (-236)}{20 - (-236)} \cdot 100 = 107.8 \, [\Omega].$$

Example 1.11 A copper winding of a non-energized transformer has a resistance of $30[\Omega]$ to $20[°C]$. However, under rated operation, the resistance increases to $35[\Omega]$. What is the temperature of the energized winding?

Solution

Explaining T_2 in Eq. 1.42 results that

$$T_2 = T_0 + \frac{R_2}{R_1}(T_1 - T_0)$$

From Table 1.2, $T_0 = -234.5[°C]$. Since $R_1 = 30[\Omega]$, $T_1 = 20[°C]$ and $R_2 = 35[\Omega]$, then,

$$T_2 = -234.5 + \tfrac{35}{30}(20 + 234.5) = 62.4 \, [°C].$$

Circuit elements are said **in series** when they are connected in series and then run by the **same current**, and **in parallel** when they are connected in parallel and then subjected to the **same voltage**.

If N resistances are connected in series, then the total resistance, called the **equivalent resistance**, is given by

$$R_{eq} = R_1 + R_2 + \cdots + R_N \tag{1.45}$$

If, however, they are connected in parallel, the equivalent resistance is obtained from

$$\frac{1}{R_{eq}} = \frac{1}{R_1} + \frac{1}{R_2} + \cdots + \frac{1}{R_N} \tag{1.46}$$

When only two resistors are in parallel, Eq. 1.46 provides that

$$R_{eq} = \frac{R_1 \cdot R_2}{R_1 + R_2}, \tag{1.47}$$

known as the **rule of the product by the sum**.

Capacitance

A **capacitor** is an electrical device basically formed by two metallic plates, separated by a thin layer of insulating material, called **dielectric**, like the one with parallel plates shown in Fig. 1.17a. Capacitors can have different shapes, but the goal, always, is to place two conductive surfaces very close and interspersed by a dielectric. For example, one type consists of a sandwich made of aluminum foil, or another metal, but separated by waxed paper. Another consists of concentric cylinders, separated by air.

The charges on the two plates must have the same module, but they must have opposite signals and produce an electrostatic field capable of storing energy, as shown in Fig. 1.17b. In it, the upper plate is charged with $+q$ and the lower plate

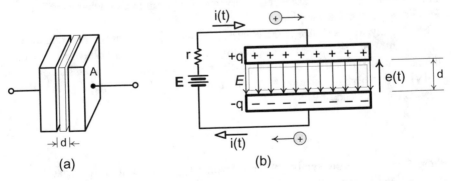

(a) (b)

Fig. 1.17 Capacitor formed by two metallic plates, separated by a thin layer of dielectric

with $-q$, producing an electrostatic field E, represented by the lines of force directed from the positive plate to the negative plate. Obviously, the lines of force between the edges of the metal plates are not shown. The energy is stored as voltage, $e(t)$, in the electrostatic field between the plates.

A battery of **electromotive force** E and internal resistance r connected to the capacitor, as shown in Fig. 1.17b, causes the capacitor to charge, with an excess of positive charges on the upper plate and an excess of negative charges on the lower plate. The positive charges of the lower plate are attracted to the negative terminal of the source, pass through it, going to its positive terminal, from where they are propelled to the upper plate of the capacitor. Due to the fact that each loss of positive charge through the lower plate is absorbed by the upper plate, the charge q is the same on both plates, although with opposite signs, as shown in Fig. 1.17b. Note that, at no time, there is a direct displacement of charges from one plate to another, as a result of the presence of the dielectric. It is this supposed displacement of positive charges that gives rise to the (conventional) electric current $i(t)$. It is necessary to keep in mind that, in reality, the charges that actually circulate are the electrons. The current i(t) will circulate until the voltage $e(t)$ at the capacitor terminals is equal to that of the source. It is said, then, that the capacitor is charged.

It is obvious that the energy stored in the electrostatic field produced in the dielectric comes from the battery, which does work in the system, by increasing the potential energy of the charges. In line with Eq. 1.21, the intensity of the electrostatic field between the capacitor plates in Fig. 1.17b, which is perpendicular to them and directed from the positive plate to the negative plate, is

$$E = \frac{e(t)}{d}$$

(1.48)

Due to the thin thickness of the dielectric, it is necessary to limit the voltage applied to the capacitors to values lower than that which could cause the dielectric to rupture, called the **breakdown voltage**. If the voltage on the capacitor is continuously increased, a point is reached where the intensity of the electric field between its plates is sufficient to break the dielectric. When this occurs, a discharge appears between the plates. As a result, the capacitors have a **working or rated voltage**, which cannot be exceeded. It decreases with frequency when sinusoidal or periodic waveforms are applied.

Electrically, **capacitance** is the electrical charge storage capacity. It can also be considered to be the property of opposing sudden changes in voltage in a circuit. Objectively, if the voltage or potential difference between the conductive plates is $e(t)[V]$, when there is a charge $+q(t)[C]$ in one and a charge $-q(t)[C]$ in the other, capacitance is defined by

$$q(t) = C.e(t),$$

(1.49)

where C is the quantity symbol of the capacitance. The *SI* unit of the capacitance is the farad, with the symbol [F]. However, the farad is too large a unit for practical applications. The microfarad $[\mu F]$, equal to $10^{-6}[F]$ and the picofarad $[pF]$, equal to $10^{-12}[F]$, are more common. According to Eq. 1.49, the farad is the capacitance that stores one coulomb of charge when the voltage at its terminals is one volt.

The capacitance of any given conductor system depends on two factors:

(1) **Geometric arrangement** of conductors. This includes the dimensions (area of the plates), the shape (squares, disks, cylinders, spheres) and the spacing of the conductors, as well as the relationships with each other.
(2) **Properties of the medium** in which the conductors are placed (vacuum, air, dielectric).

It makes it easier to understand these two items better: the fact that capacitance is manifested by itself when two conductors are energized and separated by a dielectric. This means that it is not necessary to build a capacitor to obtain capacitance.

The simplest example of a capacitor is that of two parallel plates shown in Fig. 1.17a. If each plate has an area $A[m^2]$ and if they are separated by a distance $d[m]$, which represents the thickness of the dielectric, which is generally very small compared to the linear dimensions of the plates $(d \ll \sqrt{A})$, then capacitance is given by

$$C = \varepsilon \cdot \frac{A}{d}, \tag{1.50}$$

where ε is the **permittivity** of the dielectric, whose unit is farads per meter $[F/m]$.

The permittivity ε is related to the atomic effects on the dielectric. As shown in Fig. 1.18, the charges on the capacitor plates interfere with the atoms of the dielectric, resulting in a total negative charge on the upper surface of the dielectric and a total positive charge on its lower surface. And this dielectric charge partially neutralizes the effects of the charge stored on the capacitor plates. In other words, an electric field opposite E_0 appears in the dielectric, so that the electric field between the plates now decreases to $E - E_0$. This phenomenon, called **dielectric polarization**, in turn, also reduces the voltage between the conductor plates of the capacitor, for a fixed electrical charge, and, as a consequence, leads to an increase in its capacitance. Mathematically speaking, the following occurs:

Initially, according to Eq. 1.48, $e = d \cdot E$.

Since, from Eq. 1.49, $C = q/e$, then,

$$C = \frac{q}{d \cdot E} \tag{1.51}$$

For fixed d and q, it is evident that the reduction of the electrostatic field from E to $E - E_0$ causes an increase in the value of capacitance C. It was Michael Faraday who discovered that for a fixed geometry, the capacitance of a capacitor increases when replacing air (or vacuum) by a dielectric, that is, by an insulating substance.

Fig. 1.18 Effect of dielectric on the electric field and the capacitance

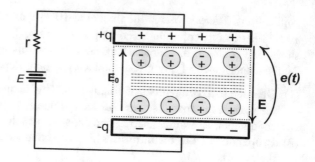

The permittivity of the vacuum is $\varepsilon_0 = 8.85 \times 10^{-12}$ [F/m]. The permittivity of other dielectric materials is related to that of vacuum by a factor called the **dielectric constant** or **relative permittivity** $K = \varepsilon_r$. Therefore,

$$K = \varepsilon_r = \frac{\varepsilon}{\varepsilon_0} \tag{1.52}$$

For vacuum $K = 1$, while for air, $K = 1.00059$, under standard conditions. In view of Eq. 1.50, Eq. 1.52 can be rewritten as

$$C = K\varepsilon_0 \frac{A}{d} = \left(8.85 \times 10^{-12}\right) \cdot \frac{KA}{d} \, [\text{F}] \tag{1.53}$$

Typical dielectric constant values are shown in Table 1.3.

Table 1.3 Approximate dielectric constants of some materials

Material	Dielectric Constant K	Temperature [°C]
Vacuum	1	–
Air	1.00059(1 atm)	20
Hydrogen	1.00026	100
Water	80.4	20
Mica	3–7	25
Quartz (fused)	3.75–4.10	20
Glass (Pyrex)	4.5	20
Glass	4–7	20
Rubber	2.94	27
Paraffin	2.0–2.5	20
Paper	4	20
Light plastic	2–3	20
Heavy plastic	4–12	20
Ceramic	7, 500	20

Example 1.12 The electric collapse of air, when it ceases to be dielectric and becomes conductive, happens whenever the electric field exceeds $30\,[kV/cm]$. What is the maximum charge that a $2\,[nF]$ parallel plate air capacitor can contain if the plates have an area of $100\,[cm^2]$ each?

Solution

The maximum load, q_{max}, is proportional to the maximum voltage, e_{max}, that can be applied between the plates, according to Eq. 1.49. That is,

$$q_{max} = C \cdot e_{max} \tag{1.54}$$

On the other hand, the electric field module between the capacitor plates, which is perpendicular to them and directed from the plate with positive charge to the plate with negative charge, as shown in Fig. 1.17b, according to Eq. 1.48, is

$$E = \frac{e}{d}$$

So, $e_{max} = d \cdot E_{max}$.
From Eq. 1.53,

$C = \varepsilon_0 \frac{A}{d}$, because $K \cong 1$. Substituting C in equation 1.54, it follows that:

$$q_{max} = \varepsilon_0 \frac{A}{d} \cdot d \cdot E_{max} = \varepsilon_0 A E_{max} = (8.85 \times 10^{-12})(10^{-2})(3 \times 10^6) = 26.55 \times 10^{-8}\,[C].$$

Note that:

$$e_{max} = \frac{q_{max}}{C} = \frac{26,55 \times 10^{-8}\,[C]}{2 \times 10^{-9}\,[F]} = 132.75[V]$$

$$d = \frac{e_{max}}{E_{max}} = \frac{132.75\,[V]}{30,000\,[V/cm]} = 4.43 \times 10^{-3}\,[cm] = 44.3\,[\mu m].$$

If the voltage between the plates exceeds $132.75\,[V]$, the air between the plates will become conductive and the loads on the plates will be neutralized. It is said, then, that the dielectric broke. When this occurs, a spark appears between the conductors.

For a capacitor formed by two hollow conductive cylinders, coaxial and long, with internal radius r_b and external radius r_a, containing air between them, as shown in Fig. 1.19, the capacitance (distributed) per unit of linear length is given by

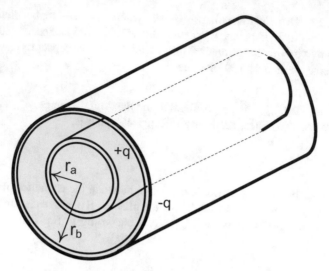

Fig. 1.19 Capacitor formed by two hollow conductive cylinders, coaxial and long

$$C/l = \frac{2\pi\varepsilon_0}{\ln(r_a/r_b)} \, [\text{F/m}] = \frac{24,18}{\log(r_a/r_b)} \, [\text{pF/m}] \qquad (1.55)$$

If the dielectric is not air, just enter the dielectric constant K in the numerator of the expressions in Eq. 1.55.

The capacitance of two hollow and concentric conducting spheres, with air between them, with internal radius r_a and external radius r_b is

$$C = \frac{4\pi\varepsilon_0 r_a r_b}{r_b - r_a}, \qquad (1.56)$$

with r_a e r_b in meters.

It is interesting to describe the capacitance of a single spherical conductor with radius r_a. To do this, just assume that the second conductor has an infinite radius. That way,

$$C = \lim_{r_b \to \infty} \frac{4\pi\varepsilon_0 r_a}{1 - r_a/r_b}$$

So,

$$C = 4\pi\varepsilon_0 r_a \qquad (1.57)$$

For an isolated conductive wire, suspended on a metal chassis, or another grounded plane, whose dimensions are sufficiently large in relation to the conductive wire, as shown in Fig. 1.20a, the capacitance per meter of wire is given by

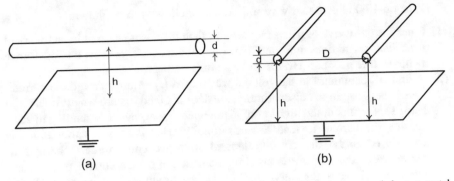

Fig. 1.20 Capacitance per linear meter in **a** isolated conductive wire, suspended on a metal chassis, or another grounded plane and **b** between two parallel conductors

$$C = \frac{24,18}{\log\left[\frac{2h}{d}\left(1 + \sqrt{1 - \frac{1}{(2h/d)^2}}\right)\right]} \, [\text{pF/m}], \qquad (1.58)$$

where $h\,[m]$ is the height in relation to the center of the wire on the ground plane or chassis.

The capacitance per linear meter between two parallel conductors, long located at a distance $h\,[m]$ on a ground plane, and whose diameter $d[m]$ is negligible compared to their length, as shown in Fig. 1.20b, is given by

$$C = \frac{12.073}{\log\left[\frac{2D}{d} \cdot \frac{1}{\sqrt{1 + (D/2h)^2}}\right]}, \qquad (1.59)$$

where $D\,[m]$ is the distance between the wires.

In circuit analysis, voltages and currents are used instead of electrical charges. And the voltage-current relationship in a time-invariant capacitance is obtained by deriving Eq. 1.49:

$$\frac{d}{dt}q(t) = C\frac{d}{dt}e(t)$$

But as the derivative of the electric charge is the current, then it turns out that

$$i(t) = C\frac{de(t)}{dt} = C \cdot \lim_{\Delta t \to 0} \frac{\Delta e}{\Delta t} \qquad (1.60)$$

Equation 1.60 allows two very important conclusions to be drawn:

(1) Since the derivative of a constant is zero, then there is no current in an invariant capacitance with time subjected to a constant voltage. Therefore, **a capacitance behaves like an open circuit for constant voltage**.

(2) A finite discontinuity in the voltage waveform $e(t)$ requires an infinite current. This is because, in the discontinuity, Δe does not tend to zero when Δt tends to zero. This results in the division of an amount $\Delta e \neq 0$ by an Δt tending to zero, causing the current to tend towards infinity. This is why **a capacitance is considered to be the circuit element that prevents sudden changes in voltage**. Of course, there is no similar restriction on the current of a capacitance. It can vary instantly, that is, suffer discontinuities or even change direction instantly, that is, change the sign.

The voltages in the capacitances do not suffer discontinuities means that the values of the voltages immediately after a switching operation, called instant $t = 0_+$, are exactly the same as immediately before it, called instant $t = 0_-$. In other words, if a changeover occurs at time $t = 0$, then $e(0_+) = e(0_-) = e(0)$. This fact is a very important fact in the transient analysis of circuits containing capacitances.

From Eq. 1.60, it can be obtained, inversely, that

$$e(t) = \frac{1}{C}q(t) = \frac{1}{C}\int i(t)dt \tag{1.61}$$

But, by virtue of Eq. 1.27, Eq. 1.61 can be rewritten as

$$e(t) = \frac{1}{C}q(t) = \frac{1}{C}\int_{-\infty}^{t} i(\lambda)d\lambda \tag{1.62}$$

Or yet,

$$e(t) = \frac{1}{C}\int_{-\infty}^{t_0} i(\lambda)d\lambda + \frac{1}{C}\int_{t_0}^{t} i(\lambda)d\lambda$$

However, according to Eq. 1.62 itself,

$$\frac{1}{C}\int_{-\infty}^{t_0} i(\lambda)d\lambda = e(t_0)$$

Finally,

$$e(t) = e(t_0) + \frac{1}{C} \int_{t_0}^{t} i(\lambda)d\lambda \qquad (1.63)$$

If $t_0 = 0$, Eq. 1.63 becomes

$$e(t) = e(0) + \frac{1}{C} \int_{0}^{t} i(\lambda)d\lambda \qquad (1.64)$$

In practice, the full past history of a capacitance is almost always unknown, but its initial voltage, that is, its voltage at $t = 0$, the moment when a circuit's operating time is usually counted, can be known; consequently, Eq. 1.64 is usually the most useful expression.

It must be clear that, in general, the voltage and current waveforms of a capacitance are not the same. It is also important to note, again, that, in general, the integral tends to smooth a waveform, while the derivation tends to introduce discontinuities. Obviously, trigonometric sine and cosine functions are exceptions. So are exponential functions.

Equation 1.60 implies that the voltage and current arrows are in opposition. Otherwise, a minus sign is required, as shown in Fig. 1.21.

Example 1.13 Analyze the issue of phase difference between voltage and current in a fixed capacitance, for sinusoidal quantities.

Solution
If the voltage at the capacitance in Fig. 1.21 is

$$e(t) = E \, sen\omega t [\text{V}], \; para \, t \geq 0, \qquad (1.65)$$

then the current will be

$i(t) = C\frac{de(t)}{dt} = \omega C E \cos \omega t [\text{A}], \; para \, t > 0.$
Since $\cos \omega t = sen(\omega t + \pi/2)$,

$$i(t) = \omega C E sen(\omega t + \pi/2) = \omega C E sen[\omega(t + \pi/2\omega)] \qquad (1.66)$$

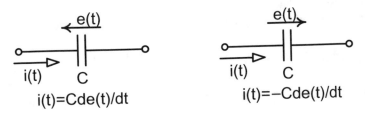

Fig. 1.21 Reference arrows for voltage between the terminal nodes and current across a Capacitance

The comparison of Eqs. 1.65 and 1.66 allows us to conclude that: for sinusoidal waveforms, in a pure capacitance, **the current $i(t)$ is leading** of $\pi/2$ [rd], or 90 [°], in relation to the voltage $e(t)$, or, equivalently, that the voltage $e(t)$ is **lagging** by $\pi/2$ [rd], or 90 [°], in relation to the current $i(t)$, when considering the two waveforms as functions of ωt [rd]. This conclusion is also supported by Eq. 1.62, because if the voltage $e(t)$ depends on the charge $q(t)$, which, in turn, depends on the area under the current curve $i(t)$, obviously, it cannot be advanced to her. In other words, it is necessary to first circulate current, depositing charges on the capacitor plates, and establishing an electric field in the dielectric, so that the voltage can arise. Thus, the current in a capacitance is always ahead of the voltage, because it is the cause and the voltage, the consequence. After studying the singular functions in Chap. 2, the understanding of these last statements, of a general character, becomes more consistent.

Taking a closer look at the last term of Eq. 1.66, we see that, **at time t [s]**, the current is ahead of the voltage of the value $\pi/2\omega$ [s]. Thus, for a frequency $f = 1$ [MHz], that is, one million cycles per second, the current is advanced in voltage for a time equal to

$$\pi/2\omega = \pi/(2 \times 2\pi f) = 1/4f = 1/(4 \times 10^6) = 0.25 \times 10^{-6} \text{ [s]} = 250 \text{ [ns]}.$$

This means that if the sinusoidal current passes through its maximum value at a certain time t_0, the voltage will pass through its maximum value 0.25 [μs] after t_0, although both exist at that time t_0 and in all previous and subsequent moments.

When a capacitance varies with time, it is denoted by $C(t)$. In this case, Eq. 1.49 should be replaced by

$$q(t) = C(t) \cdot e(t) \tag{1.67}$$

Then the current at such a capacitance will be

$$i(t) = \frac{dq(t)}{dt} = C(t)\frac{de(t)}{dt} + e(t)\frac{dC(t)}{dt} \tag{1.68}$$

In this case, a constant voltage $e(t) = E$ produces the current $i(t) = E\frac{dC(t)}{dt}$.

An example of a capacitor with variable capacitance is one composed of two sets of parallel plates, mounted on the same axis, one fixed and the other mobile. Another way to obtain variable capacitance is to vary the spacing between the capacitor's parallel plates.

The power absorbed at capacitance C in Fig. 1.21 is given by

$$p(t) = e(t) \cdot i(t) = C \cdot e(t)\frac{de(t)}{dt} = \frac{dW(t)}{dt} \tag{1.69}$$

Then the energy stored in the capacitance in an instant t is given by

$$W(t) = \int p(t)dt = \int\limits_{-\infty}^{t} p(\lambda)d\lambda \tag{1.70}$$

But Eq. 1.70 can be broken down into two, as follows:

$$W(t) = \int\limits_{-\infty}^{t_0} p(\lambda)d\lambda + \int\limits_{t_0}^{t} p(\lambda)d\lambda \tag{1.71}$$

However, according to Eq. 1.70,

$$\int\limits_{-\infty}^{t_0} p(\lambda)d\lambda = W(t_0) = \int\limits_{-\infty}^{t_0} Ce(\lambda)\frac{de(\lambda)}{d\lambda}d\lambda = C\int\limits_{-\infty}^{t_0} e(\lambda)de(\lambda) = \frac{C}{2}[e(\lambda)]^2\Big|_{-\infty}^{t_0}$$

$$= \frac{C}{2}\left\{ [e(t_0)]^2 - \lim_{\lambda \to -\infty} [e(\lambda)]^2 \right\}$$

It turns out that at $-\infty$ the capacitor had not yet been manufactured, so that $e(-\infty) = 0$. Then,

$$W(t_0) = \frac{C}{2}[e(t_0)]^2$$

So, Eq. 1.71 can be rewritten as

$$W(t) = W(t_0) + \int\limits_{t_0}^{t} p(\lambda)d\lambda = \frac{C}{2}[e(t_0)]^2 + C\int\limits_{t_0}^{t} e(\lambda)de(\lambda) = \frac{C}{2}[e(t_0)]^2$$

$$+ \frac{C}{2}[e(\lambda)]^2\Big|_{t_0}^{t} = \frac{C}{2}[e(t_0)]^2 + \frac{C}{2}[e(t)]^2 - \frac{C}{2}[e(t_0)]^2$$

Finally,

$$W(t) = \frac{C}{2}[e(t)]^2 \tag{1.72}$$

Equation 1.72 shows that the energy stored in capacitance C, at any instant t, depends only on the corresponding value of the voltage on it at that same instant, in addition to the value of the capacitance, obviously. Furthermore, that $W \geq 0$ always, since $C \geq 0$, confirming that capacitance is a passive element of the circuit.

As $e(t) = q(t)/C$, one can also write that

$$W(t) = \frac{[q(t)]^2}{2C} \tag{1.73}$$

Example 1.14 If the capacitance C of Fig. 1.21 has the voltage waveform given by the triangular pulse shown in Fig. 1.22a, find the corresponding waveforms of current, power and energy.

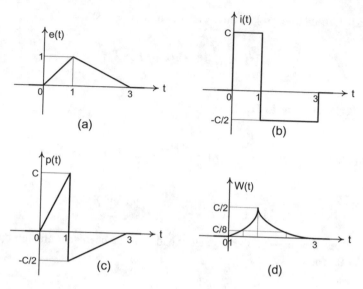

Fig. 1.22 Waveforms for the capacitance C of Fig. 1.21, for **a** voltage, **b** current, **c** power, and **d** energy

Solution

The voltage is defined analytically by

$$
e(t) = \begin{cases}
0 & for\ t \le 0 \\
t & for\ 0 \le t \le 1 \\
(-t+3)/2 & for\ 1 \le t \le 3 \\
0 & for\ t \ge 3.
\end{cases}
$$

So, the current, obtained from $i(t) = C\frac{de(t)}{dt}$, is

$$
i(t) = \begin{cases}
0 & for\ t < 0 \\
C & for\ 0 < t < 1 \\
-C/2 & for\ 1 < t < 3 \\
0 & for\ t > 3.
\end{cases}
$$

The waveform of current $i(t)$ is shown in Fig. 1.22b.
The power, obtained from $p(t) = e(t) \cdot i(t)$, is

$$p(t) = \begin{cases} 0 & for\ t \le 0 \\ Ct & for\ 0 \le t < 1 \\ C(t-3)/4 & for\ 1 < t \le 3 \\ 0 & for\ t \ge 3. \end{cases}$$

The waveform of the power $p(t)$ is recorded in Fig. 1.22c. And the energy, calculated using Eq. 1.72, is given by

$$W(t) = \begin{cases} 0 & for\ t \le 0 \\ Ct^2/2 & for\ 0 \le t \le 1 \\ C(-t+3)^2/8 & for\ 1 \le t \le 3 \\ 0 & for\ t \ge 3. \end{cases}$$

The energy at capacitance C is recorded in the curve of Fig. 1.22d. It is interesting to note that:

1. In the interval $0 < t < 1$ the current $i(t)$, with the reference indicated in Fig. 1.21, moves to the right; however, in the interval $1 < t < 3$ it is worth half the previous value and is directed to the left, because it has changed signs.
2. In the interval $0 \le t < 1$, the power $p(t)$ is positive, meaning that the capacitance is absorbing power; in interval $1 < t \le 3$ the power is negative, meaning that the capacitance is supplying power— obviously, for the rest of the circuit of which it is part.
3. In the range $0 \le t \le 1$, the capacitance is storing energy; however, in the range $1 \le t \le 3$ it is returning all the energy it has accumulated at the end of the first interval.

When N capacitors are connected in parallel, intuitively, one can imagine that another capacitor is formed, in such a way that the area of their plates is the sum of the individual areas of each plate and that, thus, an equivalent capacitance can be obtained through the sum of the individual capacitances. Therefore, for N **capacitances in parallel**, the equivalent capacitance is

$$C_{eq} = C_1 + C_2 + \cdots + C_N \tag{1.74}$$

So, the equivalent capacitance of N capacitances **in series** is obtained from

$$\frac{1}{C_{eq}} = \frac{1}{C_1} + \frac{1}{C_2} + \cdots + \frac{1}{C_N} \tag{1.75}$$

For only two capacitances in series, Eq. 1.75 provides that

$$C_{eq} = \frac{C_1 \cdot C_2}{C_1 + C_2}, \tag{1.76}$$

which is the famous rule of the product by the sum, also used to determine the equivalent resistance of two resistances in parallel.

Inductance

Magnets

The phenomena related to magnetism were discovered by the Chinese surely more than 4000 years ago. At least 3000 years ago, the magnets used in Chinese primitive compasses were pieces of an iron ore known as magnetite, or black iron oxide, Fe_3O_4, which has the property of attracting small pieces of iron. And since magnetite has such magnetic properties in its natural state, these pieces of ore are called **natural magnets**. In addition to these, the only other natural magnet that exists is the Earth itself. All other magnets are man-made and are, therefore, called **artificial magnets** or **electromagnets**.

The regions of a body where magnetism seems to be more concentrated are called **magnetic poles**. Every magnet has exactly two opposite poles, which are designated the **North pole** (N) and the **South pole** (S).

Just as with electrical charges, which repel each other when they are equal and attract each other when they are opposite, **equal magnetic poles repel each other and different poles attract each other**, as shown in Fig. 1.23.

When two north poles are placed next to each other, the magnetic force lines that come out of the north poles have opposite directions and, consequently, repel each other. This repelling force tends to drive the two magnets apart. On the other hand, when the north pole of a magnet is approached with the south pole of another, the magnetic force lines of the respective fields have concordant directions and come together to form loops of longer force lines. These long continuous lines tend to contract, and this force of attraction brings the two magnets together.

In electrostatics there is no difficulty in producing charged bodies, either positively or negatively, as demonstrated by the experiences of Benjamin Franklin. It is also possible to break an electric dipole in two and then obtain a positively charged and a negatively charged end. The magnets, on the other hand, invariably have north and south poles of exactly the same intensity. Therefore, **they are essentially dipoles**. In addition, when a bar of a magnet is cut in half, an isolated north pole or an isolated south pole is not obtained, but two halves, each with their respective north and south poles of exactly equal intensities. In other words, cutting a magnet in half gives rise to two other magnets, as shown in Fig. 1.24, and not two halves, each with its respective pole. Therefore, **there are no free magnetic charges; only in pairs**.

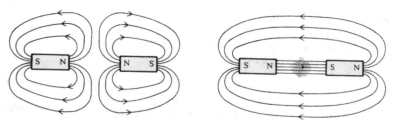

Fig. 1.23 Repulsion or attraction between poles of magnets

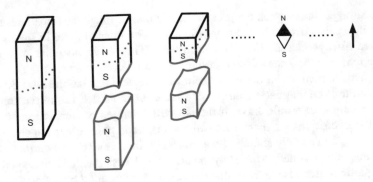

Fig. 1.24 Magnetic dipoles

The right arrow in Fig. 1.24 represents an elementary magnetic dipole, that is, at the atomic level. And the diamond on its left is usually used to represent the needle of a compass.

The magnetic forces exerted by permanently magnetized substances, such as magnetite, on objects made of iron constitute what is called a **magnetic field**. A vector field of forces that, although invisible, can be seen through the effects of its forces, spreading iron filings on a glass plate or sheet of paper placed on a bar-shaped magnet. After that, when the glass plate or paper sheet vibrates lightly and repeatedly, the grains of the filings will be distributed in a well-defined configuration, which describes the magnetic field of forces around the magnet.

Another way of delineating the **magnetic induction field B** is to observe the direction that a small, freely suspended magnetic dipole, such as a compass needle, takes when moving at different points in the vicinity of a bar-shaped magnet, as shown in Fig. 1.25.

Fig. 1.25 Magnetic induction field **B**

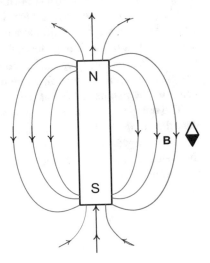

The stronger the magnet, the greater the number of magnetic force lines and the greater the space covered by the magnetic field. **The lines of force in the magnetic field are always closed, that is, continuous**. They have no point of origin or termination, like the lines of force in the electric field.

Magnetic induction **B** is defined as being a force field that has the same properties in relation to "**the magnetic charge**" that the electric field **E** has in relation to the electric charge—**or would have**, if the magnetic charge, as such, existed because, unlike the electric charge, there is no isolated magnetic pole, as can happen with the electric charge. The only source of any magnetic field is electrical current, even in natural magnets. It is defined in *SI* by the unit called Tesla, with the unit symbol [T].

Magnetic materials are those that are attracted or repelled by a magnet and that can be magnetized by it. Iron and steel are the most common magnetic materials. Magnetic materials are also nickel, cobalt and the alloy with 78.5% nickel and 21.5% iron, called permalloy. **Permanent magnets** are formed by hard magnetic materials, such as cobalt steel, which keeps its magnetism even when the magnetizing field is moved away.

Magnetic Flux

Not very strictly, the **magnetic flux** ϕ is defined as the total number of lines of force of the magnetic induction field **B** that crosses a given surface S. (It is not very accurate because each line "pierces" the surface at a different angle with it, at the point where it crosses, in addition to the fact that it is not necessarily flat). Thus, at the flat top of the north pole of the magnet in Fig. 1.25, where each line is perpendicular to it, the magnetic flux is the set, or bundle, of all magnetic force lines that emerge from the area of the top of the north pole of the magnet, and enter it by the south pole. His SI unit is Weber, with the unit symbol $[W_b]$, in honor of the German physicist Wilhelm Weber (1804–1891).

The precise definition of magnetic flux ϕ, through a given surface S, is the integral of the component of **B** normal to the surface over the area established by S.

Figure 1.26 shows any surface S and, over it, at any point P, an infinitesimal element of area *da*, in addition to a unit vector **n** perpendicular to the element of the surface at that point P. In general, the magnetic induction **B** is variable in modulus, direction and sense over the entire surface S, but, for point P, it is possible to consider the area element *da* as being flat and the vector **B** representing the local value of the magnetic field at that point P.

The component of **B** perpendicular to the surface at this point P is nothing more than the component of **B** along the direction of the normal vector **n**, whose value is $B \cdot \cos \theta$. However, since **n** has a unitary module, the scalar product $\mathbf{B} \cdot \mathbf{n}$ also has the value $B \cdot \cos \theta$. Then, the surface-normal component of **B** at point P can be written as

$$B_n = B \cdot \cos \theta = \mathbf{B} \cdot \mathbf{n} \qquad (1.77)$$

Fig. 1.26 Magnetic induction field at point P

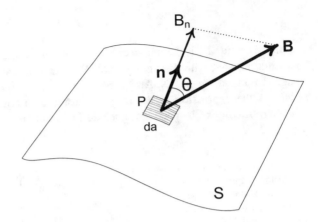

Therefore, the infinitesimal element of magnetic flux through the smallest surface area da is defined as the perpendicular component of the magnetic induction field at the surface times the area da, that is,

$$d\phi = B_n \cdot da = \boldsymbol{B} \cdot \boldsymbol{n}\, da \tag{1.78}$$

To find the total magnetic flux through the total surface S of Fig. 1.26, it is only necessary to integrate all the infinitesimal elements of area that make up the surface. That way,

$$\phi = \int_S \boldsymbol{B} \cdot \boldsymbol{n}\, da \tag{1.79}$$

Usually the task of performing the integral in Eq. 1.79 is extremely difficult and time-consuming. The work, however, is much easier if the magnetic induction $\boldsymbol{B}$ is constant in modulus, direction and sense at all points on the surface S and if, in addition, it is planar, so that the normal vector $\boldsymbol{n}$ is also constant in all points. Under these conditions, the quantity $\boldsymbol{B.n}$ is the same for all surface area elements and can therefore be written outside the integral of Eq. 1.79, that is,

$$\phi = \boldsymbol{B} \cdot \boldsymbol{n} \int_S da = (\boldsymbol{B} \cdot \boldsymbol{n})A = (B \cdot 1 \cdot \cos\theta)A$$

or

$$\phi = B \cdot A \cdot \cos\theta, \tag{1.80}$$

where A represents the total surface area S.

If, in addition, the direction of $\boldsymbol{B}$ is normal to the surface S, the angle θ between the vectors $\boldsymbol{B}$ and $\boldsymbol{n}$ will be zero and Eq. 1.80 will become

$$\phi = B \cdot A \tag{1.81}$$

Based on Eq. 1.81, the magnetic induction unit is more commonly called webers per square meter, with the unit symbol $[\mathrm{Wb/m^2}]$, instead of Tesla. Another unit of magnetic induction is $[\mathrm{N/Am}]$, resulting from $\boldsymbol{F} = q(\boldsymbol{v} \times \boldsymbol{B}) = q \cdot \boldsymbol{v} \cdot \boldsymbol{B} \cdot sen\theta$, which is the expression of the magnetic force in a conductor carrying current— charge q moving with speed $\boldsymbol{v}$, immersed in an external magnetic field $\boldsymbol{B}$, as shown in Fig. 1.27.

Fig. 1.27 Magnetic force in a conductor carrying current— charge q moving with speed $\boldsymbol{v}$, immersed in an external magnetic field $\boldsymbol{B}$

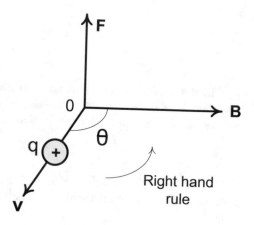

Example 1.15 Find the magnetic flux through a circular surface of 32 [cm] in diameter, whose normal makes an angle of 20° with a constant magnetic induction of 0.6 $[\mathrm{Wb/m^2}]$, as illustrated in Fig. 1.28.

Fig. 1.28 Magnetic flux through a circular surface

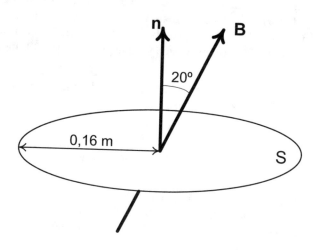

Solution

From Eq. 1.80,

$$\phi = B \cdot A \cdot \cos\theta = (0.6) \cdot \pi(0.16)^2 \cdot (0.940) = 0.045\,[\text{Wb}]$$

If the magnetic field were perpendicular to the plane of the circle then $\theta = 0°$ and

$$\phi = (0.6) \cdot \pi(0.16)^2 = 0.048\,[\text{Wb}].$$

The Oersted Experiment

In 1819 the Danish scientist Hans Christian Oersted (1777–1851), using the voltaic cell capable of establishing a direct current, discovered that the needle of a compass was deflected in the vicinity of a conductor in which an electric current was flowing. And the immediate conclusion was that **an electric current flowing through a conductor produces a magnetic field around it**. More than that, Oersted proved that magnetism and electricity are related phenomena. In Fig. 1.29a the iron filings, when forming a well-defined configuration of concentric circumferences around the conductor, through which a current circulates, show the existence of the magnetic field created by it, so that each section of the wire has around it this field of magnetic forces in a plane perpendicular to the conductor, as shown in Fig. 1.29b. The intensity of the magnetic field around a conductor carrying an electric current depends directly on the intensity of that current.

The **right hand rule** is a very convenient way of determining the relationship between the direction of current flow in a wire and the direction of the lines of force in the magnetic field around the conductor. It is as follows: suppose you are holding the wire that conducts the current with your right hand, with the four adjacent fingers around the wire and with your thumb extended. If the thumb extended along the wire indicates the direction of the current flow, then the other four fingers will indicate the direction of the magnetic force lines of the field around the conductor. This rule is illustrated in Fig. 1.30.

If the conductive wire forms a closed path, such as the circumference conducting the current i, shown in Fig. 1.31, the right hand rule should be used inverting the previous description, that is, the adjacent fingers indicating the direction of the current and the thumb, the direction of the field, in Fig. 1.31, represented by flow ϕ.

If the conductive wire is wound with N circumferences, called **turns**, forming what is called a **coil**, as shown in Fig. 1.32, and if, in addition, an iron core, called **ferromagnetic material**, is placed inside it, it increases the concentration of flux within it, that is, the lines of force are approached in the the magnetic field inside the coil. **Permeability**, with the quantity symbol μ, is a measure of this flow-intensifying property; μ creates a channel for flux ϕ. It has the unit of Henry per meter, with the symbol $[\text{H/m}]$ in the SI. The permeability of the vacuum, designated μ_0, is worth $4\pi \times 10^{-7}\,[\text{Wb/A m}] = 0.4\,[\mu\text{H/m}]$. The permeability of

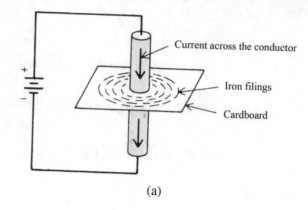

(a)

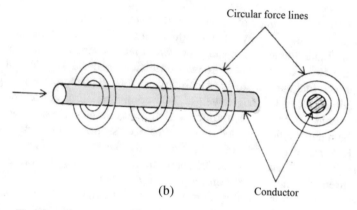

(b)

Fig. 1.29 a Electric current flowing through a conductor producing a magnetic field around it;
b Field of magnetic forces in a plane perpendicular to the conductor [9]

Fig. 1.30 Right hand rule
determining the relationship
between the direction of
current flow in a wire and the
direction of the lines of force
in the magnetic field around
the conductor [9]

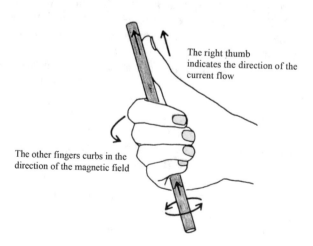

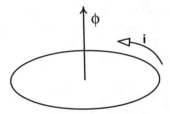

Fig. 1.31 Adjacent fingers indicating the direction of the current and the thumb, the direction of the field, for conductive wire in a closed path

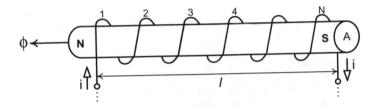

Fig. 1.32 Magnetic flux produced by a current across a coil with N turns and with ferromagnetic core

other materials is related to that of vacuum through a factor called relative permeability, with the symbol μ_r. Such a relationship is $\mu = \mu_r \cdot \mu_0$, that is, $\mu_r = \mu/\mu_0$. Most materials have a relative permeability close to 1, but pure iron has a range of 6,000–8,000 and nickel has a range of 400–1,000. The so-called permalloy has a relative permeability of around 80,000.

The relative permeability can also be expressed by $\mu_r = 1 + \chi_m$, where χ_m is the **magnetic susceptibility**.

An interesting fact that is worth remembering is that

$$\frac{1}{\sqrt{\varepsilon_0 \mu_0}} = \frac{1}{\sqrt{8.85 \times 10^{-12} \times 4\pi \times 10^{-7}}} = 299.863.380,4 \left[\frac{m}{s}\right] \cong 3 \times 10^8 \left[\frac{m}{s}\right]$$

$$= 300,000 \left[\frac{km}{s}\right],$$

which is the speed of propagation of electromagnetic phenomena in free space.

When all the N turns of a coil are associated with the same amount of flux ϕ, it is said that the coil has an **interlaced or concatenated flow**, or an **inductive coupling** $N\phi$.

Magnetomotriz force— *fmm* is the product of the current by the number of turns of a coil. It is the "pressure" required to establish a magnetic flux in a ferromagnetic material. Its unit is [At], which means ampere-turns. In fact, only amperes, because N is a dimensionless number.

$$fmm = N \cdot i \tag{1.82}$$

The **intensity of the magnetic field** is defined by the *fmm* quotient by the distance between the poles of a coil. Its unit symbol is H and its unit is [At/m]. So,

$$H = \frac{Ni}{l} \tag{1.83}$$

H is the field at the center of a solenoid air core. With an iron core, H is the magnetic field through the entire core.

Electromagnetic Induction

Shortly after the Oersted experiment, the French André Marie Ampère (1775–1836) demonstrated that there is a linear relationship, that is, a line passing through the origin, between the magnetic flux ϕ and the current i that produced it, valid for space free. In an iron core coil, however, the relationship between the magnetic flow and the current that produced it is initially almost a straight line, but as the current grows, it curves to the right, presenting the so-called saturation phenomenon, thus characterizing a non-linear relationship. As H depends on i, according to Eq. 1.83, and B is related to ϕ, according to Eq. 1.80, then, for the linear case, it can also be said that the graphical relationship between B and H is a straight line passing through the origin. And the slope of this line is exactly μ. So that, for the linear case,

$$B = \mu \cdot H \tag{1.84}$$

The next innovation step occurred around 1831, when Michael Faraday (1791–1867) and Joseph Henry (1774–1836) discovered, almost simultaneously, that a variable magnetic field could produce a voltage in a closed path - circuit, that is, that **an electric current could be generated magnetically**, but that such an effect was observed only when the magnetic flux through the circuit varied with time. This effect is referred to as **electromagnetic induction**, and the currents and electromotive forces—e.m.fs.—generated in this way are called **induced currents** and **induced e.m.fs**.

Both observed, also, that when a current that varies with time flows in a given circuit, the magnetic field produced by it acts to produce an e.m.f. in this same circuit, but whose effects are opposite to e.m.f. external that makes the current circulate, and vary, first. This effect is called **self-induction**. Therefore, when the voltage across a coil is due to the variable flow produced by the current variation in the coil itself, it is said that the voltage that appears is **self-induced** or **counter-electromotive force**—c.e.m.f.

The experimental results of Faraday and Henry, regarding the production of f.e. ms. and induced currents, can be summarized in the following statement, which constitutes the so-called **Faraday Law of induction**.

Whenever there is a magnetic flux that varies with time through a circuit, an e.m.f. is induced in it, the modulus of which is directly proportional to the rate of change of the magnetic flux in relation to time.

Therefore, from Eq. 1.79, which defines magnetic flux, Faraday's law can be mathematically written as

$$e(t) = k\frac{d\phi}{dt} = k\frac{d}{dt}\int_S \mathbf{B} \cdot \mathbf{n}\, da \qquad (1.85)$$

The integral of Eq. 1.85 is performed over the area enclosed by the circuit and, in the SI, the proportionality constant is -1. Therefore, Faraday's law becomes

$$e(t) = -\frac{d\phi}{dt} = -\frac{d}{dt}\int_S \mathbf{B} \cdot \mathbf{n}\, da \qquad (1.86)$$

Under the conditions that resulted in Eq. 1.80, then

$$e(t) = -\frac{d\phi}{dt} = -\frac{d}{dt}(BA\cos\theta)$$

$$e(t) = -A\cos\theta \cdot \frac{dB}{dt} - B\cos\theta \cdot \frac{dA}{dt} + AB\,sen\theta\,\frac{d\theta}{dt}$$

Considering $\frac{d\theta}{dt} = \omega$, angular speed with which the circuit can rotate, it comes that:

$$e(t) = -A\cos\theta \cdot \frac{dB}{dt} - B\cos\theta \cdot \frac{dA}{dt} + \omega AB\,sen\theta \qquad (1.87)$$

Equation 1.87 is only valid if the surface S is planar and the magnetic induction B is the same at all points in S.

Equation 1.87 shows that there are several ways to induce an e.m.f., according to Faraday's law. They are:

$$C_3^1 + C_3^2 + C_3^3 = \frac{3!}{1!(3-1)!} + \frac{3!}{2!(3-2)!} + \frac{3!}{3!(3-3)!} = 3+3+1 = 7$$

Of these seven possibilities, three deserve more emphasis. They are:

1. The circuit does not rotate, $\omega = 0$, the area does not vary, $A = constant$ and B varies with time. So,

$$e(t) = -A\cos\theta \cdot \frac{dB}{dt} \tag{1.88}$$

2. A and B are constant and the circuit rotates with angular velocity ω. In this case,

$$e(t) = \omega AB sen\theta \tag{1.89}$$

3. B is constant, the circuit does not rotate, $\omega = 0$, and area A varies with time. So,

$$e(t) = -B\cos\theta \cdot \frac{dA}{dt} \tag{1.90}$$

When it comes to a coil with N turns and a concatenated flow $N\phi$, Faraday's law is described mathematically by

$$e(t) = N\frac{d\phi}{dt} \tag{1.91}$$

The polarity of this voltage $e(t)$ is such that any current resulting from it produces a flow that is opposed to altering the original flow. This rule for determining the polarity of the induced voltage is known as **Lenz's Law**. As flow ϕ is produced by current i, it can also be said that the polarity of the induced voltage $e(t)$ is such that it tends to oppose the change in the current that was originally responsible for it. It can be summed up by saying that: **an induced effect is always produced in order to oppose the cause that produced it**. Therefore, in the case of a coil, Eq. 1.91 provides the magnitude of the voltage, and Lenz's law, its polarity. And to elucidate how well to apply Lenz's law, some examples are shown as follows.

Figure 1.33 shows a permanent magnet moving to the right, with its north pole approaching a coil with N turns, wound in one of the two possible directions. The flow ϕ of the permanent magnet is constant. But, as it approaches the coil, it becomes influenced by a **flow growing upwards** and, therefore, **variable in it**. According to Faraday's law, then an e.m.f. is induced in the coils, given by Eq. 1.91, which appears between its two terminals. Then Lenz's law comes in to establish the polarity of this voltage. To try to prevent the growth of the flow that is reaching it, the coil must produce another flow, ϕ_o, that is able to **oppose the growth** of the flow coming from the magnet. According to the right hand rule, with the adjacent fingers agreeing with the winding direction, this requires a current i entering the left terminal of the coil and exiting its right terminal. But this is only possible if the voltage arrow is directed to the left, that is, if the left terminal of the coil is positive and, of course, the right terminal, negative.

Figure 1.34 shows the same description as in Fig. 1.33, but with the winding direction of the coil inverted. Similar reasoning shows that polarity has now also been reversed.

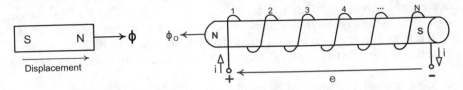

Fig. 1.33 Permanent magnet moving to the right, with its north pole approaching a coil with N turns

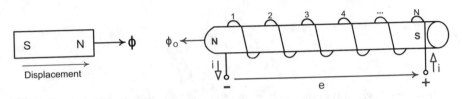

Fig. 1.34 Permanent magnet moving to the right, with its north pole approaching a coil with N turns, with the winding direction of the coil inverted

Figure 1.35 shows a ferromagnetic core of μ permeability over which N turns are wound. Assuming that the magnetic flux $\phi(t)$, clockwise, produced by an external source, not shown, is at a time of growth, then, according to Faraday's law, an e.m.f. $e(t)$ is induced and appears between the two winding terminals. Its magnitude is given by Eq. 1.91. Its polarity comes from the application of Lenz's law. To try to prevent the growth of $\phi(t)$ it is necessary to have an opposite flow $\phi_o(t)$, in a counterclockwise direction. But, according to the right hand rule for coils, this requires an electric current $i(t)$ entering the lower terminal and exiting the

Fig. 1.35 Faraday and Lenz's laws applied to a coil in an iron core

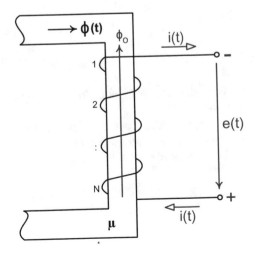

upper terminal of the winding, which determines the voltage polarity of the voltage $e(t)$. Then the voltage reference arrow should be directed downwards, as shown.

It is evident that if $\phi(t)$ is decreasing, $\phi_o(t)$ should reinforce it, also clockwise. That is, there will be an inversion of the current $i(t)$ and, consequently, the polarity of $e(t)$.

Since flow $\phi(t)$ is produced by current $i(t)$ Eq. 1.91 can be rewritten as

$$e(t) = N\frac{d\phi}{di} \cdot \frac{di}{dt} = L\frac{di}{dt} = L\left(\lim_{\Delta t \to 0}\frac{\Delta i}{\Delta t}\right), \tag{1.92}$$

where

$$L = N\frac{d\phi}{di} \tag{1.93}$$

L is called **self-inductance** or, simply, the **coil inductance** shown in Fig. 1.32. Its SI unit is Henry, with the unit symbol [H]. According to Eq. 1.92, one Henry is the amount of inductance that allows the induction of one Volt when the current varies in the ratio of one Amp per second. Typical inductance values are in the order of [mH].

A coil's inductance depends on its shape, how it is wound with one or more layers, the material of the core around which it is wound, the number of turns and the distance between them, and other factors.

Considering that $\phi = BA$ and that $i = Hl/N$, expressions obtained from Eqs. 1.81 and 1.83, we have that $d\phi = A \cdot dB$ and $di = (l/N)dH$. So, from Eq. 1.93,

$$L = N \cdot \frac{A \cdot dB}{(l/N)dH}$$

That is,

$$L = \frac{AN^2}{l} \cdot \frac{dB}{dH} \tag{1.94}$$

B is the magnetic induction in [Wb/m^2] and H is the magnetic intensity in [At/m].

The relationship between B and H depends on the material from which the core is made, and two typical curves are shown in Figs. 1.36a, b.

For the air core,

$\frac{dB}{dH} = tg\theta = \mu$, where μ is the core permeability. So, Eq. 1.94 becomes

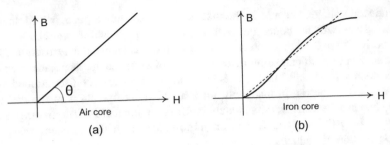

Fig. 1.36 Relationship between B and H depending on the material from which the core is made

$$L = \mu N^2 A / l \qquad (1.95)$$

As the inductance has a constant value, independent of H and, therefore, of the current throught it, it is a linear circuit element. Furthermore, Eq. 1.95 shows that the inductance L always corresponds to the product of permeability μ by a geometric factor with the dimension of length. For the linear case, $d\phi/di$ can be replaced by ϕ/i. Therefore, from Eq. 1.93, $L = N\phi/i$ or,

$$N\phi = Li \qquad (1.96)$$

For the iron core, the dB/dH value depends on the H value and, therefore, on the current, so that the inductance is a non-linear element. Even for the iron core, however, an equivalent linear inductance can be used to produce good results. In this book, the inductance L will always be considered constant.

Physically, any extension of a conductor has some inductance associated with it, when a variable current over time flows throught it. That is, it is not necessary to build an **inductor**—a coil to obtain inductance. An inductor is a device built with the purpose that inductance is its main characteristic, which is the ability to store and supply energy. One can therefore speak of a conductor's inductance or a coil's inductance.

Example 1.16 An inductance of 500 [mH] undergoes a current variation of 15 [mA] in 0.5 [μs]. What is the voltage generated at its terminals?

Solution
As the current curve waveform was not given, it can be considered that

$$\Delta e = L\Delta i / \Delta t = 500 \times 10^{-3} \times 15 \times 10^{-3}(....)/0.5 \times 10^{-6} = 15,000 \text{ [V]}$$
$$= 15 \text{[kV]}$$

Inductance can also be seen as the property of an electrical circuit to counteract sudden changes in current. According to Eq. 1.92, a sudden change in the current, that is, a finite discontinuity in the current curve $i(t)$, would require an infinite self-induced voltage, because in the discontinuity Δt tends to zero, but Δi does not. (These issues will be better understood with the help of the singular functions covered in Chap. 2 of this book). Of course, there is no similar restriction for the inductance voltage. It can suffer discontinuities or even change its polarity instantly, that is, change its signal. The currents in the inductances are not discontinued means that at $t = 0_+$, that is, immediately after a switching operation in the circuit, at $t = 0$, they are the same as immediately before switching, at $t = 0_-$. In other words, that $i(0_+) = i(0_-) = i(0)$. This is an important data in the transient analysis of circuits containing inductances.

If an attempt is made to open a branch of a circuit containing an inductance, through which an electric current is flowing and storing an energy given by $(1/2)Li^2$, an electric **arc** may appear between the switch terminals, and the energy stored in the inductance will be spent on ionizing the air in the region where the arc exists.

It is necessary to pay attention to the fact that Eq. 1.92 implies **associated references** for the voltage and current arrows in an inductance, that is, arrows in opposition; otherwise, a minus sign is required, as shown in Fig. 1.37.

Equation 1.92 also reveals that if the current in an inductance is continuous, that is, constant, then the voltage at its terminals is zero, because $di/dt = 0$. With a constant current flowing through it, but with zero voltage at its terminals, **an inductance behaves like a short-circuit in relation to the direct current**.

From Eq. 1.92, conversely,

$$i(t) = \frac{1}{L} \int e(t)dt \tag{1.97}$$

Or, according to Eq. 1.27,

$$i(t) = \frac{1}{L} \int_{-\infty}^{t} e(\lambda)d\lambda \tag{1.98}$$

Or yet,

$$i(t) = \frac{1}{L} \int_{-\infty}^{t_0} e(\lambda)d\lambda + \frac{1}{L} \int_{t_0}^{t} e(\lambda)d\lambda \tag{1.99}$$

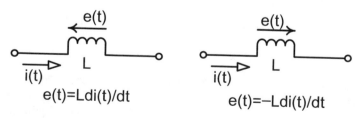

Fig. 1.37 Reference arrows for voltage between the terminal nodes and current across an indutor

But, according to Eq. 1.98,

$$\frac{1}{L} \int_{-\infty}^{t_0} e(\lambda)d\lambda = i(t_0)$$

So,

$$i(t) = i(t_0) + \frac{1}{L} \int_{t_0}^{t} e(\lambda)d\lambda \qquad (1.100)$$

Equation 1.91 can be rewritten as

$$d(N\phi) = edt$$

Or that

$$\int d(N\phi) = N\phi = \int edt \qquad (1.101)$$

Thus, the magnitude $\int edt$ represents the concatenated magnetic flow of a coil of N turns. And the comparison of Eqs. 1.97 and 1.101 allows us to conclude that

$$i(t) = \frac{1}{L}N\phi(t)$$

Or that

$$N\phi(t) = Li(t),$$

confirming Eq. 1.96
  Obviously, for $N = 1$,

$$\phi(t) = Li(t) \qquad (1.102)$$

The power absorbed at the inductance in Fig. 1.37 is given by

$$p(t) = e(t) \cdot i(t) = Li(t)\frac{di(t)}{dt} = \frac{dW(t)}{dt}$$

Then the energy stored in the inductance until any moment t is

$$W(t) = \int p(t)dt = \int_{-\infty}^{t} p(\lambda)d\lambda \qquad (1.103)$$

On the other hand, Eq. 1.103 can be rewritten as

$$W(t) = \int_{-\infty}^{t_0} p(\lambda)d\lambda + \int_{t_0}^{t} p(\lambda)d\lambda \tag{1.104}$$

But, according to Eq. 1.103,

$$\int_{-\infty}^{t_0} p(\lambda)d\lambda = W(t_0) = \int_{-\infty}^{t_0} Li(\lambda)di(\lambda) = \frac{L}{2}[i(\lambda)]^2\Big|_{-\infty}^{t_0} = \frac{L}{2}\Big\{[i(t_0)]^2 - [i(-\infty)]^2\Big\}$$

It turns out that at $-\infty$ the inductance didn't even exist yet, so that $i(-\infty) = 0$. Then,

$$\int_{-\infty}^{t_0} p(\lambda)d\lambda = \frac{L}{2}[i(t_0)]^2$$

Therefore, Eq. 1.104 becomes

$$W(t) = \frac{L}{2}[i(t_0)]^2 + \int_{t_0}^{t} Li(\lambda)di(\lambda) = \frac{L}{2}[i(t_0)]^2 + \frac{L}{2}[i(\lambda)]^2\Big|_{t_0}^{t}$$

$$= \frac{L}{2}[i(t_0)]^2 + \frac{L}{2}[i(t)]^2 - \frac{L}{2}[i(t_0)]^2$$

Finally,

$$W(t) = \frac{L}{2}[i(t)]^2 \tag{1.105}$$

Equation 1.105 shows that the energy in an inductance L, at any instant t, depends only on the value of the current at that exact moment, and the value of the inductance, obviously. Being $L \geq 0$, then always $W(t) \geq 0$, confirming that the inductance is a passive circuit element.

Example 1.17 Find the inductance of a single layer coil with 50 turns wound in a cylindrical ferromagnetic core 1.5 [cm] long and 1.5 [mm] in diameter, if the relative permeability of the material is $\mu_r = 7,000$. What is, approximately, the resistance of the copper wire that was wound to build the coil?

Solution
According to Eq. 1.95,

$$L = \frac{\mu N^2 A}{l} = \frac{\mu_r \mu_0 N^2 A}{l} = \frac{\mu_r \mu_0 N^2 \cdot \pi r_n^2}{l}$$

$\mu_0 = 4\pi \times 10^{-7} [\text{H/m}]; \mu_r = 7,000; \quad N = 50; \quad l = 1.5 [\text{cm}] = 1.5 \times 10^{-2} [\text{m}];$
$d_n = 1.5 [\text{mm}] = 1.5 \times 10^{-3} [\text{m}] \therefore r_n = 0.75 \times 10^{-3} [\text{m}].$

Therefore,

$$L = \frac{7 \times 10^3 \times 4\pi \times 10^{-7} \times 50^2 \times \pi(0.75 \times 10^{-3})^2}{1.5 \times 10^{-2}} = 2.59 [\text{mH}]$$

Disregarding the thickness of the copper wire insulation film, its diameter is approximately equal to the length of the coil divided by the number of turns, as they are juxtaposed. So,

$$d_f = \frac{l}{N} = \frac{1.5 \times 10^{-2}}{50} = 3 \times 10^{-4} [\text{m}]$$

The total length of the wire is the result of the circumference of a loop multiplied by the number of turns. So,

$$l_f \cong 2\pi r_n \cdot N = 2\pi \cdot 0.75 \times 10^{-3} \times 50 = 236 \times 10^{-3} [\text{m}]$$

The cross-sectional area of the copper wire is given by

$$A_f = \pi \frac{d_f^2}{4} = \frac{\pi}{4} (3 \times 10^{-4})^2 = 7.069 \times 10^{-8} [\text{m}^2]$$

From Eq. 1.41,

$$R = \rho \frac{l_f}{A_f}$$

From Table 1.1, $\rho = 1.72 \times 10^{-8} [\Omega\,\text{m}]$. Finally,

$R = 1.72 \times 10^{-8} \frac{236 \times 10^{-3}}{7.069 \times 10^{-8}} = 0.057 [\Omega] \cong 60 [\text{m}\Omega].$

If an inductance varies over time, it will be denoted by $L(t)$. In this case, Eq. 1.102 should be replaced by

$$\phi(t) = L(t)i(t)$$

Then the voltage induced in such an inductance will be

$$e(t) = \frac{d\phi(t)}{dt} = L(t)\frac{di(t)}{dt} + i(t)\frac{dL(t)}{dt} \qquad (1.106)$$

Equation 1.106 shows that, even for direct current, there is a voltage induced in the inductance and the component cannot be treated as a short-circuit. That is, for a direct current I, the voltage at a time-varying inductance is

$$e(t) = I\frac{dL(t)}{dt} \qquad (1.107)$$

Inductance and capacitance are two dual electrical elements, as their definition equations, $e(t) = Ldi(t)/dt$ and $i(t) = Cde(t)/dt$, are essentially the same, except that the roles of $e(t)$ and $i(t)$ are switched. Thus, to obtain all the equations of one of them, just make the appropriate changes in all the equations of the other. For example, if in Eq. 1.49, $q(t) = Ce(t)$, C is replaced by L and $e(t)$ by $i(t)$ Eq. 1.102 is obtained, $\phi(t) = Li(t)$, showing that the dual of q, the charge at capacitance, is ϕ, the flux at inductance. It is important to remember that dual is not the same as equivalent. As mentioned at the beginning of this chapter, two electrical circuits are dual when they are described by the same equations, although they have different physical functioning. The only restriction is that the duality rules are only applicable to planar circuits, that is, those where their different branches only cross at the nodes.

Table 1.4 shows a list of dual element pairs.

According to Table 1.4, the dual circuit of N capacitances in parallel is a circuit with N inductances in series. Thus, series inductances combine in the same way as capacitances in parallel. So for N series inductances,

$$L_{eq} = L_1 + L_2 + L_3 + \cdots + L_N \qquad (1.108)$$

Therefore, for N inductances in parallel,

$$\frac{1}{L_{eq}} = \frac{1}{L_1} + \frac{1}{L_2} + \frac{1}{L_3} + \cdots + \frac{1}{L_N} \qquad (1.109)$$

Table 1.4 Duality of electrical elements

Series connection	Parallel connection
Electric charge $q(t)$	Magnetic flux $\phi(t)$
Voltage $e(t)$	Current $i(t)$
Voltage source	Current source
Short-circuit	Open circuit
Resistance R	Conductance $G = 1/R$
Capacitance C	Inductance L
Mesh	Node
Mesh current	Nodal voltage
Switch that closes	Switch that opens
Mesh with N branches in series	Node with N branches in parallel

In particular, for two inductances in parallel, the rule of the product by the sum applies:

$$L_{eq} = \frac{L_1 \cdot L_2}{L_1 + L_2} \qquad (1.110)$$

Example 1.18 Make an analysis of the phase difference between voltage and current in an inductance subjected to sinusoidal quantities.

Solution
If the current at the inductance in Fig. 1.37 is

$$i(t) = I \cdot sen\omega t \, for \, t \geq 0, \qquad (1.111)$$

then the voltage at its terminals will be

$$e(t) = L\frac{d}{dt}i(t) = \omega LI \cos \omega t = \omega LIsen\left(\omega t + \frac{\pi}{2}\right) = \omega LIsen\left[\omega\left(t + \frac{\pi}{2\omega}\right)\right] \qquad (1.112)$$

Comparing the two equations, we can then conclude that: for sine waveforms, in pure inductance, the voltage is ahead of the current of $\pi/2$ radians (90°) or, equivalently, that **the current $i(t)$ is lagging by $\pi/2$[rd], (90°), in relation to the voltage $e(t)$**. This conclusion is corroborated by Eq. 1.98, because if the current depends on the area under the voltage curve, obviously, it cannot anticipate it. Therefore, the current in an inductance is **always** behind the voltage for any waveforms. This last statement is best understood after studying the singular functions, covered in the second chapter of this book. But, you see, the lag is $\pi/2$ [rd], (90°), when the reference is ωt [rd]. When the reference is t [s], the current delay is $\phi/(2\omega)$ [s]. Thus, if the sinusoidal voltage goes through its maximum value at a given time $t = \tau$, the corresponding sinusoidal current will pass through its peak value a time $\phi/(2\omega)$ [s] after τ, although both exist at that time $t = \tau$ and in all others antecedent and subsequent moments.

The following is a summary of the dual correspondence between the main equations related to capacitance and inductance.

Capacitance	Inductance
$q(t) = C \cdot e(t)$	$\phi(t) = L \cdot i(t)$
$\frac{d}{dt}q(t) = i(t) = C\frac{d}{dt}e(t)$	$\frac{d}{dt}\phi(t) = e(t) = L\frac{d}{dt}i(t)$
$q(t) = \int\limits_{-\infty}^{t} i(\lambda)d\lambda = q(t_0) + \int\limits_{t_0}^{t} i(\lambda)d\lambda$	$\phi(t) = \int\limits_{-\infty}^{t} e(\lambda)d\lambda = \phi(t_0) + \int\limits_{t_0}^{t} e(\lambda)d\lambda$
$e(t) = e(t_0) + \frac{1}{C}\int\limits_{t_0}^{t} i(\lambda)d\lambda$	$i(t) = i(t_0) + \frac{1}{L}\int\limits_{t_0}^{t} e(\lambda)d\lambda$

(continued)

(continued)

Capacitance	Inductance
$p(t) = \frac{d}{dt}W(t) = e(t)i(t) = Ce(t)\frac{d}{dt}e(t)$	$p(t) = \frac{d}{dt}W(t) = i(t)e(t) = Li(t)\frac{d}{dt}i(t)$
$W(t) = \int_{-\infty}^{t} p(\lambda)d\lambda = W(t_0) + \int_{t_0}^{t} p(\lambda)d\lambda$	$W(t) = \int_{-\infty}^{t} p(\lambda)d\lambda = W(t_0) + \int_{t_0}^{t} p(\lambda)d\lambda$
$W(t) = \frac{1}{2}C \cdot e^2(t)$	$W(t) = \frac{1}{2}L \cdot i^2(t)$
Capacitances in parallel $C = C_1 + C_2$	Inductances in series $L = L_1 + L_2$
Capacitances in series $C = \frac{C_1 \cdot C_2}{C_1 + C_2}$	Inductances in parallel $L = \frac{L_1 \cdot L_2}{L_1 + L_2}$

Example 1.19 Find the waveform of the voltage induced in a 400 [mH] coil, when the current shown in Fig. 1.38a passes through it.

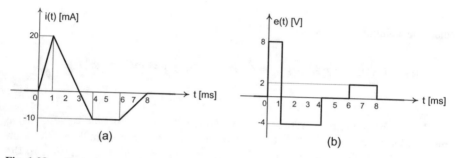

Fig. 1.38 **a** Current across and **b** Voltage between the terminals of a coil (inductor)

Solution

The induced voltage is given by

$$e(t) = L\frac{d}{dt}i(t) = 400 \times 10^{-3}\frac{d}{dt}i(t) = 0.4i'(t)$$

A good way to proceed is to find the slope of the current curve in each stretch, which, according to the geometric interpretation of the derivative concept, is $i'(t)$, and then multiply by 0.4.

For $0 < t < 1$ [ms], the slope is $20\,[\text{mA}]/1\,[\text{ms}] = 20\,[\text{A/s}]$. So, $e(t) = 0.4 \times 20 = 8$ [V].

For $1\,[\text{ms}] < t < 4\,[\text{ms}]$, the slope is $-30\,[\text{mA}]/3\,[\text{ms}] = -10\,[\text{A/s}]$. So, $e(t) = 0.4 \times (-10) = -4[\text{V}]$.

For $4\,[\text{ms}] < t < 6\,[\text{ms}]$, $i'(t) = 0$. Therefore, $e(t) = 0$ [V].

For $6\,[\text{ms}] < t < 8\,[\text{ms}]$, the slope is $10\,[\text{mA}]/2\,[\text{ms}] = 5\,[\text{A/s}]$. So, $e(t) = 0.4 \times 5 = 2$ [V].

The waveform of the voltage induced in the coil is shown in Fig. 1.38b.

Sources

As mentioned earlier, **sources are the active elements in a circuit**. They are elements of two terminals for which there is no pre-established relationship between voltage and current, as is the case with passive elements, that is, resistors, capacitors and inductors. Therefore, when one of the two quantities is given, the other cannot be calculated without knowing the rest of the circuit. They are classified into two types: independent sources and controlled sources, also called linked or dependent.

Independent sources are those for which either the voltage $e(t)$ or the current $i(t)$ is always given, either through a constant value or through a waveform that varies with time. The independent voltage source shown in Fig. 1.39a has a voltage $e(t)$ which is a specified function of time and which is independent of any external connections at its two terminals. On the other hand, the current $i(t)$ depends on the external connection that is made, and can pass through the voltage source in any of the two possible directions. Thus, with concordant reference arrows, if at some point $e(t)$ and $i(t)$ are both positive, the voltage source will be supplying and not absorbing power, that is, $p(t) < 0$. The current $i(t)$ can assume any value, so that the voltage source is theoretically capable of supplying an unlimited amount of power and energy to the rest of the circuit. As ideal elements, which only approximate the real behavior of physical devices, the sources are considered to be capable of supplying energy indefinitely, that is, without limits.

An independent voltage source that maintains a constant voltage at its terminals is called a **continuous (DC) voltage source**. And one that has a voltage that varies sinusoidally over time is called an **alternating voltage source**.

A **battery** is an independent voltage source that provides an almost constant voltage, regardless of the current being requested from it. When $e(t)$ is a constant, that is, when the voltage source is continuous, it is represented as in Fig. 1.39b, where the small straight lines suggest the plates of a battery. The greater sign must be associated with the positive sign. Thus, although the $+$ and $-$ signs represent a redundancy of notation, they are usually included. It is worth remembering that a voltage source of zero value is equivalent to a short circuit.

A generator is a machine on which electromagnetic induction is used to produce a voltage by rotating coils of conductive wire through a stationary magnetic field or by rotating a magnetic field produced by a source of direct voltage, called a

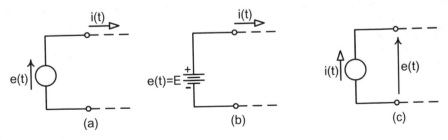

Fig. 1.39 a Independent voltage source; **b** DC voltage source; **c** Independent current source

rotating field of the rotor through the stationary stator coils. A generator that produces a sinusoidal alternating voltage is called an **alternator**. When the stator has three sets of coils: $a - a', b - b', c - c', 120°$ spatially out of phase, sinusoidal voltages of the same amplitude are also induced, also out of phase with 120°, but in the independent variable ωt [rd], where $\omega = 2\pi f$ [rd/s], with f [cycles/second] = f [Hz] the frequency, given by $f = p \cdot n/60$, with n representing the number of rotor turns in [rpm] and p, the number of pairs of magnetic poles of the rotor coils. Such a generator is called a **three-phase alternator**.

The independent current source shown in Fig. 1.39c produces a specific current $i(t)$ independent of the external connections. As the voltage $e(t)$ at its terminals depends on the rest of the circuit and, theoretically, can take on any value, the independent current source can also provide an unlimited amount of power and energy. Current source causes some surprise in those who are more used to voltage sources, but many devices can be represented by a current source in conjunction with some passive elements. In addition, inductors initially energized, that is, traversed by an initial current $i(0) = I$, can be represented by an equivalent consisting of a de-energized inductance L in parallel with a current source of value equal to I. It is evident that the dual situation also exists, that is, a capacitor with an initial voltage $e(0) = E$ can be replaced by a de-energized capacitance C in series with an independent voltage source of value E. A current source of zero value is equivalent to an open circuit.

The term **source at rest** is often used to denote a zero value source. So, a source of voltage at rest is a short-circuit, and a source of current at rest is an open circuit.

A controlled source is one whose value is not independent of the rest of the circuit of which it is a part, but is a known function of some other voltage or current in the same circuit. They are particularly important in building models of electronic components. A controlled source is not really a component of two terminals, as the voltage or current on which it depends must also be shown so that the element is fully characterized.

A symbol for a **common emitter NPN transistor** is shown in Fig. 1.40a, where b, e and c denote the base, emitter, common to the input and output, and collector terminals. This three-terminal component can be operated in approximately linear or non-linear fashion. In the linear range, the collector current $i_c(t)$ is approximately proportional to the base current $i_b(t)$, but it also depends on the voltage between the collector and the emitter $e_{ce}(t)$. The small voltage between the base and the emitter $e_{be}(t)$ depends on the base current $i_b(t)$ and, to a lesser extent, on the collector voltage to the emitter $e_{ce}(t)$. A model that includes these effects is shown in Fig. 1.40b, and contains two controlled sources.

A network that does not contain any active elements, that is, an independent source, is called a **passive network**.

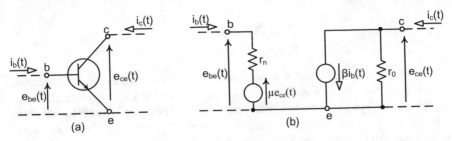

Fig. 1.40 a Common emitter NPN transistor; **b** Equivalent circuit model for a common emitter NPN transistor

1.6 Kirchhoff's Laws

The two laws dealing with the effects of the interconnection of different circuit elements were proposed by Gustav Robert Kirchhoff (1824–1887), a German physicist. Kirchhoff's laws are strictly valid only when the signal propagation time is neglected, that is, when the elements of the circuit are concentrated. In this book, any connection of elements in such a way that the dimensions of them and the circuit itself are negligible compared to the wavelength of the highest frequency of interest will be called a **lumped circuit**.

The restriction on the size of the elements is a consequence of Kirchhoff's laws being approximations of Maxwell's famous equations, which are the general laws of electromagnetism. The approach is analogous to the fact that Newton's laws of classical physics are approximations to the laws of relativistic mechanics.

A **branch** of a circuit is a single path containing a single element that connects a node to any other node. Remembering that all interconnecting wires of circuit elements are considered as short-circuits, a **node** is defined as the meeting of all points that are subject to the same potential.

Kirchhoff formulated the law of currents, also called the first law, assuming that no charge can accumulate in an element or in a node and that the charge must be conserved, that is, that a node cannot store, generate or destroy electrical charge. Therefore, the total current entering a node of any concentrated circuit, at any time, must be equal to the total current leaving it at the same time. In summary, the first law is

Kirchhoff's Currents Law

The algebraic sum of currents leaving any node in a lumped circuit, at any time, is equal to zero

Symbolically,

$$\sum_{n=1}^{k} i_n(t) = 0 \qquad\qquad (1.113)$$

When applying Eq. 1.113, it is necessary to remember that a current $i(t)$ entering a node is equivalent to a current $-i(t)$ leaving it. Furthermore, in a circuit with N nodes, the application of the law of currents will only produce **linearly independent equations** if it is applied at most to any $N - 1$ nodes in the circuit. Therefore, the law of currents should not be applied to all nodes in the circuit; one node must always be excluded. With linearly dependent equations, the number of unknowns will be greater than the number of truly useful equations. In other words, applying the law to the total number of nodes N of the circuit will result in N equations, but so that any one of them will be the negative of the sum of the other $N - 1$ and, therefore, will not add any new information. Usually, this node that is excluded in the application of the law of currents is the one taken as a reference node in the application of the nodal circuit solution method, to which the grounding symbol is assigned.

The Kirchhoff's currents law is independent of the nature of the circuit elements, that is, it applies to any circuit with lumped parameters, whether linear, non-linear, active, passive, variant or invariant over time.

A **loop** of a circuit is a set of branches that form a single closed path. So, each node in a loop must have exactly two branches of the loop attached to it. A **mesh** is a particular case of loop, which does not involve any other branch.

Kirchhoff's Voltages Law

Kirchhoff's voltages law is based on the principle of energy conservation. Based on the fact that voltage is energy per unit of charge, it establishes that

The sum of the voltages along any closed path is equal to zero

Symbolically,

$$\sum_{n=1}^{k} e_n(t) = 0 \qquad\qquad (1.114)$$

A closed path is understood as a loop or mesh.

In the application of Eq. 1.114, when the route adopted, clockwise or counter-clockwise, coincides with the voltage reference arrow, entering the tail and exiting the arrowhead, the corresponding voltage must be included positively in the equation. Otherwise, it enters the equation with a negative sign. The number of times you can apply Eq. 1.114 to a circuit also has a limit. It matches the number of loops that can be counted in a circuit, even if the equation is being done by loops and not by meshes. And the objective is always to guarantee a set of linearly independent equations, that is, one in which no equation can be obtained by linearly combining the others. Alternatively, you can calculate the maximum number of times you can apply Eq. 1.114 using the formula

$$L = B - N + 1, \tag{1.115}$$

where N is the total number of nodes in the circuit, B is the total number of branches and L is the maximum number of loops—or meshes—to which Eq. 1.114 can be applied.

Note that it is the path that needs to be closed. This means that if it includes an open switch, the voltage at the switch terminals must also be included in the sum.

1.7 Analytical and Digital Circuit Solution

The following example is a preview of the main objective of this book, which is the search for a complete solution of concentrated parameter circuits, either analytically discrete-time, that is, in the domain of continuous-time, or digitally, in the domain of discrete-time. It is, therefore, expected that the reader, at the end of this chapter, will feel sufficiently motivated to continue until the last of its pages.

Example 1.20 Find the current, voltage, power and energy waveforms in the elements of the RC series circuit, powered by a constant source, shown in Fig. 1.41, knowing that the capacitance is de-energized at $t = 0_-$. Then, perform the digital simulation, using the ATPDraw program, for the following numerical values $E = 10\,[\text{V}], R = 10\,[\Omega], C = 1{,}000\,[\mu\text{F}]$.

Fig. 1.41 RC series circuit, powered by a constant source

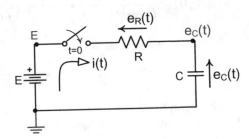

Solution
As the digital solution uses the nodal method, it will also be used in the analytical solution. With the switch closed, there are three nodes in the circuit, namely: the lower node, consisting of all points of the connection between the negative battery terminal and the lower capacitance plate; the upper left node, consisting of all the points that connect the positive terminal of the battery and the left terminal of the resistor; and the upper right node, which gathers all the connection points from the right terminal of the resistor to the upper plate of the capacitance.

As the essence of the nodal method consists of the application of the law of currents to nodes of unknown voltages in the circuit, always ignoring one of them, the lower node was taken as a reference. However, when doing this the upper left node has the source voltage, E, known, and the upper right node, with the capacitance voltage. Since there is no equation for a known voltage node, it remains, therefore, to equate only the node of unknown voltage, $e_C(t)$. Therefore, adding the two currents that supposedly leave this node and equaling to zero, it comes that

$$\frac{e_C(t) - E}{R} + C\frac{d}{dt}e_C(t) = 0$$

Or

$$e'_C(t) + \frac{1}{RC}e_C(t) = \frac{E}{RC}, \text{ with } e_C(0_-) = 0. \tag{1.116}$$

The solution of the corresponding homogeneous differential equation

$\left(D + \frac{1}{RC}\right)e_C(t) = 0$ is

$$e_{C_H}(t) = ke^{-t/RC}$$

The particular integral can be obtained using the so-called abbreviated method (Chap. 3), namely:

$$e_{C_P}(t) = \frac{1}{D + \frac{1}{RC}}\left(\frac{E}{RC}\right) = \frac{1}{D + \frac{1}{RC}}\left(\frac{E}{RC}\epsilon^{0.t}\right) = \frac{1}{0 + \frac{1}{RC}}\left(\frac{E}{RC}\epsilon^{0.t}\right) = E$$

Therefore, the general solution is

$$e_C(t) = e_{C_P}(t) + e_{C_H}(t) = E + ke^{-t/RC} \tag{1.117}$$

To determine the constant k, it is necessary to know the initial condition $e_C(0_+)$. Since the voltage at the capacitance cannot vary instantly, it is evident that $e_C(0_+) = e_C(0_-) = 0$. Taking $t = 0_+$ in Eq. 1.117 comes that

$0 = E + k$ then $k = -E$.
So,

$$e_C(t) = E - E\epsilon^{-t/RC} = E\left(1 - \epsilon^{-t/RC}\right), \quad \text{for } t \geq 0. \tag{1.118}$$

The product $RC = \tau$ is called the **RC circuit time-constant**. Theoretically, the term containing the exponential function only tends to zero when time t tends to infinity. It turns out that for $t = 5\tau = 5RC$, $\epsilon^{-t/RC} = \epsilon^{-5} = 0.007$, that is, the exponential has a value equal to just 0.7% of its initial value, which is equal to 1. Thus, it is customary to consider that for $t \geq 5\tau$ the exponential function is already given as null. With this practice, it can then be said that the expression of the voltage in the capacitance, Eq. 1.118, contains a **transient component**, the exponential, and a component of **steady-state regime**, the constant E. In other words, that in a time approximately equal to 5τ the capacitance has already been charged and reached the source voltage, E. It is possible to predict, then, that after a time greater than 5τ, there is no longer a need to circulate current to put more positive electrical charge on the upper plate of the capacitance, with the consequent equally positive repellency of the bottom plate. The voltage between the terminals of the resistance is

$$e_R(t) = E - e_C(t)$$

So,

$$e_R(t) = E\epsilon^{-t/RC}, \quad \text{for } t > 0, \tag{1.119}$$

The current flowing through the circuit elements is

$$i(t) = \frac{e_R(t)}{R} = \frac{E}{R}\epsilon^{-t/RC}, \quad \text{for } t > 0. \tag{1.120}$$

The powers in the passive elements are:

$$p_R(t) = R[i(t)]^2 = \frac{E^2}{R}\epsilon^{-2t/RC}, \quad \text{for } t > 0 \tag{1.121}$$

$$p_C(t) = e_C(t) \cdot i(t) = E\left(1 - \epsilon^{-t/RC}\right) \cdot \frac{E}{R}\epsilon^{-t/RC}$$

Therefore,

$$p_C(t) = \frac{E^2}{R}\left(\epsilon^{-t/RC} - \epsilon^{-2t/RC}\right), \quad \text{for } t \geq 0. \tag{1.122}$$

Note that the power in the capacitance has two time constants: $\tau_1 = \tau = RC$ and $\tau_2 = RC/2 = \tau_1/2$.

The energy dissipated in the resistance is obtained by applying Eq. 1.36, with $t_0 = 0_+$:

$$W_R(t) = W_R(0_+) + \int_{0_+}^{t} p_R(\lambda)d\lambda = 0 + \frac{E^2}{R}\int_{0_+}^{t} \epsilon^{-2\lambda/RC}d\lambda = \frac{E^2}{R} \cdot \frac{\epsilon^{-2\lambda/RC}}{\frac{-2}{RC}}\Big|_{0_+}^{t}$$

$$= -\frac{CE^2}{2} \cdot \epsilon^{-2\lambda/RC}\Big|_{0_+}^{t}$$

Then,

$$W_R(t) = \frac{CE^2}{2}\left(1 - \epsilon^{-2t/RC}\right), \quad \text{for } t \geq 0. \tag{1.123}$$

The energy stored in the capacitance is obtained by applying Eq. 1.72:

$$W_C(t) = \frac{C}{2}[e_C(t)]^2$$

So,

$$W_C(t) = \frac{C}{2}\left[E\left(1 - \epsilon^{-t/RC}\right)\right]^2 = \frac{CE^2}{2}\left(1 - 2\epsilon^{-t/RC} + \epsilon^{-2t/RC}\right), \quad \text{for } t \geq 0. \tag{1.124}$$

Note that

$$W_R(t) + W_C(t) = CE^2\left(1 - \epsilon^{-t/RC}\right). \tag{1.125}$$

The power at the voltage source is

$$p_F(t) = -E \cdot i(t) = -\frac{E^2}{R}\epsilon^{-t/RC} \quad \text{for } t > 0.$$

Note that $p_F(t) + p_R(t) + p_C(t) = 0$.
The energy at the voltage source is

$$W_F(t) = W_F(0_+) + \int_{0_+}^{t} p_F(\lambda)d\lambda = 0 - \frac{E^2}{R}\int_{0_+}^{t} \epsilon^{-\lambda/RC}d\lambda = -\frac{E^2}{R} \cdot \frac{\epsilon^{-\lambda/RC}}{\frac{-1}{RC}}\Big|_{0_+}^{t}$$

Therefore,

$$W_F(t) = CE^2\left(\epsilon^{-t/RC} - 1\right), \quad \text{for } t \geq 0.$$ Note that $W_F(t) + W_R(t) + W_C(t) = 0$.
Substituting the given numeric values, the following expressions result:

$$e_C(t) = 10(1 - \epsilon^{-100t}) \text{ [V]}, \ t \geq 0 \text{ [s]}; \tau = 1/100 = 10 \text{ [ms]}.$$

$$e_R(t) = 10\epsilon^{-100t} \text{ [V]}, t > 0 \text{ [s]}.$$

Note that $e_C(t) + e_R(t) = E, t > 0$ [s], satisfying Kirchhoff's voltages law.

$$i(t) = \epsilon^{-100t} \text{ [A]}, t > 0 \text{ [s]}.$$

$$p_R(t) = 10\epsilon^{-200t} \text{ [W]}, t > 0 \text{ [s]}.$$

$$p_C(t) = 10(\epsilon^{-100t} - \epsilon^{-200t}), t \geq 0 \text{ [s]}. \tag{1.126}$$

$$W_R(t) = 50(1 - \epsilon^{-200t}) \text{ [mJ]}, t \geq 0 \text{ [s]}.$$

$$W_C(t) = 50(1 - 2\epsilon^{-100t} + \epsilon^{-200t}) \text{ [mJ]}, t \geq 0 \text{ [s]}.$$

The circuit built for the digital solution using the ATPDraw program is shown in Fig. 1.42. (See Appendix A) As $\tau = 10$ [ms], $t_{max} = 70$[ms] was adopted, enough to capture the transient and steady state regimes. In the absence of a formula for determining the value of Δt, the standard ATP value was adopted, that is, $\Delta t = 1$ [µs]. A common practice is to adopt a value in the order of one hundredth of the time constant, that is, $\Delta t = \tau/100$. This practice, however, produces satisfactory results for first order circuits, in which it is already known, previously, that the transient components will be simple exponentials.

The waveforms obtained digitally are shown in Fig. 1.43.

Deriving $p_C(t)$, given by Eq. 1.124, and equaling zero, we obtain that the instant when the power in the capacitance passes through the maximum corresponds to $t = \tau \cdot ln2 = 10 \times 0.6931 = 6.931$ [ms]. At this moment, the power $p_{C_{max}} = 2.5$ [W], in agreement with the value shown in the curve of $p_C(t)$ shown in Fig. 1.43.

Fig. 1.42 *RC* series circuit, powered by a constant source, built for the digital solution using the ATPDraw program

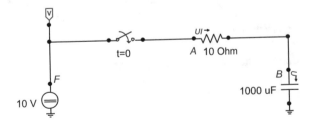

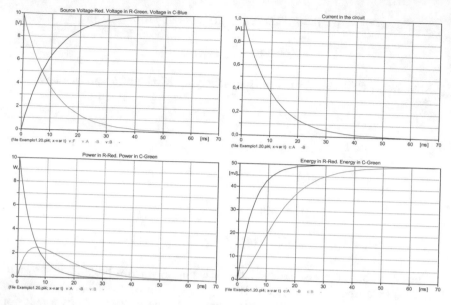

Fig. 1.43 *RC* series circuit transient response for voltage, current, power and energy with digital solution using the ATPDraw program

1.8   Conclusions and Motivations for Electromagnetic Transients Studies

From this brief introduction to the theme of transients in basic electrical circuits, it is evident that the understanding of physical phenomena requires adequate mathematical modelling and the use of computer simulation tools through digital models capable of representing accurately and with stability all components of the system under study. We think that the book is intended for engineering students who already have been successfully approved in introductory circuit analysis, calculus and ordinary differential equations courses. Our experience as educators is that, when it comes to truly understand and apply the mathematical knowledge in electrical circuits transients, the majority of students have much more difficulties. Therefore, this book supplies, in an integrated way, all the necessary knowledge and practice through extensive examples. In the next chapters of this book we aim to present an introduction to the modelling and mathematical calculation of transients in electrical circuits through an analytical solution and digital computational solution for a wide variety of basic cases (at least for circuits with concentrated parameters), using an EMTP-based program, the ATP software—Alternative Transients Program. At the end of the studies, it is expected that students have developed essential skills to understand the facilities, limitations and potential of these engineering tools applicable to the industry in the electrical sector.

References

1. Alexander CK, Sadiku MNO (2008) Fundamentals of electrical circuits, 3rd edn. McGraw-Hill do Brasil, Ltda, São Paulo
2. Boylestad RL (2012) Introduction to circuit analysis, 12th edn. Pearson Education of Brazil, São Paulo
3. Close CM (1966) The analysis of linear circuits. Harcourt, Brace & World Inc., New York
4. McKelvey JP, Grotch H (1979) Physics, vol 1 and 3. Harper & Row of Brazil Ltda, São Paulo
5. Nilson JW, Riedel SA (2009) Electrical circuits, 8th edn. Pearson Education of Brazil, São Paulo
6. Nussenzveig HM (1997) Basic physics course, vol 1 and 2, 3rd edn. Edgard Blücher,
7. O'Malley JR (1992) Circuit analysis. In: Schaum's outline, 2nd edn. McGraw-Hill, New York
8. Resnick R, Halliday D, Walker J (2010) Fundamentals of physics 3: electromagnetism; LTC, 8th edn.
9. Young HD, Freedman RA (2009) Physics III: electromagnetism, vol 3, 12th edn. Pearson Education

Chapter 2
Singular Functions

2.1 Single Step Function

This chapter introduces the singular functions, also called generalized functions or distributions, that constitute a special class of functions, because they do not fit perfectly into the conventional mathematical concept of function. Here they will be designated by $U_n(t)$, $n = 0, \pm 1, \pm 2, \pm \cdots$, in such a way that for each value of n a specific singular function will correspond. They can be considered as a family of functions that are generated from the unit impulse function, $U_0(t)$, by successive derivations or integrations. However, for simplicity and also for strictly didactic reasons, the study of them will be done starting from the unit step function, $U_{-1}(t)$. An important reason to study them is that the response of any circuit linear to an arbitrary entry can be determined, indirectly, as long as its response to a unit step or unit impulse is known. This is done through the integral convolution call. Thus, the response of a linear circuit to a unitary impulse, denoted by $h(t)$, or to a unitary step, called $r(t)$, can be used as a characteristic property of such a circuit.

Although the step and impulse functions do not actually appear as signals on any physical circuit, that is, on any realizable circuit, they can approximate many physical signals. Often, a circuit's excitation is a pulse whose duration is short compared to the circuit's time constants. Under this condition, it is then possible to approximate the desired response of the circuit, to the pulse, by responding to an impulse of the same area, acting from the same moment in which the original pulse would act. Also, the energy initially stored in elements of electrical circuits can be conveniently represented by hypothetical sources of the type of step or impulse. Furthermore, in mathematical manipulations involving the differential equations that describe a given circuit, impulses may appear, even if the source that excites it is not an impulse.

The **unit step function**, $U_{-1}(t)$, shown in Fig. 2.1a, also known as the **Heaviside Unit Function**, is defined as being null for negative values of t, unit for positive values of t, and discontinuous for $t = 0$. Then,

© The Author(s), under exclusive license to Springer Nature Switzerland AG 2021
J. C. Goulart de Siqueira and B. D. Bonatto, *Introduction to Transients in Electrical Circuits*, Power Systems, https://doi.org/10.1007/978-3-030-68249-1_2

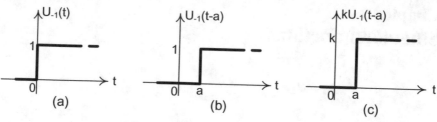

Fig. 2.1 a Unitary step function $U_{-1}(t)$; **b** Unitary step function delayed of a: $U_{-1}(t-a)$; **c** Amplitude k step function $kU_{-1}(t-a)$

$$U_{-1}(t) = \begin{cases} 0 & for\ t<0 \\ 1 & for\ t>0. \end{cases} \qquad (2.1)$$

Therefore, the unit step function is not defined for $t=0$, but is known for all values arbitrarily close to $t=0$. This behavior is indicated by writing that $U_{-1}(0_-)=0$ and $U_{-1}(0_+)=1$.

If, however, the discontinuity occurs at time $t=a>0$, and not at $t=0$, then the unit step will be denoted by $U_{-1}(t-a)$, as shown in Fig. 2.1b, for $a>0$. Therefore,

$$U_{-1}(t-a) = \begin{cases} 0 & for\ t<a \\ 1 & for\ t>a. \end{cases} \qquad (2.2)$$

It is said, then, that this is a delayed unit step of a value a.

It should also be noted that $U_{-1}(a_-)=0$ and that $U_{-1}(a_+)=1$.

When the discontinuity has a value $k \neq 1$ and occurs at time $t=a$, that is, when the step function has an amplitude equal to k and is delayed by a time a, then the function is designated by $kU_{-1}(t-a)$ and represented as shown in Fig. 2.1c.

The possibility of making delays and amplitude changes in the unit step function allows to express analytically a large number of different pulses. For example, the rectangular pulse shown in Fig. 2.2a can be expressed by adding two step functions, namely:

$$p(t) = AU_{-1}(t-a) - AU_{-1}(t-b), \qquad (2.3)$$

as explained in Fig. 2.2b.

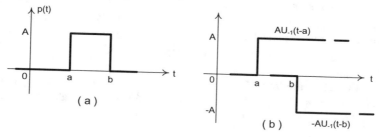

Fig. 2.2 a Pulse function; **b** Unitary step functions used to represent the pulse

Fig. 2.3 Triangular pulse

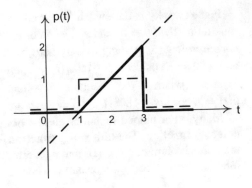

The multiplication of the rectangular, dashed, unitary pulse equation, between $t = 1$ and $t = 3$, that is, $U_{-1}(t - 1) - U_{-1}(t - 3)$, by the line defined by equation $(t - 1)$ produces the triangular pulse

$$p(t) = (t - 1)[U_{-1}(t - 1) - U_{-1}(t - 3)], \tag{2.4}$$

shown in full line in Fig. 2.3.

In the theory of electrical circuits, it is common to convene the instant of application of the excitation as the instant $t = 0$. This means that if an ideal switch is closed at time $t = 0$, in series with a source of direct voltage E, for example, by exciting the circuit from $t = 0$, as shown in Fig. 2.4a, then it is possible to eliminate the switch and take the voltage source as $e(t) = EU_{-1}(t)$, as shown in Fig. 2.4b.

In both Fig. 2.4a and Fig. 2.4b, we have that:

$$e(t) = \begin{cases} 0 & for\ t < 0 \\ E & for\ t > 0 \end{cases} = EU_{-1}(t).$$

Therefore, if the switch is closed in series with a sinusoidal voltage source $E \cos \omega t$, at time $t = 0$, then the switch can be eliminated by taking the voltage source como $e(t) = E \cos(\omega t + \theta)U_{-1}(t)$, where the phase angle θ was introduced to express the value that the voltage source $e(t)$ presents exactly at the moment of closing the switch, $t = 0$, that is, $e(0) = E \cos \theta t$. (The term sinusoidal voltage source will be applied here for both $E \cos \omega t$ and $E \cos \omega t$, as the only difference between them is a 90° lag).

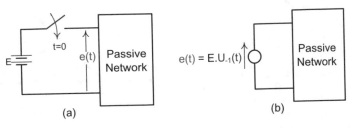

Fig. 2.4 a Switching at time $t = 0$; **b** Equivalent circuit with unit step function $U_{-1}(t)$

For example, if the switch is closed at the instant when the source voltage is maximum, then $\theta = 0^0$ and $e(t) = E\cos\omega t.U_{-1}(t)$. If the switch is closed when the source voltage is passing exactly zero, with a positive derivative, then $\theta = -90^0$ and $e(t) = E\cos(\omega t - 90^0)U_{-1}(t) = E\operatorname{sen}\omega t.U_{-1}(t)$. If, however, the switch is façade when the source voltage is minimal, $\theta = 180^0$ and $e(t) = E\cos(\omega t + 180^0)U_{-1}(t) = -E\cos\omega t.U_{-1}(t)$.

Many other unique functions can be obtained by successive integrations of the unit step function. The **unit ramp function**, $U_{-2}(t)$, for example, is the integral of the unit step function $U_{-1}(t)$, that is, it represents the area involved by the unit step function from $-\infty$ to the instant t. Then,

$$U_{-2}(t) = \int U_{-1}(t)dt = \int_{-\infty}^{t} U_{-1}(\lambda)d\lambda = \int_{-\infty}^{0_-} U_{-1}(\lambda)d\lambda + \int_{0_+}^{t} U_{-1}(\lambda)d\lambda$$

$$= \int_{-\infty}^{0_-} 0.dt + \int_{0_+}^{t} 1.d\lambda = \begin{cases} 0 & for\ t \leq 0 \\ t & for\ t \geq 0. \end{cases}$$

$$(2.5)$$

So Eq. (2.5) can also be written as

$$U_{-2}(t) = t.U_{-1}(t) = \begin{cases} 0 & for\ t \leq 0 \\ t & for\ t \geq 0. \end{cases} \qquad (2.6)$$

Figure 2.5a shows the unitary ramp function $U_{-2}(t)$, which is continuous at $t = 0$, and in Fig. 2.5b, the unitary ramp function delayed of a, namely:

$$U_{-2}(t - a) = (t - a)U_{-1}(t - a) = \begin{cases} 0 & for\ t \leq a \\ t - a & for\ t \geq a. \end{cases} \qquad (2.7)$$

From the last term in Eq. (2.5), the one containing the switch, and also from Fig. 2.5a, there is that

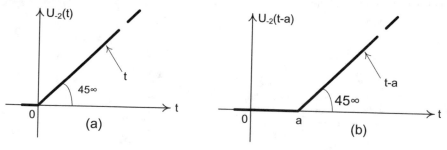

Fig. 2.5 a Unitary ramp function $U_{-2}(t)$; **b** Unitary ramp function delayed of a: $U_{-2}(t - a)$

$$U_{-1}(t) = \frac{d}{dt} U_{-2}(t). \tag{2.8}$$

In turn, the integration of the unit ramp function $U_{-2}(t)$, that is, the area comprised by the unit ramp function from $-\infty$ to time t, generates the singular **second degree unit parabola function**, $U_{-3}(t)$. So,

$$U_{-3}(t) = \int U_{-2}(t)dt = \int_{-\infty}^{t} U_{-2}(\lambda)d\lambda = \int_{-\infty}^{t} U_{-2}(\lambda)d\lambda + \int_{0}^{t} U_{-2}(\lambda)d\lambda$$

$$= \int_{-\infty}^{0} 0.dt + \int_{0}^{t} \lambda d\lambda = \begin{cases} 0 & \text{for } t \leq 0 \\ \frac{t^2}{2} & \text{for } \geq 0. \end{cases} \tag{2.9}$$

Alternatively, Eq. (2.9) can be written as

$$U_{-3}(t) = \frac{t^2}{2} U_{-1}(t) = \begin{cases} 0 & \text{for } t \leq 0 \\ \frac{t^2}{2} & \text{for } \geq 0. \end{cases} \tag{2.10}$$

The functions $U_{-3}(t)$ and $U_{-3}(t-a) = \frac{(t-a)^2}{2} U_{-1}(t-a)$ are represented in Figs. 2.6a and 2.6b, respectively.

From the last term of Eq. (2.9), the one containing the switch, and also from Fig. 2.6a, we have

$$U_{-2}(t) = \frac{d}{dt} U_{-3}(t). \tag{2.11}$$

Continuing with the successive integrations, other singular functions are being obtained, namely:

$$U_{-4}(t) = \int U_{-3}(t)dt = \int_{-\infty}^{t} U_{-3}(\lambda)d\lambda = \frac{t^3}{3!} U_{-1}(t) \tag{2.12}$$

$$U_{-3}(t) = \frac{d}{dt} U_{-4}(t) \tag{2.13}$$

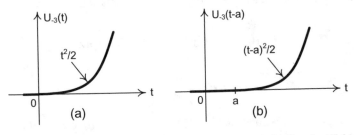

Fig. 2.6 a Unitary parabola function $U_{-3}(t)$; **b** Unitary parabola function delayed of a: $U_{-3}(t-a)$

$$U_{-5}(t) = \int U_{-4}(t)dt = \int_{-\infty}^{t} U_{-4}(\lambda)d\lambda = \frac{t^4}{4!}U_{-1}(t) \qquad (2.14)$$

$$U_{-4}(t) = \frac{d}{dt}U_{-5}(t) \qquad (2.15)$$

$$U_{-(m+1)}(t) = \int U_{-m}(t)dt = \int_{-\infty}^{t} U_{-m}(\lambda)d\lambda = \frac{t^m}{m!}U_{-1}(t) \qquad (2.16)$$

$$U_{-m}(t) = \frac{d}{dt}U_{-(m+1)}(t), m = 1, 2, 3, \ldots \qquad (2.17)$$

Generalizing, we have that:

$$U_{k-1}(t) = \int U_k(t)dt = \int_{-\infty}^{t} U_k(\lambda)d\lambda \qquad (2.18)$$

$$U_k(t) = \frac{d}{dt}U_{k-1}(t) \qquad (2.19)$$

In the process of integrating a function that is multiplied by a unit step, the unit step factor should be used to define the lower limit of the integral, but it is neither necessary nor desirable to include it in the integration process, that is, not integrating. After the integration has been carried out, the corresponding unit step factor is introduced again. This can be described mathematically as:

$$\int_{-\infty}^{t} f(\lambda)U_{-1}(\lambda)d\lambda = \left[\int_{0}^{t} f(\lambda)d\lambda\right]U_{-1}(t). \qquad (2.20)$$

$$\int_{0}^{t} f(\lambda)U_{-1}(\lambda - a)d\lambda = \left[\int_{a}^{t} f(\lambda)d\lambda\right]U_{-1}(t - a). \qquad (2.21)$$

2.2 Unitary Impulse Function

Taking $k = 0$ in Eqs. (2.18) and (2.19) give the expressions

$$U_{-1}(t) = \int U_0(t)dt = \int_{-\infty}^{t} U_0(\lambda)d\lambda \qquad (2.22)$$

$$U_0(t) = \frac{d}{dt}U_{-1}(t) \qquad (2.23)$$

But, as they were obtained, Eqs. (2.18) and (2.19) are only valid for integer and negative values of k, ($k = -1, -2, -3, -\cdots$). And you can't just assign zero to the k index and extrapolate from there. However, and despite the inconsistency pointed out when adopting $k = 0$, Eqs. (2.18) and (2.19) are absolutely correct, as we intend to show, with a wealth of details, below.

Equation (2.22) suggests the existence of a singular function $U_0(t)$, of unit area for $t > 0$ and null for $t < 0$, according to the definition of the unit step function $U_{-1}(t)$, Eq. (2.1). On the other hand, Eq. (2.23) establishes that the function $U_0(t)$ is null for $t \neq 0$, since the unit step derivative is null for $t \neq 0$, that is, derived from zero for $t < 0$ and from 1 to $t > 0$. Thus, the existence of the function $U_0(t)$ is restricted only to the time $t = 0$, which, in turn, implies that the function $U_0(t)$ is the derivative of the unit step function only at time $t = 0$, this instant in which it is discontinuous. And this, from the point of view of conventional classical mathematics, is inadmissible (in classical mathematics there is no derivative of discontinuous function). **So the function $U_0(t)$, in addition to having to exist only at time $t = 0$, must involve a unit area**, because,

$\int_{-\infty}^{t} U_0(\lambda)d\lambda = 0$, when $t < 0$, given that $U_{-1}(t) = 0$ for $t < 0$, and

$\int_{-\infty}^{t} U_0(\lambda)d\lambda = 0$, when $t > 0$, given that $U_{-1}(t) = 1$ for $t > 0$.

The impossibility of proposing a rigorous definition for the function $U_0(t)$, based on classical mathematics of the 1930s, led physicist Paul AM Dirac, who employed it for the first time in his work in Quantum Mechanics, to call it improper function. Currently, the $U_0(t)$ function is called the **unitary impulse function** or the **Dirac delta function**. Many authors denote it as $\delta(t)$.

In 1950, Laurent Schwartz published a treatise entitled "The Theory of Distributions", within which he proposed a sufficiently rigorous and satisfactory basis for Dirac's delta function. Unfortunately, Schwartz's theory proved to be too abstract for the mathematicians and physicists of the day.

In 1953, George Temple developed a more elementary theory, although no less rigorous, under the title "Theories and Applications of Generalized Functions", within which the impulse function was rigorously defined; thus, Eqs. (2.22) and (2.23) are duly proved.

In Engineering courses, the unit impulse function is usually presented in three different ways, namely: as a limit, as an equation and through a property.

■ **Definition as Limit** In this case, the unit impulse function is defined as the limit of a sequence of functions $f_n(t)$, that is:

$$\lim_{n \to \infty} f_n(t) = \delta(t) = U_0(t), \tag{2.24}$$

provided that the functions $f_n(t)$ satisfy the following conditions:

$$\int_{-\infty}^{+\infty} f_n(t)dt = 1, \text{ and} \tag{2.25}$$

$$\lim_{n \to \infty} f_n(t) = 0, \text{ for } t \neq 0. \tag{2.26}$$

(The condition imposed by Eq. (2.26) is not always necessary).

As can be seen, there are a large number of functions that satisfy Eqs. (2.25) and (2.26). Figure 2.7 shows the rectangular pulse $p_\Delta(t)$, with base Δ, between $-\Delta/2$ and $+\Delta/2$, and height $1/\Delta$. The pulse sequence formed when $\Delta \to 0$ satisfies Eq. (2.26). It also satisfies Eq. (2.25), because

$$\int_{-\infty}^{+\infty} p_\Delta(t)dt = 1,$$

regardless of the value assigned to Δ. In other words, when Δ tends to zero, the limit results in "a zero-based pulse, infinite amplitude and unit area", that is,

$$U_0(t) = \overset{lim}{\Delta \to 0} p_\Delta(t). \tag{2.27}$$

The triangular pulse shown in Fig. 2.8 also becomes a unitary pulse when a tends to zero. Like this,

$$U_0(t) = \overset{lim}{a \to 0} p_a(t).$$

Another function that generates the unit impulse is shown in Fig. 2.9. So,

$$U_0(t) = \overset{lim}{\varepsilon \to 0} \frac{\varepsilon^{-t^2/\varepsilon}}{\sqrt{\pi\varepsilon}}$$

The unit impulse function will be represented as shown in Fig. 2.10a. Figure 2.10b shows an area A pulse, delayed by time a.

Logically, to obtain a pulse of area A, starting from Fig. 2.7, for example, the amplitude of the rectangular pulse must be A/Δ. It is interesting to note, then, that **the factor A that multiplies an impulse does not represent its amplitude, which is infinite, but the area it contains**. Positive and negative impulses must be represented by arrows of the same size, drawn above and below the time axis, respectively.

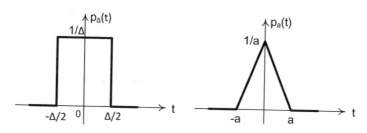

Fig. 2.7 Unitary impulse function $U_0(t)$ defined as a rectangular pulse

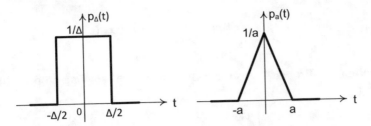

Fig. 2.8 Unitary impulse function $U_0(t)$ defined as a triangular pulse

Fig. 2.9 Unitary impulse function $U_0(t)$ defined as a gaussian function

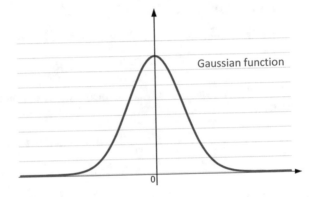

Fig. 2.10 **a** Unitary impulse function $U_0(t)$; **b** Unitary Amplitude A impulse function delayed of a: $U_0(t-a)$

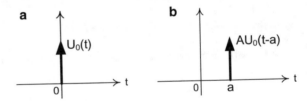

■ **Definition by Equation** The definition by the equation is simply as follows:

$$\begin{cases} U_0(t) = 0 \; for \; t \neq 0 \\ \int_{-\infty}^{+\infty} U_0(t)dt = 1. \end{cases} \tag{2.28}$$

As the unit impulse is equal to zero for $t \neq 0$, one can also write that

$$\int_{0_-}^{0_+} U_0(t)dt = 1, \tag{2.29}$$

because the total area of the unit impulse is "concentrated" at $t = 0$.

In consequence,

$$\int_{-\infty}^{0_-} U_0(t)dt = \int_{0_+}^{+\infty} U_0(t)dt = 0. \tag{2.30}$$

■ **Definition by Property** Also known as the pulse sampling property, it is the following:

$$\int_{-\infty}^{+\infty} f(t)U_0(t-a)dt = f(a), \tag{2.31}$$

provided that $f(a)$ exists.

Equation (2.31) is easily justified, since $U_0(t-a) = 0$ for $t \neq a$. So,

$$f(t)U_0(t-a) = 0 \text{ for all } t \neq a.$$

For $t = a$, where $f(t)$ is defined at this point, $f(t)U_0(t-a)$ becomes $f(a)U_0(t-a)$. Then,

$$f(t)U_0(t-a) = f(a)U_0(t-a) \tag{2.32}$$

So,

$$\int_{-\infty}^{+\infty} f(t)U_0(t-a)dt = \int_{-\infty}^{+\infty} f(a)U_0(t-a)dt = f(a)\int_{-\infty}^{+\infty} U_0(t-a)dt$$
$$= f(a).1 = f(a).$$

As a particular case of Eq. (2.31), it is obvious that:

$$\int_{-\infty}^{+\infty} f(t)U_0(t)dt = f(0). \tag{2.33}$$

One way of defining the unit impulse function also emerges from the Inverse Fourier Transform:

$$U_0(t) = \frac{1}{2\pi}\int_{-\infty}^{+\infty} e^{j\omega t}d\omega = \frac{1}{2\pi}\int_{-\infty}^{+\infty} \cos \omega t \, d\omega = \frac{1}{\pi}\int_{0}^{+\infty} \cos \omega t \, d\omega.$$

Using Eq. (2.33), we can now present a first justification for the very important Eq. (2.23).

Integrating by parts ($\int u dv = uv - \int v du$) the expression $f(t)\frac{d}{dt}U_{-1}(t)$, between the limits $-\infty$ and $+\infty$, comes that:

$$\int_{-\infty}^{+\infty} f(t)\frac{d}{dt}U_{-1}(t)dt = U_{-1}(t)f(t)\Big|_{-\infty}^{+\infty} - \int_{-\infty}^{+\infty} U_{-1}(t)\frac{d}{dt}f(t)dt$$

$$= f(+\infty) - \int_{0}^{+\infty} d[f(t)] \tag{2.34}$$

$$= f(+\infty) - f(t)\Big|_{0}^{+\infty} = f(0).$$

Comparing Eq. (2.34) with Eq. (2.33), or considering the operators that produce the same result to be the same, it is concluded that

$$U_0(t) = \frac{d}{dt}U_{-1}(t),$$

Equation (2.23) is justified.

A second way of justifying Eq. (2.23), and also (2.22), which is more intuitive and less abstract, is done with the help of Figs. 2.11a, b. It is easy to see that:

$$f_0(t) = \frac{d}{dt}f_{-1}(t) \tag{2.35}$$

$$f_{-1}(t) = \int_{-\infty}^{t} f_0(\lambda)d\lambda \tag{2.36}$$

When a tends to zero, the function $f_{-1}(t)$ converges to the function $U_{-1}(t)$, and the function $f_0(t)$, to the function $U_0(t)$. Thus, Eqs. (2.36) and (2.35) are transformed, respectively, into Eqs. (2.22) and (2.23), namely:

$$U_{-1}(t) = \int_{-\infty}^{t} U_0(\lambda)d\lambda = \int U_0(t)dt.$$

$$U_0(t) = \frac{d}{dt}U_{-1}(t).$$

It is justified, therefore, that Eqs. (2.18) and (2.19) are also valid for $k = 0$.

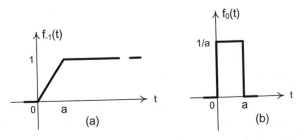

Fig. 2.11 a Approximate representation of a unitary step function; b Approximate representation of a unitary impulse function

It is also observed, from the last expression, that **the unit of the unitary impulse function is** $[s^{-1}]$, when the unit step function is dimensionless. Otherwise, the unit is second^{-1} times the unit of the step function. For example, if the step function is the electrical charge.

$$q(t) = QU_{-1}(t)[C], \text{ then, } q(0_-) = 0 \ [C], \ q(0_+) = Q[C], \text{ and}$$

$$i(t) = \frac{d}{dt}q(t) = \frac{d}{dt}[QU_{-1}(t)] = QU_0(t)\left[\frac{C}{S}\right] = QU_0(t)[A].$$

It should be noted that the area of a current impulse, Q, represents the electrical charge that the current impulse transfers instantly at $t = 0$, causing $q(0_+) = Q[C] \neq q(0_-) = 0[C]$.

If the step has amplitude A and its discontinuity occurs at time $t = a$, that is, if the step is $AU_{-1}(t-a)$, then, according to the chain derivation rule,

$$\frac{d}{dt}[AU_{-1}(t-a)] = A\frac{d}{dt}U_{-1}(t-a) = A\frac{dU_{-1}(t-a)}{d(t-a)} \cdot \frac{d(t-a)}{dt} = AU_0(t-a)$$

(2.37)

Equation (2.37) highlights a most relevant fact, namely: **The derivative of a finite discontinuity A, present in any step function, is an impulse function of area A, located exactly at the moment when the step function discontinuity occurs**. this conclusion can be expanded to any function $f(t)$, which presents a finite discontinuity at any time $t = a$. Its derivation should be done normally for the intervals $-\infty < t \leq a_-$ e $a_+ \leq t < +\infty$. Then, the impulse $[f(a_+) - f(a_-)]U_0(t - a)$ must be added, exactly at time $t = a$.

2.3   The Family of Singular Functions

When $k = 1$ is assigned to Eqs. (2.19) and (2.18), it follows that:

$$U_1(t) = \frac{d}{dt}U_0(t).$$

(2.38)

$$U_0(t) = \int_{-\infty}^{t} U_1(\lambda)d\lambda = \int U_1(t)dt.$$

(2.39)

Although again there are difficulties when taking $k = 1$ in Eqs. (2.18) and (2.19), which have been deducted for integer and negative values of k, we will no longer enter into the merits of this question, simply accepting that Eqs. (2.38) and (2.39) are correct.

The singular function $U_1(t)$ is called a **unit double**.

One way to try to visualize the unit stunt function uses Fig. 2.12. In part (a) of the figure, a triangular pulse $f_0(t)$ is shown, which, as seen above, generates a unit impulse when it tends to zero. In part (b) the function $f_1(t)$ is shown, which is the derivative of the pulse of part (a). When it tends to zero, the function $f_1(t)$ will converge to the unitary double function, $U_1(t)$. Part (c) of Fig. 2.12 shows the graphical representation used by some authors. But the graphic representation that will be used here is the one shown in part (d) of the same Fig. 2.12. That is, a single arrow will be used, not two, but writing next to it that it is the stunt function, $U_1(t)$. And the main reason for the adoption of this representation is justified in the next paragraph.

Assigning the integer and positive values $2, 3, 4, ...$, to the k index in Eq. (2.19), other singular functions are being obtained, namely: $U_2(t) = \frac{d}{dt} U_1(t); U_3(t) = \frac{d}{dt} U_2(t); U_4(t) = \frac{d}{dt} U_3(t)$, etc., which, of course, also satisfy Eq. (2.18). They will also always be represented here by a single arrow, next to which the function referred to will be noted. Otherwise, using the notation of other authors, the function $U_2(t)$ would have to be represented by three arrows, $U_3(t)$ by four, and so on, creating a lot of confusion when two or more of them are occurring at the same point. Obviously, the **impulsive functions** $U_0(t)$, $U_1(t)$, $U_2(t)$, ... must be understood in the sense of **generalized, symbolic functions** or **distributions**, and not in the classical sense of functions.

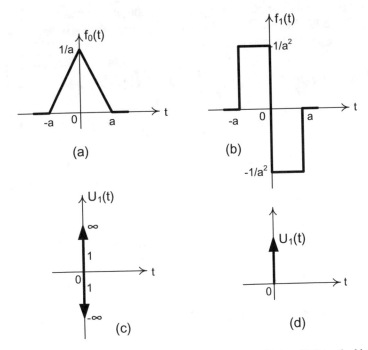

Fig. 2.12 **a** Triangular pulse function; **b** Derivative of pulse of (**a**); **c** Unitary double function, $U_1(t)$; **d** graphic representation that will be used here for a unitary double function, $U_1(t)$

It remains evident, then, that the unit impulse function, $U_0(t)$, can be considered as the starting point for obtaining the entire **family of singular functions,** through successive integrations or derivations. Seen from this point of view, the reason for choosing negative and positive indices used in the notation of singular functions is perfectly understood. Thus, negative indices mean that the singular function in question was obtained by successively integrating the unit impulse function, $U_0(t)$. And positive indexes, by successive derivation of it. Therefore, **to derive or integrate a singular function just add or subtract one to its respective index.** There is no simpler way to derive or integrate a function!

The notation $U_n(t)$ used here for singular functions also has the advantage of bringing together the Laplace Transform of all of them in a single formula, namely:

$$\mathcal{L}[U_n(t)] = s^n, n = 0, \pm 1, \pm 2, \pm \cdots \tag{2.40}$$

Then,

$$\mathcal{L}[U_0(t)] = s^0 = 1; \mathcal{L}[U_1(t)] = s; \mathcal{L}[U_2(t)] = s^2; \mathcal{L}[U_{-1}(t)] = \frac{1}{s}; \mathcal{L}[U_{-2}(t)] = \frac{1}{s^2}.$$

Figure 2.13 presents a synthetic representation of the singular functions, with the main objective of reaffirming that they constitute a family of functions, having the unitary impulse function as a generator.

The singular functions obtained by deriving the unit impulse function, on the right in Fig. 2.13, include the following properties, demonstrated through the process of integration by parts:

$$\int_{-\infty}^{+\infty} U_1(t)f(t)dt = U_0(t)f(t)\Big|_{-\infty}^{+\infty} - \int_{-\infty}^{+\infty} U_0(t)f'(t)dt$$

$$= U_0(t)f(0)\Big|_{-\infty}^{+\infty} - f'(0)\int_{-\infty}^{+\infty} U_0(t)dt = -f'(0). \tag{2.41}$$

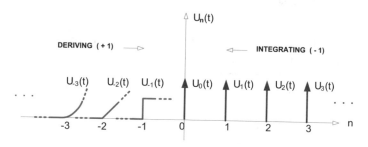

Fig. 2.13 Synthetic representation of the family of singular functions, having the unitary impulse function as a generator

$$\int_{-\infty}^{+\infty} U_2(t)f(t)dt = U_1(t)f(t)\Big|_{-\infty}^{+\infty} - \int_{-\infty}^{+\infty} U_1(t)f'(t)dt$$

$$= -\int_{-\infty}^{+\infty} U_1(t)f'(t)dt = -[-f''(0)] = f''(0). \tag{2.42}$$

Generally,

$$\int_{-\infty}^{+\infty} U_n(t)f(t)dt = -\int_{-\infty}^{+\infty} U_{n-1}(t)f'(t)dt = (-1)^n f^{(n)}(0), n = 1, 2, 3, \cdots \tag{2.43}$$

where,

$$f^{(n)}(0) = \frac{d^n}{dt^n} f(t)\Big|_{t=0}.$$

And also,

$$\int_{-\infty}^{+\infty} U_n(t-a)f(t)dt = (-1)^n f^{(n)}(a), \tag{2.44}$$

for $n = 0, 1, 2, 3, \cdots$.

Note that Eqs. (2.43) and (2.44) encompass Eqs. (2.33) and (2.31), respectively.

Example 2.1 Prove the following properties:

(a) $\qquad \int_{-\infty}^{+\infty} U_0(t-a)f(t)dt = \int_{-\infty}^{+\infty} U_0(t)f(t+a)dt = f(a).$ $\qquad$ (2.45)

(b) $\qquad \int_{-\infty}^{+\infty} U_0(at)f(t)dt = \frac{1}{|a|}\int_{-\infty}^{+\infty} U_0(t)f\left(\frac{t}{a}\right)dt = \frac{1}{|a|}f(0), a \neq 0.$ $\qquad$ (2.46)

Solution

(a) Considering $t - a = \tau$, $t = \tau + a$, $dt = d\tau$,

$$\int_{-\infty}^{+\infty} U_0(t-a)f(t)dt = \int_{-\infty}^{+\infty} U_0(\tau)f(\tau+a)d\tau = \int_{-\infty}^{+\infty} U_0(t)f(t+a)dt$$

Because of Eq. (2.33),

$$\int_{-\infty}^{+\infty} U_0(t)f(t+a)dt = f(t+a)|_{t=0} = f(a).$$

(b) Considering, now, $at = \tau, t = \tau/a, dt = (1/a)d\tau$, we have for $a > 0$, that:

$$\int_{-\infty}^{+\infty} U_0(at)f(t)dt = \frac{1}{a}\int_{-\infty}^{+\infty} U_0(\tau)f(\tau/a)d\tau = \frac{1}{a}\int_{-\infty}^{+\infty} U_0(t)f(t/a)dt$$

$$= \frac{1}{a}f(0) = \frac{1}{|a|}f(0).$$

When $a < 0$, t and τ you have opposite signs and

$$\int_{-\infty}^{+\infty} U_0(at)f(t)dt = \frac{1}{a}\int_{+\infty}^{-\infty} U_0(\tau)f(\tau/a)d\tau = -\frac{1}{a}\int_{-\infty}^{+\infty} U_0(t)f(t/a)dt$$

$$= \frac{1}{|a|}f(0).$$

Example 2.2 Prove that the unit impulse function is even.

Solution

From Eqs. (2.46) and (2.33), it follows that

$$\int_{-\infty}^{+\infty} U_0(at)f(t)dt = \frac{1}{|a|}f(0) = \frac{1}{|a|}\int_{-\infty}^{+\infty} U_0(t)f(t)dt = \int_{-\infty}^{+\infty} [U_0(t)/|a|]f(t)dt$$

From the extreme integrals, considering that $f(t)$ is any function, we conclude that

$$U_0(at) = \frac{1}{|a|}U_0(t) \tag{2.47}$$

Taking $a = -1$ in the above relationship, it follows that

$$U_0(-t) = U_0(t), \tag{2.48}$$

showing that $U_0(t)$ **is an even function.**

Example 2.3 Demonstrate that

$$\int_{-\infty}^{+\infty} U_{-1}(at)f(t)dt = \frac{1}{a|a|}\int_{-\infty}^{+\infty} U_1(t)f(t)dt = \frac{-1}{a|a|}f'(0), \ a \neq 0. \tag{2.49}$$

Solution

For $a > 0$, considering $\tau = at, t = \tau/a, dt = (1/a)d\tau$, it comes that

$$
\begin{aligned}
\int_{-\infty}^{+\infty} U_1(at)f(t)dt &= \frac{1}{a}\int_{-\infty}^{+\infty} U_1(\tau)f(\tau/a)d\tau = \frac{1}{a}\int_{-\infty}^{+\infty} U_1(t)f(t/a)dt \\
&= \frac{1}{a}\left[-\frac{d}{dt}f\left(\frac{t}{a}\right)\right]\Big|_{t=0} = \frac{1}{a}\left[-\frac{df(t/a)}{d(t/a)}\cdot\frac{(t/a)}{dt}\right]\Big|_{t=0} \\
&= -\frac{1}{a}\left[f'(t/a)\cdot\frac{1}{a}\right]\Big|_{t=0} = -\frac{1}{a^2}f'(0) = \frac{1}{a^2}\int_{-\infty}^{+\infty} U_1(t)f(t)dt,
\end{aligned}
$$

$$(2.50)$$

according to Eq. (2.41).

For $a < 0$, with the same variable change, considering that now when $t \to -\infty, \tau \to +\infty$ and vice versa, it comes that

$$
\begin{aligned}
\int_{-\infty}^{+\infty} U_1(at)f(t)dt &= \frac{1}{a}\int_{+\infty}^{-\infty} U_1(\tau)f(\tau/a)d\tau = -\frac{1}{a}\int_{-\infty}^{+\infty} U_1(\tau)f(\tau/a)d\tau \\
&= \frac{1}{-a}\int_{-\infty}^{+\infty} U_1(t)f(t/a)dt = \frac{1}{-a}\left[-\frac{d}{dt}f\left(\frac{t}{a}\right)\right]\Big|_{t=0} \\
&= \frac{1}{-a}\left[-\frac{df(t/a)}{d(t/a)}\cdot\frac{(t/a)}{dt}\right]\Big|_{t=0} = \frac{1}{a^2}|f'(t/a)|_{t=0} \\
&= \frac{1}{a^2}f'(0) = -\frac{1}{a^2}\int_{-\infty}^{+\infty} U_1(t)f(t)dt.
\end{aligned}
$$

$$(2.51)$$

With $a|a| = \begin{cases} a^2 & para\ a > 0 \\ -a^2 & para\ a < 0 \end{cases}$, Equations (2.50) and (2.51) can be combined into one:

$$
\int_{-\infty}^{+\infty} U_1(at)f(t)dt = \frac{1}{a|a|}\int_{-\infty}^{+\infty} U_1(t)f(t)dt = \frac{-1}{a|a|}f'(0), a \neq 0.
$$

Example 2.4 Prove that the unit double function is odd.

Solution

From the two integrals of Eq. (2.49) it appears that:

$$
U_1(at) = \frac{1}{a|a|}U_1(t) \tag{2.52}
$$

For $a = -1$,

$$U_1(-t) = -U_1(t),$$

confirming that the unit stunt function $U_1(t)$ **is odd**.

Example 2.5 Prove that

$$f(t)U_1(t) = f(0)U_1(t) - f'(0)U_0(t). \tag{2.53}$$

Then, calculate $tU_1(t)$.

Solution

From Eq. (2.32), for $a = 0$, we have that:

$$f(t)U_0(t) = f(0)U_0(t). \tag{2.54}$$

Deriving the two members of Eq. (2.54), it results

$$f'(t)U_0(t) + f(t)U_1(t) = f(0)U_1(t)$$

But, according to Eq. (2.54), $f'(t)U_0(t) = f'(0)U_0(t)$. So,

$$f(t)U_1(t) = f(0)U_1(t) - f'(0)U_0(t).$$

For $f(t) = t$,

$$tU_1(t) = -U_0(t). \tag{2.55}$$

Example 2.6 Establish that

$$f(t)U_2(t) = f(0)U_2(t) - 2f'(0)U_1(t) + f''(0)U_0(t). \tag{2.56}$$

Then, calculate $t^2 U_2(t)$.

Solution

Deriving Eq. (2.53), it follows that:

$$f'(t)U_1(t) + f(t)U_2(t) = f(0)U_2(t) - f'(0)U_1(t)$$

where,

$$f(t)U_2(t) = f(0)U_2(t) - f'(0)U_1(t) - f'(t)U_1(t)$$

But, because of Eq. (2.53) itself,

$$f'(t)U_1(t) = f'(0)U_1(t) - f''(0)U_0(t)$$

So, after replacing the last equation on the penultimate,

$$f(t)U_2(t) = f(0)U_2(t) - 2f'(0)U_1(t) + f''(0)U_0(t).$$

For $f(t) = t^2$, Eq. (2.56) provides that:

$$t^2 U_2(t) = 2U_0(t). \tag{2.57}$$

The generalization of the identities of examples 2.5 and 2.6 is

$$f(t)U_n(t) = \sum_{k=0}^{n} (-1)^k \frac{n!}{k!(n-k)!} f^{(k)}(0)U_{n-k}(t), \text{for } n = 1, 2, 3, \ldots \tag{2.58}$$

It can also be shown that

$$f(t)U_n(t-a) = \sum_{k=0}^{n} (-1)^k \frac{n!}{k!(n-k)!} f^{(k)}(a)U_{n-k}(t-a), \text{for } n = 1, 2, 3, \ldots$$

$$\tag{2.59}$$

Example 2.7   Prove that

$$U_{-1}(\alpha t) = \begin{cases} U_{-1}(t) & \text{for } \alpha > 0 \\ U_{-1}(-t) & \text{for } \alpha < 0. \end{cases} \tag{2.60}$$

Solution
The definition of unitary step, given by Eq. (2.1), allows to write that

$$U_{-1}(\alpha t) = \begin{cases} 0 & \text{for } \alpha t < 0 \\ 1 & \text{for } \alpha t > 0. \end{cases}$$

Or,

$$U_{-1}(\alpha t) = \begin{cases} 0 & \text{for } t < 0 \\ 1 & \text{for } t > 0 \end{cases} = U_{-1}(t) \text{ for } \alpha > 0.$$

Now let $\alpha = -\beta, \beta > 0$. From Eq. (2.1),

$$U_{-1}(\alpha t) = U_{-1}(-\beta t) = \begin{cases} 0 & \text{if} -\beta t < 0 \\ 1 & \text{if} -\beta t > 0 \end{cases} = \begin{cases} 0 & \text{if } \beta t > 0 \\ 1 & \text{if } \beta t < 0 \end{cases} = \begin{cases} 0 & \text{if } t > 0 \\ 1 & \text{if } t < 0 \end{cases}$$

$$= U_{-1}(-t) \text{ if } \alpha < 0,$$

Thus, Eq. (2.60) is proved.

Example 2.8 Prove that

$$U_n(-t) = (-1)^n U_n(t), \text{ for } n = 0, 1, 2, 3, \ldots \tag{2.61}$$

Solution

For $n = 0$, the identity has already been demonstrated, Eq. (2.48), showing that the impulse function is even, namely:

$$U_0(-t) = + U_0(t).$$

For $n = 1$, the Eq. (2.61) has already been demonstrated in example 2.4, showing that the unit double function is odd. But it's easy to prove it. From Eq. (2.19), for $k = 1$,

$$U_1(t) = \frac{d}{dt} U_0(t)$$

So,

$$U_1(-t) = \frac{d}{d(-t)} U_0(-t) = \frac{d}{-dt} U_0(-t) = -\frac{d}{dt} U_0(t)$$

$$= -U_1(t).$$

Also from Eq. (2.19), for $k = 2$,

$$U_2(t) = \frac{d}{dt} U_1(t)$$

So,

$$U_2(-t) = \frac{d}{d(-t)} U_1(-t) = \frac{d}{-dt} U_1(-t) = -\frac{d}{dt}[-U_1(t)] = \frac{d}{dt} U_1(t) = + U_2(t).$$

Still from Eq. (2.19),

$$U_3(t) = \frac{d}{dt} U_2(t)$$

So,

$$U_3(-t) = \frac{d}{d(-t)} U_2(-t) = \frac{d}{-dt} U_2(-t) = -\frac{d}{dt}[+U_2(t)] = -U_3(t).$$

Therefore, generalizing, in fact,

$$U_n(-t) = (-1)^n U_n(t), \text{for } n = 0, 1, 2, 3, \ldots$$

Example 2.9 Express the function $f(t)$ of Fig. 2.14 in a series of singular functions.

Fig. 2.14 Function of
Example 2.9

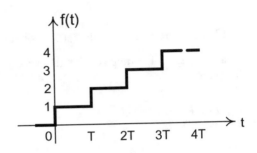

Solution
By simple inspection, it is easy to conclude that

$$f(t) = U_{-1}(t) + U_{-1}(t - T) + U_{-1}(t - 2T) + U_{-1}(t - 3T) + \cdots$$

Example 2.10 Perform the step expansion of the periodic function $f_1(t)$ shown in Fig. 2.15a.

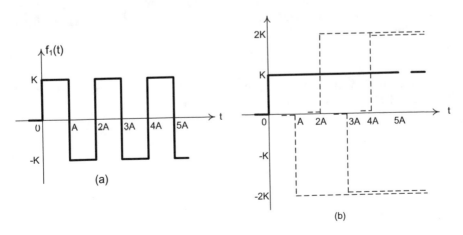

Fig. 2.15 a Periodic function $f_1(t)$; **b** Alternating rectangular pulse functions

Solution

According to Fig. 2.2 and Eq. (2.3), $f_1(t)$ can be written as the sum of alternating rectangular pulse equations, namely:

$$f_1(t) = K[U_{-1}(t) - U_{-1}(t-A)] - K[U_{-1}(t-A) - U_{-1}(t-2A)]$$
$$+ K[U_{-1}(t-2A) - U_{-1}(t-3A)] - K[U_{-1}(t-3A) - U_{-1}(t-4A)] + \ldots$$
$$= K[U_{-1}(t) - 2U_{-1}(t-A) + 2U_{-1}(t-2A) - 2U_{-1}(t-3A) + \ldots]$$

Or,

$$f_1(t) = KU_{-1}(t) + 2K \sum_{n=1}^{+\infty} (-1)^n U_{-1}(t-nA)$$

Figure 2.15b shows the first steps contained in the solution obtained.

Example 2.11 Decompose the function $g(t)$ shown in Fig. 2.16 into singular functions.

Fig. 2.16 Function $g(t)$ of Example 2.11

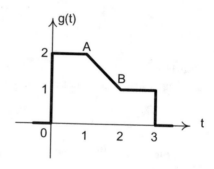

Solution

The function $g(t)$ can be considered as the juxtaposition of the next three pulses.

The rectangular pulse of amplitude 2, between $t = 0$ and $t = 1$, and which is defined by

$$2[U_{-1}(t) - U_{-1}(t-1)]$$

The equation of the line passing through points A and B is given by $-t + 3$. Therefore, the trapezoidal pulse between $t = 1$ e $t = 2$ is given by

$$(-t+3)[U_{-1}(t-1) - U_{-1}(t-2)]$$

Finally, the amplitude 1 rectangular pulse, between $t = 2$ and $t = 3$, is defined by

$$U_{-1}(t-2) - U_{-1}(t-3)$$

Considering, then, that the function $g(t)$ is the juxtaposition, in sequence, of the three pulses mentioned, it can be written that:

$$g(t) = 2[U_{-1}(t) - U_{-1}(t-1)] + (-t+3)[U_{-1}(t-1) - U_{-1}(t-2)]$$
$$+ U_{-1}(t-2) - U_{-1}(t-3) = 2U_{-1}(t) - 2U_{-1}(t-1) - (t-1)U_{-1}(t-1)$$
$$+ 2U_{-1}(t-1) + (t-2)U_{-1}(t-2) - U_{-1}(t-2) + U_{-1}(t-2) - U_{-1}(t-3)$$
$$= 2U_{-1}(t) - (t-1)U_{-1}(t-1) + (t-2)U_{-1}(t-2) - U_{-1}(t-3).$$

Referring to Eq. (2.7), we can finally write that:

$$g(t) = 2U_{-1}(t) - U_{-2}(t-1) + U_{-2}(t-2) - U_{-1}(t-3). \qquad (2.62)$$

The singular functions that make up $g(t)$ are shown, in sequence, in Fig. 2.17.

Fig. 2.17 Singular functions that make up $g(t)$

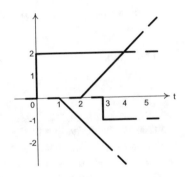

Example 2.12 Find the derivative of the functions $U_{-1}(-t)$ e $f(t)$ shown in Figs. 2.18a, b.

Solution
As can be seen from Fig. 2.18c,

$$U_{-1}(-t) = U_{-1}(t+\infty) - U_{-1}(t) = 1 - U_{-1}(t).$$

So,

$$\frac{d}{dt}U_{-1}(-t) = \frac{d}{dt}[1 - U_{-1}(t)] = -\frac{d}{dt}U_{-1}(t).$$

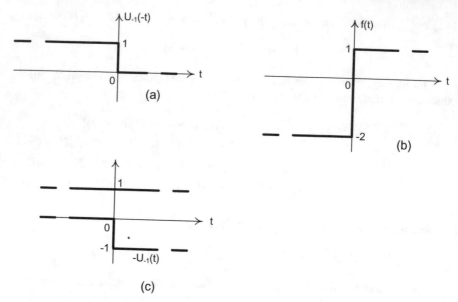

Fig. 2.18 a Unit step function $U_{-1}(-t)$; b Function $f(t)$; c Composition of functions

Therefore,

$$\frac{d}{dt}U_{-1}(-t) = -U_0(t).$$

That is, the derivative of the function $U_{-1}(-t)$ is a pulse with area equal to -1, which is exactly the value of the discontinuity present in the function $U_{-1}(-t)$, at time $t = 0$, that is: $0 - 1 = -1$.

Similar reasoning shows that the function $f(t)$, in Fig. 2.18b, can be expressed as

$$f(t) = -2U_{-1}(t+\infty) + 3U_{-1}(t) = -2 + 3U_{-1}(t).$$

So,

$$f'(t) = 3U_0(t).$$

Therefore, the derivative of $f(t)$ is an impulse of area equal to 3, which is the discontinuity value of $f(t)$ exactly at time $t = 0$, that is: $1 - (-2) = 3$.

Example 2.13 Derive the function $g(t)$ from example 2.11 and graph it.

Solution

From Eq. (2.62), we have that:

$$g'(t) = 2U_0(t) - \frac{d}{dt}U_{-2}(t-1) + \frac{d}{dt}U_{-2}(t-2) - \frac{d}{dt}U_{-1}(t-3)$$

$$= 2U_0(t) - \frac{d}{d(t-1)}U_{-2}(t-1).\frac{d(t-1)}{dt} + \frac{d}{d(t-2)}U_{-2}(t-2).\frac{d(t-2)}{dt}$$

$$- \frac{d}{d(t-3)}U_{-1}(t-3).\frac{d(t-3)}{dt}$$

$$= 2U_0(t) - U_{-1}(t-1) + U_{-1}(t-2) - U_0(t-3).$$

Or,

$$g'(t) = 2U_0(t) - [U_{-1}(t-1) - U_{-1}(t-2)] - U_0(t-3).$$

The graphical representation of $g'(t)$ is shown in Fig. 2.19. Note that the two finite discontinuities present in $g(t)$, Fig. 2.16, gave rise to the impulses contained in $g'(t)$. Furthermore, that the areas of the two pulses correspond to the values of $g(t)$, discontinuities.

Fig. 2.19 Graphical representation of $g'(t)$

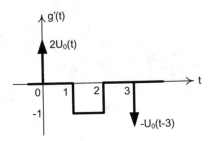

Example 2.14 Calculate the following integrals:

(a) $\int_{-\infty}^{+\infty} e^{j\omega t}.U_0(t-T)dt$

(b) $\int_{-\infty}^{+\infty} cost.U_0(t-\pi/3)dt$

(c) $\int_{-\infty}^{+\infty} U_0(t-a).U_{-1}(t-b)dt, a > b$

(d) $\int_{-\infty}^{+\infty} sent.[U_0(t-\pi/4) + U_1(t-\pi/2) + U_2(t-\pi)]dt$

(e) $I = \int_0^{2t} f(x)dx$, sendo que $f(x) = U_{-1}(x).U_{-1}(t-x), t > 0$ e parameter.

Solution

By Eq. (2.31),

(a) $\int_{-\infty}^{+\infty} e^{j\omega t}.U_0(t-T)dt = e^{j\omega t}|_{t=T} = e^{j\omega T}$

(b) $\int_{-\infty}^{+\infty} \cos t.U_0(t-\pi/3)dt = \cos \pi/3 = \cos 60° = sen\ 30° = 1/2$

(c) $\int_{-\infty}^{+\infty} U_0(t-a).U_{-1}(t-b)dt = U_{-1}(t-b)|_{t=a} = U_{-1}(a-b)$

With $a > b$, $U_{-1}(a-b) = U_{-1}(\tau)$ for $\tau > 0$, that is, $U_{-1}(a-b) = 1$. So,

$$\int_{-\infty}^{+\infty} U_0(t-a).U_{-1}(t-b)dt = 1, so\ a > b$$

(d) Under Eq. (2.44),

$$\int_{-\infty}^{+\infty} sen t.[U_0(t-\pi/4) + U_1(t-\pi/2) + U_2(t-\pi)]dt$$

$$= \int_{-\infty}^{+\infty} sen t.U_0(t-\pi/4)dt$$

$$+ \int_{-\infty}^{+\infty} sen t.U_1(t-\pi/2)dt + \int_{-\infty}^{+\infty} sen t.U_2(t-\pi)dt$$

$$= sen\pi/4 - \cos \pi/2 - sen\pi = \sqrt{2}/2$$

(e)
$$U_{-1}(x) = \begin{cases} 0 & if\ x<0 \\ 1 & if\ x>0 \end{cases}; U_{-1}(-x)$$

$$= \begin{cases} 1 & if\ x<0 \\ 0 & if\ x>0 \end{cases} \therefore U_{-1}(t-x) = U_{-1}[-(x-t)]$$

$$= \begin{cases} 1 & if\ x-t<0 \\ 0 & if\ x-t>0 \end{cases} \begin{cases} 1 & if\ x<t \\ 0 & if\ x>t \end{cases}.$$

So,

$$U_{-1}(t-x) = \begin{cases} 1 & for\ -\infty<x<t \\ 0 & for\ t<x<+\infty \end{cases} e U_{-1}(x) = \begin{cases} 0 & for\ -\infty<x<0 \\ 1 & for\ 0<x<+\infty. \end{cases}$$

Then,

$$I = \int_0^{2t} f(x)dx = \int_0^t 1.dx + \int_t^{2t} 0.dx = t.$$

Example 2.15 Synthesize the function $f(t)$ given in Eq. (2.63) and graph the result obtained.

$$f(t) = U_{-2}(t+1) - 2U_{-1}(t) - U_{-2}(t-1) \qquad (2.63)$$

Since $U_{-2}(t) = tU_{-1}(t)$, Eq. (2.6), then $U_{-2}(t \pm 1) = (t \pm 1)U_{-1}(t \pm 1)$, and Eq. (2.63) can be rewritten as

$$f(t) = (t+1)U_{-1}(t+1) - 2U_{-1}(t) - (t-1)U_{-1}(t-1)$$

Adding these three functions from the moments when they become non-zero, that is, synthesizing them, it results that:

$$f(t) = \begin{cases} 0 \; for - \infty < t \le -1 \\ t+1 \; for -1 \le t < 0 \\ t-1 \; for \; 0 < t \le 1 \\ 0 \; for \; 1 \le t < +\infty. \end{cases} \qquad (2.64)$$

The graphical representation of $f(t)$ is made with the information contained in Eq. (2.64), and is shown in Fig. 2.20.

Fig. 2.20 Graphical representation of $f(t)$

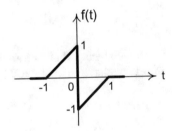

Example 2.16 To a capacitance C, initially de-energized, that is, at rest, the current $i(t)[A]$ shown in Fig. 2.21-a is applied.

(a) Calculate the charge $q[C]$ stored in it at time $t = 5[s]$, that is, $q(5)$.

(b) Find the waveform of q (t), for t > 0.

Solution

(a) For $i(t) = \frac{d}{dt}q(t)$, then $q(t) = \int i(t)dt = \int_{-\infty}^{t} i(\lambda)d\lambda$ and, from Fig. 2.21a,

$$q(5) = \int_{-\infty}^{5} i(\lambda)d\lambda = \int_{-\infty}^{1_-} 0.d\lambda + \int_{1_-}^{1_+} 3U_0(\lambda - 1)d\lambda + \int_{1_+}^{2_-} 0.d\lambda + \int_{2_+}^{3_-} (-1)d\lambda$$

$$+ \int_{3_+}^{4_-} 0.d\lambda + \int_{4_-}^{4_+} -U_0(\lambda - 4)d\lambda + \int_{4_+}^{5} 0.d\lambda.$$

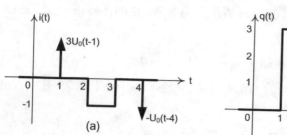

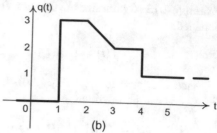

Fig. 2.21 a Current $i(t)$ applied to a capacitor; **b** Charge $q(t)$

Therefore,

$$q(5) = 0 + 3 + 0 - \lambda|_2^3 + 0 - 1 + 0 = 3 - 1 - 1 = 1 [C].$$

Considering that **the load is the net area under the current curve**, alternatively, we have that:

$q(5) =$ area of impulse $3U_0(t-1) +$ area of square pulse $+$ area of impulse $-U_0(t-4) = 3 + (-1) + (-1) = 1 [C].$

(b) The decomposition of i(t) into singular functions, from Fig. 2.21a, gives that

$$i(t) = 3U_0(t-1) - U_{-1}(t-2) + U_{-1}(t-3) - U_0(t-4).$$

Remembering that **to integrate any single function, just subtract 1 from its index**, we have to

$$q(t) = \int i(t)dt = 3U_{-1}(t-1) - U_{-2}(t-2) + U_{-2}(t-3) - U_{-1}(t-4)$$
$$= 3U_{-1}(t-1) - (t-2)U_{-1}(t-2) + (t-3)U_{-1}(t-3) - U_{-1}(t-4)$$
$$= 3U_{-1}(t-1) + (-t+2)U_{-1}(t-2) + (t-3)U_{-1}(t-3) - U_{-1}(t-4).$$

The synthesis of these unique functions gives

$$q(t) = \begin{cases} 0 \ for \ t < 1 \\ 3 \ for \ 1 < t \le 2 \\ -t+5 \ for \ 2 \le t \le 3 \\ 2 \ for \ 3 \le t < 4 \\ 1 \ for \ t > 4. \end{cases}$$

The waveform of $q(t)$ is shown in Fig. 2.21b.

Example 2.17 Calculate $I = \int f(t)dt$, where $f(t)$ is the function shown in Fig. 2.20, and which is expressed in terms of singular functions by Eq. (2.63) of example 2.15. Graph the result obtained.

Solution

As mentioned earlier, and noted in Fig. 2.13, to integrate any singular function, just subtract 1 from its index. Therefore, from Eq. (2.63),

$$I = \int f(t)dt = U_{-3}(t+1) - 2U_{-2}(t) - U_{-3}(t-1) \qquad (2.65)$$

Considering now that $U_{-3}(t) = \frac{t^2}{2}U_{-1}(t)$, Eq. (2.10), then.
$U_{-3}(t \pm 1) = \frac{(t \pm 1)^2}{2}U_{-1}(t \pm 1)$. Then,

$$I = \frac{(t+1)^2}{2}U_{-1}(t+1) - 2tU_{-1}(t) - \frac{(t-1)^2}{2}U_{-1}(t-1)$$

$$= \left(\frac{t^2}{2} + t + \frac{1}{2}\right)U_{-1}(t+1) - 2tU_{-1}(t) + \left(-\frac{t^2}{2} + t - \frac{1}{2}\right)U_{-1}(t-1)$$

$$= \begin{cases} 0 \; for - \infty < t \le -1 \\ \frac{t^2}{2} + t + \frac{1}{2} for - 1 \le t \le 0 \\ \frac{t^2}{2} - t + \frac{1}{2} for \; 0 \le t \le 1 \\ 0 \; for \; 1 \le t \le +\infty. \end{cases} \qquad (2.66)$$

The graphical representation of I is shown in Fig. 2.22. It is interesting to note that integration, in general, tends to "smooth out" the curve that has been integrated. More than that: it eliminates the finite discontinuities present in the curve that has been integrated.

Fig. 2.22 Graphical representation of current $i(t)$

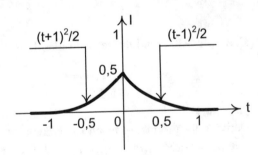

Example 2.18 Transform $f(t) = 10sen(200t + \pi/6).U_1(t)$ into a sum of singular functions.

Solution

According to Eq. (2.53),

$$f(t) = 10sen(200t + \pi/6)|_{t=0}.U_1(t) - \frac{d}{dt}[10sen(200t + \pi/6)]|_{t=0}.U_0(t)$$

$$= 10sen\pi/6.U_1(t) - 2.000\cos \pi/6 = 5U_1(t) - 1.000\sqrt{3}U_0(t).$$

Example 2.19 Eliminate the step function of the function $f(t) = U_{-1}(t^2 + 5t + 6)$.

Solution

$t^2 + 5t + 6 = 0$ has roots of -2 and -3. That is, $t^2 + 5t + 6 = (t+2)(t+3)$.

So,

$$f(t) = U_{-1}[(t+2)(t+3)] = \begin{cases} 0 & for \ (t+2)(t+3) < 0(A) \\ 1 & for \ (t+2)(t+3) > 0(B) \end{cases}$$

Study of (A)

A-1: $(t+2) < 0$ and $(t+3) > 0$ or $t < -2$ and $t > -3$; or, yet, $-3 < t < -2$.
A-2: $(t+2) > 0$ and $(t+3) < 0$ or $t > -2$ and $t < -3$; inconsistent conditions.

Study of (B)

B-1: $(t+2) > 0$ and $(t+3) > 0$ or $t > -2$ and $t > -3$; then $t > -2$.
B-2: $(t+2) < 0$ and $(t+3) < 0$ or $t < -2$ and $t < -3$; then $t < -3$.

Therefore,

$$f(t) = \begin{cases} 0 & for -3 < t < -2 \\ 1 & for \ t\langle -3 \ and \ for \ t\rangle -2 \end{cases}$$

Finally,

$$f(t) = \begin{cases} 1 \ for \ t < -3 \\ 0 \ for -3 < t < -2 \\ 1 \ for \ t > -2. \end{cases}$$

Example 2.20 Find the second order derivative of the function

$$e(t) = E\cos(\omega t + \theta).U_{-1}(t).$$

Solution

Deriving the product of functions, it comes that:

$$e'(t) = -\omega E sen(\omega t + \theta).U_{-1}(t) + E\cos(\omega t + \theta)|_{t=0}.U_0(t).$$

where,

$$e'(t) = E\cos\theta U_0(t) - \omega E sen(\omega t + \theta)U_{-1}(t).$$

Deriving again,

$$e''(t) = E\cos\theta U_1(t) - \omega^2 E\cos(\omega t + \theta)U_{-1}(t) - \omega E sen(\omega t + \theta)|_{t=0}.U_0(t)$$

So,

$$e''(t) = E\cos\theta U_1(t) - \omega E sen\theta U_0(t) - \omega^2 E\cos(\omega t + \theta)U_{-1}(t).$$

Example 2.21 Prove that

$$U_{-1}(t)U_0(t) = \frac{1}{2}U_0(t) \tag{2.67}$$

Solution

Equation (2.32) cannot be applied directly because for $t = 0$ the function $U_{-1}(t)$ is not defined.

If $f(t) = U_{-1}(t)U_0(t) = ?$

For $t < 0$, $U_{-1}(t) = 0$ and $U_0(t) = 0$. Then, $f(t) = 0.0 = 0$.

For $t > 0$, $U_{-1}(t) = 1$ and $U_0(t) = 0$. Then, $f(t) = 1.0 = 0$.

The indefinite integral of the function $f(t)$ is

$$\int f(t)dt = \int U_{-1}(t)U_0(t)dt + C.$$

The rule of integration by parts provides that $\int u dv = uv - \int v du$, is $u = U_{-1}(t), du = U_0(t)dt; dv = U_0(t)dt, v = U_{-1}(t)$. Então,

$$\int f(t)dt = [U_{-1}(t)]^2 - \int U_{-1}(t)U_0(t)dt + C = U_{-1}(t) - \left[\int f(t)dt - C\right] + C.$$

Then,

$$2\int f(t)dt = U_{-1}(t) + 2C$$

Or,

$$\int f(t)dt = \frac{1}{2}U_{-1}(t) + C.$$

Deriving this last expression, it comes that:

$$f(t) = \frac{1}{2}U_0(t)$$

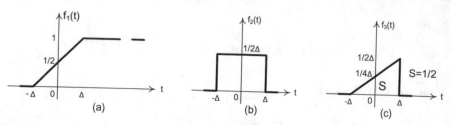

Fig. 2.23 a–c Alternative ways of conceiving Eq. (2.67)

Therefore,

$$U_{-1}(t)U_0(t) = \frac{1}{2}U_0(t).$$

It is interesting to note that $\frac{1}{2} = \frac{U_{-1}(0_+) + U_{-1}(0_-)}{2}$, that is, that the area of the resulting impulse is the mean value of the step discontinuity.

An alternative way of conceiving Eq. (2.67) uses Figs. 2.23a–c.

The function $f_1(t)$ shown in Fig. 2.23a tends to the unit step when $\to 0$. On the other hand, the function $f_2(t)$ of Fig. 2.23b tends to the unit impulse when $\to 0$. In Fig. 2.23c there is that $f_3(t) = f_1(t).f_2(t)$. The area under the triangular pulse $f_3(t)$ is $S = 1/2$. When $\Delta \to 0$ the function $f_3(t)$ tends towards a pulse with an area equal to $1/2$.

2.4   Causal Functions

A function $f(t)$ is called causal when it has the form $f(t) = \Phi(t).U_{-1}(t)$, that is:

$$f(t) = \begin{cases} 0 \ for \ t < 0 \\ \Phi(t) for \ t > 0. \end{cases}$$

If $\Phi(t)$ is continuously and successively differentiable for $t > 0$, that is, for $0_+ \leq t < +\infty$, then it is possible to show that a causal function can be expanded into an infinite series of singular functions, with decreasing indices, as explained in Eq. (2.68).

$$\Phi(t).U_{-1}(t) = \sum_{n=0}^{+\infty} \Phi^{(n)}(0_+)U_{-(n+1)}(t) \qquad (2.68)$$

The following particular cases will be useful in the future:

$$\sin(t).U_{-1}(t) = U_{-2}(t) - U_{-4}(t) + U_{-6}(t) - U_{-8}(t) + \cdots \qquad (2.69)$$

$$\cos(t).U_{-1}(t) = U_{-1}(t) - U_{-3}(t) + U_{-5}(t) - U_{-7}(t) + \cdots \qquad (2.70)$$

$$\varepsilon^{-\alpha t}.U_{-1}(t) = U_{-1}(t) - \alpha U_{-2}(t) + \alpha^2 U_{-3}(t) - \alpha^3 U_{-4}(t) + \cdots \qquad (2.71)$$

$$\begin{aligned} sen(\omega t + \theta).U_{-1}(t) = &\, sen\theta.U_{-1}(t) + \omega cos\theta.U_{-2}(t) - \omega^2 sen\theta.U_{-3}(t) \\ &- \omega^3 cos\theta.U_{-4}(t) + \omega^4 sen\theta.U_{-5}(t) + \cdots \end{aligned} \qquad (2.72)$$

$$\begin{aligned} \cos(\omega t + \theta).U_{-1}(t) = &\, cos\theta.U_{-1}(t) - \omega sen\theta.U_{-2}(t) - \omega^2 cos\theta.U_{-3}(t) \\ &+ \omega^3 sen\theta.U_{-4}(t) + \omega^4 cos\theta.U_{-5}(t) - \cdots \end{aligned} \qquad (2.73)$$

Example 2.22 Prove Eq. (2.69).

Solution
The MacLaurin series of the sine function is

$$sen(t) = t - \frac{t^3}{3!} + \frac{t^5}{5!} - \frac{t^7}{7!} + \cdots \ for \ -\infty < t < +\infty$$

Multiplying the two members by the unitary step comes that

$$sen(t).U_{-1}(t) = t.U_{-1}(t) - \frac{t^3}{3!}.U_{-1}(t) + \frac{t^5}{5!}.U_{-1}(t) - \frac{t^7}{7!}.U_{-1}(t) + \cdots$$

But,

$$t.U_{-1}(t) = U_{-2}(t); \frac{t^3}{3!}.U_{-1}(t) = U_{-4}(t); \frac{t^5}{5!}.U_{-1}(t) = U_{-6}(t); \frac{t^7}{7!}.U_{-1}(t)$$
$$= U_{-8}(t); \ldots$$

Then,

$$sent.U_{-1}(t) = U_{-2}(t) - U_{-4}(t) + U_{-6}(t) - U_{-8}(t) + \cdots$$

2.5 Derivative of an Ordinary Sectionally Continuous Function

A function is called sectionally continuous, in a given interval, when it has, at most, a finite number of finite discontinuities.

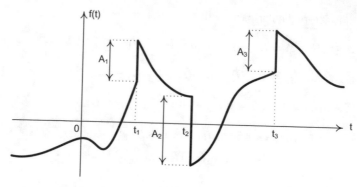

Fig. 2.24 Function $f(t)$ sectionally continuous

Consider the function $f(t)$ sectionally continuous, for a supposedly infinite interval, with its finite value discontinuities $A_1, A_2, A_3, \ldots, A_n$, occurring in instants $t_1, t_2, t_3, \ldots, t_n$, respectively, illustrated in Fig. 2.24.

Logically, $f(t)$ can be expressed in terms of steps occurring at times $t_1, t_2, t_3, \ldots, t_n$, and of a function $f_1(t)$ strictly continuous at all times in the interval, with ordinary derivative $f_1'(t)$, such that $f_1'(t) = f'(t)$, except for the discontinuity points of $f(t)$, as indicated in Eq. (2.74).

$$f(t) = f_1(t) + \sum_{k=1}^{n} A_k U_{-1}(t - t_k) \tag{2.74}$$

Deriving Eq. (2.74) results that:

$$f'(t) = f_1'(t) + \sum_{k=1}^{n} A_k U_0(t - t_k) \tag{2.75}$$

Equation (2.75) shows that the generalized derivative $f'(t)$, of an ordinary sectionally continuous function $f(t)$, is formed by the ordinary derivative $f_1'(t)$, where it exists, plus the sum of pulses occurring at t_k, discontinuity of $f(t)$, with areas equal to the amplitudes A_k of the jumps. This conclusion, already observed in examples 2.12 and 2.13, associated with the possibility of obtaining generalized derivatives of higher orders of an ordinary function, allows to establish a systematic method for the decomposition into singular functions of functions defined by polynomial sections, sectional continuous or not, and which is described below.

2.6 Decomposition of a Function into Singular Functions

Being $f(t)$ a function defined by polynomial sections, continuous or sectionally continuous with finite discontinuities, and null for t less than a certain time t_0, positive, negative or null. The decomposition of such a function into singular functions, by a systematic procedure, consists essentially of two stages:

1. Initially, the function $f(t)$ is derived as many times as necessary to obtain a result consisting only of singular functions of non-negative indices, the so-called **impulsive functions**.
2. As the indefinite integral of a singular function is always another singular function, which is obtained by simply subtracting the unit from its index, the impulsive functions obtained in the first stage are then successively integrated, as many times as there were derivations, thus recovering the original function $f(t)$, but now expanded into singular functions.

If $f(t)$ is a constant K for $t < t_0$, then, after performing the steps described, this K value must be added to the result obtained to complete the desired decomposition.

Example 2.23 Decompose the function $f(t)$ shown in Fig. 2.25 into singular functions.

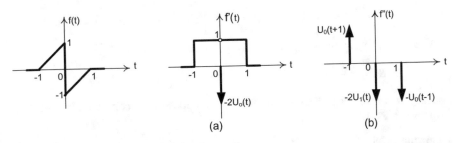

(a)   (b)

Fig. 2.25 Function $f(t)$ for Example 2.23

Solution
The first and second order derivatives of $f(t)$ are shown in parts a and b of Fig. 2.26, respectively.

Integrating, successively, twice $f''(t)$ and considering that $f(t) = 0$ for $t < t_0 = -1$, the requested expansion results:

$$f(t) = U_{-2}(t+1) - 2U_{-1}(t) - U_{-2}(t-1).$$

Equation (2.63) and Fig. 2.20 of example 2.15 confirm the result obtained.

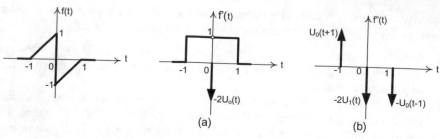

Fig. 2.26 **a** first order derivative of $f(t)$; **b** second order derivative of $f(t)$

Example 2.24 Redo example 2.11, with the procedure described in Sect. 2.6.

Solution

To make it easier, Fig. 2.16 of example 2.11 is repeated in Fig. 2.27a. Parts b and c of Fig. 2.27 show the first and second order derivatives, respectively.
    Figure 2.27c shows that

$$g''(t) = 2U_1(t) - U_0(t-1) + U_0(t-2) - U_1(t-3).$$

Consequently, integrating, successively, twice, it comes that

$$g(t) = 2U_{-1}(t) - U_{-2}(t-1) + U_{-2}(t-2) - U_{-1}(t-3).$$

As you can see, the effort is minimal to decompose $g(t)$, compared to that spent in example 2.11.

Example 2.25 Obtain the expansion in singular functions of the function $f(t)$ shown in Fig. 2.28.

Solution

In parts a and b of Fig. 2.29 the first and second order derivatives are shown, respectively. Then, integrating $f''(t)$ twice, successively, and noting that $f(t) = -1$ for $t < -2 = t_0$, the requested decomposition is obtained; to know:

$$f(t) = -1 + U_{-2}(t+2) + U_{-1}(t) - 2U_{-2}(t) + U_{-2}(t-2).$$

Example 2.26 Develop $f(t) = t^2 U_{-1}(t-2)$ in singular components.

Solution

The first order derivative of the function is

$$f'(t) = 2t.U_{-1}(t-2) + t^2|_{t=2}.U_0(t-2) = 2t.U_{-1}(t-2) + 4U_0(t-2). \text{ So,}$$

$$f''(t) = 2U_{-1}(t-2) + 2t|_{t=2}.U_0(t-2) + 4U_1(t-2)$$
$$= 2U_{-1}(t-2) + 4U_0(t-2) + 4U_1(t-2)$$

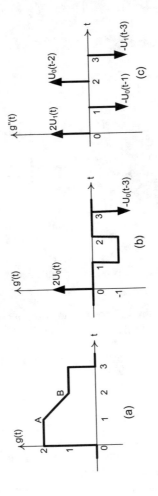

Fig. 2.27 **a** Function $g(t)$ for Example 2.24; **b** first order derivative of $g(t)$; **c** second order derivative of $g(t)$

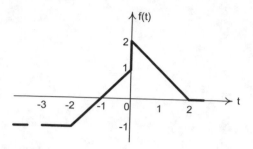

Fig. 2.28 Function $f(t)$ for Example 2.25

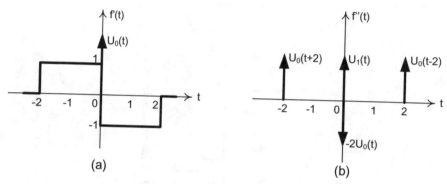

(a)   (b)

Fig. 2.29 a first order derivative of $f(t)$; **b** second order derivative of $f(t)$

Subtracting two units from the indices of the component functions of $f''(t)$, that is, integrating twice $f''(t)$ there is that:

$$f(t) = 2U_{-3}(t-2) + 4U_{-2}(t-2) + 4U_{-1}(t-2).$$

2.7 The Distribution Concept

Suppose that $f(t)$ is an ordinary function, that is, a function in the strict mathematical sense, such that the integral of Eq. (2.76) exists for any function $\varphi(t)$, having a derivative of any order and tending to zero more quickly than any det power, when $t \to \pm\infty$, called **test function**.

$$\int_{-\infty}^{+\infty} f(t)\varphi(t)dt \tag{2.76}$$

This definite integral, is, clearly, an $I_f[\varphi(t)]$ number, dependent on $\varphi(t)$.

The Generalized Distribution or Function $f(t)$ is defined as the process that associates the number $I_f[\varphi(t)]$ to the arbitrary function $\varphi(t)$. Therefore, the ordinary function $f(t)$ will be interpreted as a distribution specified by the ordinary integral of Eq. (2.76). For example, if $f(t) = U_{-1}(t)$, then

$$\int_{-\infty}^{+\infty} U_{-1}(t)\varphi(t)dt = \int_{0}^{+\infty} \varphi(t)dt = I_f[\varphi(t)] \qquad (2.77)$$

Therefore, the distribution $U_{-1}(t)$ is the process that associates an arbitrary function (test) $\varphi(t)$ with a number $I_f[\varphi(t)]$, equal to the area of $\varphi(t)$ between zero and more infinite.

The notation

$$\int_{-\infty}^{+\infty} g(t)\varphi(t)dt = N_g[\varphi(t)] \qquad (2.78)$$

is the formal extension of the process that has just been described for the case of $g(t)$ distributions **that have no meaning as ordinary functions**, such as the unitary double impulsive function.

The following are some properties and consequences related to the definition of distribution.

(a) **Linearity**

The distribution is linear, that is:

$$\int_{-\infty}^{+\infty} g(t)[a_1\varphi_1(t) + a_2\varphi_2(t)]dt = a_1 \int_{-\infty}^{+\infty} g(t)\varphi_1(t)dt + a_2 \int_{-\infty}^{+\infty} g(t)\varphi_2(t)dt$$

$$(2.79)$$

(b) **Sum of Distributions**

The sum $g(t) = g_1(t) + g_2(t)$ of two distributions is defined by:

$$\int_{-\infty}^{+\infty} g(t)\varphi(t)dt = \int_{-\infty}^{+\infty} g_1(t)\varphi(t)dt + \int_{-\infty}^{+\infty} g_2(t)\varphi(t)dt \qquad (2.80)$$

(c) **Displacement and Escalation**

The distribution $g(t - t_0)$ is given by:

$$\int_{-\infty}^{+\infty} g(t - t_0)\varphi(t)dt = \int_{-\infty}^{+\infty} g(t)\varphi(t + t_0)dt \qquad (2.81)$$

The distribution $g(at)$ is given by:

$$\int_{-\infty}^{+\infty} g(at)\varphi(t)dt = \frac{1}{|a|}\int_{-\infty}^{+\infty} g(t)\varphi(t/a)dt \qquad (2.82)$$

The demonstration of Eqs. (2.81) and (2.82) is similar to that of Eqs. (2.45) and (2.46).

(d) **Parity**

The $g(t)$ distribution is called even (or odd) if, for any test function $\varphi(t)$ odd (or even), it turns out that:

$$\int_{-\infty}^{+\infty} g(t)\varphi(t)dt = 0 \qquad (2.83)$$

(e) **Product of Distribution by Ordinary Function**

The product of a $g(t)$ distribution by an ordinary function $f(t)$ is a $g(t)f(t)$ distribution defined by:

$$\int_{-\infty}^{+\infty} [g(t)f(t)]\varphi(t)dt = \int_{-\infty}^{+\infty} g(t)[f(t)\varphi(t)]dt, \qquad (2.84)$$

which is consistent with the law of associativity. Of course, Eq. (2.84) is only true if also $f(t)\varphi(t)$ is a test function.

(f) **Derivative**

The derivative $\frac{d}{dt}g(t) = g'(t)$ of a generalized distribution or function $g(t)$ is defined by

$$\int_{-\infty}^{+\infty} g'(t)\varphi(t)dt = -\int_{-\infty}^{+\infty} g(t)\varphi'(t)dt \qquad (2.85)$$

The demonstration is carried out by applying the rule of integration by parts. Then,

$$\int_{-\infty}^{+\infty} g'(t)\varphi(t)dt = g(t)\varphi(t)|_{-\infty}^{+\infty} - \int_{-\infty}^{+\infty} g(t)\varphi'(t)dt$$

Since $\varphi(t)$ is a test function, it satisfies the $0ptlimn \to \pm\infty[g(t)\varphi(t)] = 0$ condition and, Eq. (2.85) is proved.

The generalization for the derivative of order n of a distribution $g(t)$ leads to

$$\int_{-\infty}^{+\infty} \frac{d^n}{dt^n}g(t)\varphi(t)dt = (-1)^n \int_{-\infty}^{+\infty} g(t)\frac{d^n}{dt^n}\varphi(t)dt \qquad (2.86)$$

2.8   The Unitary Impulse Function as Distribution

The biased form as the unit impulse function was introduced in Sect. 2.2, assigning zero value to the k index in Eqs. (2.18) and (2.19)—which were designed exclusively for integer and negative k values, followed by the graphical illustrations of Figs. 2.7, 2.8 and 2.9, in an attempt to give more concreteness to it, for not being sufficiently rigorous and consistent, can, in certain applications, lead to incorrect conclusions—when not contradictory. This is due to the fact that one tries to consider the unit impulse function as an ordinary function.

Fortunately, such risks are largely avoided when it is conceived of as a generalized function or distribution.

Strictly speaking, the **unit impulse function $U_0(t)$ is a distribution that associates the function test $\Phi(t)$ with the number $\Phi(t)$**. Therefore, according to Eq. (2.78),

$$\int_{-\infty}^{+\infty} U_0(t)\Phi(t)dt = \Phi(0). \tag{2.87}$$

Equation (2.87) is nothing more than Eq. (2.33), which was the form used in Sect. 2.2 to present the unit impulse function through a property.

The integral of Eq. (2.87) has no meaning as an ordinary integral. The integral and the function $U_0(t)$ are simply defined by the number $\Phi(t)$ associated with the function $\Phi(t)$.

Note that the definition given by Eq. (2.87) **does not require any grafic visualization.**

Equations (2.81) and (2.82) also show, respectively, that:

$$\int_{-\infty}^{+\infty} U_0(t-t_0)\Phi(t)dt = \int_{-\infty}^{+\infty} U_0(t)\Phi(t+t_0)dt = \Phi(t_0) \tag{2.88}$$

$$\int_{-\infty}^{+\infty} U_0(at)\Phi(t)dt = \frac{1}{|a|}\int_{-\infty}^{+\infty} U_0(t)\Phi(t/a)dt = \frac{1}{|a|}\Phi(0)$$

$$= \frac{1}{|a|}\int_{-\infty}^{+\infty} U_0(t)\Phi(t)dt \tag{2.89}$$

Equations (2.88) and (2.89) are nothing more than the confirmation, on more rigorous bases, of Eqs. (2.45) and (2.46).

Equation (2.84) shows that

$$\int_{-\infty}^{+\infty} [f(t)U_0(t)]\Phi(t)dt = \int_{-\infty}^{+\infty} U_0(t)[f(t)\Phi(t)]dt = f(0)\Phi(0)$$

$$= \int_{-\infty}^{+\infty} [f(0)U_0(t)]\Phi(t)dt$$

That is, that

$$f(t)U_0(t) = f(0)U_0(t),$$

provided that $f(0)$ exists, and that is the confirmation of Eq. (2.54). (See example 2.28).

From Eq. (2.85),

$$
\begin{aligned}
\int_{-\infty}^{+\infty} U_1(t)f(t)\Phi(t)dt &= -\int_{-\infty}^{+\infty} U_0(t)\frac{d}{dt}[f(t)\Phi(t)]dt = -\int_{-\infty}^{+\infty} U_0(t)f(t)\Phi'(t)dt \\
&\quad - \int_{-\infty}^{+\infty} U_0(t)f'(t)\Phi(t)dt = -f(0)\Phi'(0) - f'(0)\Phi(0) \\
&= f(0)\int_{-\infty}^{+\infty} U_1(t)\Phi(t)dt - f'(0)\int_{-\infty}^{+\infty} U_0(t)\Phi(t)dt \\
&= \int_{-\infty}^{+\infty} [f(0)U_1(t) - f'(0)U_0(t)]\Phi(t)dt
\end{aligned}
$$

Therefore, comparing the first with the last integral, there is that

$$f(t)U_1(t) = f(0)U_1(t) - f'(0)U_0(t),$$

confirming Eq. (2.53).

The procedure used so far, in this section, could be continued to prove other relationships and properties related mainly to the singular impulsive functions, that is, those obtained by successive derivations of the impulse function. This section is closed, however, with example 2.27.

Example 2.27 Show that if $f(t)$ is a generalized function, the rule for deriving the product:

$$\frac{d}{dt}[f(t)U_0(t)] = f(t)U_1(t) + f'(t)U_0(t)$$

remains valid.

Solution

The derivative $f'(t)$ of a generalized function or distribution $f(t)$ was defined in Eq. (2.85), and allows to write that:

$$\int_{-\infty}^{+\infty} \frac{d}{dt}[f(t)U_0(t)]\Phi(t)dt = -\int_{-\infty}^{+\infty} [f(t)U_0(t)]\Phi'(t)dt$$

$$= -\int_{-\infty}^{+\infty} U_0(t)[f(t)\Phi'(t)]dt$$

$$= -\int_{-\infty}^{+\infty} U_0(t)\left\{\frac{d}{dt}[f(t)\Phi(t)] - f'(t)\Phi(t)\right\}dt$$

$$= -\int_{-\infty}^{+\infty} U_0(t)\frac{d}{dt}[f(t)\Phi(t)]dt$$

$$+ \int_{-\infty}^{+\infty} U_0(t)[f'(t)\Phi(t)]dt$$

$$= \int_{-\infty}^{+\infty} U_1(t)[f(t)\Phi(t)]dt + \int_{-\infty}^{+\infty} [U_0(t)f'(t)]\Phi(t)dt$$

$$= \int_{-\infty}^{+\infty} [U_1(t)f(t) + U_0(t)f'(t)]\Phi(t)dt.$$

Therefore,

$$\frac{d}{dt}[f(t)U_0(t)] = f(t)U_1(t) + f'(t)U_0(t) \tag{2.90}$$

2.9 Initial Condition at t = 0₊

In Fig. 2.30 two functions are shown, namely: in dashed line, the continuous function $f_c(t)$; in full line, the function $f(t)$, discontinuous at $t = 0$, satisfying the following conditions:

$$\begin{cases} f(t) = f_c(t) \text{ for } t < 0 \\ f(t) = f_c(t) + A \text{ for } t > 0. \end{cases}$$

With the aid of the step function of amplitude A, also shown in Fig. 2.30, it is possible to write, alternatively, that:

$$f(t) = f_c(t) + AU_{-1}(t) \tag{2.91}$$

Taking $t = 0_+$ in Eq. (2.91), it follows that:

$$f(0_+) = f_c(0_+) + A$$

But $f_c(0_+) = f_c(0_-) = f(0_-)$, because $f_c(t)$ is continuous. Then,

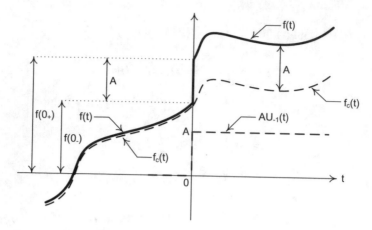

Fig. 2.30 Functions $(t), f_c(t)$, and $AU_{-1}(t)$

$$f(0_+) = f(0_-) + A \qquad (2.92)$$

From Eq. (2.92) it can be concluded, then, that:

The initial condition $f(0_+)$ is the sum of the initial condition $f(0_-)$ with the amplitude A of the step function $AU_{-1}(t)$ present in the analytical expression of $f(t)$.

Therefore, the absence of a step function, at the origin, in the analytical definition of a given function $f(t)$ implies its continuity at time $t = 0$, that is, that $f(0_+) = f(0_-)$. Or, still: **where there is discontinuity, there is a step function.**

In Fig. 2.31 an alternative way of approaching the same discontinuous function $f(t)$ of Fig. 2.30 is shown.

$$f(t) = f(0_-) + \phi(t), for\ t > 0 \qquad (2.93)$$

Fig. 2.31 Functions $f(t)$ and $\varnothing(t)$

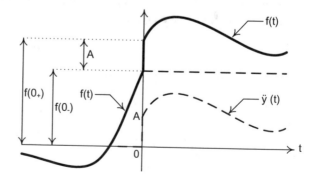

Since $\phi(t)$ is a causal function, and assuming that it is successively differentiable, it can be expanded into an infinite series of singular functions, as established in Eq. (2.68). Like this,

$$\phi(t) = C_1 U_{-1}(t) + C_2 U_{-2}(t) + C_3 U_{-3}(t) + \cdots \tag{2.94}$$

Replacing Eq. (2.94) in (2.93), it follows that:

$$f(t) = f(0_-) + C_1 U_{-1}(t) + C_2 U_{-2}(t) + C_3 U_{-3}(t) + \cdots, for\ t > 0. \tag{2.95}$$

Taking $t = 0_+$ in Eq. (2.95), it follows that:

$$f(0_+) = f(0_-) + C_1 \tag{2.96}$$

But, by Eq. (2.94) and even more by Fig. 2.31, $C_1 = \phi(0_+) = A$, then,

$$f(0_+) = f(0_-) + A,$$

which is the same Eq. (2.92).

From Eqs. (2.95) and (2.96), it can be seen, therefore, that: **once the initial condition $f(0_-)$ of the discontinuous function $f(t)$ is known, the initial condition $f(0_+)$ will be determined as soon as the C_1 coefficient is calculated of the step $U_{-1}(t)$ participating in the analytical expression of $f(t)$, for $t > 0$.**

Example 2.28 Assuming that $f(t)$ is a function with finite discontinuity at point $t = a$, prove that:

$$f(t)U_0(t-a) = \frac{f(a_+) + f(a_-)}{2} U_0(t-a). \tag{2.97}$$

Solution

The function $f(t)$ can be considered as the sum of a totally continuous function, $f_c(t)$, with a step occurring exactly at the point $t = a$; something similar to what is shown in Fig. 2.30, with a delay equal to a in the step function. That is:

$$f(t) = f_c(t) + [f(a_+) - f(a_-)]U_{-1}(t-a)$$

So,

$$f(t)U_0(t-a) = f_c(t)U_0(t-a) + [f(a_+) - f(a_-)]U_{-1}(t-a)U_0(t-a).$$

Since $f_c(t)$ is continuous, $f_c(t)U_0(t-a) = f_c(a)U_0(t-a)$. But, $f_c(a) = f(a_-)$. So,

$$f_c(t)U_0(t-a) = f(a_-)U_0(t-a).$$

On the other hand, from Eq. (2.67),

$$U_{-1}(t-a)U_0(t-a) = \frac{1}{2}U_0(t-a).$$

Since,

$$f(t)U_0(t-a) = f(a_-)U_0(t-a) + [f(a_+) - f(a_-)] \cdot \frac{1}{2}U_0(t-a).$$

Therefore,

$$f(t)U_0(t-a) = \frac{f(a_+) + f(a_-)}{2}U_0(t-a).$$

2.10 Initial Decomposition of Functions in Steps

The continuous function $f(t)$ of Fig. 2.32, considered in the interval $t_0 \leq t \leq t_0 + T$, can be approximated by the function "in stairs" $f_1(t)$, with intervals equal to $\Delta\lambda$, which, in turn, can be decomposed into a finite number, N, of steps, similar to what was done in example 2.9. Thus, $T = N.\Delta\lambda$. So,

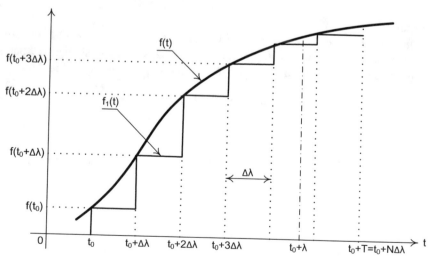

Fig. 2.32 Continuous function $f(t)$ approximate by "steps"

$$f(t) \cong f_1(t) = f(t_0)U_{-1}(t - t_0) + [f(t_0 + \Delta\lambda) - f(t_0)]U_{-1}(t - t_0 - \Delta\lambda)$$
$$+ [f(t_0 + 2\Delta\lambda) - f(t_0 + \Delta\lambda)]U_{-1}(t - t_0 - 2\Delta\lambda)$$
$$+ [f(t_0 + 3\Delta\lambda) - f(t_0 + 2\Delta\lambda)]U_{-1}(t - t_0 - 3\Delta\lambda) + \ldots$$
$$+ \{f[t_0 + (N - 1)\Delta\lambda] - f[t_0 + (N - 2)\Delta\lambda]\}$$
$$U_{-1}[t - t_0 - (N - 1)\Delta\lambda], t_0 \le t \le t_0 + T. \tag{2.98}$$

Equation (2.98) can be rewritten as:

$$f(t) \cong f(t_0)U_{-1}(t - t_0)$$
$$+ \sum_{k=1}^{N-1} \{f[t_0 + K\Delta\lambda] - f[t_0 + (k - 1)\Delta\lambda]\}U_{-1}(t - t_0 - k\Delta\lambda)$$

Multiplying and dividing each summation term by $\Delta\lambda$, we have that:

$$f(t) \cong f(t_0)U_{-1}(t - t_0)$$
$$+ \sum_{k=1}^{N-1} \frac{f[t_0 + k\Delta\lambda] - f[t_0 + (k - 1)\Delta\lambda]}{\Delta\lambda}U_{-1}(t - t_0 - k\Delta\lambda).\Delta\lambda \tag{2.99}$$

Obviously, when $\Delta\lambda \to 0, N \to \infty$, **but $T = N.\Delta\lambda$ remains, since T was previously fixed, that is, T is a parameter**, and the approximation becomes exact. But,

$$\lim_{\Delta\lambda \to 0} \frac{f[t_0 + k\Delta\lambda] - f[t_0 + (k - 1)\Delta\lambda]}{\Delta\lambda} = f'(t)|_{t=t_0+\lambda} = f'(t_0 + \lambda). <ucodeType$$
$$= ``Latin''_1 > x00DB; </ucode>$$

In addition, when $\Delta\lambda \to 0$, the sum of Eq. (2.99) becomes an integral, $\Delta\lambda$ becomes $d\lambda$, with λ ranging from 0 to T. Therefore, crossing the limit when $\Delta\lambda \to 0$ in Eq. (2.99), decomposition results

$$f(t) = f(t_0)U_{-1}(t - t_0) + \int_0^T f'(t_0 + \lambda)U_{-1}(t - t_0 - \lambda)d\lambda. \tag{2.100}$$

With the change of variable $\tau = t_0 + \lambda$, $d\lambda = d\tau$, and the extension of the upper limit T, from the interval initially considered, to infinity, it comes that:

$$f(t) = f(t_0)U_{-1}(t - t_0) + \int_{t_0}^{+\infty} f'(\tau)U_{-1}(t - \tau)d\tau, \text{for } t \ge t_0. \tag{2.101}$$

As in a defined integral, the integration variable disappears after replacing the upper and lower limits, which is why it is called a **mute variable**, it can be designated by any letter. Therefore, the integration variable τ in Eq. 2.101 can be replaced by λ, resulting, finally, that:

$$f(t) = f(t_0)U_{-1}(t - t_0) + \int_{t_0}^{+\infty} f'(\lambda)U_{-1}(t - \lambda)d\lambda, \textit{for } t \geq t_0. \qquad (2.102)$$

Comments:

(1) In the deduction of Eq. (2.102), the function $f(t)$ was assumed to be continuous at $t = t_0$, the initial instant of its decomposition in steps.

(2) When the function $f(t)$ is discontinuous at $t = t_0$, its derivative f ^ '(t) will contain the pulse $[f(t_{0_+}) - f(t_{0_-})]U_0(t - t_0)$, occurring exactly at $t = t_0$. In this case, Eq. (2.102) must be written as

$$f(t) = f(t_{0_-}).U_{-1}(t - t_0)$$

$$+ \int_{t_{0_-}}^{+\infty} \{[f(t_{0_+}) - f(t_{0_-})]U_0(\lambda - t_0) + f'(\lambda)\}U_{-1}(t - \lambda)d\lambda$$

$$= f(t_{0_-}).U_{-1}(t - t_0) + \int_{t_{0_-}}^{t_{0_+}} [f(t_{0_+}) - f(t_{0_-})]U_0(\lambda - t_0)d\lambda$$

$$+ \int_{t_{0_+}}^{+\infty} f'(\lambda)U_{-1}(t - \lambda)d\lambda = f(t_{0_-}) + [f(t_{0_+}) - f(t_{0_-})]$$

$$+ \int_{t_{0_+}}^{+\infty} f'(\lambda)U_{-1}(t - \lambda)d\lambda, \textit{for } t > t_0.$$

Therefore,

$$f(t) = f(t_{0_+}).U_{-1}(t - t_0) + \int_{t_{0_+}}^{+\infty} f'(\lambda)U_{-1}(t - \lambda)d\lambda, \textit{for } t > t_0. \qquad (2.103)$$

As $U_{-1}(t - \lambda) = 0$ for $\lambda > t$, (t parameter, variable λ), we can still write that

$$f(t) = f(t_{0_+}).U_{-1}(t - t_0) + \int_{t_{0_+}}^{+\infty} f'(\lambda)U_{-1}(t - \lambda)d\lambda, \textit{para } t > t_0. \qquad (2.104)$$

(3) In most cases, the function $f(t)$ is causal. So, $t_0 = 0$ and $f(t_{0_-}) = f(0_-) = 0$. So, from Eq. (2.102), we can show that

$$f(t) = \int_{0_-}^{\infty} f'(\lambda)U_{-1}(t - \lambda)d\lambda = \int_{0_-}^{t} f'(\lambda)U_{-1}(t - \lambda)d\lambda, \textit{for } t > 0, \qquad (2.105)$$

being understood that $f'(t)$ must include any impulse $f(0_+)U_0(t)$ that may occur at the origin, that is, at $t = 0$. Logically, if also $f(0_+) = 0$, there will be no impulse at the origin and lower limit of Eq. (2.105) must be replaced by 0.

(4) Supposing that the function $f(t)$ arose long before $t = 0$, that is, in $-\infty$, so that $t \overset{lim}{\to} -\infty f(t) = 0$, the general expression of its decomposition stepped is:

$$f(t) = \int_{-\infty}^{+\infty} f'(\lambda)U_{-1}(t - \lambda)d\lambda, \text{for all } t. \qquad (2.106)$$

Example 2.29 Establish the decomposition of the function $x(t)$ shown in Fig. 2.33 in steps (infinitesimal).

Solution
The derivatives of first and second orders of $x(t)$ are shown in Figs. 2.34 and 2.35, respectively.

Integrating the function $x''(t)$ shown in Fig. 2.35 successively, results in the decomposition of the function $x(t)$ on ramps, namely:

$$x(t) = 2U_{-2}(t) - 4U_{-2}(t - 1) + 2U_{-2}(t - 2),$$

shown in Fig. 2.36.
From Fig. 2.35, it is also possible to integrate $x''(t)$, once, which

$$x'(t) = 2U_{-1}(t) - 4U_{-1}(t - 1) + 2U_{-1}(t - 2)$$

According to Eq. (2.105), considering that there is no impulse at the origin,

$$x(t) = \int_0^t x'(\lambda)U_{-1}(t - \lambda)d\lambda$$

$$= \int_0^t [2U_{-1}(\lambda) - 4U_{-1}(\lambda - 1) + 2U_{-1}(\lambda - 2)]U_{-1}(t - \lambda)d\lambda, \qquad (2.107)$$

what expression the requested decomposition.

Fig. 2.33 Function $x(t)$ for decomposition into singular functions

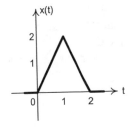

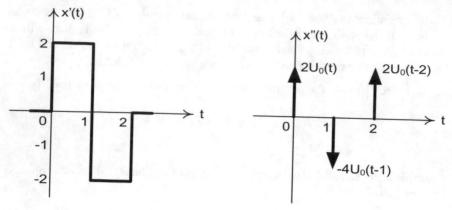

Fig. 2.34 a First order derivative of function $x(t)$; **b** second order derivative of function $x(t)$

Fig. 2.35 Function $x(t)$
expressed as a sum of singular
functions

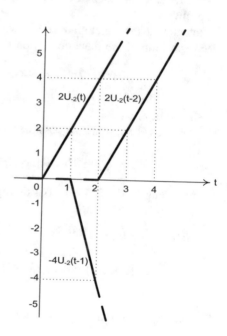

To confirm the accuracy of the result obtained, consider Fig. 2.37, where the signs $x'(\lambda)$ and $U_{-1}(t-\lambda)$ are shown; last for different ranges of values of t. Note that when t grows, the function $U_{-1}(t-\lambda)$ shifts to the right, while $x'(\lambda)$ remains fixed.

From Fig. 2.37 and Eq. (2.107), it follows that:

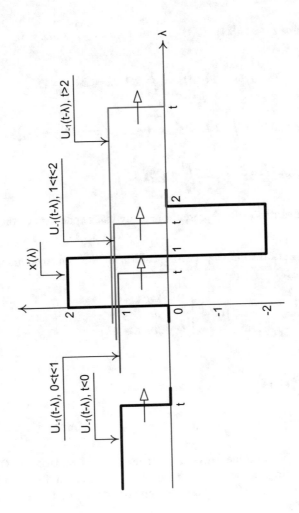

Fig. 2.36 Function $x(t)$ decomposition in singular functions

130

Fig. 2.37 Truncated ramp function

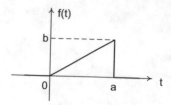

$$\textbf{For } t < 0 : x'(\lambda).U_{-1}(t - \lambda) = 0ex(t) = \int_0^t 0.d\lambda = 0.$$

$$\textbf{For } 0 \le t \le 1 : x(t) = \int_0^t 2.1.d\lambda = 2\lambda|_0^t = 2t.$$

$$\textbf{For } 1 \le t \le 2 : x(t) = \int_0^1 2.1.d\lambda + \int_1^t (-2).1.d\lambda = 2\lambda|_0^1 - 2\lambda|_1^t = -2t + 4.$$

$$\textbf{For } t \ge 2 : x(t) = \int_0^1 2.1.d\lambda + \int_1^2 (-2).1.d\lambda = 0.$$

The results obtained really represent the analytical expression of the signal $x(t)$ in Fig. 2.33.

Example 2.30 Perform the step decomposition of the truncated ramp function, defined by:

$$f(t) = \begin{cases} 0 & \text{for } t \le 0 \\ \frac{b}{a}t & \text{for } 0 \le t < a \\ 0 & \text{for } t > a. \end{cases}$$

Solution

The derivative of $f(t)$ is

$$f'(t) = \frac{b}{a}U_{-1}(t) - \frac{b}{a}U_{-1}(t - a) - bU_0(t - a), \text{for } 0_+ \le t \le a.$$

From Eq. (2.105), with the lower limit of the integral replaced by 0_+ and the upper limit by a_+, to indicate that the integration interval includes the point $\lambda = a$, in which the impulse $-bU_0(\lambda - a)$ occurs, and considering that $U_{-1}(t - \lambda) = 0$ for $\lambda > t$, we have that:

$$f(t) = \int_{0_+}^{a_+} \left[\frac{b}{a}U_{-1}(\lambda) - \frac{b}{a}U_{-1}(\lambda - a) - bU_0(\lambda - a) \right] U_{-1}(t - \lambda)d\lambda$$

$$= \int_{0_+}^{a_-} \frac{b}{a}U_{-1}(t - \lambda)d\lambda + \int_{a_-}^{a_+} -bU_0(\lambda - a)U_{-1}(t - \lambda)d\lambda.$$

To verify the correction of the result, the obtained decomposition is integrated:

For $0_+ \leq t \leq a_-$:
$$f(t) = \int_{0_+}^{t} \frac{b}{a}.1.d\lambda + \int_{t}^{a_-} \frac{b}{a}.0.d\lambda$$
$$+ \int_{a_-}^{a_+} -bU_0(\lambda - a).0.d\lambda = \frac{b}{a}t.$$

For $0_+ \leq t \leq a_-$:
$$f(t) = \int_{0_+}^{t} \frac{b}{a}.1.d\lambda + \int_{t}^{a_-} \frac{b}{a}.0.d\lambda + \int_{a_-}^{a_+} -bU_0(\lambda - a).0.d\lambda = \frac{b}{a}t.$$

For $a_- \leq t < +\infty$:
$$f(t) = \int_{0_+}^{a_-} \frac{b}{a}.1.d\lambda + \int_{a_-}^{a_+} -bU_0(\lambda - a).1.d\lambda = b - b = 0.$$

Example 2.31 Establish the formula for the decomposition of a continuous ramp function, valid for all t.

Solution
By analogy with Eq. (2.106), the formula sought must have the general form

$$f(t) = \int_{-\infty}^{+\infty} g(\lambda)U_{-2}(t - \lambda)d\lambda. \tag{2.108}$$

And the problem is solved by finding $g(\lambda)$.
Deriving the Eq. (2.108), in relation to t, comes that:

$$f'(t) = \int_{-\infty}^{+\infty} g(\lambda)\frac{d}{dt}[U_{-2}(t - \lambda)]d\lambda = \int_{-\infty}^{+\infty} g(\lambda)U_{-1}(t - \lambda)d\lambda$$

Deriving again, and remembering that the impulse function is even, that is, that $U_0(t - \lambda) = U_0(\lambda - t)$, comes that

$$f''(t) = \int_{-\infty}^{+\infty} g(\lambda)U_0(t - \lambda)d\lambda = \int_{-\infty}^{+\infty} g(\lambda)U_0(\lambda - t)d\lambda = g(\lambda)|_{\lambda=t} = g(t)$$

Then, $g(\lambda) = f''(\lambda)$ and,

$$f(t) = \int_{-\infty}^{+\infty} f''(\lambda)U_{-2}(t - \lambda)d\lambda. \tag{2.109}$$

By analogy, it can be shown that

$$f(t) = \int_{-\infty}^{+\infty} f'''(\lambda)U_{-3}(t-\lambda)d\lambda$$

is the decomposition of $f(t)$ in parables of the second degree, valid for all t.

Example 2.32 Show that the decomposition of a function $f(t)$ into exponentials of type $\varepsilon^{-t}U_{-1}(t)$ is given by

$$f(t) = \int_{-\infty}^{+\infty} [f'(\lambda) + f(\lambda)]\varepsilon^{-(t-\lambda)}U_{-1}(t-\lambda)d\lambda.$$

Solution
As in example 2.31, either

$$f(t) = \int_{-\infty}^{+\infty} g(\lambda)\varepsilon^{-(t-\lambda)}U_{-1}(t-\lambda)d\lambda.$$

Deriving in relation to t, (considering λ as a parameter), it comes that:

$$f'(t) = \int_{-\infty}^{+\infty} \left[-g(\lambda)\varepsilon^{-(t-\lambda)}U_{-1}(t-\lambda) + g(\lambda)\varepsilon^{-(t-\lambda)}U_0(t-\lambda) \right]d\lambda$$

$$= -\int_{-\infty}^{+\infty} g(\lambda)\varepsilon^{-(t-\lambda)}U_{-1}(t-\lambda)d\lambda + \int_{-\infty}^{+\infty} g(\lambda)\varepsilon^{-(t-\lambda)}U_0(t-\lambda)d\lambda$$

$$= -f(t) + \varepsilon^{-t}\int_{-\infty}^{+\infty} g(\lambda)\varepsilon^{\lambda}U_0(\lambda-t)d\lambda$$

$$= -f(t) + \varepsilon^{-t}[g(\lambda)\varepsilon^{\lambda}]|_{\lambda=t} = -f(t) + g(t).$$

Then, $g(t) = f'(t) + f(t)$, $g(\lambda) = f'(\lambda) + f(\lambda)$ and

$$f(t) = \int_{-\infty}^{+\infty} [f'(\lambda) + f(\lambda)]\varepsilon^{-(t-\lambda)}U_{-1}(t-\lambda)d\lambda, \text{for all } t.$$

2.11   Decomposition of Functions into Impulses

The function $f(t)$ shown in Fig. 2.38 can be approximated, in any finite range $-T \le t \le T$, by a finite number of 2 N rectangular pulses, juxtaposed in sequence, of width $\Delta\lambda$, occurring at $t = k\Delta\lambda$, for $k = -N, \ldots, -3, -2, -1,$ $0, 1, 2, \ldots, (N-1)$, where $N = T/\Delta\lambda$, as shown in the same figure.

Assuming, initially, a sequence of 2 N rectangular pulses of unitary area $p_{\Delta\lambda}(t - k\Delta\lambda)$, that is, a sequence of pulses of width $\Delta\lambda$ and height $1/\Delta\lambda$, logically, a generic pulse occurring at $t = k\Delta\lambda$ must be multiplied by $f(k\Delta\lambda)\Delta\lambda$ to make its

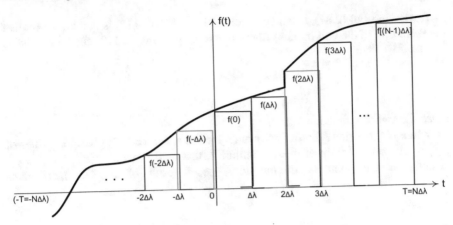

Fig. 2.38 Decomposition of function $x(t)$ into sequential rectangular pulses

amplitude equal to the value of the approximate, quantized function, in the interval $[k\Delta\lambda, (k+1)\Delta\lambda]$. Therefore, the approximation is

$$f(t) \cong \sum_{k=-N}^{N-1} f(k\Delta\lambda)p_{\Delta\lambda}(t - k\Delta\lambda)\Delta\lambda, \quad -T \le t \le T, \tag{2.110}$$

where,

$$p_{\Delta\lambda}(t - k\Delta\lambda) = \frac{1}{\Delta\lambda}\{U_{-1}[t - k\Delta\lambda] - U_{-1}[t - (k+1)\Delta\lambda]\}.$$

The approximation improves as the pulses become narrower and, as a result, increase in number. At the limit, when $\Delta\lambda \to 0$ and $N \to \infty$, keeping $T = N\Delta\lambda$ constant, however, the $k\Delta\lambda$ values of the argument of the function f of Eq. (2.110) can assume all possible values within the range $[-T, +T]$, and are therefore replaced by the continuous variable λ. But, on the other hand, this causes the $p_{\Delta\lambda}(t - k\Delta\lambda)$ pulses to become unit impulses and the sum to become an integral; and the approximation is accurate. Therefore, when $\Delta\lambda \to 0$ and $N \to \infty$, the Eq. (2.110) becomes

$$f(t) = \int_{-T}^{+T} f(\lambda)U_0(t - \lambda)d\lambda \tag{2.111}$$

To extend Eq. (2.111) to the entire domain of definition of $f(t)$, that is, $-\infty < t < +\infty$, just do $T \to \infty$, obtaining that:

$$f(t) = \int_{-\infty}^{+\infty} f(\lambda)U_0(t - \lambda)d\lambda, \textit{for all } t. \tag{2.112}$$

Reviewing Eq. (2.87), it can be concluded that Eq. (2.112) is nothing more than the definition of the unit impulse function itself, since Eq. (2.112) can be written as

$$\int_{-\infty}^{+\infty} f(\lambda)U_0(\lambda - t)d\lambda = f(\lambda)|_{\lambda=t} = f(t),$$

since the impulse function is even.

When $f(t)$ has finite discontinuity at any point a, Eq. (2.112) gives the values of $f(t)$ to the left and right of the discontinuity, that is, it provides $f(a_-)$ and $f(a_+)$. Exactly at the point of discontinuity, as a result of Eq. (2.97), it gives $[f(a_+) + f(a_-)]/2$.

2.12 Convolution Integral

The convolution $f(t)$ of any two functions $f_1(t)$ and $f_2(t)$ is defined by:

$$f(t) = f_1(t) * f_2(t) = \int_{-\infty}^{+\infty} f_1(\lambda)f_2(t - \lambda)d\lambda \qquad (2.113)$$

Comparing the Eqs. (2.112), (2.106) and (2.109) with the Eq. (2.113), it is concluded, then, that the expressions that allow to decompose a function in impulses, steps or ramps constitute integrals of convolution. For example, to decompose a function $f(t)$ in steps, its derivative $f'(t)$ is convolved with the step function $U_{-1}(t)$, that is,

$$f(t) = f'(t) * U_{-1}(t) = \int_{-\infty}^{+\infty} f'(\lambda)U_{-1}(t - \lambda)d\lambda,$$

as set out in Eq. (2.106).

In the graphical interpretation of the convolution integral, the quantities $f_1(t)$ and $f_2(t - \lambda)$ are plotted as a function of the independent variable λ, for specific ranges of t values. Then, the two curves are multiplied to form the integrand, and the value of $f(t)$ is equal to the total area under the product curve.

To form $f_1(\lambda)$, just replace t with λ in the analytical expression of $f_1(t)$. To form $f_2(t - \lambda)$, first exchange t for λ in the analytical expression of $f_2(t)$, obtaining $f_2(\lambda)$. Then exchange λ for $-\lambda$, generating $f_2(-\lambda)$. This operation corresponds to rotating 180° in f_2 (λ) about the vertical axis of the rectangular coordinate system. Then, the independent variable λ is replaced by $\lambda - t$, obtaining $f_2[-(\lambda - t)] = f_2(t - \lambda)$, which corresponds to delay $f_2(-\lambda)$ of a t value, shifting the respective curve to the right.

In summary, it can be said that the graphic procedure for carrying out a convolution integral consists of four steps.

(1) **Rotate** the function $f_2(\lambda)$ around the vertical axis, obtaining $f_2(-\lambda)$.
(2) **Shiftf** $f_2(-\lambda)$ to the right of a quantity t, generating $f_2(t - \lambda)$.
(3) **Multiply** $f_1(\lambda)$ by $f_2(t - \lambda)$, obtaining $f_1(\lambda)f_2(t - \lambda)$.
(4) **Integrate**, calculating the area under the product curve.

The graphical interpretation of the convolution integral is very useful in determining the values that should be used as limits of integration. In addition, it is an important aid in calculating convolution when one or both of the member's functions have discontinuities.

The four steps described are conveniently worked out in the next example.

Example 2.33 Calculate $f(t) = f_1(t) * f_2(t)$ (read f_1 **convoluted with** f_2), with $f_1(t)$ and $f_2(t)$ being the functions shown in Fig. 2.39a, b, respectively.

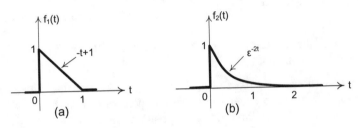

Fig. 2.39 a Function $f_1(t)$; **b** function $f_2(t)$, to be convoluted

Solution
As mentioned, $f_1(\lambda)$ is obtained from $f_1(t)$ by simply exchanging t for λ. So,

$$f_1(\lambda) = \begin{cases} 0 \; for - \infty < \lambda < 0 \\ -\lambda + 1 \; for \; 0 < \lambda \le 1 \\ 0 \; for \; 1 \le \lambda < +\infty. \end{cases}$$

It is shown in Fig. 2.41 and, as it does not depend on t, it remains static when t varies. Similarly,

$$f_2(\lambda) = \begin{cases} 0 \; for - \infty < \lambda < 0 \\ \varepsilon^{-2\lambda} for \; 0 < \lambda < +\infty. \end{cases}$$

So, **rotating** $f_2(\lambda)$ around the vertical axis, comes that:

$$f_2(-\lambda) = \begin{cases} \varepsilon^{2\lambda} \; for - \infty < \lambda < 0 \\ 0 \; for \; 0 < \lambda < +\infty. \end{cases}$$

The function $f_2(-\lambda)$ is shown in Fig. 2.40a.

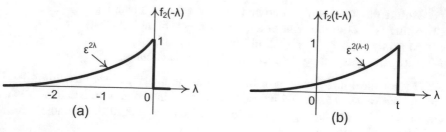

Fig. 2.40 **a** Function $f_2(t - \lambda)$; **b** function $f_2(t - \lambda)$ shifted to the right

Substituting λ for $\lambda - t$ in $f_2(-\lambda)$, it comes that:

$$f_2[-(\lambda - t)] = f_2(t - \lambda) = \begin{cases} \varepsilon^{2(\lambda-t)} \text{ for } -\infty < \lambda - t < 0 \\ 0 \text{ for } 0 < \lambda - t < +\infty. \end{cases} = \begin{cases} \varepsilon^{2\lambda}\varepsilon^{-2t}, -\infty < \lambda < t \\ 0, t < \lambda < +\infty. \end{cases}$$

The curve of the function $f_2(t - \lambda)$ is plotted in Fig. 2.40b. It is $f_2(-\lambda)$ **shifted to the right** by an amount t.

Figure 2.41 shows the curves of $f_1(\lambda)$ and $f_2(t - \lambda)$. Because it depends on t, the function $f_2(t - \lambda)$ can occupy innumerable positions along the axis λ, when t varies. However, only the three different positional possibilities of $f_2(t - \lambda)$ are shown in the figure to $f_1(\lambda)$, which remains fixed. These three positions are established by the following intervals: $-\infty < t \leq 0; 0 \leq t \leq 1; 1 \leq t < +\infty$. Steps 3 and 4 described above, that is, **multiplication and integration**, are performed simultaneously, for each of the three intervals mentioned, according to Eq. (2.113). **Note that the integration is only performed in the interval in which the product of the two functions is non-zero.**

For $-\infty < t \leq 0$:

$$f(t) = \int_{-\infty}^{t} 0.\varepsilon^{2\lambda}\varepsilon^{-2t} \cdot d\lambda = 0.$$

For $0 \leq t \leq 1$:

$$f(t) = \int_{0}^{t} (-\lambda + 1) \cdot \varepsilon^{2\lambda}\varepsilon^{-2t} \cdot d\lambda = \varepsilon^{-2t} \int_{0}^{t} (-\lambda + 1) \cdot \varepsilon^{2\lambda} \cdot d\lambda$$

Integrating by parts, are:

$u = -\lambda + 1, du = -d\lambda$ e $dv = \varepsilon^{2\lambda}d\lambda, v = \frac{\varepsilon^{2\lambda}}{2}$. So,

$$f(t) = \varepsilon^{-2t} \left[\frac{\varepsilon^{2\lambda}(-\lambda + 1)}{2} + \frac{\varepsilon^{2\lambda}}{4} \right] \Big|_{0}^{t}$$

$$= \varepsilon^{-2t} \left[\varepsilon^{2\lambda} \left(-\frac{\lambda}{2} + \frac{3}{4} \right) \right] \Big|_{0}^{t} = \frac{3}{4}(1 - \varepsilon^{-2t}) - \frac{t}{2}.$$

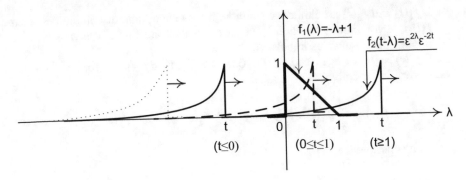

Fig. 2.41 Curves of $f_1(\lambda)$ and $f_2(t - \lambda)$

For $1 \leq t < +\infty$:

$$f(t) = \int_0^1 (-\lambda + 1) \cdot \varepsilon^{2\lambda} \varepsilon^{-2t} \cdot d\lambda = \varepsilon^{-2t} \left[\varepsilon^{2\lambda} \left(-\frac{\lambda}{2} + \frac{3}{4} \right) \right]\Big|_0^1 = \frac{\varepsilon^2 - 3}{4} \varepsilon^{-2t}.$$

Therefore,

$$f(t) = \begin{cases} 0 \ for - \infty < t \leq 0 \\ \frac{3}{4}(1 - \varepsilon^{-2t}) - \frac{t}{2} \ for \ 0 \leq t \leq 1 \\ \frac{\varepsilon^2 - 3}{4} \varepsilon^{-2t} \ for \ 1 \leq t < +\infty \end{cases}$$

2.13   Properties of the Convolution Integral

The convolution integral has the following properties:

1. Commutativity property

$$f(t) = f_1(t) * f_2(t) = f_2(t) * f_1(t), \text{where,}$$

$$f_2(t) * f_1(t) = \int_{-\infty}^{+\infty} f_2(\lambda) f_1(t - \lambda) d\lambda.$$

(2.114)

2. Distributive property in relation to the sum

$$f_1(t) * [f_2(t) + f_3(t)] = f_1(t) * f_2(t) + f_1(t) * f_3(t).$$

(2.115)

3. If $f_1(t)$ and $f_2(t)$ are positive, then $f(t)$ is also positive.

4. If $f_1(t)$ and $f_2(t)$ are limited, even though they have finite discontinuities, then $f(t)$ is a continuous function. (The fact that they are limited, obviously, also implies the absence of impulses).
5. If $f_1(t) = 0$ for $t > T_1$ and $f_2(t) = 0$ for $t > T_2$, then $f(t) = 0$ for $t > T_1 + T_2$.
6. Derivative of convolution.

$$f'(t) = \frac{d}{dt}[f_1(t) * f_2(t)] = f'_1(t) * f_2(t) = f_1(t) * f'_2(t). \tag{2.116}$$

As a consequence,

$$\begin{aligned} f(t) &= \frac{d^2}{dt^2}[f_1(t) * f_2(t)] = f'_1(t) * f'_2(t) \\ &= f_1(t) * f_2(t) = f_1(t) * f_2(t). \end{aligned} \tag{2.117}$$

7. $\int_{-\infty}^{+\infty} f(t)dt = \left(\int_{-\infty}^{+\infty} f_1(t)dt\right)\left(\int_{-\infty}^{+\infty} f_2(t)dt\right).$
8. If $f_2(t)$ is causal, that is, if $f_2(t) = 0$ for $t < 0$, then

$$f(t) = f_1(t) * f_2(t) = \int_{-\infty}^{t} f_1(\lambda)f_2(t - \lambda)d\lambda.$$

Or,

$$f(t) = f_2(t) * f_1(t) = \int_{0}^{+\infty} f_2(\lambda)f_1(t - \lambda)d\lambda.$$

9. If $f_1(t)$ is causal, then

$$f(t) = f_1(t) * f_2(t) = \int_{0}^{+\infty} f_1(\lambda)f_2(t - \lambda)d\lambda.$$

Or,

$$f(t) = f_2(t) * f_1(t) = \int_{-\infty}^{t} f_2(\lambda)f_1(t - \lambda)d\lambda.$$

10. If $f_1(t)$ and $f_2(t)$ are causal, then

$$f_1(t)*f_2(t) = \int_0^t f_1(\lambda)f_2(t-\lambda)d\lambda$$

$$= \int_0^t f_2(\lambda)f_1(t-\lambda)d\lambda = f_2(t)*f_1(t)$$

(2.118)

11. If $f_1(t)$ and $f_2(t)$ are causal with $f_1(t) = 0$ for $t > T_1$ and $f_2(t) = 0$ for $t > T_2$, then $f(t)$ is causal and $f(t) = 0$ for $t > T_1 + T_2$.

12. If $f_2(t) = kU_0(t) + f_3(t)$, where $f_1(t)$ and $f_3(t)$ are causal, then

$$f(t) = f_1(t)*f_2(t) = kf_1(t) + \int_{0_+}^t f_1(\lambda)f_3(t-\lambda)d\lambda.$$

13. If $\mathcal{F}[f_1(t)] = F_1(\omega)$ and $\mathcal{F}[f_2(t)] = F_2(\omega)$, then

$$\mathcal{F}[f_1(t)*f_2(t)] = F_1(\omega).F_2(\omega)..$$

(2.119)

The Fourier Transform of the convolution of two functions is equal to the product of the respective transforms. In other words, to **convolve in the time domain corresponds to multiply in the simple frequency domain** $(j\omega)$.

14. If $\mathcal{L}[f_1(t)] = F_1(s)$ and $\mathcal{L}[f_2(t)] = F_2(s)$, so,

$$\mathcal{L}[f_1(t)*f_2(t)] = F_1(s).F_2(s).$$

(2.120)

The Laplace transform of the convolution of two functions is equal to the product of the respective transforms. Or, **converging in the time domain corresponds to multiplying in the complex frequency domain** $(s = \sigma + j\omega)$.

15. Since $U_m(t)$ and $U_n(t)$ are any singular functions, then,

$$AU_m(t-a)*BU_n(t-b) = ABU_{m+n}(t-a-b).$$

(2.121)

16.
$$f(t-a)*U_0(t\pm b) = f(t-a\pm b).$$

(2.122)

Then,

$$f(t)*U_0(t) = \int_{-\infty}^{+\infty} f(\lambda)U_0(t-\lambda)d\lambda = f(t),$$

(2.123)

as already established in Eq. (2.112).

Decomposing a function $f(t)$ **into impulses is equivalent to convoluting it with a unitary impulse.**

17.
$$f(t) * U_1(t) = \int_{-\infty}^{+\infty} f(\lambda) U_1(t - \lambda) d\lambda = f'(t). \qquad (2.124)$$

Convoluting a function with a single stunt is the same as deriving it.

18.
$$f(t) * U_{-1}(t) = \int_{-\infty}^{t} f(\lambda) U_{-1}(t - \lambda) d\lambda = \int_{-\infty}^{t} f(\lambda) d\lambda = \int f(t) dt.$$

$$(2.125)$$

Convoluting a function with a unitary step is equivalent to integrating it.
Then,

$$f(t) * U_{-1}(t - a) = \int_{-\infty}^{t-a} f(\lambda) d\lambda. \qquad (2.126)$$

$$f(t) * [U_{-1}(t - a) - U_{-1}(t - b)] = \int_{t-b}^{t-a} f(\lambda) d\lambda, \text{ with } a < b. \qquad (2.127)$$

19. $f(t) * \frac{1}{b-a}[U_{-1}(t - a) - U_{-1}(t - b)] = \frac{1}{b-a} \int_{t-b}^{t-a} f(\lambda) d\lambda = \text{mean (run) of } f(t) \text{ in the interval } [t - b, t - a], \text{ with } a < b.$

20. $f(t) * \left[\frac{1}{2T} p_T(t)\right] = \frac{1}{2T} \int_{t-T}^{t+T} f(\lambda) d\lambda = \bar{f}(t) = \text{average (running) of } f(t) \text{ in the interval } [t - T, t + T], \text{ where}$

$$p_T(t) = U_{-1}(t + T) - U_{-1}(t - T).$$

21. The convolution of two even functions is an even function; the convolution of two odd functions is an even function; convolution of an even function with an odd function is odd

22. If $f(t) = f_1(t) * f_2(t)$, then

$$f_1(t - a) * f_2(t) = f_1(t) * f_2(t - a) = f(t - a). \qquad (2.128)$$

In consequence

$$f_1(t - a) * f_2(t + b) = f_1(t + b) * f_2(t - a) = f(t - a + b).$$

$$f_1(t - a) * f_2(t + a) = f_1(t + a) * f_2(t - a) = f(t).$$

23. The convolution of a function $f_1(t)$, not null in the interval (a, b), with a function $f_2(t)$, not null in the interval (c, d), is a function $f(t)$, not null in the range $(a + c, b + d)$.

Example 2.34 Perform the functions $f_1(t)$ and $f_2(t)$ shown in Fig. 2.42.

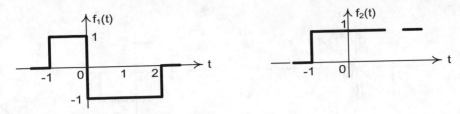

Fig. 2.42 Functions $f_1(t)$ and $f_2(t)$

Solution

In terms of singular functions, the functions are:

$f_1(t) = U_{-1}(t+1) - 2U_{-1}(t) + U_{-1}(t-2)$ and $f_2(t) = U_{-1}(t+1)$

Using properties 1, 2 and 15, it comes that:

$$f(t) = f_2(t) * f_1(t) = U_{-1}(t+1) * [U_{-1}(t+1) - 2U_{-1}(t) + U_{-1}(t-2)]$$
$$= U_{-2}(t+2) - 2U_{-2}(t+1) + U_{-2}(t-1)$$
$$= (t+2)U_{-1}(t+2) + (-2t-2)U_{-1}(t+1) + (t-1)U_{-1}(t-1)$$
$$= \begin{cases} 0 \ for -\infty < t \le -2 \\ t+2 \ for -2 \le t \le -1 \\ -t \ for -1 \le t \le 1 \\ -1 \ for \ 1 \le t < +\infty. \end{cases}$$

The convolution function $f(t)$ is shown in Fig. 2.43.

Fig. 2.43 Convolution function $f(t)$

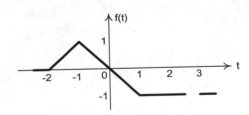

Example 2.35 Using properties, calculate $f(t) = f_1(t) * f_2(t)$, where $f_1(t)$ and $f_2(t)$ are the functions shown in Fig. 2.44.

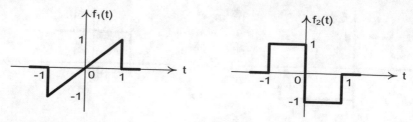

Fig. 2.44 Functions $f_1(t)$ and $f_2(t)$

Solution
According to property 6,

$$f'''(t) = f_1''(t) * f_2'(t).$$

On the other hand,

$$f_1''(t) = -U_1(t+1) + U_0(t+1) - U_1(t-1) - U_0(t-1).$$

$$f_2'(t) = U_0(t+1) - 2U_0(t) + U_0(t-1).$$

For properties 2, 16 and 15,

$$\begin{aligned}
f'''(t) &= f_1''(t) * [U_0(t+1) - 2U_0(t) + U_0(t-1)] \\
&= f_1''(t+1) - 2f_1''(t) + f_1''(t-1) \\
&= -U_1(t+2) + U_0(t+2) - U_1(t) - U_0(t) \\
&\quad + 2U_1(t+1) - 2U_0(t+1) + 2U_1(t-1) + 2U_0(t-1) \\
&\quad - U_1(t) + U_0(t) - U_1(t-2) - U_0(t-2).
\end{aligned}$$

Simply put, it results

$$\begin{aligned}
f'''(t) &= -U_1(t+2) + U_0(t+2) + 2U_1(t+1) - 2U_0(t+1) - 2U_1(t) \\
&\quad + 2U_1(t-1) + 2U_0(t-1) - U_1(t-2) - U_0(t-2).
\end{aligned}$$

Integrating, successively, three times, it comes that

$$\begin{aligned}
f(t) &= -U_{-2}(t+2) + U_{-3}(t+2) + 2U_{-2}(t+1) - 2U_{-3}(t+1) - 2U_{-2}(t) \\
&\quad + 2U_{-2}(t-1) + 2U_{-3}(t-1) - U_{-2}(t-2) - U_{-3}(t-2).
\end{aligned}$$

Or,

$$f(t) = (-t-2)U_{-1}(t+2) + \left(\frac{t^2}{2} + 2t + 2\right)U_{-1}(t+2) + (2t+2)U_{-1}(t+1)$$

$$+ (-t^2 - 2t - 1)U_{-1}(t+1) + (-2t)U_{-1}(t) + (2t-2)U_{-1}(t-1)$$

$$+ (t^2 - 2t + 1)U_{-1}(t-1) + (-t+2)U_{-1}(t-2)$$

$$+ \left(-\frac{t^2}{2} + 2t - 2\right)U_{-1}(t-2).$$

Grouping terms, we have

$$f(t) = \left(\frac{t^2}{2} + t\right)U_{-1}(t+2) + (-t^2 + 1)U_{-1}(t+1) + (-2t)U_{-1}(t)$$

$$+ (t^2 - 1)U_{-1}(t-1) + \left(-\frac{t^2}{2} + t\right)U_{-1}(t-2).$$

Finally,

$$f(t) = \begin{cases} 0 & for -\infty < t \le -2 \\ \frac{t^2}{2} + t & for -2 \le t \le -1 \\ -\frac{t^2}{2} + t + 1 & for -1 \le t \le 0 \\ -\frac{t^2}{2} - t + 1 & for 0 \le t \le 1 \\ \frac{t^2}{2} - t & for 1 \le t \le 2 \\ 0 & for 2 \le t < +\infty. \end{cases}$$

The $f(t)$ waveform is shown in Fig. 2.45. Note that $f(t)$ satisfies properties 21 and 23.

Appendix B presents a summary of the main relationships and properties of the singular functions covered in this chapter.

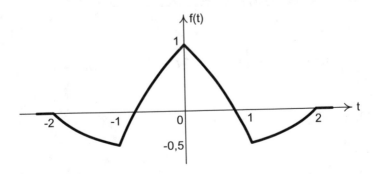

Fig. 2.45 Convolution function $f(t)$

2.14   Proposed Problems

P.2-1 Given the function $f(x) = U_{-1}(x).U_{-1}(t-x)$, where t is a parameter, determine $f'(x)$.

Answer:

$$f'(x) = U_0(x) - U_0(x-t).$$

P.2-2 Eliminate the step function at $f(t) = U_{-1}(t^2 + 3t + 2)$.

Answer:

$$f(t) = \begin{cases} 1 & for\ t < -2 \\ 0 & for\ -2 < t < -1 \\ 1 & for\ t > -1. \end{cases}$$

P.2-3 Eliminate the ramp functions of $f(t) = U_{-2}(t+1).U_{-2}(-3t-1)$.

Answer:

$$f(t) = \begin{cases} 0 & for\ t \leq -1 \\ -3t^2 - 4t - 1 & for\ -1 \leq t \leq -1/3 \\ 0 & for\ t \geq -1/3. \end{cases}$$

P.2-4 Simplify $f(t) = (\cos 4t + sen2t)U_0(t + \pi/2)$.

Answer:

$$f(t) = U_0(t + \pi/2).$$

P.2-5 Derive the function $f(t) = (\frac{3}{2}t + 1)[U_{-1}(t) - U_{-1}(t-2)]$.

Answer:

$$f'(t) = U_0(t) + \frac{3}{2}U_{-1}(t) - \frac{3}{2}U_{-1}(t-2) - 4U_0(t-2).$$

P.2-6 Calculate $\frac{d^2}{dt^2}[U_{-2}(t).U_{-1}(-t+1)]$.

Answer:

$$\frac{d^2}{dt^2}[U_{-2}(t).U_{-1}(-t+1)] = U_0(t) - U_0(t-1) - U_1(t-1).$$

P.2-7 (a) Develop $f(t) = (t^2 + 2t)U_{-1}(t - 2)$ in singular functions. (b) Same for $f(t) = (-t^3 + 4t)[U_{-1}(t+2) - U_{-1}(t - 2)]$.

Answer:

(a) $f(t) = 8U_{-1}(t - 2) + 6U_{-2}(t - 2) + 2U_{-3}(t - 2)$.
(b) $f(t) = -6U_{-4}(t+2) + 12U_{-3}(t+2) - 8U_{-2}$
    $(t+2) + 6U_{-4}(t - 2) + 12U_{-3}(t - 2) + 8U_{-2}(t - 2)$.

P.2-8 Make the following functions a sum of singular functions:

(a) $g(t) = (5t - 3)U_{-2}(t - 1)[U_{-1}(t - 2) - U_{-1}(t - 4)]$.
(b) $g(t) = (1 + 4\varepsilon^{-2t})U_1(t)$.

Answer:

(a) $g(t) = 7U_{-1}(t - 2) + 12U_{-2}(t - 2) + 10U_{-3}(t - 2) - 51U_{-1}(t - 4) - 32U_{-2}(t - 4)$
    $-10U_{-3}(t - 4)$.
(b) $g(t) = 5U_1(t) + 8U_0(t)$.

P.2-9 Prove the identities:

(a) $tU_2(t) = -2U_1(t)$.
(b) $t^2U_1(t) = 0$.
(c) $\cos t.U_3(t) = U_3(t) - 3U_1(t)$.
(d) $f(t) = \varepsilon^{2t}U_3(t) + \varepsilon^{-2t}U_2(t) = U_3(t) - 5U_2(t) + 16U_1(t) - 4U_0(t)$.

P.2-10 Synthesize the following functions:

(a) $f_1(t) = U_{-2}(t) - 3U_{-2}(t - 1) + 3U_{-2}(t - 2) - U_{-2}(t - 3)$.
(b) $f_2(t) = U_{-2}(t) + U_{-1}(t - 1) - U_{-2}(t - 1) - U_{-2}(t - 2) + U_{-2}(t - 4)$.
(c) $f_3(t) = 2U_{-3}(t) + U_{-2}(t) - U_{-3}(t - 1) - 5U_{-2}(t - 1) - U_{-3}(t - 3)$.

Answer:

(a) $f_1(t) = \begin{cases} 0 & \text{for } t \leq 0 \\ t & \text{for } 0 \leq t \leq 1 \\ -2t + 3 & \text{for } 1 \leq t \leq 2 \\ t - 3 & \text{for } 2 \leq t \leq 3 \\ 0 & \text{for } t \geq 3. \end{cases}$

(b) $f_2(t) = \begin{cases} 0 & \text{for } t \leq 0 \\ t & \text{for } 0 \leq t < 1 \\ 2 & \text{for } 1 < t \leq 2 \\ -t+4 & \text{for } 2 \leq t \leq 4 \\ 0 & \text{for } t \geq 4. \end{cases}$

(c) $f_3(t) = \begin{cases} 0 & \text{for } t \leq 0 \\ t^2+t & \text{for } 0 \leq t \leq 1 \\ t^2/2 - 3t + 9/2 & \text{for } 1 \leq t \leq 3 \\ 0 & \text{for } t \geq 3. \end{cases}$

P.2-11 Synthesize the function $f(t)$, for $0 \leq t \leq 6$, if $f(t) = 0$ for $t \leq 0$ and

$$f(t) = \sum_{n=0}^{+\infty} (-1)^n.(2n+1).U_{-2}(t-n), \text{for } t \geq 0.$$

Answer:

$$f(t) = \begin{cases} t, & 0 \leq t \leq 1 \\ -2t+3, & 1 \leq t \leq 2 \\ 3t - 7, & 2 \leq t \leq 3 \\ -4t+14, & 3 \leq t \leq 4 \\ 5t - 22, & 4 \leq t \leq 5 \\ -6t+33, & 5 \leq t \leq 6. \end{cases}$$

P.2-12 Starting from $f(t)U_0(t-a) = f(a)U_0(t-a)$ prove that:

(a) $f(t)U_1(t-a) = f(a)U_1(t-a) - f'(a)U_0(t-a)$.
(b) $f(t)U_2(t-a) = f(a)U_2(t-a) - 2f'(a)U_1(t-a) + f''(a)U_0(t-a)$.

P.2-13 Make the decomposition into singular functions of the next function, with x being a positive parameter.

$$f(t) = U_{-2}(t-x).U_{-2}(2x-t).$$

Answer:

$$f(t) = xU_{-2}(t-x) - 2U_{-3}(t-x) + xU_{-2}(t-2x) + 2U_{-3}(t-2x).$$

P.2-14 Find the value of the following integrals:

(a) $\int_1^2 e^{j2t} U_0(t-1,8)dt.$

(b) $\int_2^3 e^{j2t} U_0(t-1,8)dt.$

(c) $\int_2^3 (2t+3)[U_{-1}(t-2) - U_{-1}(t \quad 3)]U_0(t-2,5)dt.$

(d) $\int_{-2}^3 (t^3-1)U_2(\frac{t}{2}-1)dt.$

(e) $\int_3^4 \left[100\varepsilon^{-500t} + 220\cos(300t+\frac{\pi}{8})U_2(t-1)dt.\right]$

(f) $\int_{2t}^{4t} U_{-3}(4t-x)dx.$

Answer:

(a) $e^{j3.6}$,

(b) 0,

(c) 8,

(d) 96,

(e) 0,

(f) $\frac{4}{3}t^3.$

P.2-15 Make the following integrations:

(a) $\int_{-\infty}^{+\infty} U_0(at)dt.$

(b) $\int_{-\infty}^t U_0(at)dt.$

(c) $\int_{-\infty}^{+\infty} f(\lambda)U_{-1}(t-\lambda)d\lambda$, with $f(t) = \frac{1}{a}U_{-1}(t) - \frac{1}{a}U_{-1}(t-a) - U_0(t-a)$ and a a constant.

(d) $\int_{-\infty}^{+\infty} (t^2+2)U_0(\frac{t}{3}-1)dt.$

Answer:

(a) $\frac{1}{|a|}.$

(b) $\frac{1}{|a|}U_{-1}(t).$

(c) $\frac{1}{a}U_{-2}(t) - \frac{1}{a}U_{-2}(t-a) - U_{-1}(t-a) = \begin{cases} 0 & for\ t\leq 0 \\ t/a & for\ 0\leq t<a \\ 0 & for\ t>a. \end{cases}$

(d) 33.

P.2-16 Calculate the value of the integrals below.

(a) $\int_{-\pi}^{3\pi} \cos\left[\frac{3\pi}{2} + \frac{\pi}{2}U_{-1}(t-\pi)\right]U_{-2}(t-2\pi)dt.$

(b) $\int_0^{2\pi} (t+2)^2 sen(t-\pi)U_1(t-\pi)dt.$

Answer:

(a) $\frac{\pi^2}{2}.$

(b) $-(\pi+2)^2.$

P.2-17   Given   the   function $(t) = 5U_0(t) - 8\varepsilon^{-2t}U_{-1}(t)$,   calculate
$r(t) = \int h(t)dt = \int_{-\infty}^{t} h(\lambda)d\lambda$.

Answer:

$$r(t) = (1 + 4\varepsilon^{-2t})U_{-1}(t).$$

P.2-18 In a capacitor initially de-energized, the current $i(t)$ shown in Fig. 2.46 is applied.

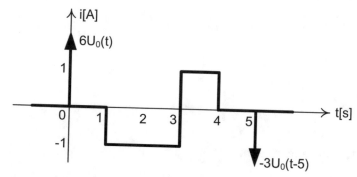

Fig. 2.46 Current $i(t)$ in a capacitor initially de-energized

(a) Find the value of the charge stored in it at time $t = 6[texts]$.
(b) Establish the analytical expression of the charge $q(t)$, for $t > 0$.

Answer:

a) $q(6) = 2[C]$.

b) $q(t) = \begin{cases} 0 & \text{for } t < 0 \\ 6 & \text{for } 0 < t \le 1 \\ -t+7 & \text{for } 1 \le t \le 3 \\ t+1 & \text{for } 3 \le t \le 4 \\ 5 & \text{for } 4 \le t < 5 \\ 2 & \text{for } t > 5. \end{cases}$

P.2-19 Develop functions $f_1(t)$ e $f_2(t)$ in singular functions shown in Figs. 2.47 and 2.48.

Answer:

$$f_1(t) = 2 - 4U_{-2}(t) + 8U_{-2}(t-1) - 8U_{-2}(t-2) + 4U_{-2}(t-3).$$
$$f_2(t) = U_{-1}(t) + U_{-2}(t) - 4U_{-1}(t-2) - U_{-1}(t-4) - U_{-2}(t-4).$$

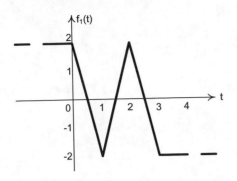

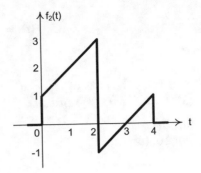

Fig. 2.47 Function $f_1(t)$

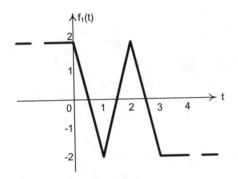

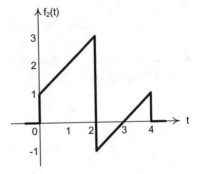

Fig. 2.48 Function $f_2(t)$

P.2-20 Prove that if the function $\varphi(t)$ is continuous and successively differentiable at $t = a$, then the decomposition into singular functions of the function $f(t) = \varphi(t)U_{-1}(t - a)$ is

$$f(t) = \sum_{n=1}^{+\infty} \varphi^{(n-1)}(a)U_{-n}(t - a),$$

where,

$$\varphi^{(n-1)}(a) = \frac{d^{n-1}}{dt^{n-1}} \varphi(t)|_{t=a}.$$

P.2-21 Knowing that

$$U_0[f(t)] = \sum_k \frac{1}{|f'(t_k)|} U_0(t - t_k),$$

t_k being the roots of $f(t) = 0$, calculate:

(a) $U_0(t^2 - a^2), a = constante.$

(b) $U_0(sent).$
(c) $2U_0(\cos 4t).$

Answer:

(a) $\frac{1}{2|a|} \cdot [U_0(t+a) + U_0(t-a)].$
(b) $\sum_{k=-\infty}^{+\infty} U_0(t - k\pi).$
(c) $\frac{1}{2} \sum_{k=1,2,3,\ldots} U_0\left[t \pm (2k+1)\frac{\pi}{8}\right].$

P.2-22 Demonstrate the formula contained in Eq. (2.57).

P.2-23 Determine $f(t)$ if $f(t) = \mathcal{F}^{-1}[F(\omega)] = \frac{1}{2\pi} \int_{-\infty}^{+\infty} F(\omega)e^{jt\omega}d\omega$ and

$$F(\omega) = \mathcal{F}[f(t)] = \frac{2+j\omega}{3+j\omega}\pi U_0(\omega).$$

Answer:

$$f(t) = \frac{1}{3}.$$

P.2-24 Show that:

(a) If $tf_1(t) = tf_2(t)$, then $f_1(t) = f_2(t) + kU_0(t)$
(b) If $t^2 f_1(t) = t^2 f_2(t)$, then $f_1(t) = f_2(t) + k_1 U_0(t) + k_2 U_1(t).$

P.2-25 Perform the decomposition in (a) steps and (b) in ramps of the function defined by

$$f(t) = \begin{cases} 0 \; for \; t \leq 0 \\ -t^2 + 2at \; for \; 0 \leq t \leq 2a \\ 0 \; for \; t \geq 2a. \end{cases}$$

Confirm the results by making the integrals obtained.

Answer:

(a) $f(t) = \int_0^{2a}(-2\lambda + 2a)U_{-1}(t - \lambda)d\lambda.$

(b) $f(t) = \int_{0_-}^{(2a)^+}[2aU_0(\lambda) - 2 + 2aU_0(\lambda - 2a)]U_{-2}(t - \lambda)d\lambda.$

P.2-26 Convolute the following functions, using the convolution definition directly.

(a) $f_1(t) = \begin{cases} 0 \quad for - \infty < t < 0 \\ 1 \quad for \; 0 < t < 1 \\ 0 \quad for \; 1 < t < +\infty. \end{cases}$ $f_2(t) = f_1(t).$

(b) $f_1(t) = \begin{cases} 0 \quad for - \infty < t < 0 \\ 1 \quad for \; 0 < t < 1 \\ -1 \quad for \; 1 < t < 2 \\ 0 \quad for \; 2 < t < +\infty. \end{cases}$ $f_2(t) = \begin{cases} 0 \quad for - \infty < t < 0 \\ 1 \quad for \; 0 < t < 2 \\ 0 \quad for \; 2 < t < +\infty. \end{cases}$

(c) $f_1(t) = \begin{cases} 0 \quad for - \infty < t \leq -1 \\ t+1 \quad for -1 \leq t \leq 0 \\ -t+1 \quad for \; 0 \leq t \leq 1 \\ 0 \quad for \; 1 \leq t < +\infty. \end{cases}$

$$f_2(t) = \sum_{k=0,1,2...} U_0(t \pm k).$$

(d) $f_1(t) = \begin{cases} 0 \quad for - \infty < t < 0 \\ 1 \quad for \; 0 < t < 1 \\ 0 \quad for \; 1 < t < +\infty. \end{cases}$ $f_2(t) = \begin{cases} 0 \quad for - \infty < t \leq 0 \\ t \quad for \; 0 \leq t \leq 1 \\ 1 \quad for \; 1 \leq t < +\infty. \end{cases}$

Answer:

(a) $f(t) = \begin{cases} 0 \quad for - \infty < t \leq 0 \\ t \quad for \; 0 \leq t \leq 1 \\ -t+2 \quad for \; 1 \leq t \leq 2 \\ 0 \quad for \; 2 \leq t < +\infty. \end{cases}$

(b) $f(t) = \begin{cases} 0 \quad for - \infty < t \leq 0 \\ t \quad for \; 0 \leq t \leq 1 \\ -t+2 \quad for \; 1 \leq t \leq 3 \\ t-4 \quad for \; 3 \leq t \leq 4 \\ 0 \quad for \; 4 \leq t < +\infty. \end{cases}$

(c) $f(t) = 1$ for $-\infty < t < +\infty$.

(d) $f(t) = \begin{cases} 0 & for -\infty < t \le 0 \\ \frac{t^2}{2} & for\ 0 \le t \le 1 \\ -\frac{t^2}{2} + 2t - 1 & for\ 1 \le t \le 2 \\ 1 & for\ 2 \le t < +\infty. \end{cases}$

P.2-27 If $f(t) = f_1(t) * f_2(t)$, calculate $f(2)$, being

$$f_1(t) = \begin{cases} 0 & for -\infty < t \le 0 \\ t & for\ 0 \le t \le 1 \\ 1 & for\ 1 \le t \le 2 \\ -t+3 & for\ 2 \le t \le 3 \\ 0 & for\ 3 \le t < +\infty. \end{cases} \qquad f_2(t) = \begin{cases} 0 & for -\infty < t \le 0 \\ 2t & for\ 0 \le t < 1 \\ \frac{t-3}{2} & for\ 1 < t \le 3 \\ 0 & for\ 3 \le t < +\infty. \end{cases}$$

Answer:

$$f(2) = \frac{7}{12}.$$

P.2-28 Using only properties, convert the following functions $f_1(t)$ and $f_2(t)$

(a) $f_1(t) = \begin{cases} 0 & for -\infty < t < -1 \\ 1 & for -1 < t < 0 \\ -1 & for\ 0 < t < 1 \\ 0 & for\ 1 < t < +\infty. \end{cases} \qquad f_2(t) = U_{-1}(t+1).$

(b) $f_1(t) = \begin{cases} -1 & for -\infty < t < -1 \\ 1 & for -1 < t < 1 \\ -1 & for\ 1 < t < +\infty. \end{cases} \qquad f_2(t) = U_0(t+1) - U_0(t-1).$

(c) $f_1(t) = \begin{cases} 0 & for -\infty < t < 0 \\ 1 & for\ 0 < t < 1 \\ 0 & for\ 1 < t < +\infty. \end{cases} \qquad f_2(t) = \begin{cases} 0 & for -\infty < t \le 0 \\ 2t & for\ 0 \le t \le 1 \\ -2t+4 & for\ 1 \le t \le 2 \\ 0 & for\ 2 \le t < +\infty. \end{cases}$

(d) $f_1(t) = \begin{cases} -2 & for -\infty < t \le -1 \\ 2t & for -1 \le t \le 1 \\ 2 & for\ 1 \le t < +\infty. \end{cases} \qquad f_2(t) = \begin{cases} 0 & for -\infty < t < -1 \\ 1 & for -1 < t < 1 \\ 0 & for\ 1 < t < +\infty. \end{cases}$

(e) $f_1(t) = \begin{cases} -1 & for -\infty < t < 0 \\ 1 & for\ 0 < t < +\infty. \end{cases} \qquad f_2(t) = \begin{cases} 0 & for -\infty < t < -1 \\ 1 & for -1 < t < 1 \\ 0 & for\ 1 < t < +\infty. \end{cases}$

(f) $f_1(t) = \begin{cases} 1 & \text{for } -\infty < t < -1 \\ 2 & \text{for } -1 < t < 0 \\ -2 & \text{for } 0 < t < 1 \\ -1 & \text{for } 1 < t < +\infty. \end{cases}$ $f_2(t) = \begin{cases} 0 & \text{for } -\infty < t < -1 \\ 1 & \text{for } -1 < t < 1 \\ 0 & \text{for } 1 < t < +\infty. \end{cases}$

Answer:

(a) $f(t) = \begin{cases} 0 & \text{for } -\infty < t \leq -2 \\ t+2 & \text{for } -2 \leq t \leq -1 \\ -1 & \text{for } -1 \leq t \leq 0 \\ 0 & \text{for } 0 \leq t < +\infty. \end{cases}$

(b) $f(t) = \begin{cases} 0 & \text{for } -\infty < t < -2 \\ 2 & \text{for } -2 < t < 0 \\ -2 & \text{for } 0 < t < 2 \\ 0 & \text{for } 2 < t < +\infty. \end{cases}$

(c) $f(t) = \begin{cases} 0 & \text{for } -\infty < t \leq 0 \\ t^2 & \text{for } 0 \leq t \leq 1 \\ -2t^2 + 6t - 3 & \text{for } 1 \leq t \leq 2 \\ t^2 - 6t + 9 & \text{for } 2 \leq t \leq 3 \\ 0 & \text{for } 3 \leq t < +\infty. \end{cases}$

(d) $f(t) = \begin{cases} -2 & \text{for } -\infty < t \leq -2 \\ \frac{t^2}{2} + 2t & \text{for } -2 \leq t \leq 0 \\ -\frac{t^2}{2} + 2t & \text{for } 0 \leq t \leq 2 \\ 2 & \text{for } 2 \leq t < +\infty. \end{cases}$

(e) $f(t) = \begin{cases} -2 & \text{for } -\infty < t \leq -1 \\ 2t & \text{for } -1 \leq t \leq 1 \\ 2 & \text{for } 1 \leq t < +\infty. \end{cases}$

(f) $f(t) = \begin{cases} 2 & \text{for } -\infty < t \leq -2 \\ t+4 & \text{for } -2 \leq t \leq -1 \\ -3t & \text{for } -1 \leq t \leq 1 \\ t-4 & \text{for } 1 \leq t \leq 2 \\ -2 & \text{for } 2 \leq t < +\infty. \end{cases}$

2.15   Conclusions

In this chapter it was presented the mathematical topic of singular functions, also called generalized functions or distributions, that constitute a special class of functions, because they do not fit perfectly into the conventional mathematical concept of function. Here they were designated by $U_n(t)$, $n = 0, \pm 1, \pm 2, \pm \ldots$, in such a way that for each value of n a specific singular function will correspond. They can be considered as a family of functions that are generated from the unit impulse function, $U_0(t)$, by successive derivations or integrations. However, for simplicity and also for strictly didactic reasons, the study of them was done starting from the unit step function, $U_{-1}(t)$. An important reason to study them is that the response of any circuit linear to an arbitrary entry can be determined, indirectly, as long as its response to a unit step or unit impulse is known. This is done through the integral convolution call. Thus, the response of a linear circuit to a unitary impulse, denoted by $h(t)$, or to a unitary step, called $r(t)$, can be used as a characteristic property of such a circuit. This is relevant for developing in the next chapter a systematic method to solve ordinary differential equations in which $x(t)$ is a singular and / or causal function. This method allows to determine the initial conditions at $t = 0_+$, from the respective initial conditions at $t = 0_-$ and the differential equation itself, without any physical reasoning about the given circuit. Therefore, a good understanding of this chapter is an indispensable condition for the good follow-up of the following chapters.

References

1. Adkins WA, Davidson MG (2012) Ordinary differential equations, undergraduate texts in mathematics. Springer
2. Boyce WE, DiPrima RC (2002) Elementary differential equations and boundary value problems, 7th ed. LTC
3. Chicone C (2006) Ordinary differential equations with applications, 22nd ed. Springer
4. Close CM (1975) Linear circuits analysis, LTC
5. Doering CI, Lopes AO (2008) Ordinary differential equations, 3rd ed. IMPA
6. Irwin JD (1996) Basic engineering circuit analysis. Prentice Hall
7. Kreider DL, Kluler RG, Ostberg DR (2002) Differential equations. Edgard Blucher Ltda
8. Leithold L (1982) The calculus with analytical geometry, vol 1, 2nd ed. Harper & How of Brasil
9. Perko L (2001) Differential equations and dynamical systems, 3rd edn. Springer, New York
10. Stewart J (2010) Calculus, vol 1, 6th ed. Thomson
11. Zill DG, Cullen MR (2003) Differential equations, 3rd ed. Makron Books

Chapter 3
Differential Equations

3.1 Introduction

When a linear system, with concentrated parameters and invariant over time, is excited with an input signal $x(t)$, the classic mathematical model for determining the output (or response) $y(t)$ is an n order ordinary linear differential equation with constant coefficients. In case the system is an electrical circuit, the order n is generally at most equal to the sum of the number of capacitances and inductances of the circuit, computing these numbers, however, only after capacitances and inductances in series and / or parallel have been replaced by their respective equivalents. This chapter presents a fundamental review about the theory and solution of ordinary differential equations. Moreover, it presents a systematic method to solve ordinary differential equations in which $x(t)$ is a singular and / or causal function. This method allows to determine the initial conditions at $t = 0_+$, from the respective initial conditions at $t = 0_-$ and the differential equation itself, without any physical reasoning about the given circuit. Therefore, a good understanding of this chapter is an indispensable condition for the good follow-up of the following chapters.

When a linear system, with concentrated parameters and invariant over time, is excited with an input signal $x(t)$, the classic mathematical model for determining the output (or response) $y(t)$ is an ordinary linear differential equation with constant coefficients [1–11], as indicated in Eq. 3.1.

$$\frac{d^n}{dt^n}y + a_1\frac{d^{n-1}}{dt^{n-1}}y + a_2\frac{d^{n-2}}{dt^{n-2}}y + a_3\frac{d^{n-3}}{dt^{n-3}}y + \ldots + a_{n-1}\frac{d}{dt}y + a_n y$$
$$= b_m\frac{d^m}{dt^m}x + b_{m-1}\frac{d^{m-1}}{dt^{m-1}}x + b_{m-2}\frac{d^{m-2}}{dt^{m-2}}x + \ldots + b_1\frac{d}{dt}x$$
$$+ b_0 x, \quad \text{with } n \geq m. \tag{3.1}$$

The a_i and b_j are constant and n is the **order** of the differential equation.

© The Author(s), under exclusive license to Springer Nature Switzerland AG 2021
J. C. Goulart de Siqueira and B. D. Bonatto, *Introduction to Transients in Electrical Circuits*, Power Systems, https://doi.org/10.1007/978-3-030-68249-1_3

In case the system is an electrical circuit, the order n is generally at most equal to the sum of the number of capacitances and inductances of the circuit, computing these numbers, however, only after capacitances and inductances in series and / or parallel have been replaced by their respective equivalents.

In Sect. 3.2 there is a wide review of the most important and necessary topics in the theory of ordinary differential equations, which take the form of Eq. 3.1.

The study of transient phenomena in linear electrical circuits, with lumped or concentrated parameters, can be done both in the time domain, by the so-called **classical method**, and in the domain of the complex frequency, by the well-known **operational method** of the Laplace transforms.

The **classical method** consists of the direct resolution, that is, in the time domain itself, of the system of ordinary differential equations that are obtained by equating the circuit. And it requires determining the arbitrary constants that appear in the general solution, either of the current, or of the voltage sought, from the initial conditions at $t = 0_+$, which must be obtained from the respective initial conditions at $t = 0_-$, a circumstance that constitutes the fundamental difficulty of this method, since it requires physical reasoning, sometimes very elaborate and quite difficult, about the given circuit.

The **operational method**, on the other hand, transforms the obtained ordinary differential equations into algebraic equations of the complex variable $s = \sigma + j\omega$, incorporating, inherently to this transformation process, the initial conditions at $t = 0_-$. This eliminates the need to determine arbitrary constants, based on the initial conditions at $t = 0_+$. However, the process of inverse transformation, that is, the return to the time domain, can become excessively laborious.

As in the following chapters it is intended to work almost exclusively with the classical method, then, in Sect. 3.2 a systematic method is developed to solve ordinary differential equations in which $x(t)$ is a singular and / or causal function. **This method allows to determine the initial conditions at $t = 0_+$, from the respective initial conditions at $t = 0_-$ and the differential equation itself, without any physical reasoning about the given circuit**. Therefore, a good understanding of Sect. 3.2 is an indispensable condition for the good comprehension of the following chapters.

3.2 Ordinary Differential Equations

Since $x(t)$ is a known function of time, the second member of Eq. 3.1, as a whole, is also a known function of time, called the **Forcing Function**, and denoted by $f(t)$. Thus, Eq. 3.1 can be rewritten as in Eq. 3.2.

$$\frac{d^n}{dt^n}y + a_1\frac{d^{n-1}}{dt^{n-1}}y + a_2\frac{d^{n-2}}{dt^{n-2}}y + a_3\frac{d^{n-3}}{dt^{n-3}}y + \ldots + a_{n-1}\frac{d}{dt}y + a_ny = f(t) \quad (3.2)$$

If $f(t)$ is identically null, that is, if $f(t) = 0$, then Eq. 3.2 becomes Eq. 3.3, known as the **Homogeneous Differential Equation**.

$$\frac{d^n}{dt^n}y + a_1\frac{d^{n-1}}{dt^{n-1}}y + a_2\frac{d^{n-2}}{dt^{n-2}}y + a_3\frac{d^{n-3}}{dt^{n-3}}y + \ldots + a_{n-1}\frac{d}{dt}y + a_ny = 0 \qquad (3.3)$$

When $f(t) \neq 0$, then Eq. 3.2 is called the **Non-homogeneous Differential Equation**.

Homogeneous Eq. 3.3, which is of order n, always presents n solutions. In the vast majority of cases these n solutions are linearly independent, which means that none of them is the linear combination of the others $n-1$, and that they are denoted by:

$$y_1(t), y_2(t), y_3(t), \ldots, y_n(t).$$

Thus, the most general solution of Eq. 3.3 is given by:

$$y_H(t) = k_1y_1(t) + k_2y_2(t) + k_3y_3(t) + \ldots + k_ny_n(t), \qquad (3.4)$$

where $k_1, k_2, k_3, \ldots, k_n$ are **arbitrary constants**.

The general solution of Eq. 3.2, which is non-homogeneous, is given by:

$$y(t) = y_H(t) + y_P(t) \qquad (3.5)$$

The term $y_H(t)$ is the solution of the corresponding homogeneous equation, as given by Eq. (3.4), and is called the **Complementary Solution**. The term $y_P(t)$ is any other solution, free of arbitrary constants, that satisfies Eq. 3.2, that is, that makes its two members identical, and is called a **Solution or Particular Integral**.

As $y_H(t)$ represents the solution of Eq. 3.2 for $f(t) = 0$, that is, the system response without external excitation, when input $x(t) = 0$, the complementary solution is also called the **Free Response** of the system. And considering that this free response depends on the general nature of the system, that is to say: the types of components present, their numerical values and the way they are interconnected, in addition to the energy stored in its non-dissipative elements, it is also called the **Natural Response** of the system. In other words, the natural response is due only to the energy initially stored in the system's own non-dissipative interior elements.

A system is considered dissipative when it transfers energy, in the form of heat, to its exterior. In an electrical circuit, for example, this role fits its resistances. And as in a dissipative system, $y_H(t)$ tends to zero after some time of operation of the system, it is still called the **Transient Response** of the system. Therefore, in a non-dissipative system, the free response cannot be called a transient response, because it does not disappear after a certain time; it has a permanent character.

On the other hand, since $y_P(t)$ depends on $f(t)$, that is, on the external source (s) that excites the system, it is called the **Forced Response or Permanent Regime (Steady-State) Response**. It is necessary to emphasize, therefore, that in a non-dissipative system, such as, for example, an electrical circuit without resistances, there is no damping of $y_H(t)$, which does not tend to zero over time. So, the natural response is also permanent, just like the forced response.

Every time a dissipative system undergoes an intervention, which forces it to move from one condition of stability to another, either through a change in the applied energy source, either through a change in its elements or even in its structure, there is a transition period, during which the variables associated with its elements change from their primitive values to new ones. This period is called the system's **Transient Regime**. In it, we have that $y(t) = y_H(t) + y_P(t)$. After the transitional regime has been extinguished, when $y_H(t)$ can already be considered null, it is said that the system is in **Permanent Regime** or **Stationary State**, when $y(t) = y_P(t)$.

In summary, it can be said that: **the complete response of a stable and dissipative system is made up of the sum of the transient response $y_H(t)$ and the forced or permanent response $y_P(t)$.** And that, after sufficient time has elapsed for the transitional regime to be extinguished, generally of the order of approximately five times the largest time constant of the system, only the permanent regime or steady-state will remain. As a consequence, one can study the forced behavior, or the permanent regime of the systems, regardless of the transient behavior. This is what is done when studying the permanent regime of electrical circuits subjected to sinusoidal sources using phasors and impedances.

Forced oscillations, in response, occur whenever a system is subjected to excitations that vary periodically over time.

3.2.1 Solution of the Homogeneous Differential Equation

Suppose that a solution of Eq. 3.3 is of the form $y(t) = \epsilon^{rt}$, where r is a constant to be determined and "e" is the Neperian number. Substituting this supposed solution in Eq. 3.3, it is obtained that:

$$\left(r^n + a_1 r^{n-1} + a_2 r^{n-2} + a_3 r^{n-3} + \ldots + a_{n-1}r + a_n\right)\epsilon^{rt} = 0$$

For this equation to be satisfied for all values of t it is necessary that

$$F(r) = r^n + a_1 r^{n-1} + a_2 r^{n-2} + a_3 r^{n-3} + \ldots + a_{n-1}r + a_n = 0 \qquad (3.6)$$

Equation 3.6 is called the **Characteristic Equation** and is an algebraic equation of order n, with constant coefficients, which contains n roots: $r_1, r_2, r_3, \ldots, r_n$, called **Characteristic Roots**. Thus, by factoring the **Polynomial Operator** $F(r)$, we have that:

$$F(r) = (r - r_1)(r - r_2)(r - r_3)\ldots(r - r_n) \qquad (3.7)$$

So, the solutions of the homogeneous differential equation are:

$$y_1(t) = \epsilon^{r_1 t}, y_2(t) = \epsilon^{r_2 t}, y_3(t) = \epsilon^{r_3 t}, \ldots, y_n(t) = \epsilon^{r_n t}.$$

And the **general solution** will fit into one of the following five cases.

(a) **If all the roots of the characteristic equation $F(r) = 0$ are real and distinct,** the solutions mentioned are independent and the general solution is the linear combination of them, namely:

$$y_H(t) = k_1 \epsilon^{r_1 t} + k_2 \epsilon^{r_2 t} + k_3 \epsilon^{r_3 t} + \ldots + k_n \epsilon^{r_n t} \tag{3.8}$$

(b) **If some roots of $F(r) = 0$ are complex and distinct,** $y_H(t)$ can be written in a different form.

When the homogeneous differential equation, 3.3, has coefficients $a_1, a_2, a_3, \ldots$, a_n real, the complex roots of $F(r) = 0$ always occur in conjugated pairs. If, for example, $r_1 = \alpha + j\beta$, where α and β are real numbers, then $r_2 = \alpha - j\beta$.
Then,

$$
\begin{aligned}
k_1 \epsilon^{r_1 t} + k_2 \epsilon^{r_2 t} &= \epsilon^{\alpha t}(k_1 e^{j\beta t} + k_2 \epsilon^{-j\beta t}) \\
&= \epsilon^{\alpha t}[k_1(\cos \beta t + j \sin \beta t) + k_2(\cos \beta t - j \sin \beta t)] \\
&= \epsilon^{\alpha t}[(k_1 + k_2) \cos \beta t + j(k_1 - k_2) \sin \beta t] \\
&= \epsilon^{\alpha t}(k_1' \cos \beta t + k_2' \sin \beta t) = k \epsilon^{\alpha t} \cos(\beta t + \phi)
\end{aligned}
$$

Therefore, if $r_1 = \alpha + j\beta$, $r_2 = \alpha - j\beta$ and $r_3, r_4, \ldots, r_n$ are real and distinct roots, then,

$$y_H(t) = \epsilon^{\alpha t}(k_1 \cos \beta t + k_2 \sin \beta t) + k_3 \epsilon^{r_3 t} + \ldots + k_n \epsilon^{r_n t} \tag{3.9}$$

or

$$y_H(t) = k \epsilon^{\alpha t} \cos(\beta t + \phi) + k_3 \epsilon^{r_3 t} + \ldots + k_n \epsilon^{r_n t}, \tag{3.10}$$

where k and ϕ are arbitrary constants; being that:

$$k = \sqrt{k_1^2 + k_2^2} \text{ and } \phi = -arctg \frac{k_2}{k_1}$$

(c) **If there are repeated real roots,** the terms of Eq. 3.8 will not all be independent and the equation will no longer represent the **general solution** of the homogeneous Eq. 3.3, since the number of arbitrary constants will be less than the order n of the differential equation. If $r_1 = r_2$, for example,

$$k_1 \epsilon^{r_1 t} + k_2 \epsilon^{r_2 t} = (k_1 + k_2) \epsilon^{r_1 t} = k_1' \epsilon^{r_1 t}$$

In this case, it can be shown that $y_1(t) = \epsilon^{r_1 t}$ and $y_2(t) = t\,\epsilon^{r_1 t}$ are independent solutions, so that the general solution of the homogeneous Eq. 3.3 is:

$$y_H(t) = (k_1 + k_2 t)\epsilon^{r_1 t} + k_3 \epsilon^{r_3 t} + \ldots + k_n \epsilon^{r_n t} \tag{3.11}$$

(d) **If the root r_1 is repeated $(m-1)$ times**, that is, if $r_1 = r_2 = r_3 = \ldots = r_m$, then the general solution is:

$$y_H(t) = \left(k_1 + k_2 t + k_3 t^2 + \ldots + k_m t^{m-1}\right)\epsilon^{r_1 t} + \ldots + k_n \epsilon^{r_n t} \tag{3.12}$$

(e) **If there are repeated complex roots**, the terms of Eqs. 3.9, or 3.10, will also not be independent and the solution, in the case of $r_1 = r_2 = \alpha + j\beta$ and $r_3 = r_4 = \alpha - j\beta$, for example, must be taken like:

$$y_H(t) = \epsilon^{\alpha t}[(k_1 + k_2 t)\cos\beta t + (k_3 + k_4 t)\sin\beta t] + \ldots + k_n \epsilon^{r_n t} \tag{3.13}$$

Example 3.1 Find the solution of the following homogeneous equations:

(a) $\frac{d^2 y}{dt^2} + \frac{dy}{dt} - 6y = 0$

(b) $(r+1)(r-2)(r+3)y = 0$

(c) $(r-1)^3(r+2)(r+3)y = 0$

(d) $\frac{d^2 y}{dt^2} - 2\frac{dy}{dt} + 10y = 0$

(e) $D(D^2 + 4)(D-1)^2 y = 0$

(f) $(D^4 + 2D^2 + 1)y = 0$

(g) $D(D^4 + 6D^2 + 9)y = 0$

D is the **Heaviside Differential Operator**, defined by:

$$D = \frac{d}{dt};\ D^2 = \frac{d^2}{dt^2};\ D^3 = \frac{d^3}{dt^3};\ldots$$

Thus, $D^2 z = \frac{d^2 z}{dt^2}$, $D^5 W = \frac{d^5 W}{dt^5}$.

Therefore, conversely:

$D^{-1} = \frac{1}{D} = \int \ldots dt$ is the **Integral Operador**.

Thus, the voltage-current relationship in an inductance, $e(t) = L\frac{d}{dt}i(t)$, can be written as $e(t) = LDi(t)$ and, conversely,

$i(t) = \frac{1}{L}\int e(t)dt$ can be written as $i(t) = \frac{1}{LD}e(t)$.

Likewise, the current-voltage relationship in a capacitance, $i(t) = C\frac{d}{dt}e(t)$ $= CDe(t)$, and $e(t) = \frac{1}{C}\int i(t)dt = \frac{1}{CD}i(t)$

Solution

(a) Using the operator D, the differential equation can be written as:

$$(D^2 + D - 6)y = 0$$

Therefore, the **polynomial operator** is $F(D) = D^2 + D - 6$, and the characteristic equation is $D^2 + D - 6 = 0$.

The characteristic roots are $r_1 = 2$ and $r_2 = -3$, and the general solution, according to Eq. 3.8, is

$$y_H(t) = k_1 e^{2t} + k_2 e^{-3t}$$

(b) The characteristic roots are $r_1 = -1$, $r_2 = 2$ e $r_3 = -3$. So,

$$y_H(t) = k_1 e^{-t} + k_2 e^{2t} + k_3 e^{-3t}$$

(c) The roots of the characteristic equation $(r-1)^3(r+2)(r+3) = 0$ are: $r_1 = r_2 = r_3 = 1$, $r_4 = -2$ and $r_5 = -3$. So, according to Eq. 3.12, the general solution of the homogeneous equation is

$$y_H(t) = (k_1 + k_2 t + k_3 t^2)e^t + k_4 e^{-2t} + k_5 e^{-3t}$$

(d) The characteristic equation is $D^2 - 2D + 10 = 0$. And its roots are $r_1 = 1 + j3$ and $r_2 = 1 - j3$.

According to Eq. 3.9,

$$y_H(t) = e^t(k_1 \cos 3t + k_2 \sin 3t)$$

or

$$y_H(t) = k e^t \cos(3t + \phi)$$

(e) The roots are $r_1 = 0$, $r_2 = j2$, $r_3 = -j2$, $r_4 = r_5 = 1$. So, by Eqs. 3.9 and 3.11,

$$y_H(t) = k_1 + k_2 \cos 2t + k_3 \sin 2t + (k_4 + k_5 t)e^t$$

(f) The roots of the characteristic equation are obtained first by doing $D^2 = z$, $D^4 = z^2$, which generates the following characteristic equation: $z^2 + 2z + 1 = (z+1)^2 = 0$. Its roots are $z_1 = z_2 = -1$. Therefore, $D = \pm\sqrt{z_1} = \pm\sqrt{-1} = \pm j$, that is, $r_1 = +j$ and $r_2 = -j$. On the other hand, also $D = \pm\sqrt{z_2} = \pm\sqrt{-1} = \pm j$, that is, $r_3 = +j$ and $r_4 = -j$. Therefore, in view of Eq. 3.13, with $\alpha = 0$,

$$y_H(t) = (k_1 + k_2 t)cost + (k_3 + k_4 t)\sin t$$

(g) The characteristic roots are $r_1 = 0$, $r_2 = r_3 = j\sqrt{3}$, $r_4 = r_5 = -j\sqrt{3}$. The general solution, therefore, is

$$y_H(t) = k_1 + (k_2 + k_3 t)cos\sqrt{3}t + (k_4 + k_5 t)\sin\sqrt{3}t$$

3.2.2 Considerations on Algebraic Equations

Obtaining the solution of the homogeneous differential equation, as seen in Sect. 3.2.1, depends essentially on determining the roots of the characteristic equation $F(D) = 0$, which is always an algebraic equation. For this reason, it is opportune to make some considerations about them.

Descartes rule

"If the coefficients of an algebraic equation $f(x) = 0$ are real, then the number of positive real roots (NPRR) that it has is equal to, or less than, an even number, to the number of sign inversions that the succession of its coefficients presents".

Applying the same rule in the algebraic equation that is obtained by exchanging x for $-y$ in the original equation $f(x) = 0$, the number of real negative roots (NNRR) of the given equation is estimated.

And the difference establishes the number of pairs of complex conjugated roots (NCCR).

Example 3.2 Apply the Descartes rule to the algebraic equation

$$f(x) = x^5 - 1.7x^4 + 2x^2 + 3.1x - 12 = 0$$

Solution
Since the equation is of fifth order, there are a total of 5 roots, and the number of sign inversions is equal to 3. Therefore, the equation must have either 1 or 3 positive real roots.

Replacing x by $-y$, we get: $-y^5 - 1.7y^4 + 2y^2 - 3.1y - 12 = 0$. Or that:

$$y^5 + 1.7y^4 - 2y^2 + 3.1y + 12 = 0.$$

The y-equation now has 2 sign inversions. So, the y-equation must have either zero or 2 positive real roots (PRR). But, as $x = -y$, positive real roots in y mean negative real roots (NRR) in x. Therefore, the original equation $f(x) = 0$ must have zero or 2 negative real roots. It is then possible to put together the following possibilities (combinations) in Table 3.1.

Table 3.1 Number of positive, negative real roots and complex conjugated roots

Cases	NPRR	NNRR	NCCR	Total
1st	1	0	4	5
2nd	1	2	2	5
3rd	3	0	2	5
4th	3	2	0	5

Example 3.3 Research the number of positive and negative real roots of the equation

$$f(x) = x^5 + 25x^4 + 230x^3 + 950x^2 + 1689x + 945 = 0.$$

Solution

Since there is no sign inversion, the equation has no positive real roots.

Taking $x = -y$, it comes that: $-y^5 + 25y^4 - 230y^3 + 950y^2 - 1689y + 945 = 0$.

There are 5 signal inversions. So, the given equation has either 1 or 3 or 5 real negative roots. Therefore, the following table of possibilities can be put together in Table 3.2.

In fact, the given equation was assembled from the five roots, making the product $(x + 1)(x + 3)(x + 5)(x + 7)(x + 9)$. That is, the roots are: $-1, -3, -5, -7$, and -9, corresponding to the 3rd case in Table 3.2.

Bolzano's theorem

"If $f(a)$ and $f(b)$ have the same sign, then $f(x) = 0$ has an even number of roots, or has none, in the range $[a, b]$. If $f(a)$ and $f(b)$ have opposite signs, then $f(x) = 0$ admits an odd number of roots in the range $[a, b]$".

Upper limit of positive real roots of an algebraic equation of degree n

If $f(x) = x^n + a_{n-1}x^{n-1} + \ldots + a_1 x + a_0 = 0$, then,

$$x < 1 + \sqrt[n-m]{A},$$

where m is the highest degree of the negative coefficient terms. A is the absolute value (modulus) of the lowest negative coefficient in the equation.

Table 3.2 Number of positive real roots and complex conjugated roots

Cases	NNRR	NCCR	Total
1st	1	4	5
2nd	3	2	5
3rd	5	0	5

Example 3.4 Establish the interval within which all the real roots of the equation $f(x) = x^5 - 1.2x^4 + x^2 + 3.4x - 8 = 0$ must be contained.

Solution

$$A = |-8| = 8, m = 4, n = 5.$$

$$x < 1 + \sqrt[5-4]{8} = 9$$

Taking $x = -y$, results that: $y^5 + 1.2y^4 - y^2 + 3.4y + 8 = 0$. So, now,

$$A = |-1| = 1, m = 2, n = 5.$$

$$y < 1 + \sqrt[5-2]{1} = 2. \text{ Then } -x < 2, \text{ or } x > -2$$

Therefore, all the real roots of $f(x) = 0$ are contained in the range $-2 < x < 9$.

Example 3.5 Find the range that contains all the real roots of the algebraic equation

$$f(x) = 3x^6 + 2x^3 - 2x^2 - 1 = 0$$

Solution

Initially, the coefficient of the highest degree term in x must be reduced to 1, according to the standard initially established for $f(x) = 0$. It follows, then, that:

$$x^6 + \frac{2}{3}x^3 - \frac{2}{3}x^2 - \frac{1}{3} = 0$$

$$A = \left| -\frac{2}{3} \right| = \frac{2}{3}, n = 6, m = 2$$

$$x < 1 + \sqrt[6-2]{\frac{1}{3}} = 1.904$$

For $x = -y$,

$$y^6 - \frac{2}{3}y^3 - \frac{2}{3}y^2 - \frac{1}{3} = 0$$

$$A = \frac{2}{3}, n = 6, m = 3$$

$$y < 1 + \sqrt[6-3]{\frac{2}{3}} = 1.874$$

Therefore, the real roots of the equation are contained in the range $-1.874 < x < 1.904$.

Newton-Raphson method

It is a powerful recurrence process, which allows starting from an initial estimate x_0, from a root of a given algebraic equation $f(x) = 0$, or any non-linear equation, and progressing, iteratively, until reaching the desired precision for the true root sought $\bar{x}$. Its general form is:

$$x_{i+1} = x_i - \frac{f(x_i)}{f'(x_i)}, \text{ for } i = 0, 1, 2, \ldots$$

Therefore, $\bar{x}$ is the limit of the succession $x_0, x_1, x_2, x_3, \ldots$

The second part of Bolzano's theorem is fundamental in the search for the referred initial estimate, x_0, since it guarantees that when $f(a)$ and $f(b)$ have opposite signs, there is at least one real root between a and b. (It is interesting to note that, digitally speaking, the signs are opposite when the product of $f(a)$ by $f(b)$ is negative). On the other hand, the knowledge of the interval where all the real roots of the equation must be located facilitates the search for the values of a and b, in which the ordinates have opposite signs, since it avoids the search without any criteria of the initial estimate. (It is worth mentioning that the Newton-Raphson method also applies to complex roots).

Depending on the initial estimate adopted, however, the Newton-Raphson method may not converge to the desired solution, that is, it will not converge to the sought $\bar{x}$ root. Studies on the convergence of the method, however, establish that:

"If $f'(x)$ does not change sign in the interval $a < x < b$, the Newton-Raphson method always converges to the root $\bar{x}$ of $f(x) = 0$, included in this interval, as long as the process is started at the end where $f(x).f''(x) > 0$. In other cases, the process may or may not converge".

In the search for two values of x with ordinates of opposite signs, especially when it comes to algebraic equations, it is necessary to build tables for different values of x and their corresponding $f(x)$ values. This is a tedious and boring job, as it requires, repeatedly, the calculation of different powers of x. However, this work can be shortened by adopting the following system, based on successive factors.

Let the following polynomial be: $f(x) = Ax^5 + Bx^4 + Cx^3 + Dx^2 + Ex + F$

Factoring $x^4: f(x) = (Ax + B)x^4 + Cx^3 + Dx^2 + Ex + F$
Factoring $x^3: f(x) = ((Ax + B)x + C)x^3 + Dx^2 + Ex + F$
Factoring $x^2: f(x) = (((Ax + B)x + C)x + D)x^2 + Ex + F$
Factoring $x: f(x) = ((((Ax + B)x + C)x + D)x + E)x + F$

Knowing that the number of parentheses that open must always be equal to the number of those that close, there is no greater problem in leaving the parentheses that open, and then write, simply, that:

$$f(x) = Ax + B)x + C)x + D)x + E)x + F$$

Proceeding in this way, the only necessary algebraic operations are multiplication and addition, alternately, without the need to repeatedly calculate increasing powers of x. Furthermore, if a given numerical value of x is stored, for example, in the memory ■ of an electronic calculator, then it follows that:

$$f(x) = A■ + B)■ + C)■ + D)■ + E)■ + F$$

Example 3.6 Given the algebraic equation $f(x) = x^3 + 2x^2 + 10x - 20 = 0$, which had one of its roots calculated in 1225, with ten correct significant numbers, by Leonardo de Pisa,

(a) apply the Descartes rule;
(b) establish the range that contains its real roots;
(c) apply Bolzano's theorem and make an initial estimate, x_0, for one of these roots;
(d) from x_0, refine by Newton-Raphson up to six correct significant numbers;
(e) by deflating the polynomial, find the other two roots of the equation.

Solution

(a) There is a sign inversion in the given equation. So, there is a real positive root. Replacing x by $-y$, we get: $-y^3 + 2y^2 - 10y - 20 = 0$. Or,

$$y^3 - 2y^2 + 10y + 20 = 0.$$

There are now two signal inversions. So, there are zero or two negative real roots. The possibilities are presented in Table 3.3

(b) $A = 20, n = 3, m = 0$

$$x < 1 + \sqrt[3]{20} = 3.714$$

From the y equation, from item (a),

$$A = 2, n = 3, m = 2$$

Table 3.3 Number of positive, negative or complex conjugated roots

Cases	NPRR	NNRR	NCCR	Total
1st	1	0	2	3
2nd	1	2	0	3

Table 3.4 Location of the range containing at least one real root

x	-3	-2	-1	0	1	2	3
$f(x)$	-59	-40	-29	-20	-7	16	55

$$y < 1 + \sqrt[3]{2} = 3.\text{or}, x > -3$$

Therefore, the real roots are contained in the range $-3 < x < 3.714$.

(c) The function $f(x)$ written in the factored form is (Table 3.4)

$$f(x) = x + 2)x + 10)x - 20.$$

Since $f(1) = -7$ and $f(2) = +16$ have opposite signs, there is certainly a root between $x = 1$ and $x = 2$.

Initial estimate:

$$f'(x) = 3x^2 + 4x + 10 \text{ and} f''(x) = 6x + 4.$$

Since the derivative $f'(x)$ does not change sign in the interval $1 < x < 2$, as it always remains positive in it, the iterative process must start at the end where $f(x).f''(x) > 0$, to ensure convergence to the sought root., Thus,

$$f(1).f''(1) = -7.10 = -70 < 0$$

$$f(2).f''(2) = 16.16 = 256 > 0$$

Therefore, the iterative process must start at the right end of the range, that is, the initial estimate must be $x_0 = 2$.

(d) The Newton-Raphson refinement is as follows:

$$x_{i+1} = x_i - \frac{f(x_i)}{f'(x_i)} = x_i - \frac{x_i^3 + 2x_i^2 + 10x_i - 20}{3x_i^2 + 4x_i + 10}$$

$$= \frac{2x_i^3 + 2x_i^2 + 20}{3x_i^2 + 4x_i + 10} = \frac{2x_i + 2)x_i)x_i + 20}{3x_i + 4)x_i + 10}$$

For $i = 0 : x_1 = \dfrac{2x_0 + 2)x_0)x_0 + 20}{3x_0 + 4)x_0 + 10} = \dfrac{2.2 + 2)2)2 + 20}{3.2 + 4)2 + 10}$

$$= \frac{44}{30} = 1.46667$$

For $i = 1$: $x_2 = \frac{30.61215}{22.32000} = 1.37151$

For $i = 2$: $x_3 = \frac{28.92181}{21.12916} = 1.36881$

For $i = 3$: $x_4 = \frac{28.87660}{21.09616} = 1.36881$

Therefore, $\overline{x_1} = 1.36881$, with six correct significant numbers, since $x_4 = x_3$.

(e) Polynomial deflation

The division of the polynomial $x^3 + 2x^2 + 10x - 20$ by $x - 1.36881$ is done next.

$$
\begin{array}{l}
x^3 + 2x^2 + 10x - 20 \quad \underline{/x - 1.368810000000000} \\
\underline{-x^3 + 1.36881x^2} \qquad\quad x^2 + 3.36881x + 14.61126 \\
3.36881x^2 + 10x \\
\underline{-3.36881x^2 + 4.61126x} \\
14.61126x - 20 \\
\underline{-14.61126x + 20.00004} \\
\cong 0
\end{array}
$$

The new equation, then, is: $x^2 + 3.36881x + 14.61126 = 0$
And the other two roots are:

$$
\overline{x_2} = -1.68441 + j3.43133
$$
$$
\overline{x_3} = -1.68441 - j3.43133.
$$

3.2.3 Solution of the Non-homogeneous Differential Equation

As highlighted by Eq. 3.5, the general solution of a non-homogeneous differential equation is composed of two parts, namely: the complementary solution $y_H(t)$, which involves a number of arbitrary constants equal to the order of the differential equation, and the particular solution $y_P(t)$, free of arbitrary constants. The purpose of this section is to recall one of the methods employed in determining $y_P(t)$.

The two main methods used to determine the particular solution $y_P(t)$ are the **Method of Variation of Parameters** and the **Method of Coefficients to be Determined**. It will be remembered here a method called **Abbreviated Method**, which is a slight variation of the method of the coefficients to be determined. The **Fasorial Method**, used in the case of sinusoidal forcing functions, will also be considered as an abbreviated method.

A particular solution of a linear differential equation with constant coefficients $F(D)y(t) = f(t)$ is given by $y_P(t) = \frac{1}{F(D)}f(t)$, where the right member must be understood as

"$\frac{1}{F(D)}$ **operating on** $f(t)$", which produces the following result:

$$y_P(t) = \epsilon^{r_1 t} \int \epsilon^{(r_2-r_1)t} \int \epsilon^{(r_3-r_2)t} \cdots \int \epsilon^{(r_n-r_{n-1})t} \int f(t)\epsilon^{-r_n t} dt^n, \qquad (3.14)$$

where $r_1, r_2, r_3, \ldots, r_n$ are the n roots of the characteristic equation $F(D) = 0$.
For $n = 1$, Eq. 3.14 is reduced to

$$y_P(t) = \frac{1}{D-r_1} f(t) = \epsilon^{r_1 t} \int f(t)\epsilon^{-r_1 t} dt \qquad (3.15)$$

For certain special forms of $f(t)$, however, the work required to calculate $\frac{1}{F(D)} f(t)$ can be greatly reduced. The shortened method will be shown for some of these more frequent special forms.

1° Case: $f(t) = A\epsilon^{\alpha t}$, where A and α are any constants. In this case,

$$y_P(t) = \frac{1}{F(D)} (A\epsilon^{\alpha t}) = \frac{A\epsilon^{\alpha t}}{F(\alpha)}, \text{ provided that } F(\alpha) \neq 0. \qquad (3.16)$$

Example 3.7 Solve the following differential equations:

(a) $(D^3 - 2D^2 - 5D + 6)y(t) = (D-1)(D+2)(D-3)y(t) = \epsilon^{4t}$
(b) $(D-1)(D+2)(D-3)y(t) = (\epsilon^{2t} + 3)^2$
(c) $(D-1)(D+2)(D-3)y(t) = \epsilon^{3t}$
(d) $(D+1)(D+2)(D+3)y(t) = 30$.

Solution

(a) The characteristic roots are $r_1 = 1$, $r_2 = -2$ and $r_3 = 3$. So, the complementary solution is:

$$y_H(t) = k_1 \epsilon^t + k_2 \epsilon^{-2t} + k_3 \epsilon^{3t}$$

The particular solution is:

$$y_P(t) = \frac{1}{(D-1)(D+2)(D-3)} \epsilon^{4t}$$

$$= \frac{1}{(4-1)(4+2)(4-3)} \epsilon^{4t} = \frac{1}{(3)(6)(1)} \epsilon^{4t} = \frac{1}{18} \epsilon^{4t}$$

And the general solution is:

$$y(t) = y_P(t) + y_H(t) = \frac{1}{18} \epsilon^{4t} + k_1 \epsilon^t + k_2 \epsilon^{-2t} + k_3 \epsilon^{3t}$$

(b) The particular solution is:

$$y_P(t) = \frac{1}{(D-1)(D+2)(D-3)}(\epsilon^{2t}+3)^2$$

$$= \frac{1}{(D-1)(D+2)(D-3)}(\epsilon^{4t}+6\epsilon^{2t}+9\epsilon^{0.t}) = \frac{\epsilon^{4t}}{(3)(6)(1)}$$

$$+ \frac{6\epsilon^{2t}}{(1)(4)(-1)} + \frac{9}{(-1)(2)(-3)} = \frac{\epsilon^{4t}}{18} - \frac{3\epsilon^{2t}}{2} + \frac{3}{2}$$

So, the general solution is:

$$y(t) = \frac{\epsilon^{4t}}{18} - \frac{3\epsilon^{2t}}{2} + \frac{3}{2} + k_1\epsilon^t + k_2\epsilon^{-2t} + k_3\epsilon^{3t}$$

(c) A particular solution is:

$$y_P(t) = \frac{1}{(D-1)(D+2)(D-3)}\epsilon^{3t}$$

In this case, $F(\alpha) = F(3) = (2).(5).(0) = 0$ and **the method cannot be applied directly**. One way to get around this difficulty is to proceed as follows, with the help of Eq. 3.15:

$$y_P(t) = \frac{1}{(D-3)}\left[\frac{1}{(D-1)(D+2)}\epsilon^{3t}\right] = \frac{1}{(D-3)}\left[\frac{\epsilon^{3t}}{(2)(5)}\right]$$

$$= \frac{1}{10}\left(\frac{1}{D-3}\epsilon^{3t}\right) = \frac{1}{10}\left(\epsilon^{3t}\int \epsilon^{3t}\epsilon^{-3t}dt\right)$$

$$= \frac{1}{10}\epsilon^{3t}\int dt = \frac{t\epsilon^{3t}}{10}$$

Therefore,

$$y(t) = \frac{t\epsilon^{3t}}{10} + k_1\epsilon^t + k_2\epsilon^{-2t} + k_3\epsilon^{3t}$$

(d)

$$y_P(t) = \frac{1}{(D+1)(D+2)(D+3)}(30\epsilon^{0.t}) = \frac{30}{(1)(2)(3)} = 5$$

$$y(t) = 5 + k_1\epsilon^{-t} + k_2\epsilon^{-2t} + k_3\epsilon^{-3t}$$

Whenever the complementary solution $y_H(t)$ contains exponential terms that participate in the forcing function $f(t)$, then $F(\alpha) = 0$ and **the method fails**. To get around this obstacle, proceed as in item c) of example 3.7, using the formula

$$\frac{1}{D+a}f(t) = \epsilon^{-at}\int f(t)\epsilon^{at}dt \qquad (3.17)$$

2° Case: $f(t) = t^n$, where n is a positive integer number.

In this case, the inverse of the polynomial operator $F(D)$ must be expanded in a series of increasing powers of D, by direct division, truncating the series in the term D^n, because from $D^{n+1}t^n$ derivatives will always be null, as indicated below:

$$y_P(t) = \frac{1}{F(D)}t^n = (a_0 + a_1D + a_2D^2 + a_3D^3 + \ldots + a_nD^n)t^n, a_0 \neq 0 \qquad (3.18)$$

Example 3.8 Solve the differential equations:

(a) $(D^2 - 4D + 3)y(t) = t^2 + 2t + 1$
(b) $(D^3 - 4D^2 + 3D)y(t) = t^2$

Solution

(a) As $r_1 = 1$ and $r_2 = 3$, the complementary solution is:

$$y_H(t) = k_1\epsilon^t + k_2\epsilon^{3t}$$

The particular integral is:

$$y_P(t) = \frac{1}{3 - 4D + D^2}(t^2 + 2t + 1)$$

The obtaining of the inverse of the expanded polynomial operator in a series of increasing powers of D, truncated in D^2, is shown below.

$$\begin{array}{r} 1 \\ \hline \end{array} \quad \underline{/\,3 - 4\,D + D^2}$$

$$-1 + \frac{4}{3}D - \frac{1}{3}D^2 \qquad \frac{1}{3} + \frac{4}{9}D + \frac{13}{27}D^2$$

$$\text{---------------------------}$$

$$+\frac{4}{3}D - \frac{1}{3}D^2$$

$$-\frac{4}{3}D + \frac{16}{9}D^2 - \frac{4}{9}D^3$$

$$\text{---------------------------}$$

$$\frac{13}{9}D^2 - \frac{4}{9}D^3$$

So,

$$
\begin{aligned}
y_P(t) &= \left(\frac{1}{3} + \frac{4}{9}D + \frac{13}{27}D^2\right)\left(t^2 + 2t + 1\right) \\
&= \frac{1}{3}\left(t^2 + 2t + 1\right) + \frac{4}{9}\frac{d}{dt}\left(t^2 + 2t + 1\right) + \\
&\quad + \frac{13}{27}\frac{d^2}{dt^2}\left(t^2 + 2t + 1\right) = \left(\frac{t^2}{3} + \frac{2}{3}t + \frac{1}{3}\right) + \left(\frac{8}{9}t + \frac{8}{9}\right) \\
&\quad + \left(\frac{26}{27}\right) = \frac{t^2}{3} + \frac{14}{9}t + \frac{59}{27}
\end{aligned}
$$

Finally,

$$y(t) = \frac{t^2}{3} + \frac{14}{9}t + \frac{59}{27} + k_1 e^t + k_2 e^{3t}$$

Obviously, the method of the coefficients to be determined can be applied directly, as indicated below.

Assuming that $y_P(t) = At^2 + Bt + C$, because the forcing function is of the second degree in t, then $y_P'(t) = 2At + B$ and $y_P''(t) = 2A$.

The given differential equation is: $y''(t) - 4y'(t) + 3y(t) = t^2 + 2t + 1$

The substitution of the assumed values for the particular integral, and its derivatives, in the differential equation results in:

$$2A - 4(2At + B) + 3(At^2 + Bt + C) = t^2 + 2t + 1$$

$$3At^2 + (-8A + 3B)t + (2A - 4B + 3C) = t^2 + 2t + 1$$

Identifying similar terms in the two members of this algebraic equation results in that

$$3A = 1 \therefore A = 1/3$$

$$-\frac{8}{3} + 3B = 2 \therefore B = 14/9$$

$$\frac{2}{3} - \frac{56}{9} + 3C = 1 \therefore C = 59/27$$

Therefore, $y_P(t) = \frac{t^2}{3} + \frac{14}{9}t + \frac{59}{27}$

(b) The complementary solution is:

$$y_H(t) = k_1 + k_2 e^t + k_3 e^{3t}$$

The particular solution is:

$$y_P(t) = \frac{1}{D(D^2 - 4D + 3)}t^2 = \frac{1}{D}\left(\frac{1}{3} + \frac{4}{9}D + \frac{13}{27}D^2\right)t^2$$

$$= \frac{1}{D}\left(\frac{t^2}{3} + \frac{8}{9}t + \frac{26}{27}\right)$$

Since $\frac{1}{D}f(t) = \int f(t)dt$,

$$y_P(t) = \int \left(\frac{t^2}{3} + \frac{8}{9}t + \frac{26}{27}\right)dt = \frac{t^3}{9} + \frac{4t^2}{9} + \frac{26t}{27}$$

So,

$$y(t) = \frac{t^3}{9} + \frac{4t^2}{9} + \frac{26t}{27} + k_1 + k_2 e^t + k_3 e^{3t}$$

3° Case: $f(t) = A\cos(\omega t + \theta)$
Initially, it is worth remembering that $\sin(\omega t + \theta) = \cos(\omega t + \theta - 90°)$.

In the case of a sinusoidal forcing function, the abbreviated method is the **Phasor Method**, namely: the forcing function $f(t)$ is replaced by its **Fasor** $\dot{F} = Ae^{j\theta} = A\underline{/\theta}$; the operator D is replaced by $j\omega$; D^2, by $(j\omega)^2 = -\omega^2$; D^3, by $(j\omega)^3 = -j\omega^3$; D^4, by $(j\omega)^4 = \omega^4$; ... Logically, the particular solution sought, $y_P(t)$, is replaced by its corresponding phasor $\dot{Y}_P$. So, for example, the differential equation

$$\left(D^4 + aD^3 + bD^2 + cD + d\right)y(t) = A\cos(\omega t + \theta)$$

is converted to the simple frequency domain as:

$$(\omega^4 - ja\omega^3 - b\omega^2 + jc\omega + d)\dot{Y}_P = \dot{F} = A\underline{/\theta}$$

Therefore,

$$\dot{Y}_P = \frac{\dot{F}}{(\omega^4 - b\omega^2 + d) + j(c\omega - a\omega^3)} = M\underline{/\varphi}$$

To return to the time domain, that is, to make explicit $y_P(t)$, the **phasor definition** is used:

$$y_P(t) = \textbf{Real Part of}\left[\dot{Y}_P.e^{j\omega t}\right]$$

So,

$$y_P(t) = RP\left[Me^{j\varphi}.e^{j\omega t}\right] = RP\left[Me^{j(\omega t + \varphi)}\right] = RP\{M[\cos(\omega t + \varphi) + j\sin(\omega t + \varphi)]\}$$

Therefore,

$$y_P(t) = M\cos(\omega t + \varphi)$$

Example 3.9 Solve the following differential equations:

(a) $(D^2 + 4)y(t) = \cos 3t$

(b) $(D^2 + 3D - 4)y(t) = \sin 2t = \cos(2t - 90°)$

Solution

(a) Since $r_1 = j2$ and $r_2 = -j2$, the complementary function is:

$$y_H(t) = k_1\cos 2t + k_2\sin 2t$$

Converting the differential equation to the simple frequency domain, with $\omega = 3\left[\frac{rd}{s}\right]$, it follows that:

$$(-9+4)\dot{Y}_P = 1/\underline{0^\circ} \therefore \dot{Y}_P = -\frac{1}{5}/\underline{0^\circ}$$

Therefore,

$$y_P(t) = -0,2\cos 3t$$

So,

$$y(t) = -0,2\cos 3t + k_1\cos 2t + k_2\sin 2t$$

(b) The characteristic roots are $r_1 = 1$, $r_2 = -4$, and the complementary solution is:

$$y_H(t) = k_1 e^t + k_2 e^{-4t}$$

In the frequency domain, we have that:

$$(-2^2 + j6 - 4)\dot{Y}_P = 1/\underline{-90^\circ}$$

$$\dot{Y}_P = \frac{1}{-8+j6}\cdot 1/\underline{-90^\circ} = \frac{1\angle -90^\circ}{10\angle 143.13^\circ} = 0,1/\underline{-233.13^\circ} = -0,1/\underline{53.13^\circ}$$

So, in the time domain,

$$y_P(t) = -0.1\cos(2t - 53.13^\circ) = -0.1\sin(2t - 53.13^\circ + 90^\circ)$$
$$= -0,1\sin(2t + 36.87^\circ) = -0,1(\sin 2t.\cos 36.87^\circ + \sin 36.87^\circ.\cos 2t)$$
$$= -0.1(0.8\sin 2t + 0.6\cos 2t) = -0.02(4\sin 2t + 3\cos 2t)$$

Therefore,

$$y(t) = -0,1\sin(2t + 36.87^\circ) + k_1 e^t + k_2 e^{-4t}$$

In the sinusoidal case, the abbreviated method itself is as follows:

$$y_P(t) = \frac{1}{F(D^2)}\cos(\omega t + \phi) = \frac{1}{F(-\omega^2)}\cos(\omega t + \phi), \text{ for } F(-\omega^2) \neq 0, \quad (3.19)$$

$$y_P(t) = \frac{1}{F(D^2)}\sin(\omega t + \phi) = \frac{1}{F(-\omega^2)}\sin(\omega t + \phi), \text{ for } F(-\omega^2) \neq 0. \quad (3.20)$$

Example 3.10 Find the particular solution of the following differential equations:

(a) $(D^2 + 4)y(t) = 10\cos 3t$

(b) $(D^4 + 20D^2)y(t) = 128\sin(4t - 45°)$

(c) $(D^2 + 3D - 4)y(t) = 100\cos(2t + 25°)$

Solution

(a) The particular solution is

$$y_P(t) = \frac{1}{D^2 + 4}10\cos 3t = \frac{1}{-(3)^2 + 4}10\cos 3t = -2\cos 3t.$$

(b)

$$y_P(t) = \frac{1}{D^2(D^2 + 20)}128\sin(4t - 45°)$$

$$= \frac{1}{-(4)^2\left[-(4)^2 + 20\right]}128\sin(2t - 45°) = -2\sin(2t - 45°).$$

(c) **The operator is not of the standard 1/F(D²) and the abbreviated method does not apply ready.** Fortunately, there are two ways to get around this difficulty. The first is

$$y_P(t) = \frac{1}{(D - 1)(D + 4)}100\cos(2t + 25°)$$

$$= \frac{(D + 1)(D - 4)}{(D^2 - 1)(D^2 - 16)}100\cos(2t + 25°) = \frac{(D^2 - 3D - 4)}{100}100\cos(2t + 25°)$$

$$= -8\cos(2t + 25°) + 6\sin(2t + 25°).$$

The second is

$$y_P(t) = \frac{1}{D^2 + 3D - 4}100\cos(2t + 25°) = \frac{1}{-(2)^2 + 3D - 4}100\cos(2t + 25°)$$

$$= \frac{1}{3D - 8}100\cos(2t + 25°) = \frac{3D + 8}{9D^2 - 64}100\cos(2t + 25°)$$

$$= -(3D + 8)\cos(2t + 25°) = -8\cos(2t + 25°) + 6\sin(2t + 25°).$$

4° Case: $f(t) = e^{kt}u(t)$, with k being a constant and $u(t)$ any function. In this case,

$$y_P(t) = \frac{1}{F(D)}e^{kt}u(t) = e^{kt}\frac{1}{F(D + k)}u(t) \qquad (3.21)$$

Example 3.11 Find the general solution of the following differential equations:

(a) $y'(t) + 4y(t) = t^2 \epsilon^{-2t}$

(b) $(D-2)^2 y(t) = \epsilon^t + t\epsilon^{2t}$.

Solution

(a) $(D+4)y(t) = t^2 \epsilon^{-2t}$

$$y_H(t) = k_1 \epsilon^{-4t}$$

$$y_P(t) = \frac{1}{D+4}(t^2 \epsilon^{-2t}) = \epsilon^{-2t} \cdot \frac{1}{(D-2)+4} t^2 = \epsilon^{-2t} \cdot \frac{1}{D+2} t^2$$

$$= \epsilon^{-2t} \left(\frac{1}{2} - \frac{1}{4}D + \frac{1}{8}D^2 \right) t^2 = \epsilon^{-2t} \left(\frac{t^2}{2} - \frac{t}{2} + \frac{1}{4} \right)$$

Therefore,

$$y(t) = k_1 \epsilon^{-4t} + \left(\frac{t^2}{2} - \frac{t}{2} + \frac{1}{4} \right) \epsilon^{-2t}$$

(b)

$$y_H(t) = (k_1 + k_2 t)\epsilon^{2t}$$

$$y_P(t) = \frac{1}{(D-2)^2}\left(\epsilon^t + t\epsilon^{2t} \right) = \frac{1}{(D-2)^2}\epsilon^t + \frac{1}{(D-2)^2}t\epsilon^{2t}$$

$$= \frac{1}{(1-2)^2}\epsilon^t + \epsilon^{2t} \cdot \frac{1}{[(D+2)-2]^2}t = \epsilon^t + \epsilon^{2t} \cdot \frac{1}{D^2}(t)$$

$$= \epsilon^t + \epsilon^{2t} \cdot \frac{1}{D}\int t\, dt = \epsilon^t + \epsilon^{2t} \cdot \int \frac{t^2}{2}\, dt = \epsilon^t + \epsilon^{2t} \cdot \frac{t^3}{6}$$

Finally,

$$y(t) = \frac{t^3}{6}\epsilon^{2t} + \epsilon^t + (k_1 + k_2 t)\epsilon^{2t}$$

3.3 Differential Equations with Singular and Causal Forcing Functions

When the forcing function $f(t)$ of a differential equation is singular and / or causal, it may happen that the solution $y(t)$ and its derivatives are discontinuous at $t = 0$. That is, for a given differential equation of order n, it may happen that

$$
\begin{aligned}
y(0_+) &\neq y(0_-) \\
y'(0_+) &\neq y'(0_-) \\
y''(0_+) &\neq y''(0_-) \\
&\cdots\cdots\cdots\cdots\cdots\cdots \\
y^{n-1}(0_+) &\neq y^{n-1}(0_-)
\end{aligned}
\tag{3.22}
$$

In electrical circuit problems, for example, the initial conditions are generally known at $t = 0_-$, **but the determination of the arbitrary n constants in the general solution, the circuit response, requires knowledge of the initial conditions at $t = 0_+$.** Therefore, one of the problems to be solved is the determination of the initial conditions at $t = 0_+$, based on their knowledge at $t = 0_-$, as they may or may not be the same.

To facilitate the treatment of such by questions, a second order differential equation will be used as a reference, which, facilitating the exposure that will be made, will not cause any loss of generality. In addition, some essentially heuristic reasoning will be made.

Suppose that the differential equation that relates the input $x(t)$ and the output $y(t)$ of a linear and time-invariant system, with concentrated parameters, is

$$
\frac{d^2y}{dt^2} + a_1 \frac{dy}{dt} + a_2 y = b_0 x(t)
$$

or

$$
y''(t) + a_1 y'(t) + a_2 y(t) = b_0 x(t)
\tag{3.23}
$$

The general solution of a differential equation of order n, as shown in Sect. 3.2, involves n essential arbitrary constants, derived from the complementary solution, and the determination of these constants is only possible when n **boundary conditions**, as they are called, are known. In the case of electrical circuits, for example, n **initial conditions at $t = 0_+$**. Consequently, **the solution of Eq. 3.23 requires knowledge of $y(0_+)$ and $y'(0_+)$.**

3.3.1   *Step Type Forcing Function $U_{-1}(t)$*

When $x(t) = U_{-1}(t)$, then, for the two members of Eq. 3.23 to be identical, at any time t, it is necessary that in some part of the first member of the differential equation the step $b_0 U_{-1}(t)$ also appear. And it is evident that this portion has to be $y''(t)$. This is because if $y'(t)$ contains the step $b_0 U_{-1}(t)$, then $y''(t)$ will contain its derivative, which is the impulse $b_0 U_0(t)$, and this will break the balance of the two members of that differential equation, as there is no impulse occurring at $t = 0$ in the second member of the same.

For similar reasoning, it can be concluded that $y(t)$ cannot contain the step $b_0 U_{-1}(t)$. For if $y(t)$ contains the step $b_0 U_{-1}(t)$, $y'(t)$ will contain its derivative, which is the impulse $b_0 U_0(t)$ and, even worse, $y''(t)$ will contain the double function $b_0 U_1(t)$, with the latter two occurring exactly at $t = 0$. It turns out that in the second member of Eq. 3.23 there is neither impulse nor double at $t = 0$.

However, if $y''(t)$ contains the step $b_0 U_{-1}(t)$, then $y'(t)$ must contain its integral, which is the ramp function $b_0 U_{-2}(t)$, and $y(t)$ must contain the ramp integral, which is the second degree parabola function $b_0 U_{-3}(t)$. And since the ramp and parabola functions are continuous at any time t, containing no step function, then, according to Eq. 1.92 and subsequent conclusion, it can be concluded, finally, that $y(0_+) = y(0_-)$ and that $y'(0_+) = y'(0_-)$.

But the presence of the step $b_0 U_{-1}(t)$ in $y''(t)$ causes $y''(0_+) \neq y''(0_-)$. It turns out, however, that the initial condition $y''(0_+)$ is not necessary, since the differential equation being second order, only the initial conditions $y(0_+)$ and $y'(0_+)$ are needed to determine the constants k_1 and k_2 that arise from the solution of the corresponding homogeneous differential equation.

Therefore, there is no difficulty in calculating the solution of Eq. 3.23, when $x(t)$ is a unit step $U_{-1}(t)$.

The particular solution is found using the abbreviated method to the equation

$$\left(D^2 + a_1 D + a_2\right) y(t) = b_0, \text{ since for } t > 0, U_{-1}(t) = 1.$$

With the values of $y(0_+)$, which is equal to $y(0_-)$, and of $y'(0_+)$, which is equal to $y'(0_-)$, the arbitrary constants k_1 and k_2 of the general solution are determined.

Although it is not necessary to know the value of $y''(0_+)$ for the complete solution of the differential equation, it can be calculated by making $t = 0_+$ in Eq. 3.23 itself. Thus,

$$y''(0_+) + a_1 y'(0_+) + a_2 y(0_+) = b_0$$

So,

$$y''(0_+) = b_0 - [a_1 y'(0_+) + a_2 y(0_+)]$$

Example 3.12 Solve the differential equation

$$y''(t) + 3y'(t) + 2y(t) = 5U_{-1}(t), \text{ with } y(0_-) = 0 \text{ and } y'(0_-) = 0.$$

Solution

The characteristic roots are $r_1 = -1$ and $r_2 = -2$, and the complementary solution is

$$y_H(t) = k_1 \epsilon^{-t} + k_2 \epsilon^{-2t}$$

The particular solution is obtained considering the forcing function equal to 5 and using the Eq. 3.16, with $\alpha = 0$. So,

$$y_P(t) = \frac{1}{D^2 + 3D + 2}(5\epsilon^{0t}) = \frac{5}{2}$$

Therefore, the general solution is

$$y(t) = \frac{5}{2} + k_1 \epsilon^{-t} + k_2 \epsilon^{-2t}, \text{ and}$$

$$y'(t) = -k_1 \epsilon^{-t} - 2k_2 \epsilon^{-2t}$$

Since the $y(t)$ and $y'(t)$ functions must be continuous at $t = 0$,

$$y(0_+) = y(0_-) = 0 \text{ and } y'(0_+) = y'(0_-) = 0.$$

Substituting these initial conditions in $y(t)$ and $y'(t)$, for $t = 0_+$, it turns out that

$$k_1 + k_2 = -\frac{5}{2}$$

$$-k_1 - 2k_2 = 0$$

Solving this system of algebraic equations, it turns out that $k_1 = -5$ and $k_2 = 5/2$. Finally,

$$y(t) = \left[-5\epsilon^{-t} + \frac{5}{2}\left(1 + \epsilon^{-2t}\right) \right].U_{-1}(t)$$

The multiplication by $U_{-1}(t)$ is to point out that the expression found for $y(t)$, enclosed in brackets, is only valid for $t > 0$, since the forcing function $5U_{-1}(t)$ only acts for $t > 0$ and, in addition, the initial conditions at $t = 0_-$ are null.

At the beginning of this section, it was stated that when a second order differential equation has a step forcing function, then the solution $y(t)$ must contain a parabola $U_{-3}(t) = (t^2/2!)U_{-1}(t)$. To confirm this statement, just expand the solution obtained using the Mac-Laurin Series of the exponential function:

$\epsilon^{\alpha t} = 1 + \frac{\alpha t}{1!} + \frac{\alpha^2 t^2}{2!} + \frac{\alpha^3 t^3}{3!} + \dots$, converging to $-\infty < \alpha t < +\infty$.

Thus,

$$y(t) = \left[-5 \left(1 - \frac{t}{1!} + \frac{t^2}{2!} - \frac{t^3}{3!} + \dots \right) + \frac{5}{2} \left(1 + 1 - \frac{2t}{1!} + \frac{4t^2}{2!} - \frac{8t^3}{3!} + \dots \right) \right] . U_{-1}(t)$$

$$= \left(5 . \frac{t^2}{2!} - 15 . \frac{t^3}{3!} + 35 . \frac{t^4}{4!} - \dots \right) . U_{-1}(t)$$

or,

$$y(t) = 5U_{-3}(t) - 15U_{-4}(t) + 35U_{-5}(t) - \dots$$

$$y'(t) = 5U_{-2}(t) - 15U_{-3}(t) + 35U_{-4}(t) - \dots$$

$$y''(t) = 5U_{-1}(t) - 15U_{-2}(t) + 35U_{-3}(t) - \dots$$

The **series expansions of singular functions** above clearly show that $y(t)$ contains the parable $5U_{-3}(t)$; $y'(t)$, the ramp $5U_{-2}(t)$; and $y''(t)$, the step $5U_{-1}(t)$. Furthermore, that the solution in the expanded form of $y(t)$, together with its two derivatives, satisfies the given differential equation. And also, that $y''(0_+) = 5$, a result that can be proved by doing $t = 0_+$ in the given differential equation itself:

$$y''(0_+) + 3y'(0_+) + 2y(0_+) = 5$$

Since $y(0_+) = y'(0_+) = 0$, it follows that $y''(0_+) = 5$.

Note, too, that the singular functions present in the expansions of $y(t)$ and $y'(t)$ are all null at $t = 0_+$, exactly because they do not have the step function $U_{-1}(t)$ and, for this very reason, they make $y(t)$ and $y'(t)$ continuous at $t = 0$.

When $x(t)$ in Eq. 3.23 is a singular forcing function of the type obtained by successive integration of the unit step, that is, ramps and parabolas, then all initial conditions are equal at $t = 0_-$ and at $t = 0_+$, i.e.,

$$y(0_+) = y(0_-); y'(0_+) = y'(0_-); y''(0_+) = y''(0_-), \qquad (3.24)$$

and the particular solution $y_P(t)$ can be obtained by using the abbreviated method directly, or the method of the coefficients to be determined, without any difficulty.

Example 3.13 A linear system is described by the differential equation.

$$y''(t) + 5y'(t) + 6y(t) = 6x(t)$$

Determine your answer $y(t)$ for an input $x(t)$ equals to a unit ramp $U_{-2}(t)$, given the initial conditions $y(0_-) = 0$ and $y'(0_-) = 1$.

Solution

The characteristic roots are $r_1 = -2$ and $r_2 = -3$. So, the complementary solution is:

$$y_H(t) = k_1 \epsilon^{-2t} + k_2 \epsilon^{-3t},$$

regardless of what the forcing function is. When it is a unitary ramp, the equation becomes:

$$(D^2 + 5D + 6)y(t) = 6U_{-2}(t) = 6tU_{-1}(t)$$

Therefore, a particular solution is determined by considering the forcing function equal to $6t$, for $0_+ \leq t < \infty$, and using the second case of the abbreviated method, that is, Eq. 3.18. Thus,

$$y_P(t) = \frac{1}{6 + 5D + D^2}(6t) = \left(\frac{1}{6} - \frac{5}{36}D\right)6t = t - \frac{5}{6}$$

So,

$$y(t) = t - \frac{5}{6} + k_1 \epsilon^{-2t} + k_2 \epsilon^{-3t}$$

$$y'(t) = 1 - 2k_1 \epsilon^{-2t} - 3k_2 \epsilon^{-3t}$$

Since the initial conditions at $t = 0_-$ and at $t = 0_+$ are the same, $y(0_+) = y(0_-) = 0$ and $y'(0_+) = y'(0_-) = 1$. Taking $t = 0_+$ in $y(t)$ and $y'(t)$ results in the following system of algebraic equations:

$$0 = -\frac{5}{6} + k_1 + k_2$$

$$1 = 1 - 2k_1 - 3k_2$$

or

$$k_1 + k_2 = 5/6$$

$$2k_1 + 3k_2 = 0$$

The solution of this system results in $k_1 = 5/2$ and $k_2 = -5/3$. Finally,

$$y(t) = t - \frac{5}{6} + \frac{5}{2}\epsilon^{-2t} - \frac{5}{3}\epsilon^{-3t}, \textbf{ for } t \geq 0.$$

Note that now the result obtained cannot be multiplied by $U_{-1}(t)$. Because if

$$y(t) = (t - \frac{5}{6} + \frac{5}{2}\epsilon^{-2t} - \frac{5}{3}\epsilon^{-3t})U_{-1}(t),$$

it turns out that $y(0_-) = 0$, due to the presence of the unitary step, confirming the initial condition given in the example statement. But on the other hand,

$$y'(t) = \frac{d}{dt}\left[\left(t - \frac{5}{6} + \frac{5}{2}\epsilon^{-2t} - \frac{5}{3}\epsilon^{-3t}\right)U_{-1}(t)\right] = \left(1 - 5\epsilon^{-2t} + 5\epsilon^{-3t}\right)U_{-1}(t)$$

$$+ \left(t - \frac{5}{6} + \frac{5}{2}\epsilon^{-2t} - \frac{5}{3}\epsilon^{-3t}\right)\Bigg|_{t=0} .U_0(t) = \left(1 - 5\epsilon^{-2t} + 5\epsilon^{-3t}\right)U_{-1}(t),$$

causing $y'(0_-) = 0$, in a clear contradiction with the statement of the example, which establishes that $y'(0_-) = 1$.

3.3.2   Forcing Function Type Impulse $U_0(t)$

If $x(t) = U_0(t)$, the differential Eq. 3.23 becomes:

$$y''(t) + a_1 y'(t) + a_2 y(t) = b_0 U_0(t) \qquad (3.25)$$

To ensure that the left side of Eq. 3.25 is identical to its right side, at any time t, $y''(t)$ must contain the impulse $b_0 U_0(t)$, $y'(t)$, in turn, contain the step $b_0 U_{-1}(t)$ and $y(t)$, the ramp $b_0 U_{-2}(t)$, as first terms in their respective expansions in infinite series of singular functions, with **decreasing indices**.

   Therefore, for the case of Eq. 3.25, which is second order, $y(t)$ does not contain the step function, causing $\mathbf{y(0_+) = y(0_-)}$. But both $y'(t)$ and $y''(t)$ contain the step $b_0 U_{-1}(t)$ and, as a consequence, $y'(t)$ and $y''(t)$ are discontinuous in $t = 0$, that is: $\mathbf{y'(0_+) \neq y'(0_-)}$ and $\mathbf{y''(0_+) \neq y''(0_-)}$.

   It is evident that if the differential equation is of third order, $y'''(t)$ will contain impulse, $y''(t)$ will contain step, $y'(t)$ will contain ramp and $y(t)$ will contain second degree parabola, in a way that $\mathbf{y'''(0_+) \neq y'''(0_-)}$, $\mathbf{y''(0_+) \neq y''(0_-)}$, $\mathbf{y'(0_+) =}$ $\mathbf{y'(0_-)}$ and $\mathbf{y(0_+) = y(0_-)}$.

   Conclusion: **for a differential equation of any order with a forcing function involving unit impulse, $U_0(t)$, the two derivatives of the highest order of $y(t)$ are discontinuous at $t = 0$.**

   Therefore, in the case of a differential equation of second order, it is necessary to calculate only $y'(0_+)$, from the knowledge of the value of $y'(0_-)$, since $y(0_+) = y(0_-)$, and it is not necessary to have knowledge of $y''(0_+)$ for the calculation of the two essential arbitrary constants that appear in the general solution of the equation.

And one way to calculate $y''(0_+)$ is to integrate Eq. 3.25 between the limits 0_- and 0_+, namely:

$$\int_{0_-}^{0_+} y''(t)dt + a_1 \int_{0_-}^{0_+} y'(t)dt + a_2 \int_{0_-}^{0_+} y(t)dt = b_0 \int_{0_-}^{0_+} U_0(t)dt$$

$$[y'(0_+) - y'(0_-)] + a_1[y(0_+) - y(0_-)] + a_2 \int_{0_-}^{0_+} y(t)dt = b_0$$

Since $y(t)$ is continuous at $t = 0$, with greater strength its integral will be continuous, resulting, therefore, that:

$$y'(0_+) - y'(0_-) = b_0$$

thus

$$y'(0_+) = b_0 + y'(0_-) \qquad (3.26)$$

Note that b_0 is the area of the impulsive forcing function, and that it is also the step amplitude in $y'(t)$.

If it were necessary to find the value of $y''(0_+)$, it would be enough to substitute t for 0_+ in Eq. 3.25, obtaining that

$$y''(0_+) = -[a_1 y'(0_+) + a_2 y(0_+)] \qquad (3.27)$$

Example 3.14 Solve the differential equation

$$y''(t) + 3y'(t) + 2y(t) = 5U_0(t),$$

with the initial conditions $y(0_-) = 0$ and $y'(0_-) = 0$.

Solution

Since the forcing function is an impulse, the series expansion of singular functions of $y''(t)$ must start with an impulse, that of $y'(t)$ with a step and that of $y(t)$ with a ramp.   Therefore, $y(0_+) = y(0_-) = 0$ and,   according   to   Eq. 3.26, $y'(0_+) = 5 + 0 = 5$. On the other hand, for $t > 0$ the equation is:

$$(D^2 + 3D + 2)y(t) = 0$$

And its general solution is

$$y(t) = \left(k_1 e^{-t} + k_2 e^{-2t}\right) U_{-1}(t)$$

Then,

$$y'(t) = -\left(k_1 e^{-t} + 2k_2 e^{-2t}\right) U_{-1}(t) + (k_1 + k_2) U_0(t)$$

Substituting the initial conditions $y(0_+) = 0$ and $y'(0_+) = 5$ in $y(t)$ and $y'(t)$, and remembering that $U_0(0_+) = 0$, results the system of equations

$$0 = k_1 + k_2$$

$$5 = -k_1 - 2k_2,$$

from which it follows that que $k_1 = 5$ and $k_2 = -5$. Finally,

$$y(t) = \left(5e^{-t} - 5e^{-2t}\right) U_{-1}(t)$$

$$y'(t) = \left(-5e^{-t} + 10e^{-2t}\right) U_{-1}(t)$$

$$y''(t) = 5U_0(t) + \left(5e^{-t} - 20e^{-2t}\right) U_{-1}(t)$$

So, checking,

$$
\begin{aligned}
y''(t) + 3y'(t) + 2y(t) = {} & 5U_0(t) + \left(5e^{-t} - 20e^{-2t}\right) U_{-1}(t) \\
& + \left(-15e^{-t} + 30e^{-2t}\right) U_{-1}(t) \\
& + \left(10e^{-t} - 10e^{-2t}\right) U_{-1}(t) = 5U_0(t)
\end{aligned}
$$

Also, from the last expressions, it is evident that $y(0_+) = 0$, $y'(0_+) = 5$ and that $y''(0_+) = -15$.
From the given differential equation itself, for $t = 0_+$, one has to

$$y''(0_+) + 3y'(0_+) + 2y(0_+) = 0, \text{ since } U_0(t) = 0 \text{ for } t \neq 0.$$

Hence, in accordance with Eq. 3.27,

$$y''(0_+) = -[3y'(0_+) + 2y(0_+)] = -15$$

Therefore,

$$y(t) = \left(5e^{-t} - 5e^{-2t}\right) U_{-1}(t)$$

it is really the solution sought.

By also replacing the exponential functions present in $y(t)$, $y'(t)$ and $y''(t)$ by their respective Mac-Laurin series, the infinite series expansions of singular functions are obtained:

$$y(t) = 5U_{-2}(t) - 15U_{-3}(t) + 35U_{-4}(t) - \dots$$

$$y'(t) = 5U_{-1}(t) - 15U_{-2}(t) + 35U_{-3}(t) - \dots$$

$$y''(t) = 5U_0(t) - 15U_{-1}(t) + 35U_{-2}(t) - \dots$$

And you can confirm the presence of the ramp $5U_{-2}(t)$, the step $5U_{-1}(t)$ and the impulse $5U_0(t)$ as first terms in the expansions in singular functions of $y(t)$, $y'(t)$ and $y''(t)$, respectively. And also, that $y(0_+) = 0$, $y'(0_+) = 5$ and $y''(0_+) = -15$, which are the amplitudes of the steps in $y(t)$, $y'(t)$ and $y''(t)$, respectively.

Note, too, that the ramp function that is present in $y(t)$, is also present in $y'(t)$ and $y''(t)$, with different coefficients, of course. And also that the step function $U_{-1}(t)$ that is contained in $y'(t)$, with amplitude 5, is also present in $y''(t)$, but with amplitude -15.

In addition, **since the initial conditions at $t = 0_-$ are all null, the initial conditions at $t = 0_+$ of y, y' e y'' are the amplitudes of the steps in the respective expansions thereof**. Logically, the expansion of y (t) in singular functions must be understood as:

$$y(t) = 0.U_{-1}(t) + 5U_{-2}(t) - 15U_{-3}(t) + 35U_{-4}(t) - \dots$$

Example 3.15 Solve example 3.14 with the initial conditions $y(0_-) = 1$ and $y'(0_-) = 2$.

Solution
The initial conditions at $t = 0_+$ are: $y(0_+) = y(0_-) = 1$ and $y'(0_+) = y'(0_-) + b_0 = 2 + 5 = 7$, according to Eq. 3.26. And the general solution is:

$$y(t) = k_1\epsilon^{-t} + k_2\epsilon^{-2t}, \textbf{ for t} > \textbf{0}.$$

In the present case, in which the initial conditions at $t = 0_-$ are different from zero, the answer cannot be multiplied by $U_{-1}(t)$, because if this were done it would be obtained that:

$$y(t) = \left(k_1\epsilon^{-t} + k_2\epsilon^{-2t}\right)U_{-1}(t)$$

and that

$$y'(t) = -\left(k_1\epsilon^{-t} + 2k_2\epsilon^{-2t}\right)U_{-1}(t) + (k_1 + k_2)U_0(t)$$

Since $y(0_+) = k_1 + k_2 = y(0_-) - 1 \neq 0$, then there would be the impulse function $(k_1 + k_2)U_0(t) = U_0(t)$ in $y'(t)$ and, consequently, the double function $U_1(t)$ in $y''(t)$, which is completely unacceptable, because it would destroy the balance of the two members of the differential equation.

In other words, as $y(0_+) = y(0_-) \neq 0$, multiplying the solution by $U_{-1}(t)$ would imply the inclusion of the step $y(0_+)U_{-1}(t)$ in $y(t)$, which would go against the requirement that $y(t)$ contain a $U_{-2}(t)$ ramp, and its successive integrals, when expanded into singular functions, but not a step function, as highlighted at the end from example 3.14. Therefore, the solution to the equation is really

$$y(t) = k_1\epsilon^{-t} + k_2\epsilon^{-2t}, \text{ for } t > 0,$$

and

$$y'(t) = -k_1\epsilon^{-t} - 2k_2\epsilon^{-2t}, \text{ for } t > 0.$$

Entering the initial conditions at $t = 0_+$, it results that:

$$1 = k_1 + k_2$$

$$7 = -k_1 - 2k_2,$$

whose solution is $k_1 = 9$ and $k_2 = -8$. Finally,

$$y(t) = 9\epsilon^{-t} - 8\epsilon^{-2t}, \text{ for } t > 0.$$

When the forcing function contains one or more impulsive functions occurring at $t = 0$, the differential equation is equal to zero for $t > 0$, since impulsive functions are null for $t \neq 0$. Therefore, the solution of the differential equation is always reduced to the solution of a homogeneous differential equation.

From the study carried out in Sect. 3.2, it appears that whatever the roots of the characteristic equation $F(D) = 0$, are the solution of the homogeneous equation can always be expressed by exponential functions, that is, decomposed into a series of powers of t by means of Mac-Laurin series. This fact forms the basis for the development of a systematic method for determining the initial conditions at $t = 0_+$, which is presented below.

3.3.3 Systematic Method for Calculating Initial Conditions at $t = 0_+$

To facilitate the understanding, example 3.15 is taken up, that is, the solution of the differential equation

$$y''(t) + 3y'(t) + 2y(t) = 5U_0(t), \text{ with } y(0_-) = 1 \text{ and } y'(0_-) = 2.$$

The need for $y''(t)$ to contain the impulse function, $y'(t)$ the step function and $y(t)$ the ramp function, together with the possibility of expressing the solution in the form of a series of powers of t, let write that

$$y(t) = 0.U_{-1}(t) + C_0U_{-2}(t) + C_1U_{-3}(t) + C_2U_{-4}(t) + C_3U_{-5}(t) + \ldots \quad (3.28)$$

$$y'(t) = C_0U_{-1}(t) + C_1U_{-2}(t) + C_2U_{-3}(t) + C_3U_{-4}(t) + \ldots \quad (3.29)$$

$$y''(t) = C_0U_0(t) + C_1U_{-1}(t) + C_2U_{-2}(t) + C_3U_{-3}(t) + \ldots \quad (3.30)$$

where C_i are coefficients to be determined.

(Obviously, Eqs. 3.28 and 3.29 contain two inconsistencies. Namely: they imply that $y(0_-) = 0$ and that $y'(0_-) = 0$, since all the singular functions contained in them are null at $t = 0_-$, which is in complete disagreement with the statement of the problem, which claims to be $y(0_-) = 1$ and $y'(0_-) = 2$. In the meantime, the comments that follow will put things in their proper terms).

Substituting Eqs. 3.28 to 3.30 in the given differential equation, it follows that:

$$[C_0U_0(t) + C_1U_{-1}(t) + C_2U_{-2}(t) + C_3U_{-3}(t) + \ldots]$$
$$+ 3[C_0U_{-1}(t) + C_1U_{-2}(t) + C_2U_{-3}(t) + C_3U_{-4}(t) + \ldots]$$
$$+ 2[C_0U_{-2}(t) + C_1U_{-3}(t) + C_2U_{-4}(t) + C_3U_{-5}(t) + \ldots] = 5U_0(t)$$

Grouping similar terms together, it follows that:

$$C_0U_0(t) + (C_1 + 3C_0)U_{-1}(t) + (C_2 + 3C_1 + 2C_0)U_{-2}(t)$$
$$+ (C_3 + 3C_2 + 2C_1)U_{-3}(t) + \ldots$$
$$= 5U_0(t) + 0U_{-1}(t) + 0.U_{-2}(t) + 0.U_{-3}(t) + \ldots$$

Identifying the two members of this equation, it is obtained that:

$$C_0 = 5 \ldots\ldots\ldots\ldots\ldots\ldots\ldots C_0 = 5$$

$$C_1 + 3C_0 = 0 \ldots\ldots\ldots\ldots\ldots C_1 = -15$$

$$C_2 + 3C_1 + 2C_0 = 0 \ldots\ldots\ldots C_2 = 35 \tag{3.31}$$

$$C_3 + 3C_2 + 2C_1 = 0 \ldots\ldots\ldots C_3 = -75$$

$$\ldots\ldots\ldots\ldots\ldots\ldots\ldots\ldots\ldots\ldots\ldots\ldots$$

If the initial conditions at $t = 0_-$ were all null, evidently the initial conditions at $t = 0_+$ would be the coefficients of the step functions of each of the Eqs. 3.28 to 3.30, namely: $y(0_+) = 0$, $y'(0_+) = C_0 = 5$ and $y''(0_+) = C_1 = -15$, as in example 3.14.

However, **as the initial conditions at $t = 0_-$ are not null, it is necessary to add to the initial conditions at $t = 0_-$ the step coefficients existing in $y(t)$, $y'(t)$ and $y''(t)$, to obtain the respective initial conditions at $t = 0_+$**, as established in Eq. 3.32. Therefore,

$$\begin{aligned}
y(0_+) &= y(0_-) + 0 = y(0_-) = 1 \\
y'(0_+) &= y'(0_-) + C_0 = 2 + 5 = 7 \\
y''(0_+) &= y''(0_-) + C_1 = y''(0_-) = -15
\end{aligned} \tag{3.32}$$

Obviously, the values found, $y(0_+) = 1$ and $y'(0_+) = 7$, agree with those obtained in example 3.15.

An additional observation, however, allows to simplify the exposed procedure considerably. The initial conditions at $t = 0_+$, listed in expressions 3.32, show that **it is necessary to calculate only the coefficients of the steps present in $y(t)$ and its derivatives.**

Then, from the series expansions of singular functions of $y(t)$, $y'(t)$ and $y''(t)$, contained in Eqs. 3.28 to 3.30, **it is sufficient to compute only the first terms of the series, truncating them in steps,** as the subsequent functions are all continuous, and null at $t = 0_+$. Thus, looking at Eqs. 3.28 to 3.30, it is possible to write only that

$$\begin{aligned}
y(t) &\doteq 0.U_{-1}(t). \\
y'(t) &\doteq C_0 U_{-1}(t). \\
y''(t) &\doteq C_0 U_0(t) + C_1 U_{-1}(t).
\end{aligned} \tag{3.33}$$

The notation $\doteq$ is used only to remember that the function $y(t)$ and its derivatives are **not** only made up of the singular functions that are indicated in each one, that is, that the expansions in singular functions, which contain an infinite number of terms, were truncated on the steps.

Therefore, replacing the expressions contained in Eq. 3.33 in the given differential equation, results:

$$C_0 U_0(t) + (C_1 + 3C_0)U_{-1}(t) \doteq 5U_0(t) + 0.U_{-1}(t)$$

So,

$$C_0 = 5$$

$$C_1 + 3C_0 = 0; C_1 = -15$$

Example 3.16 Solve the differential equation

$$y''(t) + 3y'(t) + 2y(t) = 5U_1(t), \text{ with } y(0_-) = 1 \text{ and } y'(0_-) = 2.$$

Solution

For $t > 0$, the equation is

$$y''(t) + 3y'(t) + 2y(t) = 0,$$

And its general solution is:

$$y(t) = k_1 \epsilon^{-t} + k_2 \epsilon^{-2t}$$

$$y'(t) = -k_1 \epsilon^{-t} - 2k_2 \epsilon^{-2t}$$

Since the forcing function is a double function, the balance of the two members of the given equation requires that $y''(t)$ contain a double, $y'(t)$ an impulse and $y(t)$ a step. Therefore, for the determination of the initial conditions at $t = 0_+$, one must consider that

$$y(t) \doteq C_0 U_{-1}(t).$$
$$y'(t) \doteq C_0 U_0(t) + C_1 U_{-1}(t).$$
$$y''(t) \doteq C_0 U_1(t) + C_1 U_0(t) + C_2 U_{-1}(t).$$

Substituting in the differential equation, results in

$$C_0 U_1(t) + (C_1 + 3C_0)U_0(t) + (C_2 + 3C_1 + 2C_0)U_{-1}(t) \doteq 5U_1(t) + 0.U_0(t) + 0.U_{-1}(t).$$

So,

$$C_0 = 5 \dots \dots \dots \dots \dots \dots \dots \dots C_0 = 5$$

$$C_1 + 3C_0 = 0 \dots \dots \dots \dots \dots C_1 = -15$$

$$C_2 + 3C_1 + 2C_0 = 0 \dots \dots \dots C_2 = 35$$

Therefore,

$$y(0_+) = y(0_-) + C_0 = 1 + 5 = 6$$

$$y'(0_+) = y'(0_-) + C_1 = 2 - 15 = -13$$

$$y''(0_+) = y''(0_-) + C_2 = y''(0_-) + 35$$

So, from $y(t)$ and $y'(t)$, to $t = 0_+$, it comes that

$$6 = k_1 + k_2$$

$$-13 = -k_1 - 2k_2$$

Solving this system results that $k_1 = -1$ and $k_2 = 7$.
Finally,

$$y(t) = 7\epsilon^{-2t} - \epsilon^{-t}, \text{ for } t > 0.$$

When the given differential equation is integrated between the limits 0_- and 0_+ it results that

$$[y'(0_+) - y'(0_-)] + 3[y(0_+) - y(0_-)] + 2 \int_{0_-}^{0_+} y(t)dt = 5 \int_{0_-}^{0_+} U_1(t)dt$$

The function $y(t)$ presents a finite discontinuity of 5 units at $t = 0$, since $y(0_+) = 6$ and $y(0_-) = 1$. However, **the process of integrating a function with finite discontinuity(s) always produces an absolutely continuous function**, which allows you to write that

$$\int_{0_-}^{0_+} y(t)dt = 0$$

Since the double function is odd, $\int_{0_-}^{0_+} U_1(t)dt = 0$.

Therefore,

$$y'(0_+) - y'(0_-) + 3y(0_+) - 3y(0_-) = 0.$$

This equation is satisfied by the initial conditions given and calculated.

Example 3.17   An input system $x(t)$ and output $y(t)$ is described by the differential equation $y''(t) + 3y'(t) + 2y(t) = 4x'(t) + 2x(t)$. Find the output corresponding to the input $x(t) = U_0(t)$, considering that $y(0_-) = 1$ and $y'(0_-) = 3$.

Solution

When $x(t) = U_0(t)$, the equation becomes

$$y''(t) + 3y'(t) + 2y(t) = 4U_1(t) + 2U_0(t)$$

Its solution, for $t > 0$, is

$$y(t) = k_1 \epsilon^{-t} + k_2 \epsilon^{-2t}$$

$$y'(t) = -k_1 \epsilon^{-t} - 2k_2 \epsilon^{-2t}$$

As a consequence of the second member of the differential equation, it follows that $y''(t)$ must contain a double, $y'(t)$ an impulse and $y(t)$ a step, as first terms in their respective expansions in series of singular functions. So,

$$y(t) \doteq C_0 U_{-1}(t).$$
$$y'(t) \doteq C_0 U_0(t) + C_1 U_{-1}(t).$$
$$y''(t) \doteq C_0 U_1(t) + C_1 U_0(t) + C_2 U_{-1}(t).$$

Substituting in the given equation, it results that:

$$C_0 U_1(t) + (C_1 + 3C_0)U_0(t) + (C_2 + 3C_1 + 2C_0)U_{-1}(t) \doteq 4U_1(t) + 2U_0(t) + 0.U_{-1}(t).$$

So,

$$C_0 = 4 \dots \dots \dots \dots \dots \dots \dots \dots C_0 = 4$$

$$C_1 + 3C_0 = 2 \dots \dots \dots \dots \dots C_1 = -10$$

$$C_2 + 3C_1 + 2C_0 = 0 \dots \dots \dots C_2 = 22$$

Thus,

$$y(0_+) = y(0_-) + C_0 = 1 + 4 = 5$$

$$y'(0_+) = y'(0_-) | C_1 = 3 - 10 = -7$$

Therefore,

$$k_1 + k_2 = 5$$

$$k_1 + 2k_2 = 7$$

So,
$k_1 = 3$ and $k_2 = 2$.
Finally,

$$y(t) = 3\epsilon^{-t} + 2\epsilon^{-2t}, \text{ for } t > 0.$$

Example 3.18 Solve the integral-differential equation

$$y''(t) + 8y'(t) - 7 \int_{0_-}^{t} y(\lambda)d\lambda = 3U_{-1}(t),$$

with the initial conditions $y(0_-) = -2$ and $y'(0_-) = 4$.

Solution
Considering that $y(t)$ and $y'(t)$ are continuous at $t = 0$, since there will only be a step in $y''(t)$, then, $y(0_+) = y(0_-) = -2$ and $y'(0_+) = y'(0_-) = 4$. As the differential equation is of the third order, there will be three arbitrary constants to be determined in its general solution and, as a result, it is also necessary to calculate $y''(0_+)$. And this is done by replacing t with 0_+ in the integral-differential equation itself:

$$y''(0_+) + 8y'(0_+) - 7 \int_{0_-}^{0_+} y(\lambda)d\lambda = 3U_{-1}(0_+) = 3$$

Since $y(t)$ is continuous at $t = 0$, with greater strength it will be its integral, so that $\int_{0_-}^{0_+} y(\lambda)d\lambda = 0$.

So,

$$y''(0_+) = 3 - 8y'(0_+) = 3 - 8 \times 4 = -29$$

Deriving the given integral-differential equation, we have

$$y'''(t) + 8y''(t) - 7y(t) = 3U_0(t)$$

For $t > 0$, it is

$$(D^3 + 8D^2 - 7)y(t) = 0$$

The characteristic equation is $F(D) = D^3 + 8D^2 - 7 = 0$

If there is only one signal inversion in the sequence of coefficients, there is the possibility of a single positive real root, according to Descartes' rule. Replacing D with $-D$, it becomes: $-D^3 + 8D^2 - 7 = 0$. If there are two sign inversions, then the possibility of having negative real roots is none or two. On the other hand, it is reasonably easy to see that a root is $r_1 = -1$. (If a root is not noticed to be -1, the alternative is to look for an initial estimate for r_1 and apply the Newton-Raphson method). After deflation of the polynomial, it follows that $D^2 + 7D - 7 = 0$, which gives the roots $r_2 = 0.89$ and $r_3 = -7.89$. So, the general solution of the equation is:

$$y(t) = k_1\epsilon^{-t} + k_2\epsilon^{0.89t} + k_3\epsilon^{-7.89t}$$

$$y'(t) = -k_1\epsilon^{-t} + 0.89k_2\epsilon^{0.89t} - 7.89k_3\epsilon^{-7.89t}$$

$$y''(t) = k_1\epsilon^{-t} + 0.79k_2\epsilon^{0.89t} + 62.25k_3\epsilon^{-7.89t}$$

Considering the instant $t = 0_+$, the following system results:

$$k_1 + k_2 + k_3 = -2$$

$$-k_1 + 0.89k_2 - 7.89k_3 = 4$$

$$k_1 + 0.79k_2 + 62.25k_3 = -29$$

And the solution for this system is: $k_1 = -1.00$; $k_2 = -0.56$; $k_3 = -0.44$. Finally,

$y(t) = -\epsilon^{-t} - 0.56\epsilon^{0.89t} - 0.44\epsilon^{-7.89t}$, for $t > 0$.

Example 3.19 Given the differential equation $y''(t) + 12y'(t) + 4y(t) = U_2(t) + 2U_0(t)$, with $y(0_-) = 1$ and $y'(0_-) = -2$, determine $y(0_+)$ and $y'(0_+)$.

Solution
As the most positive singular forcing function, $U_2(t)$, has to be in $y''(t)$, it comes that:

$$y(t) \doteq C_0 U_0(t) + C_1 U_{-1}(t).$$
$$y'(t) \doteq C_0 U_1(t) + C_1 U_0(t) + C_2 U_{-1}(t).$$
$$y''(t) \doteq C_0 U_2(t) + C_1 U_1(t) + C_2 U_0(t) + C_3 U_{-1}(t).$$

Substituting in the differential equation, it comes that:

$$C_0 U_2(t) + (C_1 + 12C_0)U_1(t) + (C_2 + 12C_1 + 4C_0)U_0(t)$$
$$+ (C_3 + 12C_2 + 4C_1)U_{-1}(t)$$
$$\doteq U_2(t) + 0.U_1(t) + 2U_0(t) + 0.U_{-1}(t)$$

$$C_0 = 1 \dots \dots \dots \dots \dots \dots \dots C_0 = 1$$

$$C_1 + 12C_0 = 0 \dots \dots \dots \dots \dots C_1 = -12$$

$$C_2 + 12C_1 + 4C_0 = 2 \dots \dots C_2 = 142$$

Since the initial condition at $t = 0_+$ is the sum of the initial condition at $t = 0_-$ with the amplitude of the step present in the respective function,

$$y(0_+) = y(0_-) + C_1 = 1 - 12 = -11$$

$$y'(0_+) = y'(0_-) + C_2 = -2 + 142 = 140$$

It is interesting to note that the solution of the differential equation will contain the impulse function, that is,

$$y(t) = \begin{cases} U_0(t), & for\ t = 0 \\ y_H(t), & for\ t > 0. \end{cases}$$

Example 3.20 Solve the differential equation

$$y'''(t) + 4y'(t) = 6U_1(t) + 10\,\epsilon^{-t}U_{-1}(t) = 6U_1(t)$$
$$+ 10[U_{-1}(t) - U_{-2}(t) + U_{-3}(t) - U_{-4}(t) + \ldots], \text{with}$$
$$y(0_-) = y'(0_-) = y''(0_-) = 0.$$

Solution

For $t > 0$: $D(D^2 + 4)y(t) = 10\epsilon^{-t}$

  The characteristic roots are: $r_1 = 0$; $r_2 = j2$; $r_3 = -j2$. So,

$$y_H(t) = k_1 + k_2 \cos 2t + k_3 \sin 2t$$

$$y_P(t) = \frac{1}{D^3 + 4D} 10\epsilon^{-t} = -2\epsilon^{-t}$$

And the general solution is

$$y(t) = -2\epsilon^{-t} + k_1 + k_2 \cos 2t + k_3 \sin 2t$$

$$y'(t) = 2\epsilon^{-t} - 2k_2 \sin 2t + 2k_3 \cos 2t$$

$$y''(t) = -2\epsilon^{-t} - 4k_2 \cos 2t - 4k_3 \sin 2t$$

  From the differential equation, with its second member fully expressed in singular functions, one can then write that

$$y(t) \doteq 0.U_{-1}(t).$$

$$y'(t) \doteq C_1 U_{-1}(t).$$

$$y''(t) \doteq C_1 U_0(t) + C_2 U_{-1}(t).$$

$$y'''(t) \doteq C_1 U_1(t) + C_2 U_0(t) + C_3 U_{-1}(t).$$

Substituting $y'(t)$ and $y'''(t)$ in the given differential equation, it comes that

$$C_1 U_1(t) + C_2 U_0(t) + (C_3 + 4C_1)U_{-1}(t) \doteq 6U_1(t) + 0.U_0(t) + 10U_{-1}(t).$$

Therefore,

$$C_1 = 6$$

$$C_2 = 0$$

$$C_3 + 4C_1 = 10; C_3 = -14$$

Thus,

$$y(0_+) = y(0_-) + 0 = 0$$

$$y'(0_+) = y'(0_-) + C_1 = 6$$

$$y''(0_+) = y''(0_-) + C_2 = 0$$

Taking $t = 0_+$ in $y(t)$, $y'(t)$ and $y''(t)$, it follows that:

$$0 = -2 + k_1 + k_2$$

$$6 = 2 + 2k_3$$

$$0 = -2 - 4k_2$$

Solving the algebraic system results in $k_1 = 5/2$; $k_2 = -1/2$; $k_3 = 2$.
Finally,

$$y(t) = \left[-2\epsilon^{-t} + \frac{5}{2} + \frac{1}{2}\cos 2t + 2\sin 2t \right] U_{-1}(t).$$

Example 3.21 Solve the differential equation

$y''(t) + 25y(t) = U_1(t) + 40\cos 3t U_{-1}(t) = U_1(t) + 40U_{-1}(t) - 360U_{-3}(t) + \dots$, with
the initial conditions $y(0_-) = 0$ and $y'(0_-) = -1$.

Solution
For $t > 0$, we have that

$$(D^2 + 25)y(t) = 40\cos 3t$$

So, the characteristic equation is $D^2 + 25 = 0$ and $r_1 = j5$; $r_2 = -j5$. And the
complementary solution is

$$y_H(t) = k_1 \cos 5t + k_2 \sin 5t$$

To use the phasor method in determining the particular solution of the preceding
differential equation, the following changes are required, from the time domain to
the simple frequency domain:

$$40\cos 3t \; -\, -\, -\, -\, -\, -\, \rightarrow \; 40\underline{/0^\circ}$$

$$y_P(t) \; -\, -\, -\, -\, -\, -\, \rightarrow \; \dot{Y}_P$$

$$y'_P(t) \; -\, -\, -\, -\, -\, \rightarrow \; (j3)\dot{Y}_P$$

$$y''_P(t) \; -\, -\, -\, -\, \rightarrow \; (j3)^2 \dot{Y}_P = -9\dot{Y}_P$$

Thus, the differential equation becomes the algebraic equation

$$(-9+25)\dot{Y}_P = 40/\underline{0^\circ}$$

whence

$$\dot{Y}_P = \frac{40}{16}/\underline{0^\circ} = 2.5/\underline{0^\circ}$$

So,

$$y_P(t) = 2.5\cos 3t,$$

$$y(t) = 2.5\cos 3t + k_1\cos 5t + k_2\sin 5t$$

$$y'(t) = -7.5\sin 3t - 5k_1\sin 5t + 5k_2\cos 5t$$

Observing the expansion in singular functions of the forcing function of the given differential equation, it appears that $y''(t)$ must contain the double function $U_1(t)$. So,

$$y(t) \doteq C_0 U_{-1}(t).$$

$$y'(t) \doteq C_0 U_0(t) + C_1 U_{-1}(t).$$

$$y''(t) \doteq C_0 U_1(t) + C_1 U_0(t) + C_2 U_{-1}(t).$$

Substituting in the given equation, it comes that:

$$C_0 U_1(t) + C_1 U_0(t) + (C_2 + 25C_0)U_{-1}(t) \doteq U_1(t) + 0.U_0(t) + 40U_{-1}(t).$$

$$C_0 = 1; C_1 = 0; C_2 + 25C_0 = 40, C_2 = 15.$$

So,

$$y(0_+) = y(0_-) + C_0 = 0 + 1 = 1 = 2.5 + k_1; k_1 = -1.5$$

$$y'(0_+) = y'(0_-) + C_1 = -1 + 0 = -1 = 5k_2; k_2 = -0.2$$

Finally,

$$y(t) = 2.5\cos 3t - 1.5\cos 5t - 0.2\sin 5t, \text{ for } t > 0.$$

This expression can be simplified using the trigonometric identity

$$A \cos \omega t + B \sin \omega t = \sqrt{A^2 + B^2} \cos\left(\omega t - arctg\frac{B}{A}\right).$$

But it is possible to do the same using phasors. Defining

$$z(t) = -1.5 \cos 5t - 0.2 \sin 5t = -1.5 \cos 5t - 0.2 \cos\left(5t - 90^\circ\right),$$

the corresponding phasor is

$$\dot{Z} = -1.5\underline{/0^\circ} - 0.2\underline{/-90^\circ} = -1.5 + j0.2 = 1.513\underline{/172.41^\circ}$$
$$= -1.513\underline{/172.41^\circ - 180^\circ}$$

$$\dot{Z} = -1.513\underline{/-7.59^\circ} \text{ and } z(t) = -1.513 \cos\left(5t - 7.59^\circ\right)$$

Therefore, the solution can also be written as:
$$y(t) = 2.5 \cos 3t - 1.513 \cos\left(5t - 7.59^\circ\right), \text{ for } t > 0.$$

Example 3.22 Solve the following system of differential equations:

$$x'(t) + 3x(t) - 4y(t) = U_{-1}(t)$$

$$2x(t) + y'(t) - y(t) = U_0(t),$$

with the initial conditions $x(0_-) = x'(0_-) = y(0_-) = y'(0_-) = 0$.

Solution
By introducing the differential operator $D = d/dt$, the system can be written as:

$$(D+3)x(t) - 4y(t) = U_{-1}(t)$$

$$2x(t) + (D-1)y(t) = U_0(t)$$

You can now obtain a differential equation only in $x(t)$ and another only in $y(t)$, algebraically solving the system, as follows:

$$x(t) = \frac{\begin{vmatrix} U_{-1}(t) & -4 \\ U_0(t) & D-1 \end{vmatrix}}{\begin{vmatrix} D+3 & -4 \\ 2 & D-1 \end{vmatrix}}$$

Hence, eliminating the determinant of the denominator, one obtains that

$$F(D)x(t) = \begin{vmatrix} D+3 & -4 \\ 2 & D-1 \end{vmatrix} x(t) = \begin{vmatrix} U_{-1}(t) & -4 \\ U_0(t) & D-1 \end{vmatrix}$$

Calculating the determinants results

$$(D^2 + 2D + 5)x(t) = 5U_0(t) - U_{-1}(t) \tag{3.34}$$

Similarly,

$$F(D)y(t) = (D^2 + 2D + 5)y(t) = \begin{vmatrix} D+3 & U_{-1}(t) \\ 2 & U_0(t) \end{vmatrix}$$

Whence

$$(D^2 + 2D + 5)y(t) = U_1(t) + 3U_0(t) - 2U_{-1}(t) \tag{3.35}$$

For $t > 0$,

$$(D^2 + 2D + 5)x(t) = -1$$

$$(D^2 + 2D + 5)y(t) = -2$$

The characteristic roots are obtained from the algebraic equation $F(D) = D^2 + 2D + 5 = 0$, and are $r_1 = -1 + j2$ and $r_2 = -1 - j2$. So, the complementary solutions are:

$$x_H(t) = \epsilon^{-t}(a_1 \cos 2t + a_2 \sin 2t)$$

$$y_H(t) = \epsilon^{-t}(b_1 \cos 2t + b_2 \sin 2t)$$

And the particular solutions,

$$x_P(t) = \frac{1}{D^2 + 2D + 5}(-1\epsilon^{0.t}) = -\frac{1}{5}$$

$$y_P(t) = \frac{1}{D^2 + 2D + 5}(-2\epsilon^{0.t}) = -\frac{2}{5}$$

Therefore,

$$x(t) = -\frac{1}{5} + \epsilon^{-t}(a_1 \cos 2t + a_2 \sin 2t) \tag{3.36}$$

$$y(t) = -\frac{2}{5} + \epsilon^{-t}(b_1 \cos 2t + b_2 \sin 2t) \tag{3.37}$$

In view of the second members of Eqs. 3.34 and 3.35, we have that:

$$x(t) \doteq 0.U_{-1}(t).$$

$$x'(t) \doteq C_0 U_{-1}(t).$$

$$x''(t) \doteq C_0 U_0(t) + C_1 U_{-1}(t).$$

$$y(t) \doteq D_0 U_{-1}(t).$$

$$y'(t) \doteq D_0 U_0(t) + D_1 U_{-1}(t).$$

$$y''(t) \doteq D_0 U_1(t) + D_1 U_0(t) + D_2 U_{-1}(t).$$

Replacing in Eqs. 3.34 and 3.35, it comes that:

$$C_0 U_0(t) + (C_1 + 2C_0) U_{-1}(t) \doteq 5 U_0(t) - U_{-1}(t)$$

$$D_0 U_1(t) + (D_1 + 2D_0) U_0(t) + (D_2 + 2D_1 + 5D_0) U_{-1}(t) \doteq U_1(t) + 3 U_0(t)$$
$$- 2 U_{-1}(t)$$

Thus,

$$C_0 = 5 \dots\dots\dots\dots\dots C_0 = 5 \qquad\qquad D_0 = 1 \dots\dots\dots\dots D_0 = 1$$

$$C_1 + 2C_0 = -1 \dots\dots\dots C_1 = -11 \qquad\qquad D_1 + 2D_0 = 3 \dots\dots\dots D_1 = 1$$

Since all initial conditions at $t = 0_-$ are null, the initial conditions at $t = 0_+$ are given only by the coefficients of the step functions. Therefore,

$$x(0_+) = 0; x'(0_+) = 5; y(0_+) = 1; y'(0_+) = 1$$

From Eqs. 3.36 and 3.37, we have that:

$$x'(t) = -\epsilon^{-t}(a_1 \cos 2t + a_2 \sin 2t) + \epsilon^{-t}(-2a_1 \sin 2t + 2a_2 \cos 2t) \qquad (3.38)$$

$$y'(t) = -\epsilon^{-t}(b_1 \cos 2t + b_2 \sin 2t) + \epsilon^{-t}(-2b_1 \sin 2t + 2b_2 \cos 2t) \qquad (3.39)$$

Entering the initial conditions at $t = 0_+$ in Eqs. 3.36 to 3.39, it follows that:

$$0 = -\tfrac{1}{5} + a_1 \qquad 1 = -\tfrac{2}{5} + b_1$$
$$5 = -a_1 + 2a_2 \qquad 1 = -b_1 + 2b_2$$

So,

$$a_1 = 1/5; a_2 = 13/5; b_1 = 7/5; b_2 = 6/5.$$

Finally,

$$x(t) = \frac{\epsilon^{-t}}{5}(\cos 2t + 13\sin 2t) - \frac{1}{5} = 2.608\epsilon^{-t}\cos(2t - 85.6°) - 0.2 \quad for\, t \geq 0.$$

$$y(t) = \frac{\epsilon^{-t}}{5}(7\cos 2t + 6\sin 2t) - \frac{2}{5} = 1.844\epsilon^{-t}\cos(2t - 40.6°) - 0.4 \quad for\, t \geq 0.$$

Proposed problems

P.3.1. Prove the theorems:

(a) $F(D)\epsilon^{kt} = F(k)\epsilon^{kt}$, where k is any constant. (This theorem results in the first case of the abbreviated method).

(b) If k is a constant and $u(t)$ is an arbitrary function of t, then:

$F(D)\epsilon^{kt}u(t) = \epsilon^{kt}F(D+k)u(t)$. (This theorem results in the fourth case of the abbreviated method).

P.3.2. Using the theorems of P.3.1, calculate:

(a) $(D-2)\epsilon^{3t}$
(b) $(D^2 - D - 2)\epsilon^{-2t}$
(c) $(D^5 - 6D^3 + 9D)\epsilon^{-t}$
(d) $(D^2 - D + 2)t\epsilon^{-2t}$
(e) $(D^2 - 13D + 36)\epsilon^{-t}\sin 10t$

Answer

(a) ϵ^{3t}
(b) $4\epsilon^{-2t}$
(c) $-4\epsilon^{-t}$
(d) $(8t-5)\epsilon^{-2t}$
(e) $\epsilon^{-t}(-150\cos 10t - 50\sin 10t)$

P.3.3. Calculate:

(a) $\frac{1}{D+1}(t^2 + 1)$
(b) $\frac{1}{D+2}(\epsilon^t + 3t^2)$
(c) $\frac{1}{D+2}(\epsilon^{-2t}\sin t)$
(d) $\frac{1}{D+2}\sin 3t$

Answer

(a) $t^2 - 2t + 3$

(b) $\frac{e^t}{3} + \frac{3t^2}{2} - \frac{3t}{2} + \frac{3}{4}$

(c) $-e^{-2t} \cos t$

(d) $\frac{1}{13}(2 \sin 3t - 3 \cos 3t)$

P.3.4. Find the general solution of the following homogeneous differential equations:

(a) $D(D+1)(D+2)(D+3)y(t) = 0$

(b) $D^2(D^2 + 16)(D - 5)^3 y(t) = 0$

(c) $\frac{d^4 h}{dt^4} + \frac{d^3 h}{dt^3} - 2\frac{d^2 h}{dt^2} = 0$

(d) $\frac{d^2 z}{dt^2} + 2\frac{dz}{dt} + 10z = 0$

(e) $(D^4 + 2D^2 + 10)z(t) = 0$

(f) $(D^5 + 6D^3 + 9D)y(t) = 0$

(g) $(D^3 + 2.7D^2 + 0.15D - 1.771)x(t) = 0$

Answer

(a) $y(t) = k_1 + k_2 e^{-t} + k_3 e^{-2t} + k_4 e^{-3t}$

(b) $y(t) = k_1 + k_2 t + k_3 \cos 4t + k_4 \sin 4t + (k_5 + k_6 t + k_7 t^2)e^{5t}$

(c) $h(t) = k_1 + k_2 t + k_3 e^t + k_4 e^{-2t}$

(d) $z(t) = e^{-t}(k_1 \cos 3t + k_2 \sin 3t)$

(e) $z(t) = e^{1.04t}(k_1 \cos 1.44t + k_2 \sin 1.44t) + e^{-1.04t}(k_3 \cos 1.44t + k_4 \sin 1.44t)$

(f) $y(t) = k_1 + (k_2 + k_3 t) \cos \sqrt{3}t + (k_4 + k_5 t) \sin \sqrt{3}t$

(g) $x(t) = k_1 e^{0.7t} + k_2 e^{-2.3t} + k_3 e^{-1.1t}$

P.3.5. Given the algebraic equation $f(x) = x^7 + 10x^6 - 8x^5 + 3x^3 - 9x + 20 = 0$,

(a) specify the possibilities regarding the number of real positive, real negative and complex conjugated roots;

(b) establish the interval within which the real roots of the equation must necessarily be.

Answer

(a) There are six possibilities: $0, 1, 6; 0, 3, 4; 2, 1, 4; 2, 3, 2; 4, 1, 2; 4, 3, 0.$

(b) $-11 < x < 4$.

P.3.6. Find the general solution of the following non-homogeneous differential equations:

(a) $D(D-1)(D-3)(D+2)y(t) = 40$

(b) $(D-1)(D-3)(D+2)y(t) = 20e^{2t}$

(c) $(D-1)(D-3)(D+2)y(t) = 20e^{-2t}$

(d) $(D^2 + 7D + 10)y(t) = 200U_{-3}(t) + 50U_{-2}(t)$

(e) $(D^3 + 7D^2 + 10D)y(t) = t^3$
(f) $(D^2 - 4D + 4)h(t) = t^3\epsilon^{2t} + t\epsilon^{2t}$
(g) $(D^2 + 4D + 4)i(t) = (D+1)100\cos 3t$
(h) $(D^2 - 2D)w(t) = \epsilon^{-t}\sin t$
(i) $y'''(t) + 20y''(t) + 100y'(t) + 800y(t) = 400\cos(6t + 35°)$
(j) $(D^2 - 1)y(t) = \sin^3 t$

Answer

(a) $y(t) = \frac{20}{3}t + k_1 + k_2\epsilon^t + k_3\epsilon^{3t} + k_4\epsilon^{-2t}$
(b) $y(t) = -5\epsilon^{2t} + k_1\epsilon^t + k_2\epsilon^{3t} + k_3\epsilon^{-2t}$
(c) $y(t) = \frac{4}{3}t\epsilon^{-2t} + k_1\epsilon^t + k_2\epsilon^{3t} + k_3\epsilon^{-2t}$
(d) $y(t) = 10t^2 - 9t + 4.3 + k_1\epsilon^{-2t} + k_2\epsilon^{-5t}, t > 0$
(e) $y(t) = 0.025t^4 - 0.070t^3 + 0.117t^2 + 0.1218 + k_1 + k_2\epsilon^{-2t} + k_3\epsilon^{-5t}$
(f) $h(t) = \left(\frac{t^5}{20} + \frac{t^3}{6}\right)\epsilon^{2t} + (k_1 + k_2 t)\epsilon^{2t}$
(g) $i(t) = 24.33\cos(3t - 41.1°) + (k_1 + k_2 t)\epsilon^{-2t}$
(h) $w(t) = \frac{\epsilon^{-t}}{10}(2\cos t + \sin t)$
(i) $y(t) = k_1\epsilon^{-16.88t} + \epsilon^{-1.56t}(k_2\cos 6.71t + k_3\sin 6.71t) + 1.02\cos(6t - 43.2°)$
(j) $y(t) = -\frac{3}{8}\sin t + \frac{1}{40}\sin 3t + k_1\epsilon^t + k_2\epsilon^{-t}$

P.3.7. Find the solution of the equation: $i''(t) + 4i'(t) + i(t) = 100U_1(t) + 400U_0(t)$, with zero initial conditions at $t = 0_-$.

Answer

$$i(t) = 107.737\epsilon^{-0.268t} - 7.737\epsilon^{-3.732t}, \text{ for } t > 0.$$

P.3.8. Solve the differential equation $y''(t) + y(t) = U_2(t)$, with $y(0_-) = y'(0_-) = 0$.

Answer

$$y(t) = U_0(t) - \sin t.U_{-1}(t).$$

P.3.9. Given the differential equation $y''(t) + 2y'(t) = 2U_2(t) + U_0(t) + 2U_{-1}(t)$, with $y(0_-) = 1, y'(0_-) = 0$ and $y''(0_-) = 1$, determine:

(a) $y(0_+), y'(0_+)$ and $y''(0_+)$.
(b) the solution $y(t)$, for $t \geq 0$.

Answer

(a) $y(0_+) = -3$, $y'(0_+) = 9$ and $y''(0_+) = -15$.

(b) $y(t) = \begin{cases} 2U_0(t), for\ t = 0 \\ 1 + t - 4\epsilon^{-2t}, for\ t > 0. \end{cases}$

P.3.10. Given the differential equation $y''(t) + 2y'(t) = 4U_2(t-1)$, with the initial conditions $y(0_-) = 1$ and $y'(0_-) = 2$, find $y(1_+)$ and $y'(1_+)$.

Answer

$$y(1_+) = 2 - \epsilon^{-2} - 8 = -6.135; y'(1_+) = 2\epsilon^{-2} + 16 = 16.271.$$

P.3.11. Given the differential equation $y''(t) + 2y'(t) = U_1(t-2)$, with the initial conditions $y(0_-) = 1$ and $y'(0_-) = 2$, find $y(2_+)$ and $y'(2_+)$.

Answer

$$y(2_+) = 3 - \epsilon^{-4} = 2.982 \text{ and } y'(2_+) = -2 + 2\epsilon^{-4} = -1.963.$$

P.3.12. Solve the differential equation

$$y'''(t) + 9y'(t) = 2U_2(t) + 400\cos(2t - 30°)U_{-1}(t), \text{ with } y(0_-) = y'(0_-) = y''(0_-) = 0.$$

Answer

$$y(t) = 11.111 + 10.889\cos 3t - 23.094\sin 3t + 40\cos(2t - 120°), for\ t > 0.$$

P.3.13. Solve the following system of differential equations, knowing that all initial conditions are null at $t = 0_-$, and that $e(t) = 100cos300tU_{-1}(t)$.

$$(2D+1)i_1(t) - i_2(t) = De(t)$$

$$-i_1(t) + (2D^2 + 4D + 1)i_2(t) = 0.$$

Answer

$$i_1(t) = 50\cos 300t - 50\epsilon^{-t} + 50\epsilon^{-1.5t}, for\ t > 0.$$

$$i_2(t) = -2.78 \times 10^{-4}cos300t + 8.34 \times 10^{-4}\epsilon^{-t} - 5.56 \times 10^{-4}\epsilon^{-1.5t}, for\ t \geq 0.$$

3.4 Conclusions

In this chapter it was made a revision of the principal methods of solution of homogeneous and non-homogeneous differential equations. It was also presented a systematic method to solve ordinary differential equations in which $x(t)$ is a singular and / or causal function. This method allows to determine the initial conditions at $t = 0_+$, from the respective initial conditions at $t = 0_-$ and the differential equation itself, without any physical reasoning about the given circuit. Therefore, a good understanding of this chapter is an indispensable condition for the good comprehension of the following chapters.

References

1. Adkins, William A, Davidson MG (2012) Ordinary Differential Equations, Undergraduate Texts in Mathematics, Springer
2. Boyce WE, Diprima RC (2002) Elementary differential equations and boundary value problems, LTC, 7th ed
3. Chicone C (2006) Ordinary differential equations with applications, Springer, 22nd ed
4. Close CM (1975) Linear circuits analysis, LTC
5. Doering CI, Lopes AO (2008) Ordinary differential equations, IMPA, 3rd ed
6. Irwin JD (1996) Basic engineering circuit analysis, Prentice Hall
7. Kreider DL, Kuller RG, Ostberg DR (2002) Differential equations, Edgard Blucher Ltda
8. LEITHOLD, Louis., The calculus with analytical geometry, vol.1, Harper & How of Brasil, 2nd. ed., (1982)
9. Perko L (2001) Differential equations and dynamical systems, New York: Springer, 3rd ed
10. Stewart J (2010) Calculus. vol 1, Thomson, 6th. ed
11. Zill DG, Cullen MR (2003) Differential equations, Makron Books, 3rd ed

Chapter 4
Digital Solution of Transients in Basic Electrical Circuits

4.1 Introduction

This chapter presents the fundamental algorithm of the digital computational solution of electromagnetic transients in electrical circuits used in programs based on the EMTP-Electromagnetics Transients Program. Since the nodal method is used in the fundamental EMTP algorithm, therefore, the discretization of circuit elements to concentrated parameters (inductances and capacitances) is implemented through the trapezoidal integration method, resulting in Norton Equivalent Digital Circuits for these elements. It is emphasized that such digital models depend on the appropriate choice of the integration step (Δt) so that the results of the computer simulation present the acceptable precision for the phenomena under study. The simplicity for computational implementation of the trapezoidal integration method when compared to other numerical integration methods, its precision and stability justified its use in EMTP and programs based on it. Because of the trapezoidal integration method's own discretization algorithm, under certain conditions numerical oscillations may arise in the computational solution. Therefore, it is essential to understand its origin to evaluate the different methods proposed for its solution in different computer programs based on EMTP.

The use of computer simulation programs based on the EMTP—Electromagnetic Transients Program [1, 2] is applied in several areas of Electrical Engineering, such as, for example, studies on the calculation of energization transients of transmission lines, transformers, reactors, capacitor banks or various loads. Thus, the study of electromagnetic transients in electrical power systems is of paramount importance, as in these conditions, electrical equipment is subjected to very stressful situations, which can exceed their rating specifications. Even though electrical systems operate most of the time in a permanent or steady-state regime, it is essential that these systems can be able to withstand the transient regimes caused by short-circuits, switching, lightning strikes, among others. The electrical system constantly presents a dynamic behavior that integrates generation, loads, control and protection systems.

J. C. Goulart de Siqueira and B. D. Bonatto, *Introduction to Transients in Electrical Circuits*, Power Systems, https://doi.org/10.1007/978-3-030-68249-1_4

The seminal paper [1] by Prof. Dr.-Ing. Hermann W. Dommel (the creator of EMTP) presents the digital computational solution of electromagnetic transients in single-phase and multiphase networks, along with the digital computational models for basic circuit elements such as resistors, inductors and capacitors and also transmission lines. Several technical-scientific articles and other publications [3–10] followed this development, resulting in several models and studies of practical applications that enshrined different versions of programs based on the EMTP, such as Microtran [11], EMTP-RV [12], ATP (Alternative Transients Program) [13, 14], PSCAD [15], DIGSILENT [16], SABER [17], eMEGAsim [18], HYPERsim [19], PSIM [20] among others. This universal platform satisfactorily applies to transient analysis of electrical systems, usually offline. For real time simulations (Real Time Digital Simulations), especially for tests on protection and control equipment and devices (HIL—Hardware in the Loop Simulations), hardware and software such as OPAL [21] and RSCAD constituting an RTDS [22] have been increasingly used. A well-known computational platform used in academia is MatLab [23], mainly for prototyping and model testing. There is also free software with similar applications such as Scilab [24], etc. as well as software dedicated to dynamic simulations of electronic circuits such as Pspice [25], Multisim [26], etc.

The modeling and study of Electromagnetic Transients is based on fundamental concepts of analysis of electrical circuits, in which concentrated elements are described by ordinary differential equations, and transmission lines with distributed parameters are described by partial differential equations. Various other elements and components of electrical systems are represented by combinations of basic element models.

Mathematical models of electrical systems, such as differential equations, are often difficult to solve analytically, especially when they involve a large number of energy storage circuit elements, switching, non-linear components, among others. To overcome this difficulty, digital computational modeling of the elements is used, in which the differential equations are discretized over time, involving the adoption of numerical integration methods and thus obtaining approximate solutions, but with acceptable precision and numerical stability, for complex engineering problems.

4.2 Basic Algorithm for Computational Solution of EMTP-Based Programs

The solution algorithm for calculating electromagnetic transients in electrical power systems [1, 2] uses the Nodal Method, which allows to obtain the voltage in each of the nodes of an electrical circuit. The systematization of the resolution is presented in a simplified form below:

$$[\mathbf{G}].[e(\mathbf{t})] = [\mathbf{I}] \tag{4.1}$$

$$[I] = [i(t)] + [I_h(t)] \tag{4.2}$$

where:

[G] is the N × N matrix of nodal conductance;
$[e(t)]$ is the vector of nodal voltages;
$[i(t)]$ are the known current sources; and
$[I_h(t)]$ are "historical" sources that depend on the initial conditions of the system and the used discretization rule.

The matrix [G] is real, symmetrical and remains constant as long as there is no change in the switches status and in the time integration step (Δt). The formation of the conductance matrix follows the same rules as the formation of the nodal admittance matrix in steady-state analysis.

The system of equations represented before can be arranged as follows, using partitioning techniques for the nodal conductance matrix and voltage and current vectors:

$$\begin{bmatrix} [G_{AA}] & [G_{AB}] \\ [G_{BA}] & [G_{BB}] \end{bmatrix} \begin{bmatrix} [e_A(t)] \\ [e_B(t)] \end{bmatrix} = \begin{bmatrix} [I_A] \\ [I_B] \end{bmatrix} \tag{4.3}$$

where:

$[e_A(t)]$ is the vector of unknown nodal voltages;
$[e_B(t)]$ is the vector of known nodal voltages, in nodes with connected voltage sources.

To calculate the voltages at the nodes of interest, the Gaussian elimination method is used in the matrix system:

$$[G_{AA}][e_A(t)] = [I_A] - [G_{AB}][e_B(t)] = [RHS]. \tag{4.4}$$

Thus, we have the values of nodal voltages for a given time t. Then the time t is increased with the step Δt, updating the values of the *Right-Hand-Side* [RHS] vector and calculating the nodal voltages, branch voltages and branch currents. The very simplified block diagram presented in Fig. 4.1 represents the basic algorithm for calculation and computational solution in the time domain of transients in electrical circuits [5]. A simplified source code for the base EMTP algorithm implemented in MATLAB is available at https://github.com/aptis.

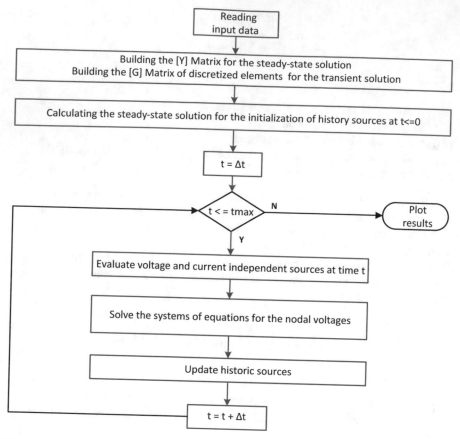

Fig. 4.1 Simplified block diagram of the basic algorithm for solving programs based on the EMTP [5]

4.3 Solution of Differential Equations via Digital Integration

In the analysis of electromagnetic transients of electrical power systems, it is common to encounter complex differential equations that describe the physical behavior of the system. The solution of this type of equation can be obtained numerically through Numerical Methods for solution of Differential Equations such as, for example, Backward Euler and Trapezoidal.

The trapezoidal integration method uses the simplification of such problems by transforming a set of differential equations into an equivalent set of algebraic equations, through discretization.

Fig. 4.2 Trapezoidal
integration method

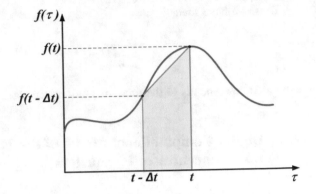

This method is used in most EMTP-based programs and consists of approximating the area below the graph curve by a trapezoid, as shown in Fig. 4.2.

The trapezoid area is given by:

$$Area = \frac{(small\ base + big\ base) * hight}{2}.$$

(4.5)

Thus, we have that:

$$\int_{t-\Delta t}^{t} f(\tau)d\tau \approx \frac{\Delta t}{2} * [f(t - Dt) + f(t)]$$

(4.6)

where Δt corresponds to the time integration step size.

The Backward Euler integration method consists of approximating the area below the graph curve with a rectangle, as shown in Fig. 4.3.

The area of the rectangle is given by: Area = Base * Height.

Fig. 4.3 Backward Euler
integration method

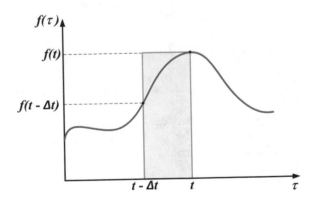

Thus, we have that:

$$\int_{t-\Delta t}^{t} f(\tau)d\tau \approx \Delta t * f(t) \tag{4.7}$$

where Δt corresponds to the time integration step size.

4.4   Digital Computational Model of the Elements with Concentrated Parameters

In order to build a digital computer solution for the calculation of electromagnetic transients in basic circuits, it is necessary to build digital models for fundamental circuit components.

4.4.1   Resistance R

Considering that the resistor is a linear and time-invariant element, the relationship between the voltage at its terminals and the current that flows through it is given by an algebraic equation. Thus, the discrete equivalent circuit model of the resistor is itself, as shown in Fig. 4.4.

$$v(t) = R.i(t). \tag{4.8}$$

where:

$$v(t) = v_{km}(t) = e_k(t) - e_m(t)$$

$$i(t) = i_{km}(t)$$

4.4.2   Inductance L

Consider that the inductor is a linear and time-invariant element, as shown in Fig. 4.5.

The relationship between the voltage at its terminals and the current that flows through it is given by the following characteristic equation, which defines the behavior of the element:

$$v(t) = L\frac{di(t)}{dt}. \tag{4.9}$$

Fig. 4.4 Linear and
time-invariant resistor

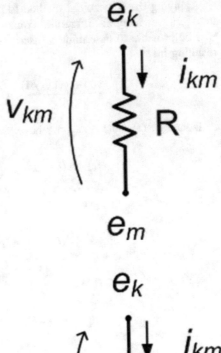

Fig. 4.5 Linear and
time-invariant inductor

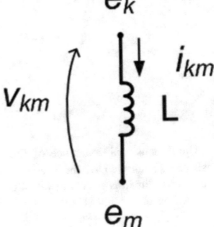

Multiplying the equation by dt on both sides, we have:

$$v(t)\, dt = L\, di(t). \tag{4.10}$$

Integrating the equation in the range from $t - \Delta t$ to t, we have:

$$\int_{t-\Delta t}^{t} v(\tau)\, d\tau = L \int_{t-\Delta t}^{t} di(\tau). \tag{4.11}$$

Applying the trapezoidal method to the solution of Eq. (4.11), it is possible to transform an integral-differential equation into an algebraic equation, and thus it will be possible to determine a discrete model for the inductor in a time interval, resulting in:

$$\frac{v(t) + v(t - \Delta t)}{2} \Delta t = L[i(t) - i(t - \Delta t)]. \tag{4.12}$$

Isolating $v(t) + v(t - \Delta t)$, we have:

$$v(t) + v(t - \Delta t) = \frac{2L}{\Delta t} i(t) - \frac{2L}{\Delta t} i(t - \Delta t). \tag{4.13}$$

Now isolating $v(t)$, we have:

$$v(t) = \frac{2L}{\Delta t} i(t) + \left[-v(t - \Delta t) - \frac{2L}{\Delta t} i(t - \Delta t) \right] \tag{4.14}$$

where,

$$eh_L(t) = \left[-v(t - \Delta t) - \frac{2L}{\Delta t} i(t - \Delta t) \right]. \tag{4.15}$$

Thus,

$$v(t) = \frac{2L}{\Delta t} i(t) + eh_L(t). \tag{4.16}$$

The dimensional coherence of Eq. (4.16) requires that $\left(\frac{2L}{\Delta t}\right)$ have the unit of resistance. Thus, we have the Thévenin Equivalent Digital Model for the inductor in the discrete time, as shown in Fig. 4.6.

Looking at the equivalent circuit in Fig. 4.6, the inductance is represented by an equivalent digital resistance $\left(\frac{2L}{\Delta t}\right)$, in series with a historical voltage source $(eh_L(t))$ that contains the past history, because at each new time interval, the source contributes with a value corresponding to data from the previous interval.

To obtain the Norton Equivalent Digital Model for the inductor in the discrete time, it is necessary to isolate $i(t) - i(t - \Delta t)$ in Eq. (4.12). Thus:

$$i(t) - i(t - \Delta t) = \frac{\Delta t}{2L} v(t) + \frac{\Delta t}{2L} v(t - \Delta t). \tag{4.17}$$

Fig. 4.6 Thévenin equivalent digital model of the inductor, discretized using the trapezoidal integration method

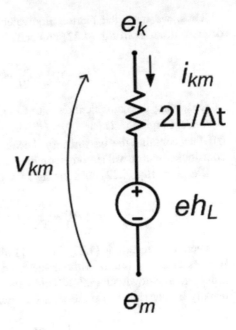

Isolating $i(t)$, we have that:

$$i(t) = \frac{\Delta t}{2L} v(t) + \frac{\Delta t}{2L} v(t - \Delta t) + i(t - \Delta t) \qquad (4.18)$$

where,

$$ih_L(t) = \left[i(t - \Delta t) + \frac{\Delta t}{2L} v(t - \Delta t) \right]. \qquad (4.19)$$

Thus,

$$i(t) = \frac{\Delta t}{2L} v(t) + ih_L(t). \qquad (4.20)$$

However, in order to maintain source conversion coherence, considering the voltage polarity of the Thévenin Equivalent Digital Model, it is necessary to invert the sign of Eq. (4.19), resulting in:

$$ih_L(t) = \left[-i(t - \Delta t) - \frac{\Delta t}{2L} v(t - \Delta t) \right]. \qquad (4.21)$$

Thus, we have the Norton Equivalent Digital Model for the inductor in the discrete time, as in Eq. (4.22) and shown in Fig. 4.7:

$$i(t) = \frac{\Delta t}{2L} v(t) - ih_L(t).$$ (4.22)

Observing the equivalent circuit of Fig. 4.7, the inductance is represented by an equivalent digital resistance $R = \left(\frac{2L}{\Delta t}\right)$, in parallel with a historical current source (ih (t)) that contains the past history, because at each new time interval, the source contributes with a value corresponding to data from the previous interval.

Based on Eq. (4.22) substituting t for $t - \Delta t$ one can write (4.23):

$$i(t - \Delta t) = \frac{\Delta t}{2L} v(t - \Delta t) - ih_L(t - \Delta t).$$ (4.23)

Therefore, inserting (4.23) in (4.21) allows the calculation of historic current sources at every time t without the need of calculating the current through the inductor, as shown in (4.24). This can save computational processing time, as usually implemented in professional software codes.

$$ih_L(t) = \left[-2\frac{\Delta t}{2L} v(t - \Delta t) + ih_L(t - \Delta t) \right].$$ (4.24)

Fig. 4.7 Norton equivalent digital model of the inductor, discretized using the trapezoidal integration method

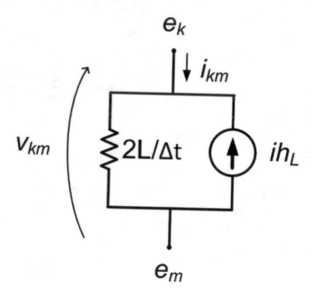

4.4.3 *Capacitance C*

Consider that the capacitor is a linear and time-invariant element, as shown in Fig. 4.8.

The relationship between the voltage at its terminals and the current that flows through it is given by the following characteristic equation, which defines the behavior of the element:

$$i(t) = C\frac{dv(t)}{dt}.$$
(4.25)

Multiplying Eq. (4.25) by dt on both sides, we have:

$$i(t)dt = Cdv(t).$$
(4.26)

Integrating the equation in the range from $t - \Delta t$ to t, we have:

$$\int_{t-\Delta t}^{t} i(\tau)d\tau = C \int_{t-\Delta t}^{t} dv(\tau).$$
(4.27)

Applying the trapezoidal integration method for the solution of Eq. (4.27), an integral-differential equation can be transformed into an algebraic equation. Thus, it will be possible to determine a discrete model for the capacitor in a period of time, resulting in:

$$\frac{i(t) + i(t - \Delta t)}{2}\Delta t = C[v(t) - v(t - \Delta t)].$$
(4.28)

Fig. 4.8 Linear and time-invariant capacitor

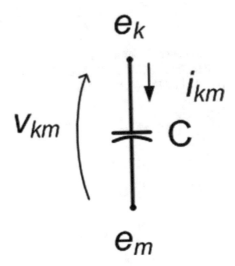

Isolating $v(t)$, we have:

$$v(t) = \frac{\Delta t}{2C} i(t) + \left[v(t - \Delta t) + \frac{\Delta t}{2C} i(t - \Delta t) \right] \tag{4.29}$$

where,

$$eh_C(t) = v(t - \Delta t) + \frac{\Delta t}{2C} i(t - \Delta t). \tag{4.30}$$

Thus,

$$v(t) = \frac{\Delta t}{2C} i(t) + eh_C(t). \tag{4.31}$$

Thus, we have the Thévenin Equivalent Digital Model for capacitor in the discrete time, as shown in Fig. 4.9.

For the Norton Equivalent Digital Model, the same previous steps are followed, however, isolating $i(t)$ in Eq. (4.28), that is:

$$i(t) = \frac{2C}{\Delta t} v(t) + \left[-\frac{2C}{\Delta t} v(t - \Delta t) - i(t - \Delta t) \right] \tag{4.32}$$

where,

$$ih_C(t) = -\frac{2C}{\Delta t} v(t - \Delta t) - i(t - \Delta t). \tag{4.33}$$

Thus,

$$i(t) = \frac{2C}{\Delta t} v(t) + \left[-\frac{2C}{\Delta t} v(t - \Delta t) - i(t - \Delta t) \right]. \tag{4.34}$$

However, in order to maintain source conversion coherence, considering the voltage polarity of the Thévenin Equivalent Digital Model, it is necessary to invert the sign of Eq. (4.33). Thus:

$$ih_C(t) = \frac{2C}{\Delta t} v(t - \Delta t) + i(t - \Delta t). \tag{4.35}$$

Fig. 4.9 Thévenin equivalent digital model of the capacitor, discretized using the trapezoidal integration method

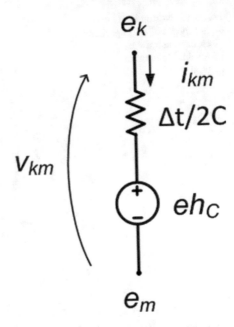

Thus, we have the Norton Equivalent Digital Model for the capacitor in the discrete time, as in Eq. (4.36) and shown in Fig. 4.10:

$$i(t) = \frac{2C}{\Delta t}v(t) - ih_C(t). \tag{4.36}$$

Observing the equivalent circuit of Fig. 4.10, the capacitance is represented by an equivalent digital resistance $\left(\frac{\Delta t}{2C}\right)$, in parallel with a historical current source $(ih(t))$ that contains the past history, because at each new time interval, the source contributes with a value corresponding to data from the previous interval.

Based on Eq. (4.36) substituting t for $t - \Delta t$ one can write (4.37):

$$i(t - \Delta t) = \frac{2C}{\Delta t}v(t - \Delta t) - ih_C(t - \Delta t). \tag{4.37}$$

Therefore, inserting (4.37) in (4.35) allows the calculation of historic current sources at every time t without the need of calculating the current through the capacitor, as shown in (4.38). This can save computational processing time, as usually implemented in professional software codes.

$$ih_C(t) = \left[2\frac{2C}{\Delta t}v(t - \Delta t) - ih_C(t - \Delta t)\right]. \tag{4.38}$$

Fig. 4.10 Norton equivalent digital model of the capacitor, discretized using the trapezoidal integration method

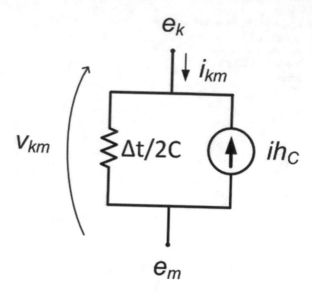

From the previous results, it is clear that the digital models depend on values associated with historical information of the inductor and capacitor. Therefore, it is possible to model these elements as digital equivalent resistors in parallel or series with historical sources, depending on the equivalent digital model chosen.

Since the nodal method is used in the fundamental EMTP algorithm, Norton Equivalent Digital Circuits are used in the discretization of circuit elements of concentrated parameters, using the trapezoidal integration method. Note also that such digital models depend on the appropriate choice of the integration time-step (Δt) so that the results of the computer simulation present the acceptable precision for the phenomena under study. Intrinsically, due to the discretization algorithm itself, the trapezoidal integration method has good precision, but in certain situations, numerical oscillations may occur, which need to be adequately addressed in computational simulation. Table 4.1 presents a comparative synthesis of precision and stability of integration methods for discretizing integral-differential equations.

4.5 Numerical Oscillations in EMTP Due to the Trapezoidal Integration Method

The simplicity for computational implementation of the trapezoidal integration method when compared to other numerical integration methods, its precision and stability justified its use in the EMTP and programs based on it. Because of the trapezoidal integration method's own discretization algorithm, under certain conditions numerical oscillations may arise in the computational solution [26, 27]. Therefore, it is essential to understand its origin to evaluate the different methods proposed for its solution in different computer programs based on the EMTP. According to [28] there are possible techniques to deal with the problem of numerical oscillations caused by the trapezoidal method:

- Output Averaging
- Restart the simulation (re-solve)
- Insert parallel resistor with inductor or series resistor with capacitor—ATP (factors *Kp* and *Ks*)
- Use Backward Euler integration rule along all simulations
- CDA—Critical Damping Adjustment
- THTA—Trapezoidal History Term Averaging.

The numerical oscillation problem in digital computational simulation occurs when there is a discontinuity in variables both in the inductor and in the capacitor modeled in a digital way. In the inductor it occurs when there is a step in the current variation, since the current cannot change instantly. And in the capacitor there is a numerical oscillation when there is a sudden voltage variation, since its voltage cannot vary instantly. To illustrate this problem, the example shown in Fig. 4.11 is used, where a current step source through an ideal inductor is applied.

Using the trapezoidal integration rule for the inductor, by Eq. (4.39) it is possible to calculate the solutions at each discrete time point as shown in Table 4.2, and in Fig. 4.12.

$$v(t) = \frac{2L}{\Delta t}i(t) + \left[-v(t - \Delta t) - \frac{2L}{\Delta t}i(t - \Delta t)\right]. \qquad (4.39)$$

Using the Backward Euler integration rule for the inductor, by Eq. (4.40) it is possible to calculate the solutions at each discrete time point as shown in Table 4.3, and in Fig. 4.13.

$$v(t) = \frac{L}{\Delta t}i(t) + \left[-\frac{L}{\Delta t}i(t - \Delta t)\right]. \qquad (4.40)$$

In the transient simulation with CDA—Critical Damping Adjustment [27, 28], the trapezoidal method is normally used and if a discontinuity occurs (such as in a switch opening or closing, or when changing the region of linearization by parts in

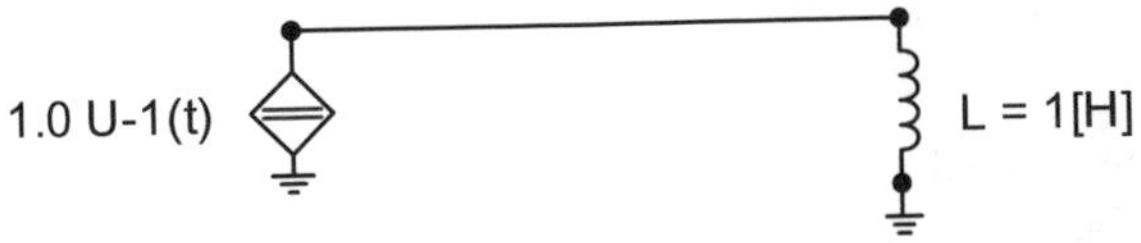

Fig. 4.11 Application of a current step source through an ideal inductor

1.0 U-1(t)   L = 1[H]

Table 4.1 Comparative synthesis of precision and stability of integration methods for discretizing integral-differential equations [5]

Integration method	Order	Precision	Stability
Trapezoidal	1st	Good	Average
Backward Euler	1st	Average	Good
Forward Euler	1st	Average	Bad
Simpson	2nd	Very good	Bad
Gear second order	2nd	Average	Average

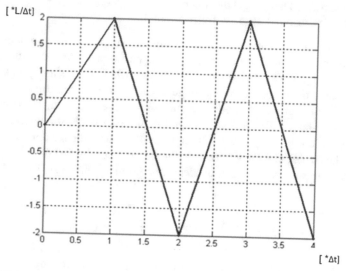

Fig. 4.12 Digital solutions at each discrete time point for voltage and current through the discretized inductor using the trapezoidal integration method

Table 4.2 Digital solutions at each discrete time point for voltage and current through the discretized inductor using the trapezoidal integration method

t	$i(t)$	$v(t)$
0	0	0
Δt	1	$+\frac{2L}{\Delta t}$
$2\Delta t$	1	$-\frac{2L}{\Delta t}$
$3\Delta t$	1	$+\frac{2L}{\Delta t}$
$4\Delta t$	1	$-\frac{2L}{\Delta t}$

saturation curves of inductors or lightning arresters, etc.), the internal algorithm of the program automatically changes the integration method to the Backward Euler and two steps of the solution are carried out, but with an integration time-step size of $\Delta t/2$. Then, the Trapezoidal method is used again until the next discontinuity occurs, as shown in Table 4.4, and in Fig. 4.14.

Table 4.3 Digital solutions at each discrete time point for voltage and current through the discretized inductor using the Backward Euler integration method

t	$i(t)$	$v(t)$
0	0	0
Δt	1	$+\frac{2L}{\Delta t}$
$2\Delta t$	1	0
$3\Delta t$	1	0
$4\Delta t$	1	0

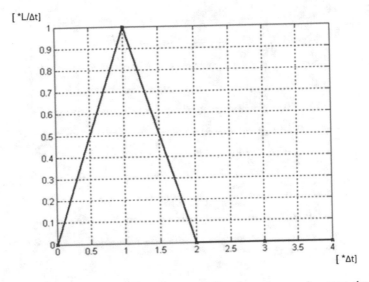

Fig. 4.13 Digital solutions at each discrete time point for voltage and current through the discretized inductor using the Backward Euler integration method

Some EMTP-based software use different approaches to minimize the problem of numerical oscillations caused by the trapezoidal integration rule. For example, the ATP software is capable of reducing the numerical oscillations through the use of an artificial damping resistance, either in series with capacitances or in parallel with inductors. This is shown in the last two elements drawn in Fig. A.3 of Appendix A. Reference [29] presents "Comparative solutions of numerical oscillations in the trapezoidal method used by EMTP-based programs". Reference [30] presents "A discussion about optimum time step size and maximum simulation time in EMTP-based programs", where it is proposed a heuristic approach to choose the proper values for Δt and t_{max} when running time domain simulation of electrical circuit transients.

Table 4.4 Digital solutions at each discrete time point for voltage and current through the discretized inductor using the CDA—Critical Damping Adjustment technique

t	$i(t)$	$v(t)$
0	0	0
$\Delta t/2$	1	$+\frac{2L}{\Delta t}$
$1\Delta t$	1	0
$2\Delta t$	1	0
$3\Delta t$	1	0
$4\Delta t$	1	0

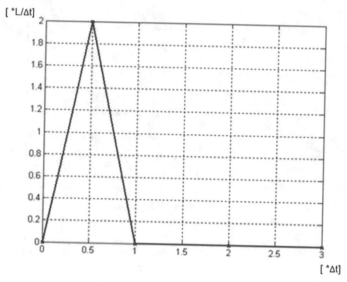

Fig. 4.14 Digital solutions at each discrete time point for voltage and current through the discretized inductor using the CDA—Critical Damping Adjustment technique

It is worth to mention the work of Scoth Meyer and Lou Van der Sluis in supporting the ATP—Alternative Transients Program [13] and of H. K. Høidalen, who created the ATPDraw™—The graphical preprocessor to ATP [14], and many other important researchers on EMTP-based programs in many countries. The ATP program has, therefore, been used all over the world by so many engineers for advanced studies and simulations of electromagnetic transients in power systems. Classical and advanced books [31–37] on electromagnetic transients in power systems has also become available, along with such a large collection of technical papers published in conferences and scientific journals. The IPST—International Power System Transients Conference is a permanent forum for such advanced development [38].

4.6 Conclusions

The proper use of software based on the EMTP—Electromagnetic Transients Program requires a basic knowledge of the computational models implemented, their potentials and limitations, since the computational modeling of a phenomenon is a simplified abstraction from the complex physical reality. Therefore, the computer simulations are validated within the scope of the simplifications and hypotheses adopted. The next chapters will present a large number of simulation cases illustrating the use of digital computer simulation for the calculation of transients in fundamental circuits.

References

1. Dommel HW (1969) Digital computer solution of electromagnetic transients in single- and multiphase networks. IEEE Trans Power Appar Syst PAS-88:388–399
2. Dommel HW (1992) EMTP theory book, Second edn. Microtran Power System Analysis Corporation, Vancouver, British Columbia, Canada (latest update 1996)
3. Dommel HW (1971) Non linear and time-varying elements in digital simulations of electromagnetic transients. Power system engineering committee of the IEEE power group for presentation at the 1971 PICA Conference, Boston, Massachusetts
4. Martí JR, Martí L, Dommel HW (1993) Transmission line models for steady-state and transients analysis. Athens Power Tech, 1993. APT 93. Proceedings Joint International Power Conference, vol 2, pp 744–750
5. Martí JR (2018) Background on electric transient programs. UBC EECE560 graduate course notes
6. MicroTran Reference Manual, Transients Program for Power and Power Electronics Circuits, Microtran Power System Analysis Corporation, Vancouver, B.C., Canada (1997). Available in: http://www.microtran.com/downloads.htm
7. Zanetta Jr LC (2003) Electromagnetic Transients in Power Systems. University of Sao Paulo, Sao Paulo
8. Araújo AEA, Neves WLA (2005) Calculation of electromagnetic transients in energy systems. UFMG, Ilust., p 261
9. Bonatto BD, Dommel HW (2010) EMTP modelling of control and power electronic devices —electromagnetic transients programs helping the analysis of the power interaction on either the load or the network side. ISBN:978-3-8383-2790-7. LAP—Lambert Academic Publishing AG & Co, p 175
10. IPST—International Power System Transients conference—http://www.ipst.org/
11. MICROTRAN—http://www.microtran.com
12. EMTP-RV—http://www.emtp.com/index.html
13. ATP—https://www.emtp.org/
14. ATPDraw—http://www.ece.mtu.edu/atp/
15. PSCAD—www.pscad.com
16. DIGSILENT—http://www.digsilent.de/
17. SABER—www.synopsys.com
18. eMEGAsim—https://www.opal-rt.com/system-emegasim/
19. HYPERsim—https://www.opal-rt.com/systems-hypersim/
20. PSIM—https://powersimtech.com/
21. OPAL—www.opal-rt.com

22. RTDS—RSCAD—http://www.rtds.com/software/rscad/rscad.html
23. MATLAB—www.mathworks.com
24. SCILAB—https://www.scilab.org/
25. PSPICE—https://www.pspice.com/
26. MULTISIM—https://www.multisim.com/
27. Martí JR, LIN J (1989) Suppression of numerical oscillations in the EMTP. IEEE Trans Power Syst 4(2):739–747
28. Lin J, Martí JR (1990) Implementation of the CDA procedure in the EMTP. Power Syst IEEE Trans 5(2):394–402
29. Ferreira LFR, Bonatto BD, Cogo JR, Jesus NC, Dommel HW, Martí JR (2015) Comparative solutions of numerical oscillations in the trapezoidal method used by EMTP-based programs. In: International conference on power systems transients, 2015, Cavtat, Croatia
30. Siqueira JCG, Bonatto BD, Martí JR, Hollmann JA, Dommel HW (2015) A discussion about optimum time step size and maximum simulation time in EMTP-based programs. Int J Electr Power Energy Syst 72:24–32
31. Greenwood A (1991) Electrical transients in power systems, 2nd edn. JohnWiley & Sons, Hoboken
32. Van der Sluis L (2001) Transients in power systems, 1st edn. John Wiley & Sons, Hoboken
33. Martinez-Velasco JA (2009) Power system transients: parameter determination. CRC Press, Boka Raton
34. Ametani A, Nagaoka N, et al (2016) Power system transients: theory and applications, 2nd edn. CRC Press, Boka Raton
35. Haginomori E, Koshiduka T, et al (2016) Power system transient analysis: theory and practice using simulation programs (ATP-EMTP). Wiley, Hoboken
36. Martinez-Velasco JA (2019) Transient analysis of power systems: a practical approach Wiley —IEEE
37. Watson N, Arrillaga J (2019) Power systems electromagnetic transients simulation (Energy Engineering), IET
38. IPST—https://www.ipstconf.org/

Chapter 5
Transients in First Order Circuits

5.1 Introduction

This chapter presents transients in first order circuits. Extensive number of examples are provided with their analytical and digital solution using the software ATP—Alternative Transients Program, using the graphical interface ATPDraw. For didactic reasons, the study is organized as a large collection of various case studies, where the fundamentals of circuit analysis, mathematical time domain solution of ordinary differential equations with singular forcing functions, and physical principles are presented in details. For example, "if all voltages and currents in a circuit remain finite, then the voltage at the terminals of a capacitance and the current through an inductance cannot be varied instantly". Therefore, a continuous knowledge is acquired along this book, integrating fundamental electrical engineering concepts in transient circuit analysis and illustrating potential applications.

All linear electrical circuit powered by a constant external source, starting from $t = 0$, or containing only energy initially stored at $t = 0_-$, and without an external source, or even with an impulsive source applied at $t = 0$, but always containing a single capacitance or a single inductance is always described by a first order differential equation, for $t > 0$, in the general form shown in Eq. 5.1.

$$y'(t) + ay(t) = A \qquad (5.1)$$

Its homogeneous solution is:

$$y_H(t) = k\epsilon^{-at}$$

And its particular solution is:

$$y_P(t) = \frac{1}{D+a}\left(A\epsilon^{0.t}\right) = \frac{A}{a}$$

So, the general solution is:

$$y(t) = y_P(t) + y_H(t) = \frac{A}{a} + k\epsilon^{-at}$$

Then, $y(0_+) = \frac{A}{a} + k$ and $k = y(0_+) - \frac{A}{a}$.

However, $\frac{A}{a} = y(\infty)$, which is the circuit's steady-state solution. Therefore, the general solution can be written as

$$y(t) = y(\infty) + [y(0_+) - y(\infty)]\epsilon^{-at}, t > 0.$$

Or else,

$$y(t) = y(\infty) + [y(0_+) - y(\infty)]\epsilon^{-t/T}, t > 0. \tag{5.2}$$

where $T = 1/a$ is called **the circuit time constant**.

It can be said, therefore, that the voltage in any element of a linear circuit with a single capacitance, or a single inductance, powered by a constant external source from $t = 0$, is always of the form.

$$e(t) = e(\infty) + [e(0_+) - e(\infty))]\epsilon^{-t/T}, t > 0. \tag{5.3}$$

And that the same goes for the current of any element of the circuit, that is, that

$$i(t) = i(\infty) + [i(0_+) - i(\infty))]\epsilon^{-t/T}, t > 0. \tag{5.4}$$

Therefore, in first-order direct current (DC) circuits, all voltages and currents that change do so by varying exponentially from their initial values $e(0_+)$ **and** $i(0_+)$ **to their final constant values** $e(\infty)$ and $i(\infty)$.

It is logical that the above also applies to capacitances in series or in parallel, since they can be replaced by a single equivalent capacitance. And the same goes for series or parallel inductances.

(a) The time constant of an RC circuit, with a single capacitance, is the product of the capacitance by the Thevenin equivalent resistance, seen from the capacitance terminals:

$$T = R_{TH}.C \tag{5.5}$$

(b) The time constant of an RL circuit, with a single inductance, is the quotient of the inductance by the Thevenin equivalent resistance, seen from the inductance terminals:

$$T = L/R_{TH} \tag{5.6}$$

5.2 Continuity Theorem

If all voltages and currents in a circuit remain finite, then the voltage at the terminals of a capacitance and the current through an inductance cannot be varied instantly.

This means that $e_C(0_+) = e_C(0_-)$ and that $i_L(0_+) = i_L(0_-)$; moreover, that voltage $e_C(t)$ and current $i_L(t)$ will be continuous signals in time. Consequently, the electric charge of a capacitance $q(t) = C.e_C(t)$ and the magnetic flux of an inductance $\phi(t) = L.i_L(t)$, also will be continuous signals in time.

A very common procedure for obtaining any initial voltages or currents at $t = 0_+$ in a given circuit, is to replace the capacitance with a voltage source of the same value as that at its terminals at $t = 0_+$; and inductance, by a current source of the same value as that passing through it at time $t = 0_+$. Obviously, if $e_C(0_+) = 0[V]$, the capacitance will be replaced by a short circuit (voltage source of zero value); and if $i_L(0_+) = 0[A]$, the inductance will be replaced by an open circuit (current source of zero value). The other initial conditions at $t = 0_+$, of voltages and currents, are obtained using the usual methods of analysis of resistive circuits.

In a direct current (DC) electrical circuit, that is, one excited only by constant sources, in the permanent regime (steady-state), when $t \to \infty$, the final values $e_C(\infty)$ and $i_L(\infty)$ can be obtained by replacing the capacitance with an open circuit $[i_C(\infty) = 0]$, and the inductance for a short circuit $[e_L(\infty) = 0]$. These conclusions result from the facts that, in the permanent direct current regime of a linear and stable circuit, all voltages and currents are equally constant. In a capacitance

$$i_C(t) = C \frac{d}{dt} e_C(t) \tag{5.7}$$

If the source is constant, in the permanent regime also $e_C(\infty) = $ constant and, according to Eq. 5.7, $i_C(\infty) = 0$.

Likewise, in an inductance

$$e_L(t) = L \frac{d}{dt} i_L(t) \tag{5.8}$$

If the source is constant, in the permanent regime also $i_L(\infty) = $ constant and, according to Eq. 5.8, $e_L(\infty) = 0$.

The other final conditions, for $t \to \infty$, of voltages and currents, are obtained with the usual techniques of analysis of resistive circuits.

More general forms of Eqs. 5.3 and 5.4 are as follows:

$$e(t) = e(\infty) + \left[e(t_{0_+}) - e(\infty) \right] \epsilon^{-(t-t_0)/T}, t > t_0. \tag{5.9}$$

$$i(t) = i(\infty) + \left[i(t_{0_+}) - i(\infty) \right] \epsilon^{-(t-t_0)/T}, t > t_0. \tag{5.10}$$

where t_0 is the switching instant.

Example 5.1 Find the time constant T of the circuit in Fig. 5.1.

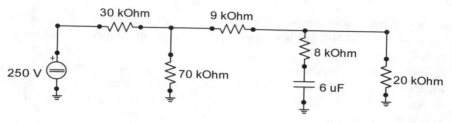

Fig. 5.1 Electric circuit of Example 5.1

Solution

The calculation of R_{TH}, the resistance seen by the capacitance, is made from the circuit of Fig. 5.2, with the 250[V] voltage source shorted, of course.

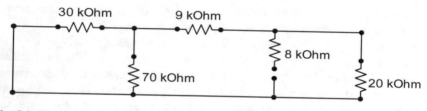

Fig. 5.2 Electric Circuit of Example 5.1 for the calculation of R_{TH}

To emphasize, the capacitance was replaced by an open circuit, in order to visualize a purely resistive circuit.

$$\frac{30 \times 70}{30 + 70} = 21[k\Omega]; 21 + 9 = 30[k\Omega]. \text{ Then,}$$

$$R_{TH} = 8 + \frac{20 \times 30}{20 + 30} = 20[k\Omega] = 20.10^3[\Omega], \text{ and}$$

$$T = R_{TH}.C = 20.10^3 \times 6.10^{-6} = 120.10^{-3}[s] = 120[ms].$$

Example 5.2 Find all initial conditions at $t = 0_+$ and final, for $t \to \infty$, in the circuit of Fig. 5.3. Capacitances are de-energized at $t = 0_-$.

Solution

(a) Calculation of initial conditions.

By the Continuity Theorem, $e_1(0_+) = e_1(0_-) = 0[V]$ and $e_4(0_+) = e_4(0_-) = 0[V]$. Then, the capacitances are converted into short-circuits, and the circuit is converted to the one in Fig. 5.4.

$$i_2(0_+) = i_3(0_+) = 100/25 = 4[A] \text{ and } i_1(0_+) = 0[A]; \; i_4(0_+) = 100/50 = 2[A].$$

$$e_2(0_+) = e_3(0_+) = 100[V].$$

(b) Calculation of final conditions.

For the calculation of the final conditions the capacitors are replaced by open circuits, as shown in Fig. 5.5.

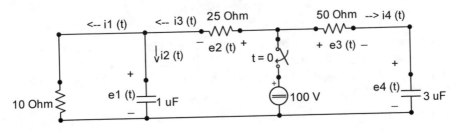

Fig. 5.3 Electric circuit of Example 5.2

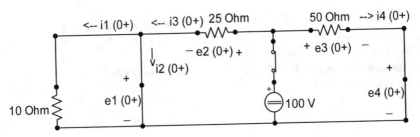

Fig. 5.4 Electric circuit of Example 5.2 for analytical solution at $t = 0_+$

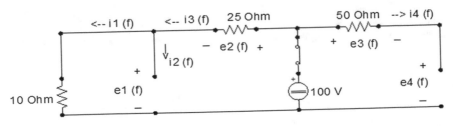

Fig. 5.5 Electric circuit of Example 5.2 for analytical solution at $t \to \infty$

$$i_2(\infty) = i_4(\infty) = 0[A]$$

$$i_1(\infty) = i_3(\infty) = 100/35 = 2.86[A]$$

$$e_1(\infty) = 10.i_1(\infty) = 28.6[V]$$

$$e_2(\infty) = 25.i_3(\infty) = 71.5[V]$$

$$e_3(\infty) = 50.i_4(\infty) = 0[V]$$

$$e_4(\infty) = 100[V].$$

Example 5.3 A capacitor of 2 [μF] capacitance, initially charged with a voltage of 300 [V], is discharged through a resistor of 270 [kΩ] , as illustrated in Fig. 5.6. What is the voltage of the capacitor 0.25 [s] after it starts to discharge? How much is the current worth right now?

Solution

$$\text{Data: } C = 2.10^{-6}[F], R = 270.10^3[\Omega], e(0_-) = 300[V].$$

$e(0_+) = e(0_-) = 300[V]$ and $e(\infty) = 0[V]$, when the capacitor is discharged.
   Then, at $t = 0_+$ the circuit is that one shown in Fig. 5.7, with the capacitor replaced by a 300 [V] voltage source.
   The time constant is $T = RC = 270.10^3.2.10^{-6} = 0.54[s]$.
   By Eq. 5.3,

$$e(t) = e(\infty) + [e(0_+) - e(\infty)]\epsilon^{-t/T} = 0 + (300 - 0)\epsilon^{-t/0.54} = 300\epsilon^{-1.85t}[V].$$

$$e(0.25) = 300\epsilon^{-1.85\times0.25} = 188.9[V]$$

Fig. 5.6 Electric circuit of Example 5.3

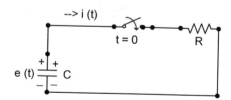

Fig. 5.7 Electric circuit of Example 5.3 for analytical solution at $t = 0_+$

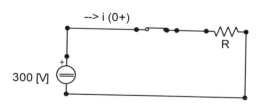

From the circuit in Fig. 5.7,

$$i(0_+) = \frac{300}{270.10^3} = 1.11[\text{mA}].$$

In the end, $i(\infty) = 0[\text{A}]$, when the capacitor behaves like a short circuit, as it has zero voltage, that is, completely discharged.

By Eq. 5.4,

$$i(t) = i(\infty) + [i(0_+) - i(\infty))]\epsilon^{-t/T} = 0 + (1.11 - 0)\epsilon^{-1.85t} = 1.11\epsilon^{-1.85t}[\text{mA}]$$

$$i(0.25) = 0.70[\text{mA}].$$

The numerical solution using the **ATPDraw program** (see Appendix A—Processing at the ATP) was performed using the circuit shown in Fig. 5.8, and the result is in Fig. 5.9, with the voltage and current values recorded on the vertical axes, on the left and to the right, respectively.

Fig. 5.8 Electric circuit of Example 5.3 for digital solution

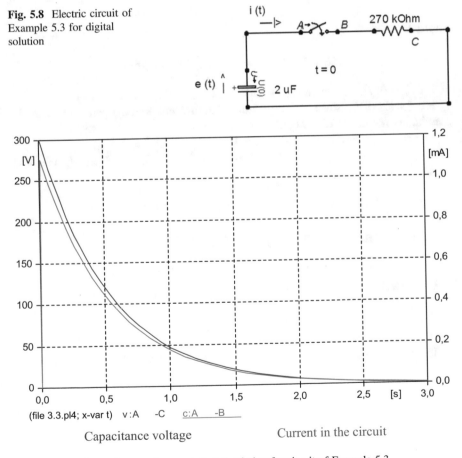

Fig. 5.9 Transient voltage and current digital solution for circuit of Example 5.3

From Fig. 5.9 it is possible, with the aid of the cursor incorporated in the ATPDraw plotter, to record that: $i(0.252) = 7.0065 \times 10^{-4}[A] = 0.70065[mA]$; $e(0.252) = 189.18[V]$.

Example 5.4 Closing a switch, at $t = 0$, connects in series a 200 [V] voltage source, a 2 [MΩ] resistor and a discharged 0.1 [µF] capacitor, as shown in Fig. 5.10. Find the capacitor current and voltage 0.1 [s] after the switch closes. Establish the general expressions of power and energy in the capacitor.

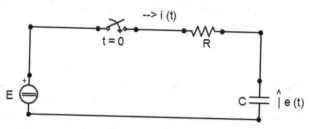

Fig. 5.10 Electric circuit of Example 5.4

Solution

$$E = 200[V], R = 2[MΩ], C = 0.1[µF].$$

$e(0_-) = 0[V]$, whence $e(0_+) = 0[V]$.
$e(\infty) = 200[V]$, since the capacitor is equivalent to an open circuit in the continuous regime of direct current, with $i(\infty) = 0[A]$.
The time constant is $T = RC = 2 \times 10^6 \times 0.1 \times 10^{-6} = 0.2[s]$.
If $e(0_+) = 0[V], i(0_+) = E/R = 200/(2.10^6) = 0.1[mA]$, where $i(0_-) = 0[A]$.
So, by Eqs. 5.3 and 5.4,

$$e(t) = 200 + (0 - 200)\epsilon^{-5t} = 200(1 - \epsilon^{-5t})[V], \text{ for } t \geq 0[s].$$

$$i(t) = 0 + (0.1 - 0)\epsilon^{-5t} = 0.1\epsilon^{-5t}[mA], \text{ for } t > 0[s].$$

Therefore, $e(0.1) = 78.69[V]$ and $i(0.1) = 60.65[µA]$.
The power in the capacitor is $p(t) = e(t).i(t) = 200(1 - \epsilon^{-5t}).10^{-4}.\epsilon^{-5t}$. So,

$$p(t) = 0.02(\epsilon^{-5t} - \epsilon^{-10t})[W], \text{ for } t \geq 0[s].$$

The general expression for calculating energy is: $W(t) = W(t_0) + \int_{t_0}^{t} p(\lambda)d\lambda$.

Fig. 5.11 Electric circuit of
Example 5.4 for digital
solution

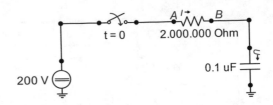

For $t_0 = 0$, it becomes: $W(t) = W(0) + \int_0^t p(\lambda)d\lambda$. Considering that the
capacitor is initially de-energized $W(0) = 0[J]$, and,

$$W(t) = \int_0^t p(\lambda)d\lambda = \int_0^t 0.02\left(\epsilon^{-5\lambda} - \epsilon^{-10\lambda}\right)d\lambda = 2 - 4\epsilon^{-5t} + 2\epsilon^{-10t}\,[mJ]\,t \geq 0[s]$$

The numerical solution for $i(t)$ and $e(t)$, by ATPDraw, using the circuit shown in
Fig. 5.11, with $\Delta t = T/100 = 2$ [ms] and $t_{max} = 5\,T = 1$ [s], is shown in Fig. 5.12.
From Fig. 5.12, with the plotter cursor, it is determined that: $i(0.101) = 60.502[\mu A]$ and $e(0.101) = 78.997[V]$.
The numerical solution for $p(t)$ and $W(t)$ is obtained from the circuit shown in
Fig. 5.13, and is shown in Fig. 5.14.
With the cursor, we get that $p_{max} = 5[mW]$, occurring when $t = 0.14[s]$, rati-
fying the values $t = Tln2 = 0.2 \times 0.69 = 0.14[s]$ and $p_{max} = 5[mW]$, obtained
from the analytical expression of $p(t)$, using the concept of maximums and mini-
mums of Differential Calculus.

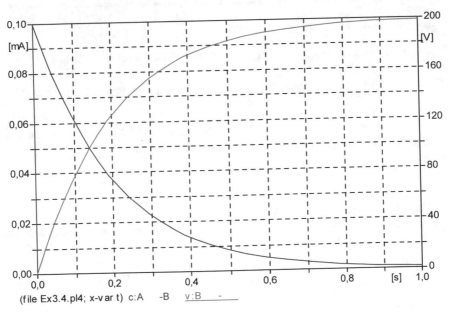

(file Ex3.4.pl4; x-v ar t) c:A   -B   v:B   -

Fig. 5.12 Transient voltage e(t) and current i(t) digital solution for circuit of Example 5.4

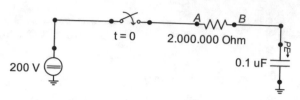

Fig. 5.13 Electric circuit of Example 5.4 for digital solution

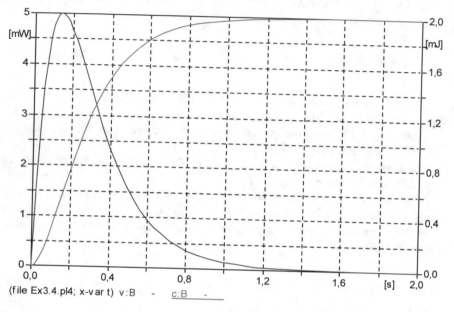

(file Ex3.4.pl4; x-var t) v:B - c:B -

Fig. 5.14 Transient power p(t) and energy W(t) digital solution for circuit of Example 5.4

Example 5.5 Determine the voltage $e_0(t)$ in the circuit of Fig. 5.15, for $t > 0$.

Solution

The Thevenin resistance seen by the inductance, with the switch closed and the voltage source short-circuited, is

$$R_{TH} = R + \frac{R}{2} = \frac{3R}{2}.$$

The time constant is $T = L/R_{TH} = 2L/3R$.

As $i_L(0_-) = 0$, also $i_L(0_+) = 0$, because at first the inductance behaves like an open circuit. So, $e_0(0_+) = 0$. When $t \to \infty$, the inductance behaves like a short circuit and the circuit changes to that of Fig. 5.16.

Using **source conversion**, the circuit becomes the one in Fig. 5.17, now with an independent current source.

Then, $e_0(\infty) = R.\frac{1}{3R} = \frac{1}{3}[V]$.

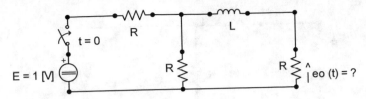

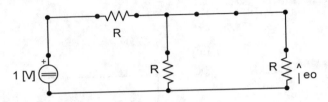

Fig. 5.15 Electric circuit of Example 5.5

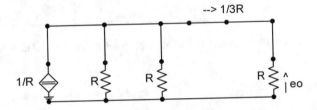

Fig. 5.16 Electric circuit of Example 5.5 for analytical solution at $t \to \infty$

Fig. 5.17 Source conversion applied to electric circuit of Example 5.5 for analytical solution at $t \to \infty$

By Eq. 5.3,

$$e_0(t) = e_0(\infty) + [e_0(0_+) - e_0(\infty)]\epsilon^{-t/T} = \frac{1}{3} + \left(0 - \frac{1}{3}\right)\epsilon^{-3Rt/2L}$$

Therefore,

$$e_0(t) = \frac{1}{3}\left(1 - \epsilon^{\frac{-3Rt}{2L}}\right)[V], t \geq 0[s]$$

Example 5.6 The circuit shown in Fig. 5.18 was operating steadily when switch k was opened at $t = 0$. Find the voltage expression $e_k(t)$, at the switch terminals, for $t > 0$.

Solution
Instant $t = 0_-$: the inductor is short-circuited and switch k is still closed. So $i(0_-) = E/R$.
   Instant $t = 0_+$: the inductor is replaced by an independent source of current equal to a $i(0_+) = i(0_-) = E/R$, as shown in Fig. 5.19, and switch k is already open.

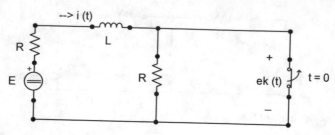

Fig. 5.18 Electric circuit of Example 5.6

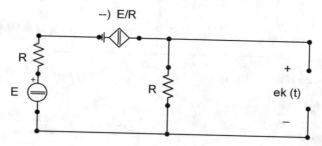

Fig. 5.19 Electric circuit of Example 5.6 for analytical solution at $t = 0_+$

$$e_k(0_+) = R.i(0_+) = E.$$

For $t \to \infty$: there is a new permanent regime, and the inductor again behaving like a short circuit, as shown in Fig. 5.20.

$i(\infty) = E/2R$, where $e_k(\infty) = E/2$.

The resistance seen by the inductance with the k switch open and the short-circuit voltage source is $R_{TH} = 2R$. So, the time constant is $T = L/R_{TH} = L/2R$. Like this,

$$e_k(t) = e_k(\infty) + [e_k(0_+) - e_k(\infty)]\epsilon^{\frac{-t}{T}} = \frac{E}{2} + \left(E - \frac{E}{2}\right)\epsilon^{\frac{-2Rt}{L}}$$

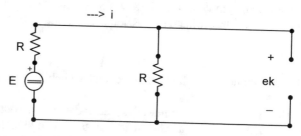

Fig. 5.20 Electric circuit of Example 5.6 for analytical solution at $t \to \infty$

Therefore,

$$e_k(t) = \frac{E}{2}\left(1 + \epsilon^{\frac{-2Rt}{L}}\right), \text{ for } t > 0[\text{s}].$$

The expression of $e_k(t)$ only applies to $t > 0$ because $e_k(0_+) = E$ and $e_k(0_-) = 0$.

Example 5.7 The circuit shown in Fig. 5.21 was operating in a steady state when, at time $t = 0$, switch k was closed. Find $i_2(t)$, for $t > 0$.

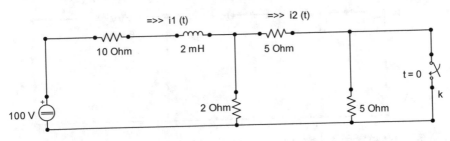

Fig. 5.21 Electric circuit of Example 5.7

Solution
Instant $t = 0_-$: switch k is still open, the circuit operates in a steady state and the inductance can be exchanged for a short circuit, as shown in Fig. 5.22.

Considering the hourly mesh currents $i_1(0_-)$ and $i_2(0_-)$, the equations are obtained:

$$12i_1(0_-) - 2i_2(0_-) = 100$$

$$-2i_1(0_-) + 12i_2(0_-) = 0.$$

Solving the system of equations, it turns out that $i_1(0_-) = 60/7 = 8.57[\text{A}]$ and $i_2(0_-) = 10/7 = 1.43[\text{A}]$.

Instant $t = 0_+$: switch k is already closed and the inductance is replaced by an independent source of current equal to $i_1(0_-)$, as in the circuit of Fig. 5.23.

By **current divider**: $i_2(0_+) = \frac{2 \times 60/7}{5+2} = 120/49 = 2.45[\text{A}]$. (Note that $i_2(0_+) \neq i_2(0_-)$).

When $t \rightarrow \infty$: again the circuit is in steady state, with the configuration shown in Fig. 5.24, in which the inductance has been replaced again by a short circuit.

Converting the voltage source to a current source, we have the circuit of Fig. 5.25.

Replacing the resistors that are in parallel with their equivalent and applying a current divider

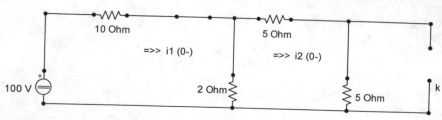

Fig. 5.22 Electric circuit of Example 5.7 for analytical solution at $t = 0_-$

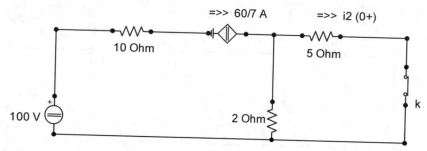

Fig. 5.23 Electric circuit of Example 5.7 for analytical solution at $t = 0_+$

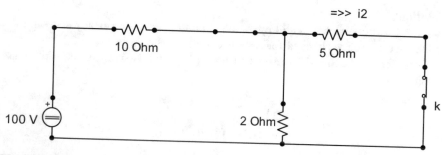

Fig. 5.24 Electric circuit of Example 5.7 for analytical solution at $t \to \infty$

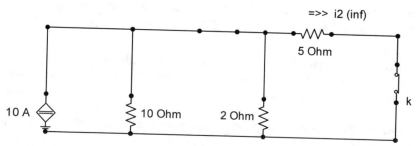

Fig. 5.25 Source conversion at electric circuit of Example 5.7 for analytical solution at $t \to \infty$

$$i_2(\infty) = \frac{10 \times \left(\frac{20}{12}\right)}{5 + \left(\frac{20}{12}\right)} = 2.5[A]$$

Time constant: opening the current source in Fig. 5.25, the inductance sees the following equivalent Thevenin resistance:

$$R_{TH} = 10 + \frac{2 \times 5}{2 + 5} = \frac{80}{7} = 11.43[\Omega]$$

$$T = \frac{L}{R_{TH}} = \frac{2.10^{-3}}{80/7} = \frac{7}{40} = 0.175[ms] = 175[\mu s] \text{ and } 1/T = 5.714,3[s^{-1}]$$

According to Eq. 5.4,

$$i_2(t) = i_2(\infty) + [i_2(0_+) - i_2(\infty)]e^{-t/T} = 2.5 + (2.45 - 2.5)e^{-5714,3t}$$

$$i_2(t) = 2.5 - 0.05e^{-5714,3t}[A], \text{ for } t > 0[s].$$

The numerical solution, using ATPDraw, performed based on the circuit shown in Fig. 5.26, in which the inductance is depicted by the model that includes the initial condition, is shown in Figs. 5.27 and 5.28. In Fig. 5.29 the current curve $i_1(t)$ is plotted.

As the initial values $i_2(0_+) = 2.45[A]$ and final values $i_2(\infty) = 2.50[A]$ are very close, the result shown in Fig. 5.27 was not very illustrative. For this reason, it was necessary to change the scale of values of the ordinate of the graph, and the result is what is shown in Fig. 5.28.

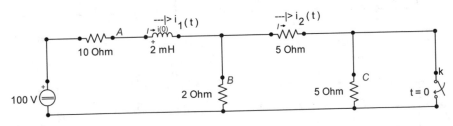

Fig. 5.26 Electric circuit of Example 5.7 for digital solution

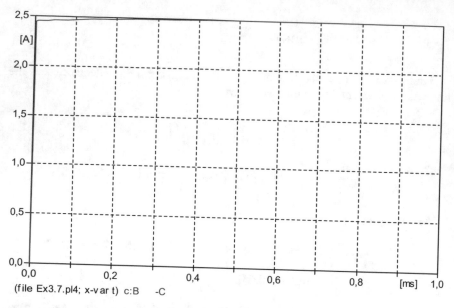

Fig. 5.27 Transient current $i_2(t)$ digital solution for circuit of Example 5.7

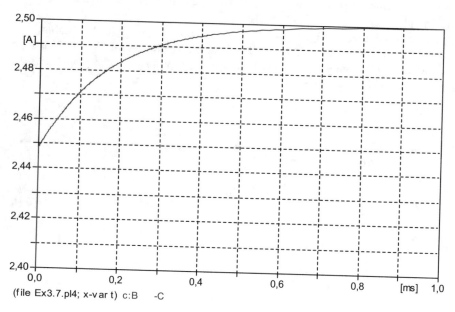

Fig. 5.28 Zoom in transient current $i_2(t)$ digital solution for circuit of Example 5.7

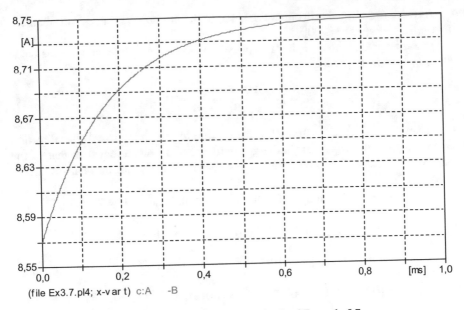

(file Ex3.7.pl4; x-v ar t) c:A   -B

Fig. 5.29 Transient current $i_1(t)$ digital solution for circuit of Example 5.7

Example 5.8 The capacitor of the circuit shown in Fig. 5.30 has an initial charge $q(0_-) = 800[\mu C]$. Obtain the expressions of the current $i(t)$ circulating in the circuit and the charge $q(t)$ in the capacitor, for $t > 0[s]$.

Solution

$q(t) = Ce(t)$. Therefore, $e(0_-) = \frac{q(0_-)}{C} = \frac{800[\mu C]}{4[\mu F]} = 200[V]$.

Fig. 5.30 Electric circuit of Example 5.8

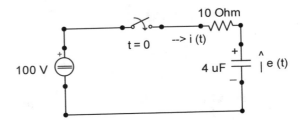

Fig. 5.31 Electric circuit of Example 5.8 for analytical solution at $t = 0_+$

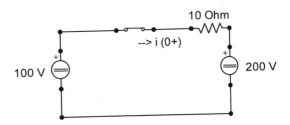

As there is no impulse in current $i(t)$, due to the presence of the resistor of $10[\Omega]$, and $e(0_+) = e(0_-) = 200[V]$.

Instant $t = 0_+$: the capacitor is replaced by a voltage source with a value equal to $e(0_+)$, as in Fig. 5.31.

$$i(0_+) = \frac{100 - 200}{10} = -10[A]$$

When $t \to \infty$: $e(\infty) = 100[V]$, because, due to the counterclockwise direction of the current, the capacitor discharges until it has a voltage equal to that of the source and $i(\infty) = 0[A]$, because in the steady state the capacitor behaves like an open circuit.

Time constant: $T = RC = 10 \times 4.10^{-6} = 40[\mu s]$ and $1/T = 25,000[s^{-1}]$.

$$i(t) = i(\infty) + [i(0_+) - i(\infty)]e^{\frac{-t}{T}} = 0 + (-10 - 0)e^{-25,000t}$$

$$i(t) = -10\epsilon^{-25,000t}[A], \text{ for } t > 0[s].$$

$$e(t) = 100 + (200 - 100)\epsilon^{-25,000t} = 100(1 + \epsilon^{-25,000t})[V], \text{ for } t \geq 0[s].$$

$$q(t) = Ce(t) = 4.10^{-6}.100(1 + \epsilon^{-25,000t})$$

$$q(t) = 400(1 + \epsilon^{-25,000t})[\mu C], \text{ for } t \geq 0[s].$$

But the load can also be obtained by integrating the current:

$$q(t) = q(0_+) + \int_{0_+}^{t} i(\lambda)d\lambda = q(0_-) + \int_{0_+}^{t} i(\lambda)d\lambda = 800.10^{-6} + \int_{0_+}^{t} -10\epsilon^{-25,000\lambda}d\lambda$$

$$q(t) = 800.10^{-6} + \frac{-10\epsilon^{-25,000\lambda}}{-25,000}\Big|_{0_+}^{t} = 800.10^{-6} + 400.10^{-6}(\epsilon^{-25,000t} - 1).$$

Then,

$$q(t) = 400(1 + \epsilon^{-25,000t})[\mu C], \text{ for } t \geq 0[s].$$

The numerical solution by the ATPDraw program, performed through the circuit of Fig. 5.32, in which the capacitance is shown by the model that includes the initial voltage, is shown in Fig. 5.33.

Fig. 5.32 Electric circuit of Example 5.8 for digital solution

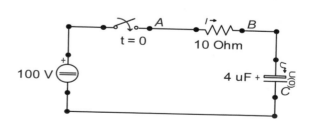

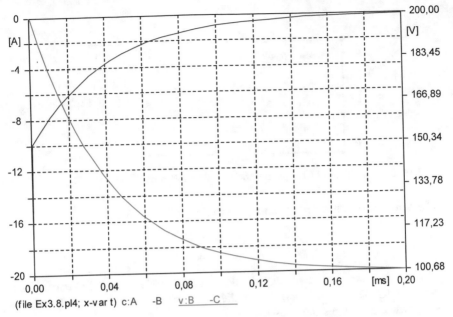

(file Ex3.8.pl4; x-var t) c:A   -B   v:B   -C

Fig. 5.33 Transient voltage and current digital solution for circuit of Example 5.8

Example 5.9 Determine the voltage expressions for $e_1(t)$ and $e_2(t)$, for $t > 0$, in the circuit shown in Fig. 5.34, knowing that $e_1(0_-) = E_1$ and $e_2(0_-) = E_2 \neq E_1$.

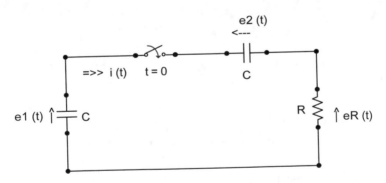

Fig. 5.34 Electric circuit of Example 5.9

Solution
As there is no impulse in the hourly current, due to the presence of resistance R, then $e_1(0_+) = e_1(0_-) = E_1$ and $e_2(0_+) = e_2(0_-) = E_2$. Without resistance R, necessarily $e_1(0_+) = e_2(0_+)$, requiring current impulse at $t = 0$, as shown in Example 5.10.

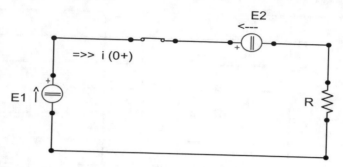

Fig. 5.35 Electric circuit of Example 5.9 for analytical solution at $t = 0_+$

Instant $t = 0_+$: the circuit can be viewed as in Fig. 5.35. So,

$$i(0_+) = \frac{E_1 - E_2}{R}$$

When $t \to \infty$: the two capacitances being equal, $e_1(\infty) = e_2(\infty)$ and $i(\infty) = 0$. (Capacitances in series have the same charge. If the capacitances are the same, then the voltages must also be).

Time constant:

$C_{eq} = \frac{C \times C}{C + C} = \frac{C}{2}$. Therefore, $T = R.C_{eq} = RC/2$.

$$i(t) = i(\infty) + [i(0_+) - i(\infty)]e^{\frac{-t}{T}}$$

$$i(t) = \frac{E_1 - E_2}{R} \epsilon^{\frac{-2t}{RC}}, \text{ for } t > 0.$$

$$e_1(t) = e_1(0_+) - \frac{1}{C} \int_{0_+}^{t} i(\lambda)d\lambda = E_1 - \frac{E_1 - E_2}{RC} \int_{0_+}^{t} \epsilon^{\frac{-2\lambda}{RC}} d\lambda$$

So,

$$e_1(t) = \frac{E_1 + E_2}{2} + \frac{E_1 - E_2}{2} \epsilon^{\frac{-2t}{RC}}, \text{ for } t \geq 0.$$

$$e_R(t) = Ri(t) = (E_1 - E_2)\epsilon^{\frac{-2t}{RC}}, \text{ for } t > 0.$$

$$e_2(t) = e_1(t) - e_R(t)$$

Therefore,

$$e_2(t) = \frac{E_1 + E_2}{2} - \frac{E_1 - E_2}{2} \epsilon^{\frac{-2t}{RC}}, \text{ for } t \geq 0.$$

Fig. 5.36 Electric circuit of Example 5.9 for digital solution

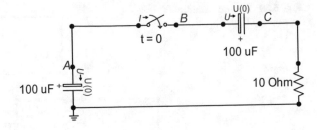

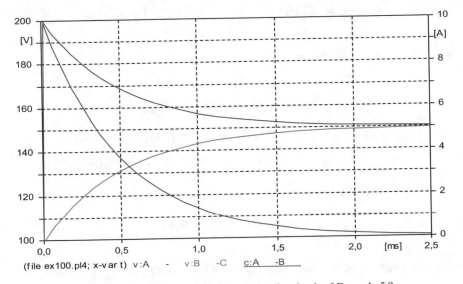

(file ex100.pl4; x-v ar t) v:A   -   v:B   -C   c:A   -B

Fig. 5.37 Transient voltage and current digital solution for circuit of Example 5.9

Alternatively, $e_2(t)$ could be obtained from:

$$e_2(t) = e_2(0_+) + \frac{1}{C} \int\limits_{0_+}^{t} i(\lambda)d\lambda = E_2 + \frac{1}{C} \int\limits_{0_+}^{t} i(\lambda)d\lambda.$$

The numerical solution, through ATPDraw, for $e_1(0_-) = 200[\text{V}]$, $e_2(0_-) = 100[\text{V}]$, $R = 10\ [\Omega]$ and $C = 100\ [\mu\text{F}]$ is shown in Figs. 5.36 and 5.37.

Example 5.10 If the capacitances C_1 and C_2 of the circuit shown in Fig. 5.38 have initial voltages $e_1(0_-) = E_1$ and $e_2(0_-) = E_2$, respectively, find the voltage $e_1(0_+) = e_2(0_+) = E$, for all $t > 0$. Also find the current $i(t)$. Consider $E_1 > E_2$.

Solution
At $t = 0_-$ the voltage at the open switch terminals is $E_1 - E_2$, with positive polarity at its left terminal, since $E_1 > E_2$. At $t = 0_+$, the circuit is shown in Fig. 5.39.

For linear and time-invariant capacitances, $q(t) = Ce(t)$. Therefore,

Fig. 5.38 Electric circuit of Example 5.10

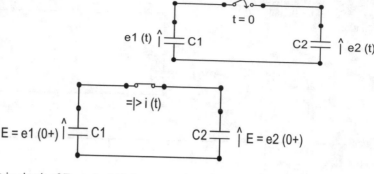

Fig. 5.39 Electric circuit of Example 5.10 for analytical solution at $t = 0_+$

$$q_1(0_-) = C_1 e_1(0_-) = C_1 E_1 \quad q_1(0_+) = C_1 e_1(0_+) = C_1 E$$

$$q_2(0_-) = C_2 e_2(0_-) = C_2 E_2 \quad q_2(0_+) = C_2 e_2(0_+) = C_2 E$$

$$q_{total}(0_-) = q_1(0_-) + q_2(0_-) \quad q_{total}(0_+) = q_1(0_+) + q_2(0_+)$$

By the **Law of Conservation of Load**, $q_{total}(0_+) = q_{total}(0_-)$. So:

$$C_1 E + C_2 E = C_1 E_1 + C_2 E_2$$

Then,

$$E = e_1(0_+) = e_2(0_+) = \frac{C_1 E_1 + C_2 E_2}{C_1 + C_2}, \text{ for } 0_+ \leq t < \infty \text{ and } C_1 + C_2 \triangleq C_p. \text{ Therefore,}$$

$$q_1(0_+) = C_1 \frac{C_1 E_1 + C_2 E_2}{C_1 + C_2} = C_1 \frac{C_1 E_1 + C_2 E_2}{C_p}$$

$$q_2(0_+) = C_2 \frac{C_1 E_1 + C_2 E_2}{C_1 + C_2} = C_2 \frac{C_1 E_1 + C_2 E_2}{C_p}$$

As $C_p = C_1 + C_2$ and $q_1(0_+) \neq q_2(0_+)$, the two capacitances are in parallel from $t = 0_+$, and not in series, not least because the circuit is closed and there is no electric current flowing for $t > 0$.

When $C_1 = C_2 = C$, then $E = \frac{E_1 + E_2}{2}$ and $q_1(0_+) = q_2(0_+) = C\frac{E_1 + E_2}{2}$.

Supposing that the capacitances were already charged well before $t = 0$, we can write, for example, that:

$$q_2(t) = q_2(0_-) U_{-1}(-t) + q_2(0_+) U_{-1}(t)$$

$$= C_2 E_2 U_{-1}(-t) + C_2 \frac{C_1 E_1 + C_2 E_2}{C_1 + C_2} U_{-1}(t)$$

So, the clockwise current flowing through the circuit is

$$i(t) = \frac{d}{dt}q_2(t) = C_2 E_2 U_0(-t).(-1) + C_2 \frac{C_1 E_1 + C_2 E_2}{C_1 + C_2} U_0(t)$$

As the impulse function is even, $U_0(-t) = U_0(t)$ and it can be written that

$$i(t) = \left[C_2 \frac{C_1 E_1 + C_2 E_2}{C_1 + C_2} - C_2 E_2 \right] U_0(t)$$

Or,

$$i(t) = \frac{C_1 C_2}{C_1 + C_2}(E_1 - E_2)U_0(t), \text{ at } t = 0.$$

The area of the current pulse, $C_1 C_2(E_1 - E_2)/(C_1 + C_2)$, represents the amount of electric charge that is transferred instantly from capacitance C_1 to the capacitance C_2, at time $t = 0$, to equalize the voltages at $t = 0_+$.
This can be confirmed by calculating $q_1(0_-) - q_1(0_+)$.

$$q_1(0_-) - q_1(0_+) = C_1 E_1 - C_1 \frac{C_1 E_1 + C_2 E_2}{C_1 + C_2} = \frac{C_1 C_2}{C_1 + C_2}(E_1 - E_2).$$

Therefore,

$$i(t) = QU_0(t), \text{ where}$$

$$Q = q_1(0_-) - q_1(0_+).$$

It is interesting to observe that the quotient $C_1 C_2/(C_1 + C_2) = C_s$ represents the equivalent of capacitances connected in series, which is absolutely consistent with the fact that at moment $t = 0$ an impulsive current circulates in the circuit. It can be said, therefore, that at time $t = 0$ the capacitances are in series, allowing the circulation of the impulsive current $i(t) = QU_0(t)$. And that for $t > 0$ the capacitances are in parallel, with the same voltage E, no more current circulating in the circuit, since $i(t) = 0$ for $t > 0$, and constant voltage does not add current in capacitances.
It is also possible to confirm the results obtained by calculating, for example, the voltage $e_1(t)$ from the current $i(t)$:

$$e_1(t) = \frac{1}{C_1} \int_{-\infty}^{t} i(\lambda)\, d\lambda = \frac{1}{C_1} \int_{-\infty}^{0_-} i(\lambda)\, d\lambda - \frac{1}{C_1} \int_{0_-}^{t} i(\lambda)d\lambda = \frac{q_1(0_-)}{C_1} - \frac{1}{C_1} \int_{0_-}^{t} i(\lambda)d\lambda$$

Or

$$e_1(t) = \frac{1}{C_1} \int_{-\infty}^{t} i(\lambda)d\lambda = \frac{1}{C_1} \int_{-\infty}^{0_-} i(\lambda)d\lambda + \frac{1}{C_1} \int_{0_-}^{t} -i(\lambda)d\lambda = \frac{q_1(0_-)}{C_1} + \frac{1}{C_1} \int_{0_-}^{t} -i(\lambda)d\lambda$$

It should be noted that the current that carries capacitance C_1, between $-\infty$ and 0_-, has the opposite sense of the one that causes it to discharge; hence the negative sign for the current in the range between 0_- and t. Moving on,

$$e_1(t) = e_1(0_-) - \frac{1}{C_1} \int_{0_-}^{t} i(\lambda)d\lambda = E_1 - \frac{1}{C_1} \cdot \frac{C_1 C_2}{C_1 + C_2}(E_1 - E_2) \int_{0_-}^{0_+} U_0(\lambda)d\lambda$$

$$= E_1 - \frac{C_2}{C_1 + C_2}(E_1 - E_2) = \frac{C_1 E_1 + C_2 E_2}{C_1 + C_2}, \text{ for all } t > 0. \text{ So,}$$

$$e_1(0_+) = \frac{C_1 E_1 + C_2 E_2}{C_1 + C_2} = E.$$

This Example 5.10 establishes the opportune moment to state the following corollary to the Continuity Theorem:

Corollary 5.1 A unitary current impulse, $i(t) = U_0(t)[A]$, circulating through a capacitance $C[F]$ produces an instantaneous variation in its voltage equal to $1/C[V]$.

This means that if the capacitance voltage at the initial time $t = 0_-$ is $e(0_-)[V]$, then the voltage at time $t = 0_+$ will be $e(0_+) = e(0_-) \pm 1/C[V]$. Being that the $1/C$ signal will be positive or negative, depending on the direction in which the current will circulate, that is, if the current will charge the capacitor even more, or if it will act to decrease its load and, consequently, its voltage. In the present case, in which the current is a unitary impulse, the **variation in the electric charge is 1 [C]**.

Obviously, **the impulsive current $i(t) = QU_0(t)[A]$ transfers $Q[C]$ of electric charge, instantly**. Considering, on the other hand, that $[A]=[C/s]$, it is evident, once again, that **the unit of $U_0(t)$ is [s^{-1}]**, being t in seconds.

For a capacitance C with voltage and current arrows from opposite directions, the current is given by:

$$i(t) = C\frac{d}{dt}e(t).$$

If $e(t)$ is discontinuous, then the current $i(t)$ will contain impulse, and also the power, because,

$$p(t) = e(t).i(t) = Ce(t)\frac{d}{dt}e(t).$$

Since $p(t) = \frac{d}{dt}W(t)$, then the energy is given by

$$W(t) = \int p(t)dt = \int_{-\infty}^{t} p(\lambda)d\lambda = \int_{-\infty}^{t} Ce(\lambda)\frac{de(\lambda)}{d\lambda}d\lambda = C\int_{-\infty}^{t} e(\lambda)de(\lambda)$$

$$= C\frac{e^2(\lambda)}{2}\Big|_{-\infty}^{t} = \frac{C}{2}\left[e^2(t) - e^2(-\infty)\right]$$

Whereas at $t \to -\infty$ the capacitor, of capacitance C, has just left the factory completely de-energized, so $q(-\infty) = e(-\infty) = 0$. So,

$$W(t) = \frac{C}{2}e^2(t).$$

Therefore, $W(0_-) = (C/2)e^2(0_-)$, regardless of the waveform of the current that flowed through the capacitance from $-\infty$ to $t = 0_-$. On the other hand, it can also be written that:

$$W(t) = \int_{-\infty}^{t} p(\lambda)d\lambda = \int_{-\infty}^{0_-} p(\lambda)d\lambda + \int_{0_-}^{t} p(\lambda)d\lambda$$

$$= W(0_-) + C\int_{0_-}^{t} e(\lambda)\frac{de(\lambda)}{d\lambda}d\lambda = \frac{C}{2}e^2(0_-) + C\frac{e^2(\lambda)}{2}\Big|_{0_-}^{t}$$

$$= \frac{C}{2}e^2(0_-) + C\frac{e^2(\lambda)}{2}\Big|_{0_-}^{t}$$

$$= \frac{C}{2}e^2(0_-) + \frac{C}{2}\left[e^2(t) - e^2(0_-)\right] = \frac{C}{2}e^2(t).$$

Therefore, regardless of the instant considered, when calculated through the voltage, the energy in a capacitance is always given by $W(t) = (C/2)e^2(t)$. But, directly from the power, you can also use the formula

$$W(t) = W(t_0) + \int_{t_0}^{t} p(\lambda)d\lambda.$$

Under no circumstances is it assumed that the energy stored in a capacitance is infinite. And since the energy is given by $W_C = (1/2)Ce_C^2$, this means that the voltage at the terminals of a capacitance can never be an impulse, as it has infinite amplitude. In an equivalent way, considering that $i_C = Cde_C/dt$, **the current $i_C(t)$ through a capacitance can never be a double.**

Example 5.11 The circuit shown in Fig. 5.40 is, strictly, the dual of that of example 5.10. Considering that it is possible that the inductances L_1 and L_2 have initial currents $i_1(0_-) = I_1$ and $i_2(0_-) = I_2$, find the current $i_1(0_+) = i_2(0_+) = I$, for all $t > 0$. Also find the voltage $e(t)$ at the open switch terminals. Consider that $I_1 > I_2$.

Fig. 5.40 Electric circuit of
Example 5.11

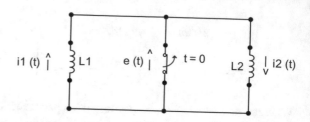

Solution

At $t = 0_-$ the current through the closed switch is $I_1 - I_2$, downwards, since $I_1 > I_2$. **For** $t > 0$, the circuit is shown in Fig. 5.41.

For any linear and time-invariant inductance, $\phi(t) = Li(t)$. So,

$$\phi_1(0_-) = L_1 i_1(0_-) = L_1 I_1 \quad \phi_1(0_+) = L_1 i_1(0_+) = L_1 I$$

$$\phi_2(0_-) = L_2 i_2(0_-) = L_2 I_2 \quad \phi_2(0_+) = L_2 i_2(0_+) = L_2 I$$

$$\phi_{total}(0_-) = \phi_1(0_-) + \phi_2(0_-) \quad \phi_{total}(0_+) = \phi_1(0_+) + \phi_2(0_+)$$

By the **Flow Conservation Law**, $\phi_{total}(0_+) = \phi_{total}(0_-)$. So,

$$L_1 I + L_2 I = L_1 I_1 + L_2 I_2$$

Then,

$I = i_1(0_+) = i_2(0_+) = \frac{L_1 I_1 + L_2 I_2}{L_1 + L_2}$, for $0_+ \le t < \infty$ and $L_1 + L_2 = L_s$. Therefore,

$$\phi_1(0_+) = L_1 \frac{L_1 I_1 + L_2 I_2}{L_1 + L_2} = L_1 \frac{L_1 I_1 + L_2 I_2}{L_s}$$

$$\phi_2(0_+) = L_2 \frac{L_1 I_1 + L_2 I_2}{L_1 + L_2} = L_2 \frac{L_1 I_1 + L_2 I_2}{L_s}$$

As $L_s = L_1 + L_2$ and $\phi_1(0_+) \ne \phi_2(0_+)$, the two inductances are in series starting at $t = 0_+$.

When $L_1 = L_2 = L$, $I = \frac{I_1 + I_2}{2}$ and $\phi_1(0_+) = \phi_2(0_+) = L\frac{I_1 + I_2}{2}$.

Whereas

Fig. 5.41 Electric circuit of
Example 5.11 for analytical
solution at $t = 0_+$

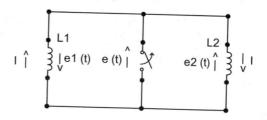

$$\phi_2(t) = \phi_2(0_-)U_{-1}(-t) + \phi_2(0_+)U_{-1}(t) = L_2I_2U_{-1}(-t) + L_2\frac{L_1I_1 + L_2I_2}{L_1 + L_2}U_{-1}(t),$$

$$e_2(t) = \frac{d}{dt}\phi_2(t) = L_2I_2U_0(-t) \times (-1) + L_2\frac{L_1I_1 + L_2I_2}{L_1 + L_2}U_0(t)$$

Since $U_0(-t) = U_0(t)$,

$$e_2(t) = \left[L_2\frac{L_1I_1 + L_2I_2}{L_1 + L_2} - L_2I_2\right]U_0(t)$$

or

$$e_2(t) = \frac{L_1L_2(I_1 - I_2)}{L_1 + L_2}U_0(t) = e(t) = -e_1(t), \text{ only at } t = 0.$$

The area of the voltage impulse, $L_1L_2(I_1 - I_2)/(L_1 + L_2)$, represents the amount of magnetic flux that was transferred instantly from inductance L_1 to inductance L_2, at time $t = 0$, to equal the currents at $t = 0_+$.
This can be proved by calculating $\phi_1(0_-) - \phi_1(0_+)$.

$$\phi_1(0_-) - \phi_1(0_+) = L_1I_1 - L_1\frac{L_1I_1 + L_2I_2}{L_1 + L_2} = \frac{L_1L_2(I_1 - I_2)}{L_1 + L_2}.$$

Therefore,

$e(t) = \Phi U_0(t)$, where

$$\Phi = \phi_1(0_-) - \phi_1(0_+)$$

It is interesting to note that the quotient $L_1L_2/(L_1 + L_2) = L_p$ is the equivalent of parallel inductances, which agrees with the fact that at moment $t = 0$ a voltage impulse appears in the circuit. Therefore, it can be said that at time $t = 0$ the inductances are in parallel, allowing the presence of the voltage $e(t) = \Phi U_0(t)$. And that, for $t > 0$, they are in series, with the same current I, but with $e(t) = 0$, because constant current does not induce electromotive force in inductances.

Corollary 5.2 A unitary voltage impulse, $e(t) = U_0(t)[V]$, applied to the terminals of an $L[\,H]$ inductance causes an instantaneous change in its current value equal to $1/L[A]$.

This corollary states that if the inductance current at $t = 0_-$ is $i(0_-)[A]$, then the current at $t = 0_+$ will be $i(0_+) = i(0_-) \pm 1/L[A]$. Being that the $1/L$ signal depends on the polarity of the applied voltage, that is, if it will increase the current, or if it will act to decrease it. In this case, where the voltage is a unitary impulse, the magnetic flux variation is 1 $[Wb]$.

Obviously, the impulsive voltage $e(t) = \Phi U_0(t)[V]$ transfers Φ $[Wb]$ of magnetic flux, instantly. Remembering, on the other hand, that $[V] = [Wb/S]$, it is evident, once again, that the unit of $U_0(t)$ is $[s^{-1}]$, being t in seconds.

Under no circumstances is it assumed that the energy stored in an induc-tance is infinite. And since the energy given by $W_L = (1/2)Li_L^2$, this means that the current through an inductance can never be a impulse, which has infinite amplitude. On the other hand, considering that $e_L = Ldi_L/dt$, then **the voltage e_L at the ends of an inductance can never be a double**.

Example 5.12 The circuit shown in Fig. 5.42 was operating steadily when switch k was opened at time $t = 0$ [s]. Find the current $i_1(t)$ for $t > 0$ [s].

Solution

At $t = 0_-$ switch k is still closed and the circuit is operating steadily. So the inductances behave like short circuits, as shown in Fig. 5.43.

Therefore,

$$i_1(0_-) = \frac{E}{R} = I_1; i(0_-) = \frac{E/2}{2R} = \frac{E}{4R}; i_2(0_-) = i_1(0_-) + i(0_-) = \frac{5E}{4R}.$$

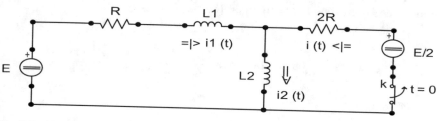

Fig. 5.42 Electric circuit of Example 5.12

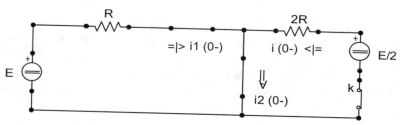

Fig. 5.43 Electric circuit of Example 5.12 for analytical solution at $t = 0_-$

Fig. 5.44 Electric circuit of Example 5.12 for analytical solution at $t = 0_+$

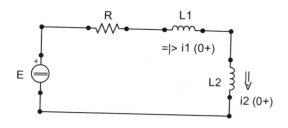

At $t = 0_+$ the switch k is already open and $i_1(0_+) = i_2(0_+)$, as in the circuit in Fig. 5.44.

From the result of Example 5.11,

$$I = i_1(0_+) = i_2(0_+) = \frac{L_1 I_1 + L_2 I_2}{L_1 + L_2} = \frac{\frac{L_1 E}{R} + \frac{5L_2 E}{4R}}{L_1 + L_2} = \frac{4L_1 + 5L_2}{4(L_1 + L_2)} \cdot \frac{E}{R}.$$

For $t \to \infty$ the inductances are again replaced by short circuits and $i_1(\infty) = \frac{E}{R} = i_2(\infty)$. The time constant is: $T = (L_1 + L_2)/R$.

As the inductances are in series, the circuit is of the first order and it can be written that:

$$i_1(t) = i_1(\infty) + [i_1(0_+) - i_1(\infty)]e^{\frac{-t}{T}}$$

$$i_1(t) = \frac{E}{R} + \left[\frac{4L_1 + 5L_2}{4(L_1 + L_2)} \cdot \frac{E}{R} - \frac{E}{R} \right] e^{\frac{-Rt}{L_1 + L_2}}$$

Finally,

$$i_1(t) = i_2(t) = \frac{E}{R}\left[1 + \frac{L_2}{4(L_1 + L_2)} e^{\frac{-Rt}{L_1 + L_2}} \right], \text{ for } t > 0.$$

To complete, the inductance voltages are calculated, assuming associated references, that is, with opposite voltage and current arrows.

As there is discontinuity in the current $i_1(t)$, since $i_1(0_-) = \frac{E}{R}$, $i_1(0_+) = \frac{4L_1 + 5L_2}{4(L_1 + L_2)} \cdot \frac{E}{R}$ and $e_1(t) = L_1 \frac{d}{dt} i_1(t)$, then the voltage $e_1(t)$ presents an impulse at $t = 0$, of value

$$e_1(t) = L_1[i_1(0_+) - i_1(0_-)]U_0(t) = \frac{L_1 L_2 E}{4R(L_1 + L_2)} U_0(t), \ t = 0.$$

For $t > 0$,

$$e_1(t) = L_1 \frac{d}{dt}\left\{ \frac{E}{R}\left[1 + \frac{L_2}{4(L_1 + L_2)} e^{\frac{-Rt}{L_1 + L_2}} \right] \right\} = -\frac{L_1 L_2}{(L_1 + L_2)^2} \cdot \frac{E}{4} e^{\frac{-Rt}{L_1 + L_2}}, \ t > 0.$$

Therefore, it can be written, alternatively, that:

$$e_1(t) = \frac{L_1 L_2 E}{4R(L_1 + L_2)} U_0(t) - \frac{L_1 L_2 E}{4(L_1 + L_2)^2} e^{\frac{-Rt}{L_1 + L_2}} U_{-1}(t).$$

Since $i_2(0_-) = \frac{5E}{4R}$ and $i_2(0_+) = \frac{4L_1 + 5L_2}{4(L_1 + L_2)} \cdot \frac{E}{R} = \left(4 + \frac{L_2}{L_1 + L_2} \right) \frac{E}{4R} < i_2(0_-)$, there is also a discontinuity in $i_2(t)$. Since $e_2(t) = L_2 \frac{d}{dt} i_2(t)$, then the voltage $e_2(t)$ also presents an impulse at $t = 0$, of value

$$e_2(t) = L_2[i_2(0_+) - i_2(0_-)]U_0(t) = -\frac{L_1 L_2 E}{4R(L_1 + L_2)} U_0(t), t = 0.$$

It is interesting to note that the impulses present in $e_1(t)$ and $e_2(t)$ satisfy Kirchhoff's voltage law, that is, that $e_1(t) + e_2(t) = 0$, at time $t = 0$, as there is no impulse in the voltage of the resistance R at any time, although there is discontinuity in the current $i_1(t)$ circulating through it.

For $t > 0$,

$$e_2(t) = L_2 \frac{d}{dt} \left\{ \frac{E}{R} \left[1 + \frac{L_2}{4(L_1 + L_2)} \epsilon^{\frac{-Rt}{L_1 + L_2}} \right] \right\} = -\frac{L_2^2}{(L_1 + L_2)^2} \cdot \frac{E}{4} \epsilon^{\frac{-Rt}{L_1 + L_2}}, t > 0.$$

Putting the results together for $t = 0$ and $t > 0$,

$$e_2(t) = -\frac{L_1 L_2 E}{4R(L_1 + L_2)} U_0(t) - \frac{L_2^2 E}{4(L_1 + L_2)^2} \epsilon^{\frac{-Rt}{L_1 + L_2}} U_{-1}(t).$$

5.3   Response to Impulse and Response to Step

The response of a circuit, without energy initially stored at time $t = 0_-$, to a unitary impulse source $U_0(t)$ or to a unitary step source $U_{-1}(t)$, for its importance, is called $h(t)$ or $r(t)$, respectively. It is important to note that the terms **Impulse Response** and **Response to Step** always refer to singular unitary functions, that is, unitary area impulse at $t = 0$ and unitary amplitude step for $t > 0$. And more: that the circuit does not contain any energy stored in its capacitors and inductors before the application of these singular functions. Thus, if the input and output of a circuit are denoted by $x(t)$ and $y(t)$, respectively, then:

$y(t) = h(t)$ where $x(t) = U_0(t)$.

$y(t) = r(t)$ where $x(t) = U_{-1}(t)$.

It is also worth noting that, being $U_0(t)$ the derivative of $U_{-1}(t)$, then, as a consequence,

$$h(t) = \frac{d}{dt} r(t). \tag{5.11}$$

And that, conversely,

$$r(t) = \int h(t) dt = \int_{-\infty}^{t} h(\lambda) d\lambda = \int_{0_-}^{t} h(\lambda) d\lambda. \tag{5.12}$$

Example 5.13 Determine $r(t)$ and $h(t)$ for the circuit in Fig. 5.45.

Solution

(a) Where $i(t) = U_{-1}(t)$, $e(t) = r(t) = ?$

According to the Continuity Theorem, the voltage $e(t)$ that is at the capacitance terminals, and that is zero for $t<0$, cannot be changed instantly, as the current source is unitary step finite. Therefore, $e(0_+) = 0$. On the other hand, at $t = 0_+$, when $i(0_+) = 1[A]$, there can still be no current going down through the resistance, because, if it did, the voltage at the capacitance terminals, which is the same in resistance terminals, it would jump to a non-zero value equal to $R \times 1 = R$. Thus, all current from the source, which is $1[A]$, is initially directed to capacitance, which begins to accumulate electrical charge and voltage. But as the charge and voltage increase in capacitance, the current going down through the resistance, $i(t) = e(t)/R$, also increases, and the current going down through the capacitance decreases, which in turn instead, it slows the growth of tension $e(t)$. When the current in the resistor reaches 1 [A], the current in the capacitance becomes zero and there is no more voltage growth e (t), which reaches its maximum value $e(\infty) = R$. As the circuit time constant is $T = RC$, according to Eq. 5.3, it can be written that.
$r(t) = e(t) = e(\infty) + [e(0_+) - e(\infty)]\epsilon^{-t/RC} = R + (0 - R)\epsilon^{-t/RC}$. So,

$$r(t) = R\left(1 - \epsilon^{-\frac{t}{RC}}\right)U_{-1}(t).$$

This result can be confirmed by writing the **nodal circuit equation**:

$$C\frac{d}{dt}e(t) + \frac{1}{R}e(t) = i(t)$$

Then,

$r'(t) + \frac{1}{RC}r(t) = \frac{1}{C}U_{-1}(t)$, with $r(0_-) = 0$.
For $t > 0$,

$$\left(D + \frac{1}{RC}\right)r(t) = \frac{1}{C}$$

$$r_H(t) = k\epsilon^{-t/RC}$$

$$r_P(t) = \frac{1}{D + \frac{1}{RC}}\left(\frac{1}{C}\epsilon^{0\times t}\right) = R$$

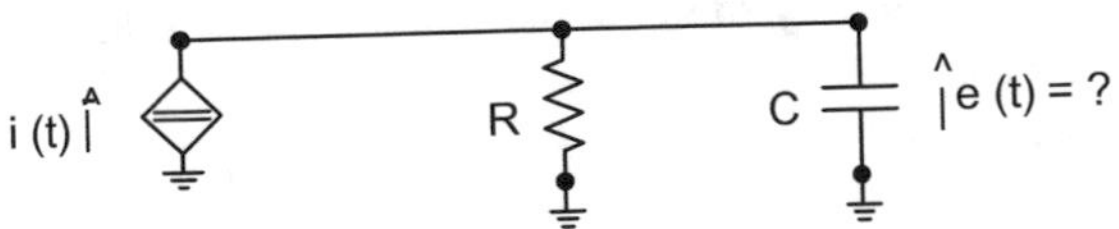

Fig. 5.45 Electric circuit of Example 5.13

$$r(t) = r_P(t) + r_H(t) = R + k\epsilon^{-t/RC}$$

Since $r(t)$ does not contain a step, since the step of the second member of the differential equation[1-11] must appear in $r'(t)$, so $r(0_+) = r(0_-) = 0 = R + k$ and $k = -R$. Soon,

$$r(t) = R\left(1 - \epsilon^{-\frac{t}{RC}}\right)U_{-1}(t).$$

(b) Where $i(t) = U_0(t), e(t) = h(t) = ?$

As the energy in the capacitor, $(1/2)Ce^2$, cannot be infinite, because this would require a double in its current, which is prohibited, the current impulse of the source, which happens at $t = 0$, cannot circulate through the circuit resistance, as this would create the voltage $RU_0(t)$, an infinite voltage, in the resistance and also in the capacitance. Thus, at $t = 0$, the current impulse of the source circulates through the capacitance. But, by Corollary 5.1, the passage of this unitary current impulse through the capacitor at $t = 0$ causes $e(0_+) = 1/C$. At $t = 0_+$, on the other hand, the impulsive current source is already zero, which means that the current source has become an open circuit for $t > 0$. Therefore, the energy $(1/2)C.(1/C)^2 = 1/2C$, existing at $t = 0_+$ in the capacitance, will be entirely dissipated in the resistance, through an anti-clockwise current in the RC loop, causing $e(\infty) = 0$. As $T = RC$, then, according to Eq. 5.3,

$$h(t) = e(t) = e(\infty) + [e(0_+) - e(\infty)]\epsilon^{-t/T} = 0 + (1/C - 0)\epsilon^{-t/RC}$$

Then,

$$h(t) = \frac{1}{C}\epsilon^{-t/RC}U_{-1}(t).$$

The impulse response can also be obtained using Eq. 5.11, that is, deriving the response to the step:

$$h(t) = \frac{d}{dt}\left[R\left(1 - \epsilon^{-\frac{t}{RC}}\right)U_{-1}(t)\right] = R\left[\frac{1}{RC}\epsilon^{-\frac{t}{RC}}U_{-1}(t) + \left(1 - \epsilon^{-\frac{t}{RC}}\right)\Big|_{t=0}U_0(t)\right]$$

Whence

$$h(t) = \frac{1}{C}\epsilon^{-t/RC}U_{-1}(t).$$

Another way to obtain $h(t)$ is the nodal equation of the circuit:

$$C\frac{d}{dt}e(t) + \frac{1}{R}e(t) = i(t)$$

Then,

$h'(t) + \frac{1}{RC}h(t) = \frac{1}{C}U_0(t)$, with $h(0_-) = 0$.

For $t > 0$,

$$\left(D + \frac{1}{RC}\right)h(t) = 0$$

$$h(t) = ke^{-t/RC}$$

Since the impulse $(1/C)U_0(t)$ must appear in $h'(t)$, then $h(t)$ must contain the step $(1/C)U_{-1}(t)$. Thus, $h(0_+) = h(0_-) + 1/C = 1/C = k$. Soon,

$$h(t) = \frac{1}{C}e^{-t/RC}U_{-1}(t).$$

Example 5.14 In the circuit shown in Fig. 5.46 the capacitance is de-energized at $t = 0_-$. (a) If the voltage source is a unitary impulse, find the expressions for current $i(t)$ and voltages $e_R(t)$ and $e_C(t)$. (b) Establish the expressions of the power $p_C(t)$ and the energy $W_C(t)$ in the capacitance.

Solution

(a) The circuit mesh equation is

$$Ri(t) + \frac{1}{C}\int i(t)dt = e(t)$$

Or,

$$i'(t) + \frac{1}{RC}i(t) = \frac{1}{R}e'(t)$$

For $e(t) = U_0(t)$, it comes that.

Fig. 5.46 Electric circuit of Example 5.14

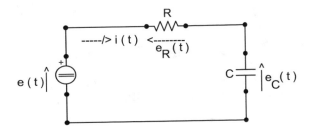

$i'(t) + \frac{1}{RC}i(t) = \frac{1}{R}U_1(t)$, with $i(0_-) = 0$.

For $t > 0$,

$$i(t) = k_1\epsilon^{-t/RC}$$

The presence of the double as a forcing function allows to write that:

$$i(t) \doteq C_0 U_0(t) + C_1 U_{-1}(t).$$

$$i'(t) \doteq C_0 U_1(t) + C_1 U_0(t) + C_2 U_{-1}(t).$$

Substituting in the differential equation, it comes that

$$C_0 U_1(t) + \left(C_1 + \frac{C_0}{RC}\right)U_0(t) + \left(C_2 + \frac{C_1}{RC}\right)U_{-1}(t) \doteq \frac{1}{R}U_1(t) + 0.U_0(t) + 0.U_{-1}(t).$$

Comparing the two sides of the above identity, we have that:

$$C_0 = \frac{1}{R}; C_1 + \frac{C_0}{RC} = C_1 + \frac{1}{R^2C} = 0 \therefore C_1 = -\frac{1}{R^2C}; C_2 - \frac{1}{R^3C^2} = 0 \therefore C_2 = \frac{1}{R^3C^2}.$$

By the conclusion drawn about the initial condition at $t = 0_+$ in Sect. 2.9 of Chap. 2,

$$i(0_+) = i(0_-) + C_1 = 0 - \frac{1}{R^2C} = -\frac{1}{R^2C}.$$

$$i'(0_+) = i'(0_-) + C_2 = 0 + \frac{1}{R^3C^2} = \frac{1}{R^3C^2}.$$

Then,

$$i(0_+) = -\frac{1}{R^2C} = k_1.$$

Therefore,

$$i(t) = \frac{1}{R}U_0(t) - \frac{1}{R^2C}\epsilon^{-t/RC}U_{-1}(t).$$

$$e_R(t) = Ri(t) = U_0(t) - \frac{1}{RC}\epsilon^{-t/RC}U_{-1}(t).$$

$$e_C(t) = e(t) - e_R(t) = \frac{1}{RC}\epsilon^{-t/RC}U_{-1}(t).$$

Or, alternatively,

$$e_C(t) = e_C(0_-) + \frac{1}{C}\int_{0_-}^{t} i(\lambda)d\lambda = 0 + \frac{1}{C}\int_{0_-}^{0_+} i(\lambda)d\lambda + \frac{1}{C}\int_{0_+}^{t} i(\lambda)d\lambda$$

$$= \frac{1}{C}\int_{0_-}^{0_+} \frac{1}{R}U_0(\lambda)d\lambda + \frac{1}{C}\int_{0_+}^{t} -\frac{1}{R^2C}\epsilon^{-\lambda/RC}d\lambda = \frac{1}{RC} + \frac{1}{RC}\epsilon^{-\lambda/RC}\Big|_{0_+}^{t}$$

$$= \frac{1}{RC}\epsilon^{-t/RC}U_{-1}(t).$$

(b) The power at the capacitance is

$$p_C(t) = e_C(t).i(t) = \frac{1}{RC}\epsilon^{-t/RC}U_{-1}(t).\left[\frac{1}{R}U_0(t) - \frac{1}{R^2C}\epsilon^{-t/RC}U_{-1}(t)\right]$$

$$= \frac{1}{R^2C}\epsilon^{-t/RC}U_{-1}(t)U_0(t) - \frac{1}{R^3C^2}\epsilon^{-2t/RC}[U_{-1}(t)]^2.$$

Since $[U_{-1}(t)]^2 = U_{-1}(t)$ and $U_{-1}(t)U_0(t) = \frac{1}{2}U_0(t)$, according to Eq. 2.67,

$$p_C(t) = \frac{1}{R^2C}\epsilon^{-t/RC}\Big|_{t=0}.\frac{1}{2}U_0(t) - \frac{1}{R^3C^2}\epsilon^{-2t/RC}U_{-1}(t)$$

Then,

$$p_C(t) = \frac{1}{2R^2C}U_0(t) - \frac{1}{R^3C^2}\epsilon^{-2t/RC}U_{-1}(t).$$

The energy in the capacitance is

$$W_C(t) = W_C(0_-) + \int_{0_-}^{t} p_C(\lambda)d\lambda = 0 + \int_{0_-}^{0_+} \frac{1}{2R^2C}U_0(\lambda)d\lambda + \int_{0_+}^{t} -\frac{1}{R^3C^2}\epsilon^{-2\lambda/RC}d\lambda$$

$$= \frac{1}{2R^2C} + \frac{1}{2R^2C}\epsilon^{-2\lambda/RC}\Big|_{0_+}^{t} = \frac{1}{2R^2C}\epsilon^{-2t/RC}U_{-1}(t).$$

Or,

$$W_C(t) = \frac{C}{2}[e_C(t)]^2 = \frac{C}{2}\left[\frac{1}{RC}\epsilon^{-t/RC}U_{-1}(t)\right]^2 = \frac{1}{2R^2C}\epsilon^{-2t/RC}U_{-1}(t).$$

On the other hand,

$$p_C(t) = \frac{d}{dt} W_C(t) = \frac{d}{dt}\left[\frac{1}{2R^2C}\epsilon^{-2t/RC}U_{-1}(t)\right] = \frac{1}{2R^2C}U_0(t) - \frac{1}{R^3C^2}\epsilon^{-2t/RC}U_{-1}(t),$$

which confirms the result previously obtained, to which the Eq. 2.67 contributed a lot.

Note that it is not possible to calculate the power either in the resistance or at the source and, consequently, also the energy. This is because in the processing of the calculation of such powers the impulse squared function, $[U_0(t)]^2$, appears, which is not known what this means. Sometimes, this is the impasse that is brought on due to the over-idealization of the circuit.

Example 5.15 Find the step response of the circuit indicated in Fig. 5.47.

Solution
Replacing the circuit seen by the capacitance with its Norton equivalent, we obtain the circuit of Fig. 5.48. (The Norton equivalent was obtained first by converting the voltage source $e(t)$, in series with the resistance R, into a current source and then using a current divider to find the Norton current).

The circuit in Fig. 5.48 has the same topology as in Example 5.13, that is, a current source in parallel with a resistance and a capacitance. When $e(t) = U_{-1}(t)$, $e_0(t) = r(t)$. Then, according to the homogeneity property of linear circuits, when multiplying the input by $1/3R$, the output is multiplied by the same factor. So, replacing the resistance R of Example 5.13 with $3R/2$ and multiplying his response by $1/3R$, it comes that:

$$r(t) = \left(\frac{1}{3R}\right)\left(\frac{3R}{2}\right)\left(1 - \epsilon^{\frac{-t}{3RC}}\right)U_{-1}(t) = \frac{1}{2}(1 - \epsilon^{-2t/3RC})U_{-1}(t).$$

This result can be confirmed using the nodal method in the circuit of Fig. 4.48:

$$C\frac{d}{dt}e_0(t) + \frac{2}{3R}e_0(t) = \frac{1}{3R}e(t)$$

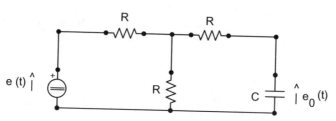

Fig. 5.47 Electric circuit of Example 5.15

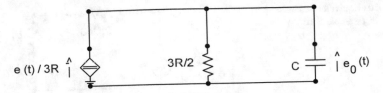

Fig. 5.48 Equivalent electric circuit of Example 5.15

For $e(t) = U_{-1}(t)$, $e_0(t) = r(t)$ and the equation becomes

$$r'(t) + \frac{2}{3RC} r(t) = \frac{1}{3RC} U_{-1}(t), \text{ with } r(0_-) = 0.$$

For $t > 0$,

$$\left(D + \frac{2}{3RC}\right) r(t) = \frac{1}{3RC}$$

$$r_H(t) = k\epsilon^{-2t/3RC}$$

$$r_P(t) = \frac{1}{D + 2/3RC} \left(\frac{1}{3RC} \epsilon^{0 \times t}\right) = \frac{1}{2}$$

$$r(t) = r_P(t) + r_H(t) = \frac{1}{2} + k\epsilon^{-2t/3RC}$$

By the differential equation,

$$r(0_+) = r(0_-) = 0 = \frac{1}{2} + k \text{ and } k = -\frac{1}{2}. \text{ Then,}$$

$$r(t) = \frac{1}{2}(1 - \epsilon^{-2t/3RC}) U_{-1}(t).$$

Example 5.16 Find the impulse response $h(t)$ of the circuit in Fig. 5.49, for $e_0(t)$ as an output.

Fig. 5.49 Electric circuit of Example 5.16

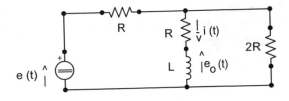

Fig. 5.50 Equivalent electric circuit of Example 5.16

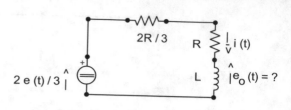

Solution

Replacing the circuit seen by the $R - L$ branch with its Thevenin equivalent, it results in Fig. 5.50.

Equating the mesh, it comes that:

$$L\frac{d}{dt}i(t) + \frac{5R}{3}i(t) = \frac{2}{3}e(t)$$

$$i'(t) + \frac{5R}{3L}i(t) = \frac{2}{3L}e(t) = \frac{2}{3L}U_0(t), \text{ with } i(0_-) = 0.$$

For $t > 0$,

$$(D + 5R/3L)i(t) = 0$$

$$i(t) = k\epsilon^{-5Rt/3L}$$

Analyzing the two members of the differential equation, it is concluded that $i'(t)$ must contain the impulse $\frac{2}{3L}U_0(t)$. Therefore, $i(t)$ must contain the step $\frac{2}{3L}U_{-1}(t)$. So $i(0_+) = \frac{2}{3L} = k$ and $i(t) = \frac{2}{3L}\epsilon^{-5Rt/3L}U_{-1}(t)$.

$$e_0(t) = h(t) = L\frac{d}{dt}i(t) = \frac{2}{3}\frac{d}{dt}\left[\epsilon^{-5Rt/3L}U_{-1}(t)\right]$$

$$= \frac{2}{3}\left[\frac{-5R}{3L}\epsilon^{-5Rt/3L}U_{-1}(t) + \epsilon^{-5Rt/3L}\Big|_{t=0}\cdot U_0(t)\right]$$

$$h(t) = \frac{2}{3}U_0(t) - \frac{10R}{9L}\epsilon^{-5Rt/3L}U_{-1}(t).$$

Example 5.17 Find the impulse response $h(t)$ of the circuit containing voltage-controlled current source in Fig. 5.51.

Solution

By the nodal method, with the lower node taken as a reference, the equation of the voltage node e_2 is:

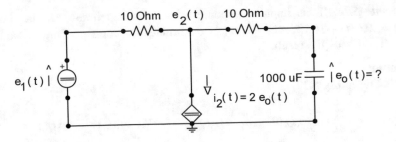

Fig. 5.51 Electric circuit of Example 5.17

$$\frac{e_2 - e_1}{10} + \frac{e_2 - e_0}{10} + i_2 = 0$$

And the e_0 voltage node equation is:

$$\frac{e_0 - e_2}{10} + 10^{-3}\frac{d}{dt}e_0 = 0$$

Replacing i_2 with $2e_0$ and grouping the terms of the two equations, and simplifying, comes that:

$$2e_2 + 19e_0 = e_1$$

$$-100e_2 + (D + 100)e_0 = 0$$

$$\begin{vmatrix} 2 & 19 \\ -100 & D+100 \end{vmatrix} e_0(t) = \begin{vmatrix} 2 & e_1(t) \\ -100 & 0 \end{vmatrix}$$

$(2D + 2100)e_0(t) = 100\, e_1(t)$

Considering that $e_1(t) = U_0(t)[\text{V}]$ and $e_0(t) = h(t)[\text{V}]$, the equation becomes:

$h'(t) + 1050h(t) = 50U_0(t)$, with $h(0_-) = 0$.

For $t > 0$,

$$(D + 1050)h(t) = 0 \text{ and } h(t) = ke^{-1050t}$$

Analyzing the two members of the differential equation, we conclude that $h(0_+) = 50 = k$.
Finally,

$$h(t) = 50\epsilon^{-1050t}U_{-1}(t)[\text{V}].$$

Example 5.18 If the impulse response of a linear and time-invariant circuit is $h(t) = 2U_0(t) + 3(\epsilon^{-4t} - 2\epsilon^{-t})U_{-1}(t)$, determine the corresponding response to step $r(t)$. Check the result.

Solution

According to Eq. 5.12,

$$r(t) = \int_{0_-}^{t} h(\lambda)d\lambda = \int_{0_-}^{t} \left[2U_0(\lambda) + 3(\epsilon^{-4\lambda} - 2\epsilon^{-\lambda})U_{-1}(\lambda)\right]d\lambda$$

$$= 2\int_{0_-}^{0_+} U_0(\lambda)d\lambda + 3\int_{0_+}^{t} (\epsilon^{-4\lambda} - 2\epsilon^{-\lambda})d\lambda = 2 + \left(6\epsilon^{-\lambda} - \frac{3}{4}\epsilon^{-4\lambda}\right)\Big|_{0_+}^{t}$$

$$= 2 + 6\epsilon^{-t} - \frac{3}{4}\epsilon - 6 + \frac{3}{4} = -\frac{13}{4} + 6\epsilon^{-t} - \frac{3}{4}\epsilon^{-4t}, \text{ for } t > 0.$$

As $r(t) = 0$ for $t < 0$,

$$r(t) = \frac{1}{4}\left(-13 + 24\epsilon^{-t} - 3\epsilon^{-4t}\right)U_{-1}(t).$$

Verification:

$$h(t) = \frac{d}{dt}r(t) = (-6\epsilon^{-t} + 3\epsilon^{-4t})U_{-1}(t) + \frac{1}{4}\left(-13 + 24\epsilon^{-t} - 3\epsilon^{-4t}\right)\Big|_{t=0}.U_0(t)$$

$$= 2U_0(t) + 3(\epsilon^{-4t} - 2\epsilon^{-t})U_{-1}(t).$$

Example 5.19 Determine the current $i(t)$ in the circuit of Fig. 5.52, knowing that the inductance is de-energized at $t = 0_-$, and that the source voltage is the rectangular pulse.

$$e(t) = \begin{cases} 0[V[, \text{ for } t < 0[s] \\ 1[V], \text{ for } 0[s] < t < 1[s] \\ 0[V], \text{ for } t > 1[s] \end{cases}$$

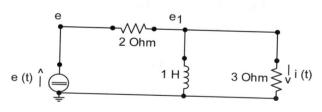

Fig. 5.52 Electric circuit of Example 5.19

Solution

The source voltage, decomposed in steps, is:

$$e(t) = U_{-1}(t) - U_{-1}(t-1)[V].$$

As the circuit is linear and time-invariant, when the input is delayed by a certain time t_0, the output is also delayed by the same t_0, and the Superposition Theorem allows to establish that:

$$i(t) = r(t) - r(t-1)[A].$$

Therefore, the current $i(t)$ will be determined as soon as the step response $r(t)$ of the given circuit is calculated.

The tension node equation $e_1(t)$ is:

$$\frac{e_1 - e}{2} + \int e_1 dt + \frac{e_1}{3} = 0$$

$$e_1'(t) + \frac{6}{5}e_1(t) = \frac{3}{5}e'(t)$$

Considering that $e(t) = U_{-1}(t), e'(t) = U_0(t)$, and the differential equation becomes:

$$e_1'(t) + \frac{6}{5}e_1(t) = \frac{3}{5}U_0(t), \text{ with } e_1(0_-) = 0.$$

$$\left(D + \frac{6}{5}\right)e_1(t) = 0, \text{ for } t > 0.$$

$$e_1(t) = ke^{-6t/5}$$

Upon inspection of the differential equation, it is concluded that $e_1(0_+) = \frac{3}{5} = k$. Therefore,

$$e_1(t) = \frac{3}{5}\epsilon^{-6t/5}U_{-1}(t)[V].$$

So, for $e(t) = U_{-1}(t)$,

$$r(t) = \frac{e_1(t)}{3} = \frac{1}{5}\epsilon^{-6t/5}U_{-1}(t)[A].$$

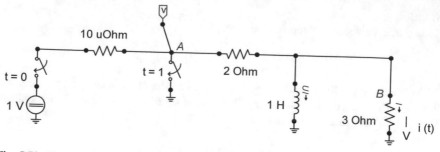

Fig. 5.53 Electric circuit of Example 5.19 for digital solution

So:

$$i(t) = r(t) - r(t-1) = \frac{1}{5}\epsilon^{-6t/5}U_{-1}(t) - \frac{1}{5}\epsilon^{-6(t-1)/5}U_{-1}(t-1)$$

$$= \frac{1}{5}\epsilon^{-6t/5}U_{-1}(t) - \frac{1}{5}\epsilon^{6/5}.\epsilon^{-6t/5}U_{-1}(t-1)$$

$$= 0.20\epsilon^{-1.2t}U_{-1}(t) - 0.66\epsilon^{-1.2t}U_{-1}(t-1)[A]$$

Summarizing this result, it comes that:

$$i(t) = \begin{cases} 0[A],\ for\ t<0[s] \\ 0.20\epsilon^{-1.2t},\ for\ 0[s]<t<1[s] \\ -0.46\epsilon^{-1.2t},\ for\ t>1[s]. \end{cases}$$

Then, $i(0_+) = 0.20[A]; i(1_-) = 0.06[A]; i(1_+) = -0.14[A]; T = 0.83[s]$.
    The circuit constructed for the numerical solution by ATPDraw is shown in Fig. 5.53.
    The small resistance of 10 [μΩ] was introduced to prevent the two switches from having a common terminal. Otherwise, the program would not produce the expected result. The results obtained are recorded in Figs. 5.54 and 5.55.

Example 5.20 A linear and time-invariant system, of input $x(t)$ and output $y(t)$, has the impulse response $h(t) = U_0(t) - \epsilon^{-2t}U_{-1}(t)$. Find the output waveform $y(t)$ for the input signal $x(t) = U_{-1}(t-1) - U_{-1}(t-2)$.

Solution

$$r(t) = \int_{-\infty}^{t} h(\lambda)d\lambda = \int_{-\infty}^{t} U_0(\lambda)d\lambda + \left[\int_{0_+}^{t} -\epsilon^{-2\lambda}d\lambda\right]U_{-1}(t)$$

$$= U_{-1}(t) + \left[\frac{\epsilon^{-2\lambda}}{2}\right]\Bigg|_{0_+}^{t} U_{-1}(t).$$

$$r(t) = \frac{1}{2}\left(1 + \epsilon^{-2t}\right)U_{-1}(t).$$

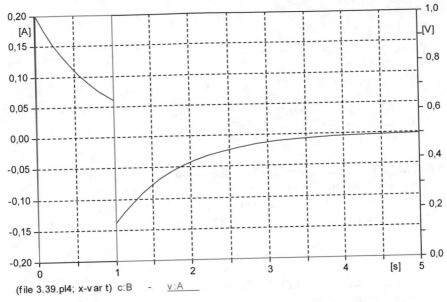

Fig. 5.54 Transient voltage and current digital solution for circuit of Example 5.19

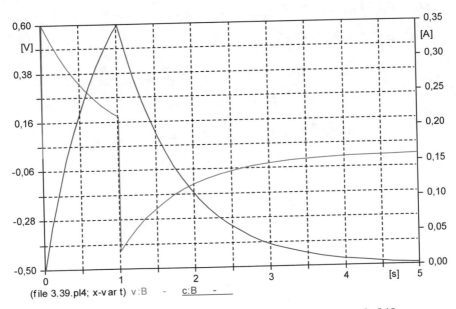

Fig. 5.55 Transient voltage and current digital solution for circuit of Example 5.19

Therefore,

$$y(t) = r(t-1) - r(t-2) = \frac{(1+\epsilon^2\epsilon^{-2t})}{2}U_{-1}(t-1) - \frac{(1+\epsilon^4\epsilon^{-2t})}{2}U_{-1}(t-2)$$

$$y(t) = \begin{cases} 0 \ for \ t \leq 1 \\ \frac{(1+\epsilon^2\epsilon^{-2t})}{2} \ for \ 1 \leq t \leq 2 \\ \frac{(\epsilon^2-\epsilon^4)\epsilon^{-2t}}{2} \ for \ t \geq 2. \end{cases}$$

Example 5.21 If all capacitances in the circuit shown in Fig. 5.56 are de-energized before $t = 0[s]$, this time when the switch is closed, (a) find the current and voltage in each capacitance for $t \geq 0[s]$; (b) find the voltages in the capacitances without first calculating the respective currents; (c) proceed to the numerical solution of the problem.

Solution

(a) The circuit can be remodeled with the introduction of the step function in the voltage source and the adoption of the mesh currents $i_1(t)$ and $i_2(t)$, as shown in Fig. 5.57.

Mesh 1: $\frac{10^6}{6}\int i_1(t)dt + \frac{10^6}{12}\int i_1(t)dt + \frac{10^6}{5}\int[i_1(t) - i_2(t)]dt = 100U_{-1}(t)$

Mesh 2: $\frac{10^6}{1}\int i_2(t)dt + \frac{10^6}{5}\int[i_2(t) - i_1(t)]dt = 0$

$$27i_1(t) - 12i_2(t) = 6.10^{-3}U_0(t)$$

$$-2i_1(t) + 12i_2(t) = 0.$$

Solving the system for the currents, it results that:

$$i_1(t) = 240.10^{-6}U_0(t)[A]$$

$$i_2(t) = 40.10^{-6}U_0(t)[A]$$

$$i_3(t) = i_1(t) - i_2(t) = 200.10^{-6}U_0(t)[A].$$

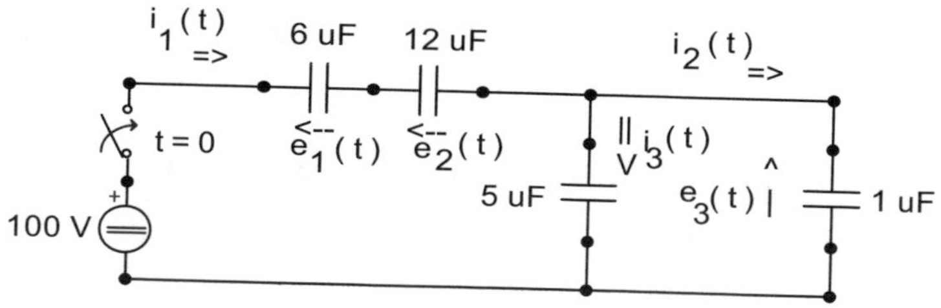

Fig. 5.56 Electric circuit of Example 5.21

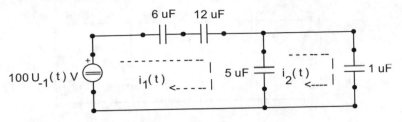

Fig. 5.57 Electric circuit of Example 5.21 for analytical solution

$$e_1(t) = e_1(0_-) + \frac{10^6}{6} \int_{0_-}^{t} i_1(\lambda)d\lambda = 0 + \frac{10^6}{6} \int_{0_-}^{t} 240.10^{-6}U_0(\lambda)d\lambda = 40U_{-1}(t)[V]$$

$$e_2(t) = e_2(0_-) + \frac{10^6}{12} \int_{0_-}^{t} i_1(\lambda)d\lambda = 0 + \frac{10^6}{12} \int_{0_-}^{t} 240.10^{-6}U_0(\lambda)d\lambda = 20U_{-1}(t)[V]$$

From the original circuit: $e_3(t) = 100U_{-1}(t) - e_1(t) - e_2(t)$. So,

$$e_3(t) = 40U_{-1}(t)[V].$$

(b) To obtain the voltages in the capacitances directly, without first calculating the respective currents, the equivalent capacitance must first be found. From there, find the charge and then use the charge to find the voltages $e_1(t)$ and $e_2(t)$ in the capacitors $6[\mu F]$ and $12[\mu F]$, which have the same charge, because they are in series, and in series with the voltage source. Note that the equivalent of capacitances in parallel is the sum of them.

$$C_{eq} = \frac{1}{1/6 + 1/12 + 1/6} = 2.4[\mu F]$$

The charge at $t = 0_+$ is: $q(0_+) = C_{eq}.e(0_+) = 2.4 \times 10^{-6} \times 100 = 240[\mu C]$.

This is the same charge on the 6 $[\mu F]$ and $12[\mu F]$ capacitors. So,

$e_1(0_+) = q(0_+)/C_1 = \frac{240 \times 10^{-6}}{6 \times 10^{-6}} = 40[V]$, that is, $e_1(t) = 40U_{-1}(t)[V]$

$e_2(0_+) = q(0_+)/C_2 = \frac{240 \times 10^{-6}}{12 \times 10^{-6}} = 20[V]$, that is, $e_2(t) = 20U_{-1}(t)[V]$

$e_3(0_+) = e(0_+) - e_1(0_+) - e_2(0_+) = 40[V]$, that is, $e_3(t) = 40U_{-1}(t)[V]$.

(c) The solution of this example by the ATPDraw program, through the circuit of Fig. 5.58, presented the results shown in Figs. 5.59, 5.60 and 5.61.

The graph of $i_1(t)$ explains the impossibility of the referred program to represent the current impulse $240.10^{-6}U_0(t)[A]$, present in the current $i_1(t)$ Obviously, this should not be surprising, due to the fact that "the impulse has a Point base, infinite

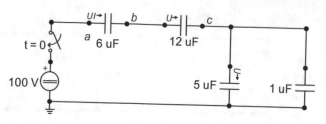

Fig. 5.58 Electric circuit of Example 5.21 for digital solution

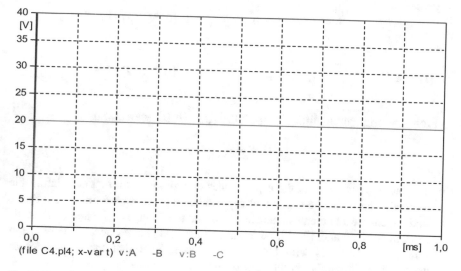

Fig. 5.59 Transient voltage and current digital solution for circuit of Example 5.21

height and an area of $240.10^{-6}[C]$"A zoom in this graph, close to $t = 0[s]$, on the right, for $\Delta t = 2$ [µs], shows the result shown in Fig. 5.62, in which what is called Numerical Oscillation is evident.

It is extremely important to point out that although the program cannot represent the current impulse $240.10^{-6}U_0(t)[A]$, it allows to obtain the value of the electric charge carried by this current impulse, since it is simply the area of the first base right triangle equal to $\Delta t = 2$ [µs], that is: $2.10^{-6}[s] \times 240[A]/2 = 240.10^{-6}[C]$.

Circuits containing controlled sources can, eventually, present indefinitely increasing responses. In these cases, the equivalent Thevenin resistance seen from the capacitor or inductor terminals is negative. This makes the time constant of the circuit also negative and, because of that, the resulting voltages and currents increase indefinitely. In a real circuit, however, there comes a time when the response reaches a limit value at which a component ends up being destroyed or enters a state of saturation, which prevents any increase in voltage or current.

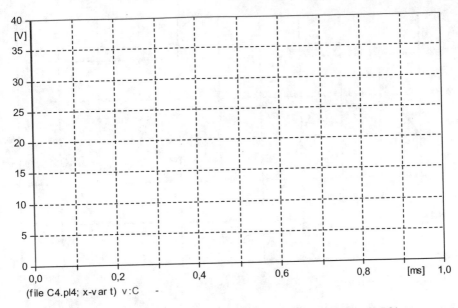

(file C4.pl4; x-var t) v:C   -

Fig. 5.60 Transient voltage and current digital solution for circuit of Example 5.21

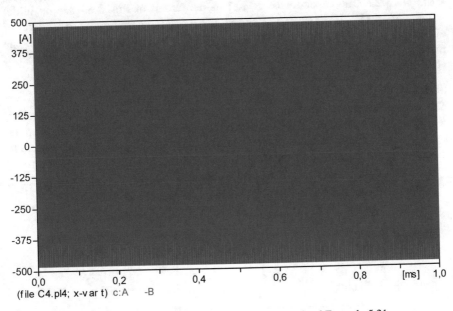

(file C4.pl4; x-var t) c:A   -B

Fig. 5.61 Numerical oscillation in the digital solution for circuit of Example 5.21

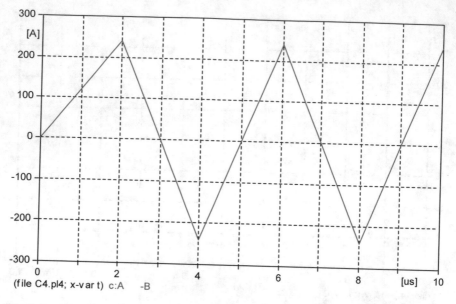

Fig. 5.62 Zoom at numerical oscillation in the digital solution for circuit of Example 5.21

Example 5.22 The capacitor of $10[\mu F]$ in the circuit of Fig. 5.63 has an initial voltage of 15 [V]. If the switch is closed at t = 0 [s], determine the expression of $e(t)$, for $t \geq 0[s]$. Assuming the capacitor's dielectric collapses when its voltage reaches 120 [V], determine how many milliseconds will elapse before the capacitor short-circuits.

Solution

The determination of the equivalent Thèvenin resistance seen by the capacitor, $R_{TH}[k\Omega]$, due to the presence of the controlled source, has to be done, for example, by injecting an independent current source $i_1[mA]$ in place of the capacitor, determining the corresponding voltage $e_1[V]$ at her terminals and calculating:

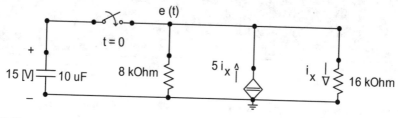

Fig. 5.63 Electric circuit of Example 5.22

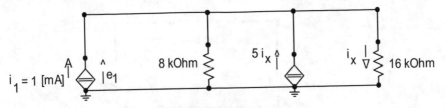

Fig. 5.64 Electric circuit of Example 5.22 for analytical solution

$$R_{TH}[k\Omega] = e_1[V]/i_1[mA].$$

Arbitrating, however, that $i_1 = 1[mA]$, $R_{TH}[k\Omega] = e_1[V/mA]$. This procedure is illustrated in Fig. 5.64.

The nodal method allows you to write that

$$i_1 = 1 = \frac{e_1}{8} + \frac{e_1}{16} - 5 \times \frac{e_1}{16} = \frac{2+1-5}{16}.e_1 = -\frac{1}{8}.e_1$$

Therefore,

$$e_1 = -[k\Omega] = R_{TH}.$$

After calculating the Thevenin resistance, seen by the capacitor, the original circuit can be exchanged for the one in Fig. 5.65.

The nodal equation is

$$10.10^{-6}\frac{de}{dt} + \frac{e}{-8.10^3} = 0$$

$e'(t) - 12.5e(t) = 0$, with $e(0_+) = e(0_-) = 15[V]$, whose solution is

$$e(t) = 15\epsilon^{12.5t}[V], \text{ for } t \geq 0[s].$$

For $e(t) = 120[V]$, it turns out that $120 = 15\epsilon^{12.5t}$ or $12.5t = ln8$ and $t = 166.36[ms]$.

For the numerical solution by ATPDraw, the circuit shown in Fig. 5.66 was used, with $\Delta t = T/100 = 0.8$ [ms] and $t_{max} = 200[ms]$. The voltage $e(t)$ is shown in Fig. 5.67.

Fig. 5.65 Equivalent electric circuit of Example 5.22 for analytical solution

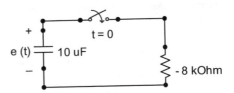

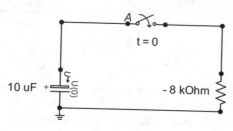

Fig. 5.66 Electric circuit of Example 5.22 for digital solution

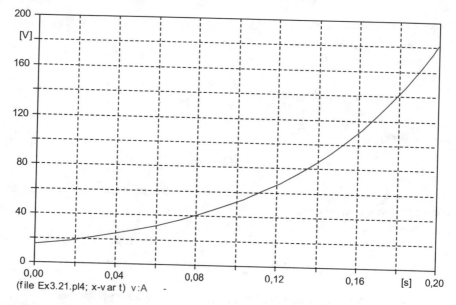

Fig. 5.67 Transient voltage digital solution for circuit of Example 5.22

With the plotter cursor, it can be detected that when $t = 208 \times \Delta t = 166.4$ [ms], $e = 119.47[V]$, and that when $t = 209 \times \Delta t = 167.2[ms]$, $e = 120.67[V]$.

Example 5.23 The circuit shown in Fig. 5.68 has an initial electrical charge of 25 $[\mu C]$ in the capacitor, with the polarity as shown. The switch is closed at $t = 0[s]$, applying the voltage $e(t) = 100\sin(1,000t + 30°)[V]$. Find the current $i(t)$, for $t > 0[s]$.

Fig. 5.68 Electric circuit of Example 5.23

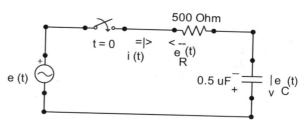

Solution

The charge is a continuous function of time, because there is no impulse in the current. So,

$$q(0_+) = Ce_C(0_+) = q(0_-)$$

So,

$$e_C(0_+) = \frac{q(0_-)}{C} = \frac{25 \times 10^{-6}}{0.5 \times 10^{-6}} = 50[V].$$

$$e(0_+) = 100 \sin 30° = 50[V].$$

$$e_R(0_+) = e_C(0_+) + e(0_+) = 100[V].$$

$$i(0_+) = \frac{e_R(0_+)}{R} = \frac{100}{500} = 0.2[A] = 200[mA].$$

The mesh equation for the circuit is:

$$Ri(t) + \frac{1}{C}\int i(t)dt = e(t)$$

$$i'(t) + \frac{1}{RC}i(t) = \frac{1}{R}e'(t)$$

$$i'(t) + 4,000i(t) = 200\cos(1,000t + 30°)$$

The solution to the homogeneous equation is:

$$i_H(t) = ke^{-4,000t}[mA]$$

The particular solution is obtained by the phasor method:

$$(j1,000 + 4,000)\dot{I}_P = 200\angle 30°$$

$$\dot{I}_P = \frac{200\angle 30°}{4,123.11\angle 14.04°} = 48.51\angle 15.96° \text{ [mA]}$$

$$i_P(t) = 48.51\cos(1,000t + 15.96°)[mA]$$

$$i(t) = i_H(t) + i_P(t) = ke^{-4,000t} + 48.51\cos(1,000t + 15.96°)$$

$$i(0_+) = 0.2[A] = 200[mA] = k + 48.51\cos 15.96°$$

$$k = 153.36[mA]$$

$$i(t) = 153.36e^{-4.000t} + 48.51\cos(1,000t + 15.96°)[mA].$$

For the numerical solution by ATPDraw, the circuit of Fig. 5.69 was used.

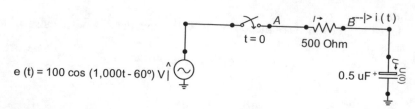

Fig. 5.69 Electric circuit of Example 5.23 for digital solution

As the program operates with the cosine function, it was necessary to convert the voltage source to $e(t) = 100\cos(1,000t + 30° - 90°) = 100\cos (1,000t - 60°)[V]$.

Where $\omega = 2\pi f = 1,000 \left[\frac{rad}{s}\right]$, $f = 159.15[Hz]$ and $T = 1/f = 6.28[ms]$.

$$T_C = RC = 500 \times 0.5 \times 10^{-6} = 0.25[ms] = 1/4,000.$$

Considering that $T/100 = 62.8[\mu s]$ and $T_C/100 = 2.5[\mu s]$, $\Delta t = 2.5[\mu s]$ was adopted.

As $5T_C = 1.25[ms] << T$, $t_{max} = 20[ms] \approx 3T$ was adopted to be able to visualize, in the plot, both the transient and the permanent regime.

To enter the initial voltage of 50[V], down, in capacitance, the upper plate, with the + mark, received the value of −50[V], and the lower plate, which is grounded, of course, received the zero value. The simulation result is recorded in Fig. 5.70.

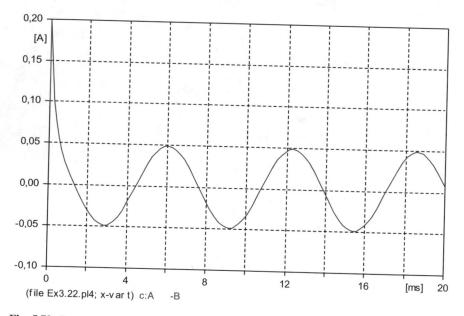

(file Ex3.22.pl4; x-var t) c:A   -B

Fig. 5.70 Transient current digital solution for circuit of Example 5.23

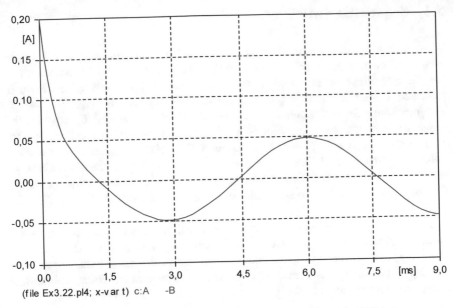

Fig. 5.71 Zoom at transient current digital solution for circuit of Example 5.23

As the transient component has a very short duration, a zoom was applied in the vicinity of $t = 0[s]$, on the right, and the result is shown in Fig. 5.71.

Figure 5.72 shows the source voltages, $e(t)$, and the capacitance voltage $e_C(t)$.

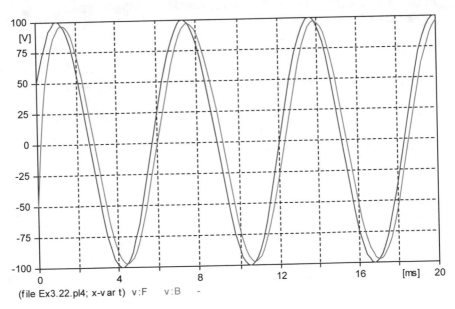

Fig. 5.72 Transient voltage digital solution for circuit of Example 5.23

5.4   Sequential Switching

When switching occurs more than once in the same circuit, there is what is called sequential switching. A two-position switch, for example, can be turned on at $t = 0[s]$ and turned off at $t = t_0[s]$, or several switches can be opened or closed in sequence. Logically, the time reference for all switches can no longer be $t = 0[s]$. The procedure for determining voltages and currents generated by a switching sequence uses the same techniques as the previous examples: calculating the expressions of $e(t)$ and $i(t)$ in a circuit element, for a given position of the switches, valid for a first interval time, for example $0[s] < t < t_0[s]$, then these solutions are used to determine the initial conditions for the next position of the switches, that is, the conditions at $t = t_{0_+}$. In sequential switching problems, Eqs. 5.9 and 5.10 are used.

It is necessary to keep in mind that any electrical quantity of a real circuit, except inductive currents and capacitive voltages, can vary instantly at the moment of switching, that is, undergo discontinuity. Thus, first calculating inductive currents and capacitive voltages is even more important in sequential switching problems. In many problems, it helps to design the circuit valid for each time interval.

Example 5.24   The three-position switch of the circuit shown in Fig. 5.73 is closed at position 1 at $t = 0[s]$ and then it is instantly moved to position 2 after exactly one time constant, that is, $t_0 = T[s]$. Obtain the expression of the current $i(t)$, for $t > 0[s]$, knowing that the capacitor is discharged at $t = 0_-[s]$.

Solution

The time constant is: $T = RC = 500 \times 0.5 \times 10^{-6}[s] = 250[\mu s]$; $1/T = 4,000[s^{-1}]$.
   For $0 < t < T$:

$$e(0_+) = e(0_-) = e(0) = 0[V].$$

$e(\infty) = 20[V]$, if the switch was not moved to position 2 at $t = T$.

$$\mathbf{e(t)} = e(\infty) + [e(0_+) - e(\infty)]\epsilon^{-t/T} = \mathbf{20 - 20\epsilon^{-4,000t}[V], \ for\ 0 \le t \le T[s].}$$

$$i(t) = \frac{20 - e(t)}{500} = 40\epsilon^{-4,000t}[mA], \ for\ 0 < t < T[s]; i(T_-) = 14.72[mA].$$

Fig. 5.73 Electric circuit of Example 5.24

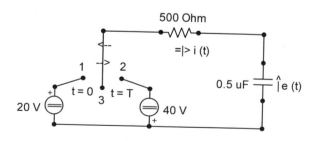

Fig. 5.74 Electric circuit of Example 5.24 for digital solution

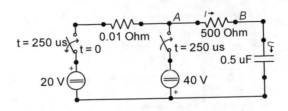

For $t = T_-$:

$$e(T_-) = 20 - 20\epsilon^{-1} = 12.642[\text{V}]$$

For $t > T$:

$$e(T_+) = e(T_-) = 12.642[\text{V}]; e(\infty) = -40[\text{V}]; T = 250[\mu s]; 1/T = 4,000[s^{-1}].$$

$$e(t) = e(\infty) + [e(T_+) - e(\infty)]\epsilon^{-(t-T)/T} = -40 + (12.642 + 40)\epsilon^1\epsilon^{-4,000t}$$

$$e(t) = -40 + 143.1\epsilon^{-4,000t}[\text{V}], \text{ for } t \geq T[s].$$

$$e(t) + 500i(t) + 40 = 0$$

$$i(t) = \frac{-40 - e(t)}{500} = -286.2\epsilon^{-4,000t}[\text{mA}], \text{ for } t > T[s]; i(T_+) = -105.29[\text{mA}].$$

The current circulates for a time approximately equal to $6T = 1.5[ms]$.

Figure 5.74 shows the circuit built in ATPDraw to perform the numerical solution of this Example 5.24. It is very important to note the need to introduce a small resistance of 0.01 [Ω], to prevent the two switches from having a common terminal, in which case the program does not process the solution.

The waveforms of current $i(t)$ and voltage $e(t)$ are recorded in the plot in Fig. 5.75.

Example 5.25 The two circuit switches shown in Fig. 5.76 were closed a long time ago, which means that it is operating in a steady state. At $t = 0[s]$, switch 1 is opened. After $60[ms]$, switch 2 is opened.

(a) Determine $i(0_-)$ and $e(0_-)$.
(b) Determine $i(t)$ and $e(t)$ for $0 < t < 60[ms]$.
(c) Determine $i(t)$ and $e(t)$ for $t > 60[ms]$.
(d) Calculate the percentage of the initial energy stored in the $120[mH]$ inductor that was dissipated in the $6[\Omega]$ resistor, on the right.
(e) Find the voltage $e_{ch1}(t)$ at the terminals of switch1, for $t > 0[s]$, with positive polarity at the left terminal.

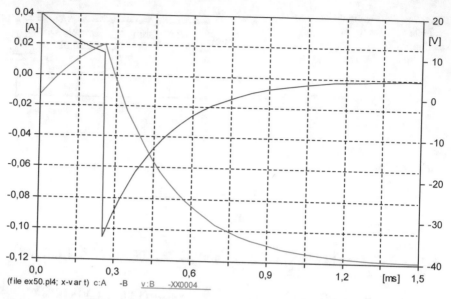

Fig. 5.75 Transient voltage and current digital solution for circuit of Example 5.24

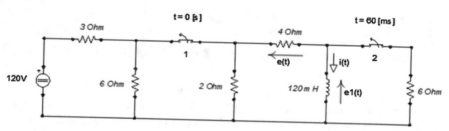

Fig. 5.76 Electric circuit of Example 5.25

Solution

(a) At $t = 0_-$ [s], switches 1 and 2 are still closed and the circuit operates in a steady state, which means that the inductor, replaced by a short circuit (because the voltage source is constant), it is also shorting the resistance of 6 [Ω] on the right, reducing the circuit to that indicated in Fig. 5.77.

By means of source transformation and combination of resistors in parallel, the circuit of Fig. 5.77 is reduced to that shown in Fig. 5.78, containing a Norton equivalent.

By current divider, $i(0_-) = \frac{1 \times 40}{1+4} = 8$ [A] and

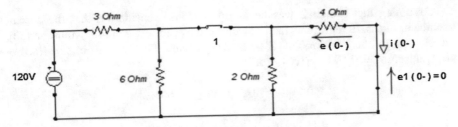

Fig. 5.77 Electric circuit of Example 5.25 for analytical solution at $t = 0_-$

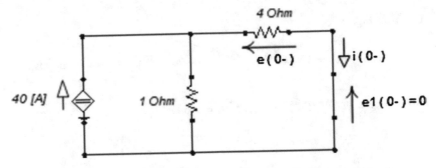

Fig. 5.78 Equivalent electric circuit of Example 5.25 for analytical solution at $t = 0_-$

$$e(0_-) = 4i(0_-) = 32[V].$$

(b) At $t = 0_+$ [s], $i(0_+) = i(0_-) = 8[A]$, switch 1 is already open and switch 2 is still closed, allowing to change the original circuit of Figs. 5.76 and 5.79, with the inductance replaced by the current source of 8[A].

Therefore, $e(0_+) = 4 \times 4 = 16[V]$ and $e_1(0_+) = -6 \times 4 = -24[V]$.

$$R_{TH} = \frac{6 \times 6}{6+6} = 3[\Omega]; T = \frac{L}{R_{TH}} = \frac{120 \times 10^{-3}}{3} = 40[ms] \text{ and } 1/T = 25[s^{-1}].$$

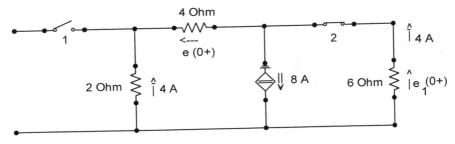

Fig. 5.79 Electric circuit of Example 5.25 for analytical solution at $t = 0_+$

Assuming that switch 2 was never opened, and considering that the current source in Fig. 5.79 represents the inductance, the energy stored in it would be fully dissipated in the resistors of 2, 4 and 6 Ohms. This would imply $i(\infty) = 0[A]$ and $e(\infty) = 0[V] = e_1(\infty)$. So,

$$i(t) = 8\epsilon^{-25t}[A], \text{ for } 0 \le t \le 60[ms].$$

$$e(t) = 16\epsilon^{-25t}[V], \text{ for } 0 < t < 60[ms].$$

$$e_1(t) = -24\epsilon^{-25t}[V], \text{ for } 0 < t < 60[ms].$$

Or, from Fig. 5.76,

$$e_1(t) = L\frac{d}{dt}i(t) = 120.10^{-3}.\frac{d}{dt}8\epsilon^{-25t} = -24\epsilon^{-25t}[V], \text{ for } 0 < t < 60[ms].$$

(c) $i(60ms_-) = 8\epsilon^{-25 \times 60 \times 10^{-3}} = 1.785[A] = i(60ms_+).$

$$e(60ms_-) = 16\epsilon^{-25 \times 60 \times 10^{-3}} = 3.570[V].$$

$$e_1(60ms_-) = -24\epsilon^{-25 \times 60 \times 10^{-3}} = -5.355[V].$$

For $t > 60[ms]$, switches 1 and 2 are open and the circuit is reduced to that of Fig. 5.80.

As $i(\infty) = e(\infty) = 0$, and the new time constant is $T' = \frac{L}{(R_{TH})'} = \frac{120 \times 10^{-3}}{6} = 20[ms]$ according to the expression (5.10),

$$i(t) = 1.785\epsilon^{-(t-60 \times 10^{-3})/20 \times 10^{-3}} = 1.785\epsilon^{-50(t-60 \times 10^{-3})} = 1.785\epsilon^3\epsilon^{-50t}$$

$$i(t) = 35.853\epsilon^{-50t}[A], \text{ for } t \ge 60[ms]; \ i(60ms_+) = 1.785[A].$$

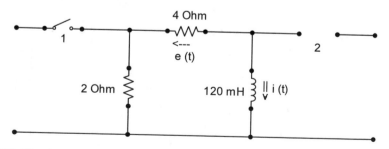

Fig. 5.80 Electric circuit of Example 5.25 for analytical solution

$$e(t) = 4i(t) = 143.412\epsilon^{-50t}[V], \text{ for } t > 60[ms]; \; e(60ms_+) = 7.140[V].$$

$$e_1(t) = -6i(t) = -215.118\epsilon^{-50t}[V], \text{ for } t > 60[ms]; \; e_1(60ms_+) = -10.71[V].$$

(d) The 6 [Ω] resistor on the right in Fig. 5.76 is in the circuit only during the initial 60[ms] of its operation, because for $t < 0[s]$ it is short-circuited by the inductance, and for $t > 60[ms]$, switch 2, open, removes it from the circuit. During this interval, the voltage across the resistor is

$$e_1(t) = -24\epsilon^{-25t}[V], \text{ for } 0 < t < 60[ms].$$

The power dissipated in said resistor is

$$p_1(t) = \frac{[e_1(t)]^2}{6} = 96\epsilon^{-50t} [W], \text{ for } 0 < t < 60[ms].$$

Thus, the energy dissipated in the resistor is

$$W = \int\limits_{0}^{0.060} 96\epsilon^{-50t} dt = \frac{96}{-50} \epsilon^{-50t}\Big|_0^{0.060} = 1.92(1 - \epsilon^{-3}) = 1,824.41[mJ].$$

The initial energy stored in the 120[mH] inductor is

$$W_L = \frac{1}{2}Li(0)^2 = \frac{1}{2}(0.120)(8)^2 = 3840[mJ].$$

Therefore, $\frac{1824.41}{3840} \times 100 = 47.51\%$ of the initial energy stored in the 120[mH] inductor is dissipated in the 6 [Ω] resistor on the right.

(e) For $t < 0$, the switch 1 is closed and $e_{ch1}(t) = 0[V]$.

For $0 < t < 60[ms]$, switch 1 is already open and the voltage on resistor 2 [Ω] is worth $2 \times e(t)/4 = (1/2) \times 16\epsilon^{-25t} = 8\epsilon^{-25t}$ [V], with positive polarity at its lower terminal. On the other hand, the voltage in the 6 [Ω] resistor from the left, with switch 1 open, calculated by voltage divider, is 80 [V], positive at the left terminal of switch 1. So, $e_{ch1}(t) = 80 + 8\epsilon^{-25t}[V]$, so that: $e_{ch1}(0_+) = 88[V]$ and $e_{ch1}(60ms_-) = 81.785[V]$.

For $t > 60[ms]$, switch 2 is also already open and the voltage at the left pole of switch 1 remains at 80[V]. But the voltage at the 2[Ω] resistor is now worth $2 \times e(t)/4 = (1/2) \times 143.4122\epsilon^{-50t} = 71.706\epsilon^{-50t}[V]$, which is the negative of the voltage at the right pole of switch1. So, $e_{ch1}(t) = 80 + 71.706\epsilon^{-50t}[V]$, so that $e_{ch1}(60ms_+) = 83.570[V]$ and $e_{ch1}(\infty) = 80[V]$.

Example 5.26 Switch 1 in the circuit of Fig. 5.81 is closed at time $t = 0$. After 300[ms], switch 2 is opened and, 60[ms] later, switch 3 is opened. Using the

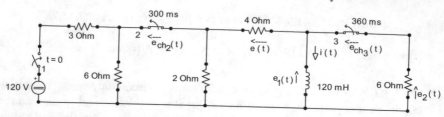

Fig. 5.81 Electric circuit of Example 5.26

ATPDraw program, with $\Delta t = 10^{-5}$ [s] and maximum simulation time $t_{max} = 0.5$ [s], determine $e(t)$, $e_1(t)$, $i(t)$, $e_2(t)$, $e_{ch2}(t)$ and $e_{ch3}(t)$, for $t > 0$.

Solution

The circuit used to determine the requested quantities is shown in Fig. 5.82. Figures 5.83, 5.84, 5.85 and 5.86 show the results obtained.

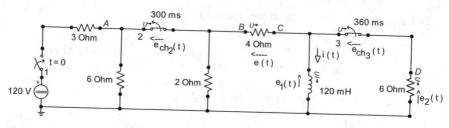

Fig. 5.82 Electric circuit of Example 5.26 for digital solution

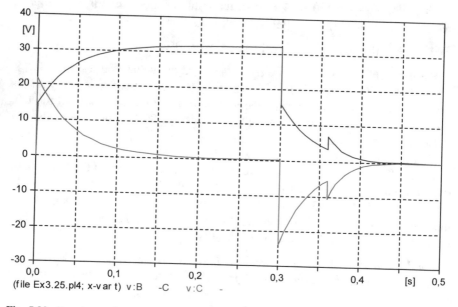

Fig. 5.83 Transient voltage and current digital solution for circuit of Example 5.26

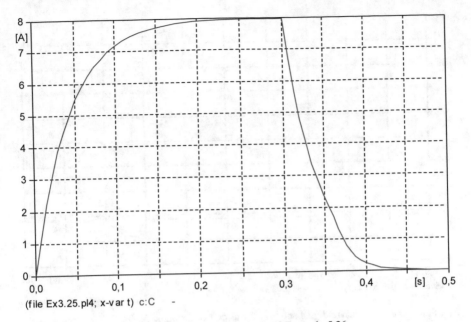

(file Ex3.25.pl4; x-var t) c:C   -

Fig. 5.84 Transient current digital solution for circuit of Example 5.26

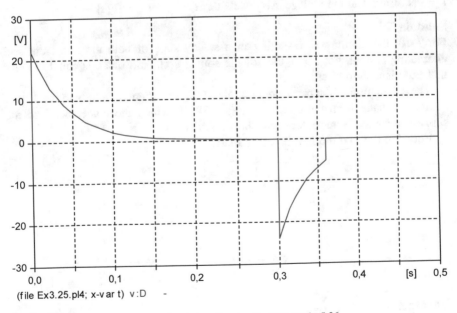

(file Ex3.25.pl4; x-var t) v:D   -

Fig. 5.85 Transient voltage digital solution for circuit of Example 5.26

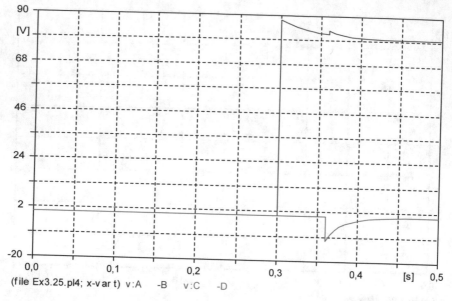

Fig. 5.86 Transient voltage digital solution for circuit of Example 5.26

Example 5.27 The switch in the circuit shown in Fig. 5.87 was opened at time $t = 0[s]$. Calculate the voltage $e_k(t)$ at its terminals, for $t > 0$ [s].

Solution

For $t < 0$, the circuit was operating in a steady state of both direct current and alternating current. For $t > 0$, the voltage source $e(t)$, sinusoidal, is equal to zero, that is, it is a short circuit.

The calculation of the permanent regime of direct current, due only to the $10[V]$, source, is made from the circuit of Fig. 5.88, in which the inductance and the sinusoidal source were replaced by short circuits.

Equating the two meshes, it comes that:

$$5i + i_k = 0$$

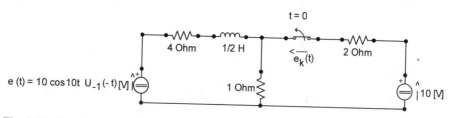

Fig. 5.87 Electric circuit of Example 5.27

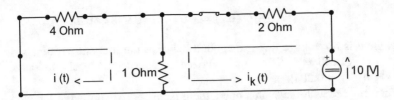

Fig. 5.88 Electric circuit of Example 5.27 for DC analytical solution at $t = 0_-$

$$i + 3i_k = 10.$$

$$i = -5/7 = -0.714[A], \text{ for } -\infty < t \leq 0_- [s].$$

Therefore, $i_{dc}(0_-) = -0.714[A]$.

The calculation of the permanent alternating current regime, due only to the sinusoidal source, is made from the circuit in the simple frequency domain of Fig. 5.89.

Since $\omega = 10[rad/s]$, $jX_L = j\omega L = j10.(1/2) = j5[\Omega]$.

$$\dot{I} = \frac{10\angle 0°}{4 + j5 + \frac{2}{3}} = \frac{30\angle 0°}{14 + j15} = \frac{30\angle 0°}{20.518\angle 46.975°} = 1.462\angle - 46.975°[A]$$

$$i(t) = 1.462\cos(10t - 46.975°)[A]. \text{ For } -\infty < t \leq 0_- [s].$$

Therefore,

$$i_{ac}(0_-) = 1.462\cos(-46.975°) = 0.998[A].$$

Overlapping the initial conditions,

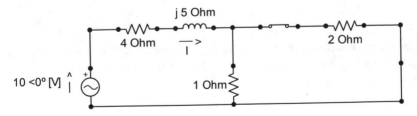

Fig. 5.89 Electric circuit of Example 5.27 for AC analytical solution at $t = 0_-$

$$i(0_-) = i_{dc}(0_-) + i_{ac}(0_-) = -0.714 + 0.998 = 0.284[A].$$

As there is no voltage impulse applied between the inductor terminals, in contrast to Corollary 5.2, its current cannot be discontinued and $i(0_+) = i(0_-) = 0.284[A]$.

When t→ ∞, with the switch open, there is no source applied to the circuit and $i(\infty) = 0[A]$.

The circuit time constant, for $t > 0[s]$, is: $T = L/R_{TH} = 0.5/5 = 0.1[s]$.
Therefore,

$$i(t) = i(\infty) + [i(0_+) - i(\infty)]\epsilon^{-t/T} = 0.284\epsilon^{-10t}, \text{ for } t \geq 0[s].$$

On the closed path involving the open switch,

$$10 + e_k(t) - (1).i(t) = 0$$

Finally,

$$e_k(t) = i(t) - 10 = (0.284\epsilon^{-10t} - 10) U_{-1}(t)[V].$$

$$e_k(0_-) = 0[V]; e_k(0_+) = -9.716[V]; e_k(\infty) = -10[V].$$

Example 5.28   There is no energy stored in the capacitor of the circuit of Fig. 5.90, when the switch k_1 closes at $t = 0$ [s]. Twenty-four microseconds later, switch k_2 is closed. Determine the voltage and current in the capacitor for $t \geq 0$ [s].

$$R_1 = 2[k\Omega], R_2 = 8[k\Omega], C = 6[nF].$$

Solution

In the first time interval: $0 < t < 24$ [μs], the circuit time constant is

$$T_1 = R_1C = 2.10^3.6.10^{-9} = 12[\mu s].$$

In the second time interval: $t > 24$ [μs], the equivalent resistance seen by the capacitance, R_{eq}, with the short-circuit voltage source and the open-circuit current source, is

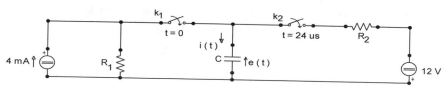

Fig. 5.90 Electric circuit of Example 5.28

$$R_{eq} = \frac{R_1 \times R_2}{R_1 + R_2} = \frac{2 \times 8}{2 + 8} = 1.6[k\Omega].$$

So, the new circuit time constant is

$$T_2 = R_{eq}C = 1.6 \times 10^3 \times 6 \times 10^{-9} = 9.6[\mu s].$$

From example 5.13, for $0 < t < 24.10^{-6}$ [s],

$$e(t) = 4.10^{-3}R_1\left(1 - \epsilon^{-t/T_1}\right) = 4.10^{-3}.2.10^3\left(1 - \epsilon^{-10^6 t/12}\right)$$

$$e(t) = 8\left(1 - \epsilon^{-10^6 t/12}\right)[V], \text{ for } 0 < t < 24.10^{-6} \text{ [s]}.$$

Then, $e(24\mu s_-) = 6.917[V]$.

$$i(t) = C\frac{d}{dt}e(t) = 6.10^{-9}.8.\frac{10^6}{12}.\epsilon^{-10^6 t/12} = 4\epsilon^{-10^6 t/12}[mA].$$

Therefore, $i(24\mu s_-) = 0.541[mA]$.

Thus, for $t > 24$ [μs], the circuit is shown in Fig. 5.91, with the two switches closed.

In Fig. 5.91, the time instant $t = 24$ [μs] was provisionally taken as a new $t = 0$. The circuit's nodal equation is:

$$\frac{e(t)}{R_1} + C\frac{d}{dt}e(t) + \frac{e(t) + 12U_{-1}(t)}{R_2} = 4 \times 10^{-3}U_{-1}(t)$$

Rearranging and simplifying, it follows that:

$$e'(t) + \frac{1}{T_2}e(t) = \frac{1}{C}\left(4 \times 10^{-3} - \frac{12}{R_2}\right)U_{-1}(t)$$

Substituting the remaining numerical values, the differential equation is:

$e'(t) + \frac{10^6}{9.6}e(t) = \frac{5}{12}.10^6.U_{-1}(t)$, with $e(0_-) = 6,917[V] = e(24\mu s_-)$.

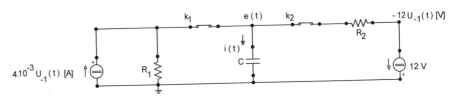

Fig. 5.91 Electric circuit of Example 5.28 for analytical solution

$$e_H(t) = k_1\varepsilon^{-10^6 t/9.6}$$

$$e_P(t) = \frac{1}{D + 10^6/9.6}\left(\frac{5}{12}\cdot 10^6\cdot \epsilon^{0.t}\right) = 4$$

$$e(t) = 4 + k_1\varepsilon^{-10^6 t/9.6}.$$

Since $(0_+) = e(0_-) = 6.917 = 4 + k_1, k_1 = 2.917$. Therefore,

$$e(t) = 4 + 2.917\varepsilon^{-10^6 t/9.6}, \text{ for } t \geq 0 \, [\mathrm{s}].$$

Returning to the original reference of the time, that is, replacing t with $t - 24[\mu s]$,

$$e(t) = 4 + 2.917\epsilon^{-10^6(t - 24.10^{-6})/9.6} = 4 + 2.917.\epsilon^{2.5}.\epsilon^{-10^6 t/9.6}$$

$$e(t) = 4 + 35.536\varepsilon^{-10^6 t/9.6}, \text{ for } t \geq 24.10^{-6}[\mathrm{s}].$$

This last result can also be obtained using Eq. (5.9), with
$e(t_{0_+}) = e(24.10^{-6}_+) = 6.917[\mathrm{V}]; e(\infty) = 4[\mathrm{V}]; T_2 = 9.6[\mu s]$.
Then,

$$e(t) = 4 + (6.917 - 4)\epsilon^{-10^6(t - 24\times 10^{-6})/9.6} = 4 + 2.917.\epsilon^{2.5}.\varepsilon^{-10^6 t/9.6}$$

$$= 4 + 35.536\varepsilon^{-10^6 t/9.6}, \text{ for } t \geq 24.10^{-6}[\mathrm{s}].$$

The current is:
$i(t) = C\frac{d}{dt}e(t) = 6.10^{-9}.\frac{d}{dt}\left(4 + 35.536\varepsilon^{-10^6 t/9.6}\right)$. The result is

$$i(t) = -22.21\varepsilon^{-10^6 t/9.6}[\mathrm{mA}], \text{ for } t > 24[\mu s]; i(24\mu s_+) = -1.823[\mathrm{mA}].$$

For the numerical solution of this example through the ATPDraw program, the circuit shown in Fig. 5.92 was assembled.

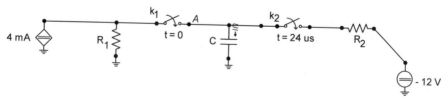

Fig. 5.92 Electric circuit of Example 5.28 for digital solution

As $T_2 = 9.6[\mu s]$,   is   considered   that $t_{max} = 80[\mu s] > 24 + 5T_2$ and
$\Delta t = 10^{-7}[s] \approx T_2/100$.

$$R_1 = 2[k\Omega], \ R_2 = 8[k\Omega], \ C = 6[nF].$$

The results obtained are shown in Figs. 5.93 and 5.94.

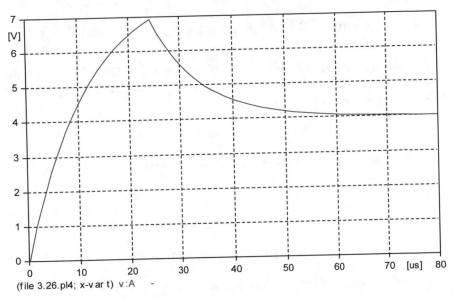

(file 3.26.pl4; x-v ar t) v:A   -

Fig. 5.93 Transient voltage digital solution for circuit of Example 5.28

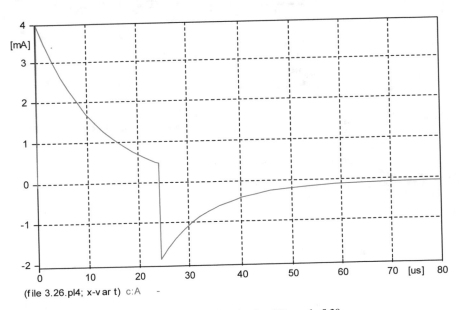

(file 3.26.pl4; x-v ar t) c:A   -

Fig. 5.94 Transient current digital solution for circuit of Example 5.28

Example 5.29 Find the current $i(t)$ in the circuit of Fig. 5.95, for $t > 0$.

$$e(t) = \left[50 + 100 \sin^2 40t\right]U_{-1}(t)[V]$$

Solution

Remembering that $\sin^2 \alpha = \frac{1 - \cos 2\alpha}{2}$, the voltage source can be rewritten as:

$$e(t) = \left[50 + 100\frac{1 - \cos 80t}{2}\right]U_{-1}(t) = 100U_{-1}(t) - 50\cos80tU_{-1}(t)[V].$$

Considering the hourly loop current $i_1(t)$, circulating in the left loop, and the hourly loop current $i(t)$, circulating in the right loop of the circuit in Fig. 5.95, the following equations result

$$5i_1 - 3i + 800\int(i_1 - i)dt = e.$$

$$-3i_1 + 9i + 800\int(i - i_1)dt = 0.$$

Or,

$$(5D + 800)i_1 - (3D + 800)i = De.$$

$$-(3D + 800)i_1 + (9D + 800)i = 0.$$

Solving this system for $i(t)$, it follows that:

$$i'(t) + \frac{1,600}{9}i(t) = \frac{1}{12}e'(t) + \frac{200}{9}e(t).$$

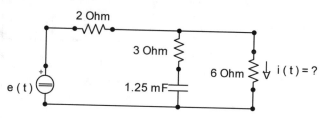

Fig. 5.95 Electric circuit of Example 5.29

1. For $e(t) = 100U_{-1}(t)[V]$, $e'(t) = 100U_0(t)[V/s]$, and the differential equation becomes:

$$\left(D + \frac{1,600}{9}\right)i(t) = \frac{25}{3}U_0(t) + \frac{20,000}{9}U_{-1}(t), \text{ with } i(0_-) = 0.$$

$$i_H(t) = k_1\epsilon^{-1,600t/9} \text{ and } i_P(t) = \frac{1}{D + \frac{1600}{9}}\left(\frac{20,000}{9}\epsilon^{0\times t}\right) = \frac{25}{2}. \text{ So,}$$

$$i(t) = \frac{25}{2} + k_1\epsilon^{-1,600t/9}$$

Since $i(0_+) = \frac{25}{3}$, then $\frac{25}{2} + k_1 = \frac{25}{3}$ and $k_1 = -\frac{25}{6}$. Finally,

$$i(t) = \frac{25}{2} - \frac{25}{6}\epsilon^{-1,600t/9} = 12.500 - 4.167\epsilon^{-177.778t}[A], \text{ for} t > 0.$$

2. If $e(t) = -50\cos 80tU_{-1}(t)[V]$, $e'(t) = -50U_0(t) + 4,000\sin 80tU_{-1}(t)[V/s]$, and the preceding differential equation becomes:

$$\left(D + \frac{1,600}{9}\right)i(t) = \frac{1}{12}[-50U_0(t) + 4,000\sin 80tU_{-1}(t)] + \frac{200}{9}[-50\cos 80tU_{-1}(t)]$$

$$= -\frac{25}{6}U_0(t) + \frac{3,000}{9}\sin 80tU_{-1}(t) - \frac{10,000}{9}\cos 80tU_{-1}(t).$$

$$z(t) \triangleq -\frac{10,000}{9}\cos 80t + \frac{3,000}{9}\sin 80t$$

$$= \frac{1}{9}[-10,000\cos 80t + 3,000\cos(80t - 90°)]$$

So, the corresponding phasor is

$$\dot{Z} = \frac{1}{9}(-10,000 - j3,000) = -\frac{1,000}{9}(10 + j3) = -\frac{1,000}{9} \times 10.44∠16.7°$$
$$= -1,160∠16.7°$$

Thus, $z(t) = -1,160\cos(80t + 16.7°)$, and the differential equation can be rewritten as:

$$\left(D + \frac{1,600}{9}\right)i(t) = -\frac{25}{6}U_0(t) - 1,160\cos(80t + 16.7°).U_{-1}(t)$$

$$= -\frac{25}{6}U_0(t) - 1,111.074U_{-1}(t) + \ldots, \text{ with } i(0_-) = 0.$$

$$i_H(t) = k_2\epsilon^{-1,600t/9}$$

$$(177.778 + j80)\dot{I}_P = -1,160\ \underline{/16.7°} \text{ and } \dot{I}_P = -5.950\underline{/} - \underline{/7.53°}$$

$$i_P(t) = -5.95 \cos(80t - 7.53°)$$

$$i(t) = k_2 \epsilon^{-1,600t/9} - 5.95 \cos(80t - 7.53°)$$

From differential equation,

$$i(0_+) = -\frac{25}{6} = -4.167 = k_2 - 5.95 \cos(-7.53°).$$

So, $k_2 = 1.732$.
Finally,

$$i(t) = 1.732\epsilon^{-177.778t} - 5.95 \cos(80t - 7.53°)[\text{A}], \text{ for } t > 0.$$

Superimposing the currents $i(t)$ obtained in items 1 and 2, finally results in the current circulating in the resistance of $6[\Omega]$ of the circuit of Fig. 5.95:

$$i(t) = 12.500 - 2.435\epsilon^{-177.778t} - 5.95 \cos(80t - 7.53°)[\text{A}], \text{ for } t > 0.$$

$i(0_+) = 4.166[\text{A}]$ and the steady-state current fluctuates between a maximum of $12.50 + 5.95 = 18.45[\text{A}]$ and a minimum of $12.50 - 5.95 = 6.55[\text{A}]$.

The numerical solution was performed by replacing the voltage source e (t) with two equivalent current sources, as shown in Fig. 5.96, with $\Delta t = 10^{-5}[\text{s}]$ and $t_{max} = 0.35[\text{s}]$.

Note, in red in Fig. 5.96, the current probe, to record the total current from the two current sources. The responses obtained are recorded in Figs. 5.97 and 5.98.

Figures 5.99 and 5.100 show the voltage and current waveforms at the capacitance of 1250 [µF]. And Fig. 5.101 shows a zoom in the capacitance current.

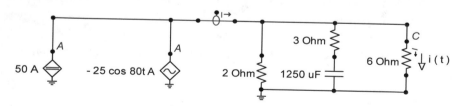

Fig. 5.96 Equivalent electric circuit of Example 5.29

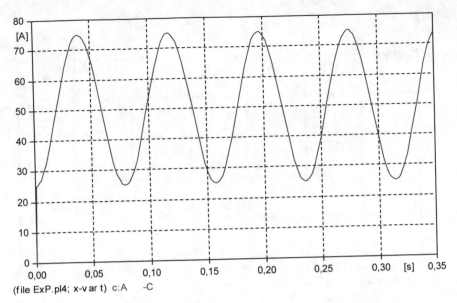

(file ExP.pl4; x-v ar t) c:A   -C

Fig. 5.97 Transient total current source digital solution for circuit of Example 5.29

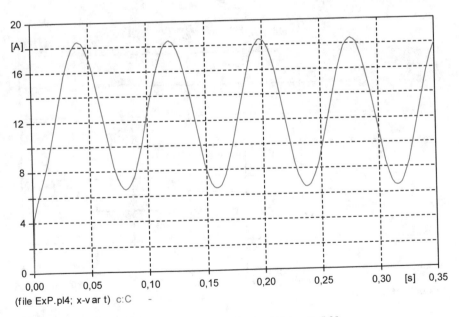

(file ExP.pl4; x-v ar t) c:C   -

Fig. 5.98 Transient current digital solution for circuit of Example 5.29

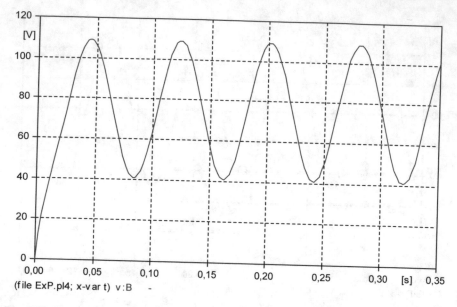

Fig. 5.99 Transient capacitor voltage digital solution for circuit of Example 5.29

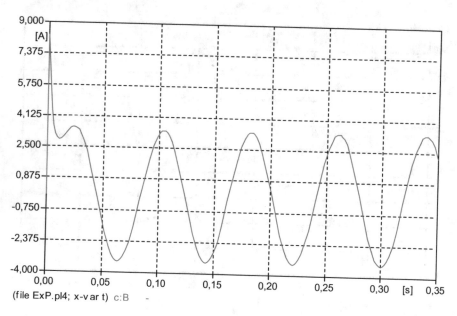

Fig. 5.100 Transient capacitor current digital solution for circuit of Example 5.29

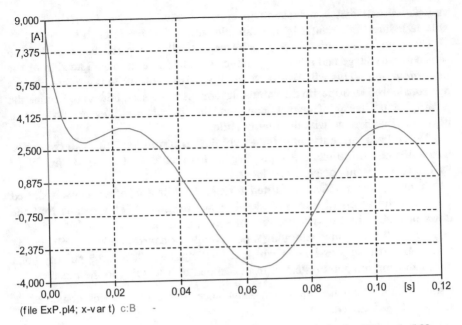

(file ExP.pl4; x-var t) c:B   -

Fig. 5.101 Zoom at transient capacitor current digital solution for circuit of Example 5.29

5.5 Magnetically Coupled Circuits

When two circuits are very close to each other, so that the variable current that circulates in one of them induces voltage (electromotive force) in the other, and vice versa, there is what is called magnetically coupled circuits. In the case of coils, this magnetic coupling can be considerably increased if one coil is wound on top of another, thus increasing the phenomenon of **mutual induction**. Moreover, the addition of a ferromagnetic core will establish a pathway for the magnetic flux, maximizing the coupling, although the iron core may introduce non-linearities.

The self-inductance L_1, as we know, is the parameter that relates the rate of change with the time of the current $i_1(t)$, which circulates in circuit1, with the voltage $e_1(t)$ induced in itself, that is: $e_1(t) = L_1 di_1(t)/dt$. Likewise, the L_2 self-inductance is the proportionality parameter between the rate of change of the current $i_2(t)$, which circulates in circuit 2, and the voltage $e_2(t)$ induced in it, that is: $e_2(t) = L_2 di_2(t)/dt$. In both cases, it is implied that the voltage reference arrow is in opposition to the current reference arrow, that is: that the current enters the inductance through the terminal towards which the voltage arrow is directed. (Otherwise, a minus sign must be introduced in the second member of the previous expressions of $e_1(t)$ and $e_2(t)$).

Mutual inductance M is the parameter that relates the voltage induced in one circuit with the rate of change with the time of the variable current circulating in the other circuit. Thus, $e_{1_M}(t) = M di_2(t)/dt$ is the voltage induced in circuit 1 by the variable current that is circulating in circuit 2. And $e_{2M}(t) = M di_1(t)/dt$ is the voltage induced in circuit 2 by the variable current flowing in circuit 1.

Therefore, when two circuits are magnetically coupled, there is both a self-induced voltage and a mutually induced voltage in each one. **The sign of the auto-induction terms will always be positive if the voltage and current arrows are considered in opposite directions**. It remains to be seen how to determine the polarity of the terms of mutual induction, that is, whether they are positive or negative. This is done with the **Points Rule**:

"**When both currents enter (or leave) dotted terminals (with asterisks), the terms of mutual induction are positive**". This means that the flows produced by the two currents are in agreement.

"**If one current enters a dotted terminal while the other exits a dotted terminal, the terms of mutual induction are negative.**" This means that the flows produced by the two currents are in opposition.

It is worth noting that: **changing a polarity point corresponds to replacing M with $-M$ in the expressions in which it participates**. Changing a polarity point means moving the point (asterisk) from one end of a winding to the other.

Example 5.30 Find the equivalent inductance of the inductances connected in series of Fig. 5.102a and b.

Solution

(a) The same current $i(t)$ passes through both windings. In addition, the current arrow enters the windings through the dotted terminals (with *), so that:

$$e = e_1 + e_2 = \left(L_1 \frac{di}{dt} + M \frac{di}{dt}\right) + \left(L_2 \frac{di}{dt} + M \frac{di}{dt}\right) = (L_1 + L_2 + 2M)\frac{di}{dt}$$

Fig. 5.102 Electric circuits of Example 5.30: **a** additive polarity; **b** subtractive polarity

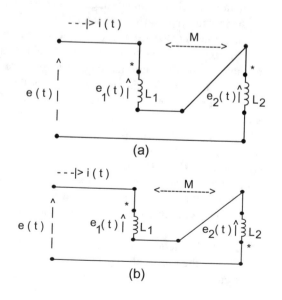

(a)

(b)

Therefore, $L_{eq} = L_1 + L_2 + 2M$.

(b) In Fig. 5.102b, the current $i(t)$ enters the first winding through the dotted terminal and exits through the dotted terminal of the second winding, which makes the terms in M negative. In addition, Fig. 5.102b is the same as Fig. 5.102a, only with the change of point (*) from the beginning to the end of the second winding. Then, the equivalent inductance is obtained from the previous expression by simply exchanging M for $-M$. Thus,

$$L_{eq} = L_1 + L_2 - 2M.$$

Example 5.31 Find the equivalent inductance of the parallel connection in Fig. 5.103.

Solution
Injecting the voltage source $e(t)$ and considering the mesh current $i_1(t)$ and the loop current $i_2(t)$, Fig. 5.104 is obtained.
If $i_1(t) + i_2(t) = i(t)$.
Consider that the corresponding equations are:

$$L_1 Di_1(t) - MDi_2(t) = e(t)$$

$$-MDi_1(t) + L_2 Di_2(t) = e(t)$$

Resolving for $i_1(t)$, it comes that:

$$\begin{vmatrix} L_1 D & -MD \\ -MD & L_2 D \end{vmatrix} i_1(t) = \begin{vmatrix} e(t) & -MD \\ e(t) & L_2 D \end{vmatrix}$$

$$(L_1 L_2 - M^2)Di_1(t) = (L_2 + M)e(t)$$

Solving for $i_2(t)$, it comes that:

$$(L_1 L_2 - M^2)Di_2(t) = (L_1 + M)e(t)$$

Therefore, adding the two differential equations, it follows that

$$(L_1 L_2 - M^2)Di(t) = (L_1 + L_2 + 2M)e(t)$$

Fig. 5.103 Electric circuit of Example 5.31

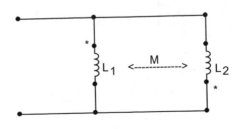

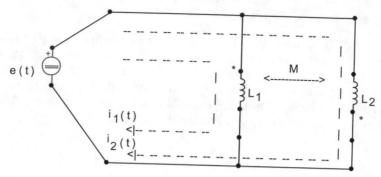

Fig. 5.104 Electric circuit of Example 5.31 for analytical solution

So,

$$e(t) = \frac{L_1 L_2 - M^2}{L_1 + L_2 + 2M} \cdot \frac{d}{dt} i(t) = L_{eq} \cdot \frac{d}{dt} i(t)$$

Then,

$$L_{eq} = \frac{L_1 L_2 - M^2}{L_1 + L_2 + 2M}.$$

If one of the polarity points of either of the two windings of the connection in Fig. 5.103 is reversed, M will be replaced by $-M$, and the new equivalent inductance will be:

$$L_{eq} = \frac{L_1 L_2 - M^2}{L_1 + L_2 - 2M}.$$

It is worth remembering that, in any situation, $M = k \sqrt{L_1 L_2}$, with $0 \leq k \leq 1$. When $k = 0$ there is no magnetic coupling between the windings, and there is no mutually induced voltage. And when $k = 1$ the magnetic coupling between the windings is maximum.

Example 5.32 There is no energy initially stored in the circuit of Fig. 5.105 when the switch is closed at $t = 0$. Find $i_1(t)$, $i_2(t)$, $i(t)$ and $e(t)$ for $t > 0$.

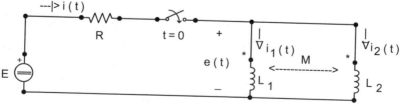

Fig. 5.105 Electric circuit of Example 5.32

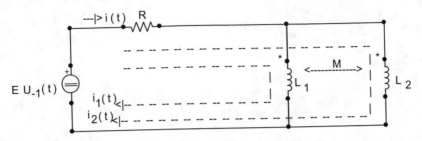

Fig. 5.106 Electric circuit of Example 5.32 for analytical solution

Solution

The switch, closing at $t = 0$, in series with the constant voltage source E can be replaced by the source $EU_{-1}(t)$. Introducing the mesh current $i_1(t)$ and the loop current $i_2(t)$, results in the circuit of Fig. 5.106.

As the polarity is additive (the two currents enter the dotted terminals), the corresponding equations are:

$$R(i_1 + i_2) + L_1 Di_1 + MDi_2 = EU_{-1}(t)$$

$$R(i_1 + i_2) + L_2 Di_2 + MDi_1 = EU_{-1}(t).$$

Or,

$$(L_1 D + R)i_1 + (MD + R)i_2 = EU_{-1}(t)$$

$$(MD + R)i_1 + (L_2 D + R)i_2 = EU_{-1}(t).$$

Solving this system for $i_1(t)$, the differential equation results

$$\left[D + \frac{R(L_1 + L_2 - 2M)}{L_1 L_2 - M^2} \right] i_1(t) = \frac{L_2 - M}{L_1 L_2 - M^2} EU_{-1}(t), \text{ with } i_1(0_-) = 0. \quad (5.13)$$

$$i_{1_H}(t) = k_1 \epsilon^{-\frac{R(L_1 + L_2 - 2M)}{L_1 L_2 - M^2} t}$$

$$i_{1_P}(t) = \frac{\frac{L_2 - M}{L_1 L_2 - M^2}}{D + \frac{R(L_1 + L_2 - 2M)}{L_1 L_2 - M^2}} \left(E\epsilon^{0 \times t} \right) = \frac{L_2 - M}{R(L_1 + L_2 - 2M)} E$$

$$i_1(t) = \frac{L_2 - M}{R(L_1 + L_2 - 2M)} E + k_1 \epsilon^{-\frac{R(L_1 + L_2 - 2M)}{L_1 L_2 - M^2} t}$$

From the Eq. (5.13), $i_1(0_+) = i_1(0_-) = 0$. So,

$$\frac{L_2 - M}{R(L_1 + L_2 - 2M)}E + k_1 = 0 \text{ and } k_1 = -\frac{(L_2 - M)E}{R(L_1 + L_2 - 2M)}$$

Therefore,

$$i_1(t) = \frac{(L_2 - M)E}{R(L_1 + L_2 - 2M)}\left[1 - \epsilon^{-\frac{R(L_1 + L_2 - 2M)}{L_1 L_2 - M^2}t}\right]U_{-1}(t). \tag{5.14}$$

Solving the system for $i_2(t)$, the differential equation results

$$\left[D + \frac{R(L_1 + L_2 - 2M)}{L_1 L_2 - M^2}\right]i_2(t) = \frac{L_1 - M}{L_1 L_2 - M^2}EU_{-1}(t), \text{ with } i_2(0_-) = 0. \tag{5.15}$$

Due to the similarity between differential Eqs. 5.15 and 5.13, we have that:

$$i_2(t) = \frac{(L_1 - M)E}{R(L_1 + L_2 - 2M)}\left[1 - \epsilon^{-\frac{R(L_1 + L_2 - 2M)}{L_1 L_2 - M^2}t}\right]U_{-1}(t). \tag{5.16}$$

Since $i(t) = i_1(t) + i_2(t)$, the sum of Eqs. 5.14 and 5.16 gives that:

$$i(t) = \frac{E}{R}\left[1 - \epsilon^{-\frac{R(L_1 + L_2 - 2M)}{L_1 L_2 - M^2}t}\right]U_{-1}(t).$$

$e(t) = L_1 Di_1(t) + MDi_2(t)$, then the requested voltage is

$$e(t) = E\epsilon^{-\frac{R(L_1 + L_2 - 2M)}{L_1 L_2 - M^2}t}U_{-1}(t).$$

Logically, if the polarity point of the right winding, for example, is changed to the lower terminal of the coil, just change M to –M to obtain the new expressions of $i_1(t)$, $i_2(t)$, $i(t)$ and $e(t)$ for $t > 0$.

Example 5.33 Find the impulse response $h(t)$ of the circuit in Fig. 5.107.

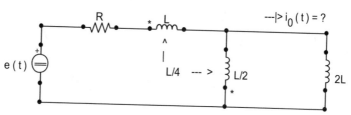

Fig. 5.107 Electric circuit of Example 5.33

Solution

Introducing the hourly mesh currents: $i_1(t)$, on the left, and $i_0(t)$, on the right, then the current going down through the inductance $L/2$ is $i_1(t) - i_0(t)$, and the corresponding equations are:

$$Ri_1 + LDi_1 - \frac{L}{4}D(i_1 - i_0) + \frac{L}{2}D(i_1 - i_0) - \frac{L}{4}Di_1 = e(t)$$

$$2LDi_0 - \frac{L}{2}D(i_1 - i_0) + \frac{L}{4}Di_1 = 0.$$

By manipulating and simplifying, the following system results:

$$(4LD + 4R)i_1 - LDi_0 = 4e(t)$$

$$-Di_1 + 10Di_0 = 0.$$

Solving for $i_0(t)$, comes the first order differential equation

$$i_0'(t) + \frac{40R}{39L}i_0(t) = \frac{4}{39L}e(t)$$

For $e(t) = U_0(t)$, $i_0(t) = h(t)$, and the equation becomes

$$h'(t) + \frac{40R}{39L}h(t) = \frac{4}{39L}U_0(t), \text{ with } h(0_-) = 0.$$

For $t > 0$,

$$\left(D + \frac{40R}{39L}\right)h(t) = 0$$

$$h(t) = k_1 e^{-40Rt/39L}$$

As $h(0_+) = 4/39L$, $k_1 = 4/39L$ and finally

$$h(t) = \frac{4}{39L}e^{-40Rt/39L} \cdot U_{-1}(t).$$

Example 5.34 Find the step response $r(t)$ of the circuit shown in Fig. 5.108.

Solution

The equations of the two meshes are:

$$L_2\left(i_1' - i_2'\right) + Mi_1' + L_1i_1' + M\left(i_1' - i_2'\right) = e_1(t)$$

$$L_2\left(i_1' - i_2'\right) + Mi_1' - R_2i_2 = 0.$$

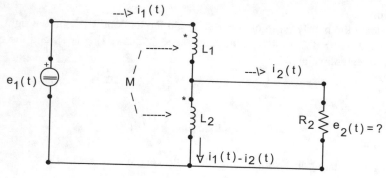

Fig. 5.108 Electric circuit of Example 5.34

Or,

$$(L_1 + L_2 + 2M)Di_1(t) - (L_2 + M)Di_2(t) = e_1(t)$$

$$-(L_2 + M)Di_1(t) + (L_2D + R_2)i_2(t) = 0.$$

Solving this algebraic system for $i_2(t)$, the following differential equation results:

$$i_2'(t) + \frac{R_2L}{L_1L_2 - M^2}i_2(t) = \frac{L_2 + M}{L_1L_2 - M^2}e_1(t) = \frac{L_2 + M}{L_1L_2 - M^2}U_{-1}(t), i_2(0_-) = 0,$$

where $L \triangleq L_1 + L_2 + 2M$.
For $t > 0$,

$$i_2'(t) + \frac{R_2L}{L_1L_2 - M^2}i_2(t) = \frac{L_2 + M}{L_1L_2 - M^2}$$

The solution to the homogeneous equation is:

$$i_{2_H}(t) = k_1\epsilon^{-\frac{R_2L}{L_1L_2 - M^2}t}$$

The particular integral is:

$$i_{2_P}(t) = \frac{1}{D + \frac{R_2L}{L_1L_2 - M^2}}\left(\frac{L_2 + M}{L_1L_2 - M^2}.\epsilon^{0\times t}\right) = \frac{L_2 + M}{R_2L}$$

And the general solution is:

$$i_2(t) = \frac{L_2 + M}{R_2L} + k_1\epsilon^{-\frac{R_2L}{L_1L_2 - M^2}t}$$

Since $i_2(0_+) = 0$, it follows that:

$$0 = \frac{L_2 + M}{R_2 L} + k_1 \text{ and } k_1 = -\frac{L_2 + M}{R_2 L}.$$

Therefore,

$$i_2(t) = \frac{L_2 + M}{R_2 L} \left[1 - \epsilon^{-\frac{R_2 L}{L_1 L_2 - M^2}t} \right] U_{-1}(t)[A].$$

The steady state current is:

$$i_2(\infty) = \frac{L_2 + M}{R_2 L} [A]. \tag{5.17}$$

As $e_2(t) = r(t) = R_2 i_2(t)$, finally

$$r(t) = \frac{L_2 + M}{L} \left[1 - \epsilon^{-\frac{R_2 L}{L_1 L_2 - M^2}t} \right] U_{-1}(t) \, [V]$$

Then, the steady state voltage at resistance R_2, that is, $e_2(\infty)$, in response to a unitary step voltage source at $e_1(t)$, is worth:

$$r(\infty) = \frac{L_2 + M}{L} [V]. \tag{5.18}$$

Equations 5.17 and 5.18 apparently contradict what has been stated since the beginning of this chapter, namely: "in the permanent regime of a circuit powered by a constant source, all of its inductances behave like short circuits". Therefore, in the permanent circuit regime of Fig. 5.108 there should be neither voltage nor current in resistor R_2, since it is in parallel with a short circuit—which is being contradicted by Eqs. 5.17 and 5.18. In the present case, what is happening is that the left mesh current, $i_1(t)$, is not constant in the steady state, and the previous rule does not apply to circuits in which the steady state voltages and currents are not limited. In other words, the current $i_1(t)$ grows without limits. To confirm, the current $i_1(t)$ is calculated.

Solving the previous algebraic system for $i_1(t)$, the following differential equation is obtained:

$$i_1''(t) + \frac{R_2 L}{L_1 L_2 - M^2} i_1'(t) = \frac{L_2}{L_1 L_2 - M^2} U_0(t) + \frac{R_2}{L_1 L_2 - M^2} U_{-1}(t), \text{ with } i_1(0_-)$$
$$= i_1'(0_-) = 0.$$

For $t > 0$, it is:

$$D\left[D + \frac{R_2 L}{L_1 L_2 - M^2}\right]i_1(t) = \frac{R_2}{L_1 L_2 - M^2}$$

The solution to the homogeneous equation is:

$$i_{1_H}(t) = k_2 + k_3 \epsilon^{-\frac{R_2 L}{L_1 L_2 - M^2}t}$$

And the particular integral,

$$i_{1_P}(t) = \frac{1}{D}\frac{1}{D + \frac{R_2 L}{L_1 L_2 - M^2}} \cdot \left(\frac{R_2}{L_1 L_2 - M^2} \cdot \epsilon^{0 \times t}\right) = \frac{1}{D}\left[\frac{1}{L}\right] = \int \frac{1}{L}dt = \frac{1}{L}t$$

Therefore, the general solution is:

$$i_1(t) = \frac{1}{L}t + k_2 + k_3 \epsilon^{-\frac{R_2 L}{L_1 L_2 - M^2}t}$$

and

$$i_1'(t) = \frac{1}{L} - k_3 \frac{R_2 L}{L_1 L_2 - M^2} \epsilon^{-\frac{R_2 L}{L_1 L_2 - M^2}t}$$

From the differential equation, by inspection, there is that:

$$i_1(0_+) = 0$$

$$i_1'(0_+) = \frac{L_2}{L_1 L_2 - M^2}.$$

So,

$$k_2 + k_3 = 0$$

$$\frac{1}{L} - \frac{R_2 L}{L_1 L_2 - M^2}k_3 = \frac{L_2}{L_1 L_2 - M^2}$$

Then,

$$k_2 = -k_3 = \frac{(L_2 + M)^2}{R_2(L_1 + L_2 + 2M)^2}$$

Finally,

$$i_1(t) = \left[\frac{1}{L}t + \frac{(L_2+M)^2}{R_2L^2}\left(1 - \epsilon^{-\frac{R_2L}{L_1L_2-M^2}t}\right)\right]U_{-1}(t)[A].$$

Or,

$$i_1(t) = \frac{1}{L}U_{-2}(t) + \frac{(L_2+M)^2}{R_2L^2}\left[1 - \epsilon^{-\frac{R_2L}{L_1L_2-M^2}t}\right]U_{-1}(t)[A].$$

In fact, the presence of the ramp in the expression of the current $i_1(t)$ confirms that $i_1(\infty) = \infty$. Logically, this would not happen if the left mesh of the circuit contained a resistance R_1, that is, if the autotransformer of Fig. 5.108 was not so idealized.

In conclusion, the responses of permanent regime are:

$$i_1(t) = \frac{1}{L}t + \frac{(L_2+M)^2}{R_2L^2} \text{ (a crescent line); } i_2(t) = \frac{L_2+M}{R_2L}; \ e_2(t) = \frac{L_2+M}{L}.$$

Example 5.35

(a) Find currents $i_1(t)$ and $i_2(t)$ in the transformer shown in Fig. 5.109, which is excited by a sinusoidal source, for $t > 0$ [s]. It is known that the coupling coefficient is $k = 1$ and that the numerical values of the parameters are:

$$E = 100 \text{ [V]}, \ \omega = 20\pi[\text{rad/s}], \ R_1 = 1[\Omega], \ L_1 = 0.02[H], \ R_2 = 10[\Omega] \text{ and } L_2 = 0.08[H].$$

(b) Using the T equivalent of the transformer, proceed to the numerical solution of the problem using the ATPDraw program.
(c) Find the analytical expressions of the voltages $e_{L_1}(t)$ and $e_{L_2}(t)$, for $t > 0$ [s].
(d) Using the T equivalent, find the numerical solution of the voltages $e_{L_1}(t)$ and $e_{L_2}(t)$.

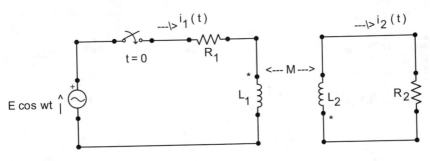

Fig. 5.109 Electric circuit of Example 5.35

Solution

(a) Since $k = 1$, $M = \sqrt{L_1 L_2} = \sqrt{0.02 \times 0.08} = 0.04 [H]$.

The primary equation is:

$$R_1 i_1(t) + L_1 i_1'(t) + M i_2'(t) = e(t) \tag{5.19}$$

The secondary equation is:

$$M i_1'(t) + L_2 i_2'(t) + R_2 i_2(t) = 0 \tag{5.20}$$

Or,

$$(L_1 D + R_1) i_1(t) + M D i_2(t) = e(t)$$

$$M D i_1(t) + (L_2 D + R_2) i_2(t) = 0$$

Then,

$$\begin{vmatrix} L_1 D + R_1 & MD \\ MD & L_2 D + R_2 \end{vmatrix} i_1(t) = \begin{vmatrix} e(t) & MD \\ 0 & L_2 D + R_2 \end{vmatrix}$$

$$\left(L_1 L_2 D^2 + L_1 R_2 D + L_2 R_1 D - M^2 D^2 + R_1 R_2 \right) i_1(t) = L_2 e'(t) + R_2 e(t)$$

As $M^2 = L_1 L_2$, it follows that:

$$\left(D + \frac{R_1 R_2}{L_1 R_2 + L_2 R_1} \right) i_1(t) = \frac{L_2}{L_1 R_2 + L_2 R_1} e'(t) + \frac{R_2}{L_1 R_2 + L_2 R_1} e(t)$$

For $e(t) = E \cos \omega t U_{-1}(t)$, $e'(t) = E U_0(t) - \omega E \sin \omega t U_{-1}(t)$.
Substituting and entering numerical values, the differential equation becomes:

$$(D + 35.714) i_1(t) = 28.571 U_0(t)$$
$$- 1{,}795.196 \sin 20\pi t U_{-1}(t) + 3{,}571.429 \cos 20\pi t U_{-1}(t).$$

Or,

$$(D + 35.714) i_1(t) = 28.571 U_0(t) + 3{,}997.228 \cos(20\pi t + 26.687°) U_{-1}(t); \ i_1(0_-)$$
$$= 0.$$

For $t > 0$,

$$(D+35.714)i_1(t) = 3{,}997.228\cos(20\pi t + 26.687°)$$

$$i_{1_H}(t) = k_1\epsilon^{-35.714t}$$

$$(35.714+j62.832)\dot{I}_{1_P} = 3{,}997.228/26.687°$$

$$\dot{I}_{1_P} = \frac{3{,}997.228\angle 26.687°}{72.273\angle 60.386°} = 55.307\angle -33.699°$$

$$i_{1_P}(t) = 55.307\cos(20\pi t - 33.699°)$$

$$i_1(t) = k_1\epsilon^{-35.714t} + 55.307\cos(20\pi t - 33.7°)$$

From the original differential equation of $i_1(t)$, we have that $i_1(0_+) = 28.571[\text{A}]$. So,

$$28.571 = k_1 + 55.307\cos(-33.7°) = k_1 + 46.013 \text{ and } k_1 = -17.442.$$

Therefore,

$$i_1(t) = -17.442\epsilon^{-35.714t} + 55.307\cos(20\pi t - 33.7°)[\text{A}], \text{ for } > 0[\text{s}].$$

Solving the system of differential equations for $i_2(t)$, it follows that:

$$\left(D + \frac{R_1 R_2}{L_1 R_2 + L_2 R_1}\right)i_2(t) = \frac{-M}{L_1 R_2 + L_2 R_1}e'(t)$$

By making the numerical substitutions, the equation becomes:

$$(D+35.714)i_2(t) = -14.286U_0(t) + 897.598\sin 20\pi t U_{-1}(t), \text{ with } i_2(0_-) = 0.$$

For $t > 0$,

$$(D+35.714)i_2(t) = 897.598\sin 20\pi t$$

$$i_{2_H}(t) = k_2\epsilon^{-35.714t}$$

$$(35.714+j62.832)\dot{I}_{2_P} = 897.598\angle 0°$$

$$\dot{I}_{2_P} = \frac{897.598\angle 0°}{72.273\angle 60.386°} = 12.420\angle -60,4°$$

$$i_{2_P}(t) = 12.420\sin(20\pi t - 60.4°) = 12.420\cos(20\pi t - 60.4° - 90°)$$

$$= 12.420 \cos(20\pi t - 150.4°) = -12.420\cos(20\pi t + 29.6°)$$

$$i_2(t) = k_2 \epsilon^{-35.714t} - 12.420\cos(20\pi t + 29.6°)$$

$$i_2(0_+) = -14.286 = k_2 - 12.420 \cos 29.6° = k_2 - 10,799 \text{ and } k_2 = -3.487.$$

Finally,

$$i_2(t) = -3.487\epsilon^{-35.714t} - 12.420\cos(20\pi t + 29.6°)[A], \text{ for } t > 0[s].$$

The time constant, both for the primary circuit and the secondary circuit, is $T = 1/35.714 = 28[ms]$. And the period of the sinusoidal permanent regime component is $T_s = 2\pi/\omega = 2\pi/20\pi = 100[ms]$.

(b) In terms of the transformer T equivalent, in this case where the polarity is additive, that is, in which the two currents enter the windings through the dotted terminals, the equivalent circuit that produces the same equations, that is, Eqs. 5.19 and 5.20, is shown in Fig. 5.110.

Then, the circuit for the numerical solution by ATPDraw is the one shown in Fig. 5.111.

After typing RUN, with $\Delta t = 10[\mu s]$, the program apparently ran. But, when having the currents $i_1(t)$ and $i_2(t)$ plotted, it issued the following message: "Invalid floating point operation". To explain what happened, it is necessary to remember that the initial values of the currents are: $i_1(0_+) = 28.571[A]$ and $i_2(0_+) = -14.286[A]$. Therefore, the currents in the three inductances in Fig. 5.110 show discontinuities at $t = 0[s]$, that is: at the inductance of $60[mH]$ the current jumps from 0 [A] at $t = 0_-[s]$ to 28.571[A] at $t = 0_+$; at the inductance of 120 [mH] it changes from 0 [A] at $t = 0_-[s]$ to $-14.286[A]$ at $t = 0_+[s]$; and at the inductance of $-40[mH]$ it varies from 0 [A] at $t = 0_-[s]$ to $i_1(0_+) - i_2(0_+) = 42.857[A]$ at $t = 0_+[s]$. However, according to Eq. (5.8), or in the face of Corollary 5.2, discontinuities in the waveforms of the inductance currents generate impulses in the voltages at their terminals, that is, infinite voltages. It is these infinite tensions that in the ATPDraw program, unable to represent them, generate so-called numerical oscillations. And that in the present case resulted in the message

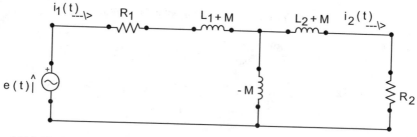

Fig. 5.110 Equivalent electric circuit of Example 5.35

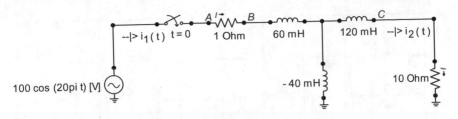

Fig. 5.111 Electric circuit of Example 5.35 for digital solution

"Invalid floating point operation". In fact, in ATPDraw, that is, in discrete time, the initial values of the currents are $i_1(\Delta t) = i_1(10^{-5}) = 37.868[A]$ and $i_2(\Delta t) = i_2(10^{-5}) = -15.905[A]$, but which generate impulses in the inductance voltages in the same way and, consequently, the mentioned difficulties.

To circumvent this difficulty, what the ATPDraw program provides is an "artificial damping", through the introduction of large resistances in parallel with the pure inductances, expressed by $R = k.2L/\Delta t$, allowing to choose $k = 5$ or $k = 10$. Thus, using $k = 10$, for the inductance of $60[mH]$, the parallel resistance is $10 \times 2 \times 60.10^{-3}/10^{-5} = 0.12[M\Omega]$; for the $120[mH]$ inductance, the resistance is $0.24 [M\Omega]$; and for $-40[mH]$, the resistance is $-0.08 [M\Omega]$.

Thus, the new circuit for the numerical solution by ATPDraw is the one shown in Fig. 5.112. It is interesting to note, in Fig. 5.112, that the representation of the inductances now differs from that of the inductances in Fig. 5.111.

In Fig. 5.113 the current $i_1(t)$ is plotted and in Figs. 5.114 and 5.115, two zooms in the same current, in the vicinity of $t = 0[s]$, on the right. In Fig. 5.115, with the help of the plotter cursor, it is possible to capture that $i_1(10\mu s) = i_1(\Delta t) = 28.585[A]$, which, logically, differs from 37.868 [A], since this last value refers to to the case where the circuit still did not have the built-in parallel resistors. However, $28.585[A] \cong i_1(0_+) = 28.571[A]$, the true, that is, the initial value obtained analytically from the original circuit of Fig. 5.109.

In Fig. 5.116 the current $i_2(t)$ is plotted. And in Fig. 5.117 a very strong zoom is shown in the current $i_2(t)$ in the vicinity of $t = 0$ [s], on the right. It is possible to extract from it, with the help of the plotter cursor, that $i_2(\Delta t) = i_2(10\mu s) = -14.283[A] \cong i_2(0_+) = -14.286[A]$, the true initial current, obtained analytically from the original circuit of Fig. 4.109.

The current that descends through the $-40[mH]$ inductance is shown in Fig. 5.118.

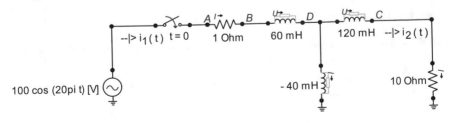

Fig. 5.112 Electric circuit of Example 5.35 for digital solution

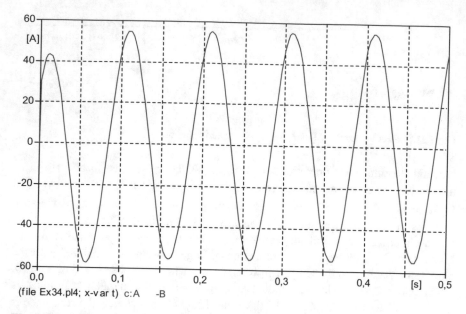

Fig. 5.113 Transient current $i_1(t)$ digital solution for circuit of Example 5.35

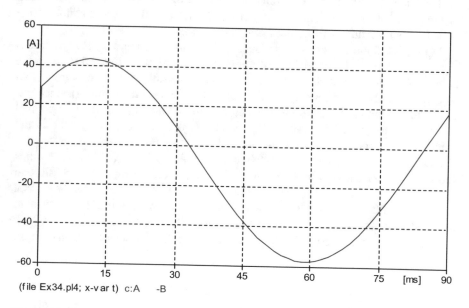

Fig. 5.114 Zoom at transient current $i_1(t)$ digital solution for circuit of Example 5.35

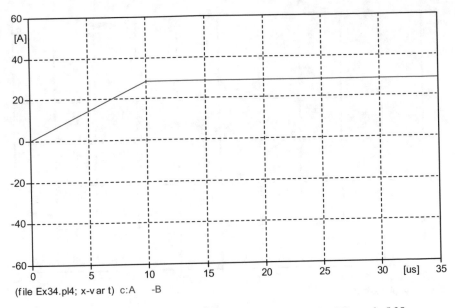

Fig. 5.115 Zoom at transient current $i_1(t)$ digital solution for circuit of Example 5.35

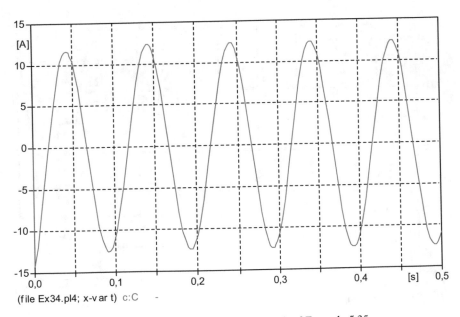

Fig. 5.116 Transient current $i_2(t)$ digital solution for circuit of Example 5.35

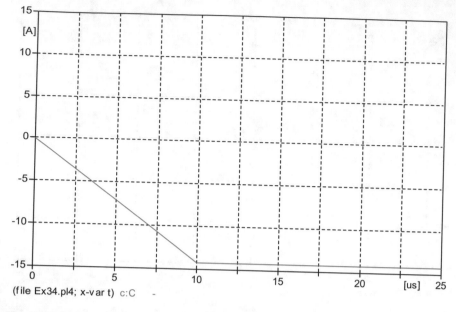

Fig. 5.117 Zoom at transient current $i_2(t)$ digital solution for circuit of Example 5.35

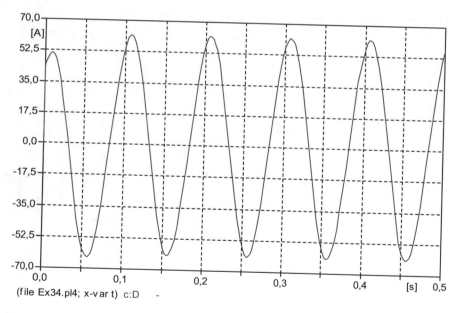

Fig. 5.118 Transient current $i_L(t)$ digital solution for circuit of Example 5.35

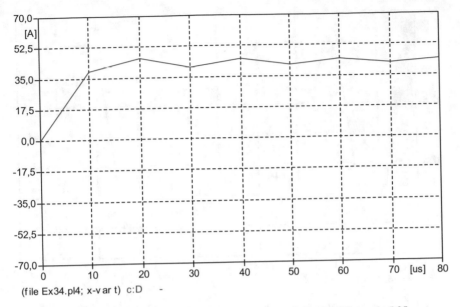

Fig. 5.119 Zoom at transient current $i_L(t)$ digital solution for circuit of Example 5.35

It is interesting to note that this current at the inductance of $-40[mH]$ presents oscillation (numeric damped), as illustrated in Fig. 5.119, which shows a zoom of the same, in the vicinity of $t = 0$ [s]. This fact, however, is not relevant, because this current does not even exist in the original circuit of Fig. 5.109.

It is also worth noting that, because the inductances of the T equivalent in Fig. 5.111 have a common point (node D in Fig. 5.112), all three must receive the addition of large resistances in parallel. Otherwise, the primary and secondary currents will have damped numerical oscillations. For example, when only the inductance of $-40[mH]$ receives the incorporation of the resistance of -0.08 [$M\Omega$] in parallel, the currents $i_1(t)$ and $i_2(t)$ present damped numerical oscillations, as shown in Figs. 5.120 and 5.121, 5.122 and 5.123. Obviously, these oscillations are due to the digital process of solution of the circuit, because, as seen in the analytical solution, they are neither part of the current $i_1(t)$, nor the current $i_2(t)$.

(c) From Eq. 5.19, the voltage at inductance L_1 is given by:

$$e_{L_1}(t) = L_1 i_1'(t) + M i_2'(t) = 0.02 i_1'(t) + 0.04 i_2'(t)$$

$$= 0.02 \frac{d}{dt} \left\{ \left[-17.442 \epsilon^{-35.714t} + 55.307 \cos(20\pi t - 33.7°) \right] U_{-1}(t) \right\}$$

$$+ 0.04 \frac{d}{dt} \left\{ \left[-3.487 \epsilon^{-35.714t} - 12.420 \cos(20\pi t + 29.6°) \right] \right\} U_{-1}(t).$$

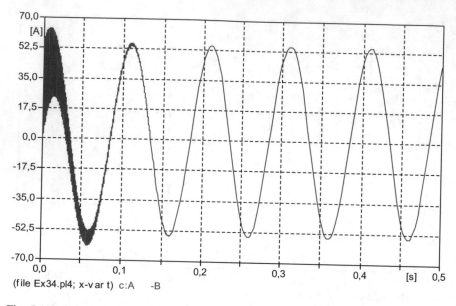

Fig. 5.120 Transient current $i_1(t)$ digital solution for circuit of Example 5.35 with damped numerical oscillations

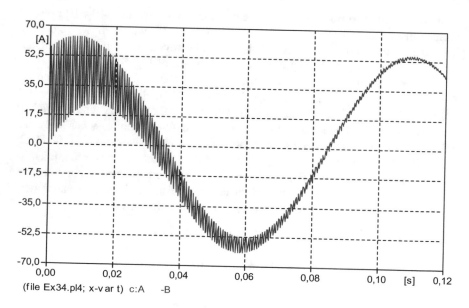

Fig. 5.121 Zoom at transient current $i_1(t)$ digital solution for circuit of Example 5.35 with damped numerical oscillations

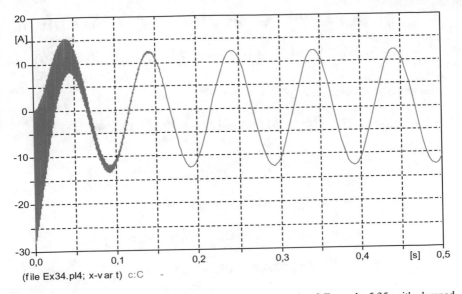

(file Ex34.pl4; x-var t) c:C   -

Fig. 5.122 Transient current $i_2(t)$ digital solution for circuit of Example 5.35 with damped numerical oscillations

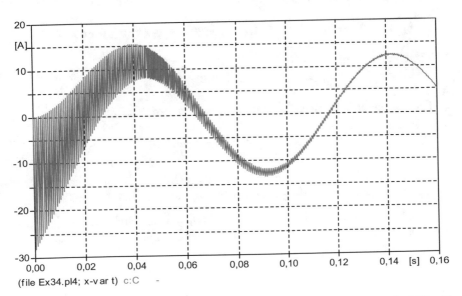

(file Ex34.pl4; x-var t) c:C   -

Fig. 5.123 Zoom at transient current $i_2(t)$ digital solution for circuit of Example 5.35 with damped numerical oscillations

$$e_{L_1}(t) = \left[12.458\epsilon^{-35.714t} - 69.501\sin(20\pi t - 33.7°)\right]U_{-1}(t) + 0.571U_0(t)$$

$$+ \left[4.981\epsilon^{-35.714t} + 31.215\sin(20\pi t + 29.6°)\right]U_{-1}(t) - 0.571U_0(t).$$

Simply put, it comes that:

$$e_{L_1}(t) = 17.439\epsilon^{-35.714t} + 62.090\cos(20\pi t + 29.613°)[V], \textbf{ for } t > 0[s].$$

Therefore,

$$e_{L_1}(0_+) = 71.419[V].$$

From Eq. 5.20, the voltage at inductance L_2 is given by:

$$e_{L_2}(t) = Mi_1'(t) + L_2 i_2'(t) = 0.04i_1'(t) + 0.08i_2'(t)$$

$$= \left[24.917\epsilon^{-35.714t} - 139.002\sin(20\pi t - 33.7°)\right]U_{-1}(t) + 1.143U_0(t)$$

$$+ \left[9.963\epsilon^{-35.714t} + 62.430\sin(20\pi t + 29.6°)\right]U_{-1}(t) - 1.143U_0(t).$$

Simply put, it follows that:

$$e_{L_2}(t) = 34.880\epsilon^{-35.714t} + 124.180\cos(20\pi t + 29.6°)[V], \textbf{ for } t > 0[s].$$

Therefore,

$$e_{L_2}(0_+) = 142.854[V].$$

The voltage at inductance L_2 could also be obtained from

$$e_{L_2}(t) = -e_{R_2}(t) = -R_2 i_2(t).$$

(d) For the numerical solution of the voltages in L_1 and L_2 using the ATPDraw program, the corresponding circuit is shown in Fig. 5.124.

Initially, the voltages in the inductances of $60[mH]$, $120[mH]$ and $-40[mH]$, are represented in Figs. 5.125, 5.126 and 5.127, respectively.

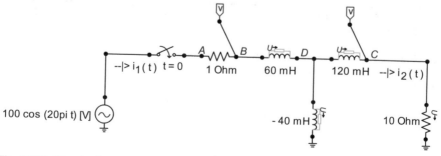

Fig. 5.124 Electric circuit of Example 5.35 for digital solution

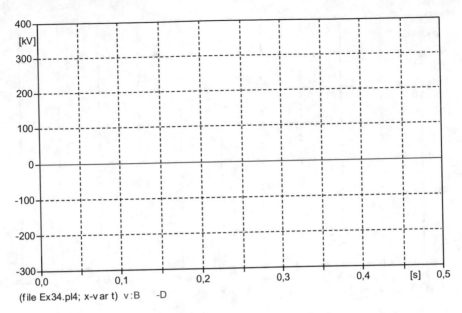

(file Ex34.pl4; x-v ar t) v :B   -D

Fig. 5.125 Transient digital voltage in the inductance of $60[mH]$

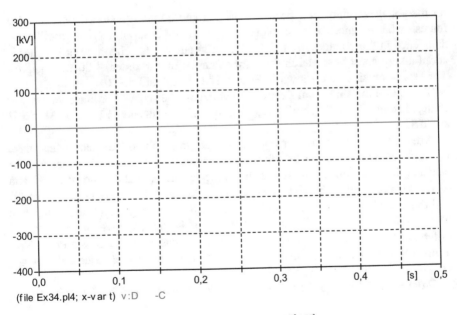

(file Ex34.pl4; x-v ar t) v :D   -C

Fig. 5.126 Transient digital voltage in the inductance of $120[mH]$

Apparently "nothing came out". It turns out that the ordinate scale is in $[kV]$ and, in addition, the transient regime lasts approximately $5T = 5 \times 28[ms] = 140[ms]$. Thus, the observation time $t_{max} = 0.5[s]$ is extremely long to be able to see the transient (exponential) component. On the other hand, along the analytical solution

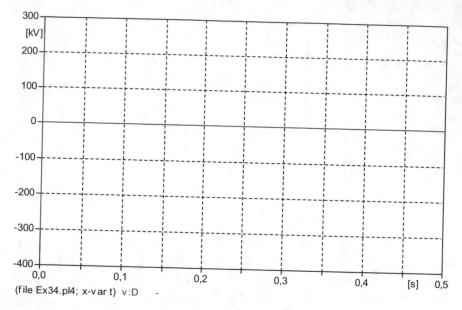

Fig. 5.127 Transient digital voltage in the inductance of $-40[mH]$

for the voltages, impulses appeared in the self and mutually induced electromotive forces, which canceled each other out. This suggests numerical oscillations dampened in the voltages in Figs. 5.125, 5.126 and 5.127. Therefore, it is better to adopt a very short observation time; for example, on the order of $20\Delta t = 0.2[ms]$. The results obtained are shown in Figs. 5.128, 5.129 and 5.130.

To capture the sinusoidal steady state voltages, the scale of ordinate values was changed from $[kV]$ to $[V]$, taking $t_{max} = 0.3[s]$. The results are in Figs. 5.131, 5.132 and 5.133.

After detailing the waveforms of the voltages in the three inductances that make up the T equivalent of the transformer, contained in Fig. 5.124, it remains to be clarified that they do not represent the requested numerical solutions: $e_{L_1}(t)$ and $e_{L_2}(t)$. This is because the voltage $e_{L_1}(t)$ is the potential difference between node B and the reference node (Earth) in Fig. 5.124, and that voltage $e_{L_2}(t)$ is the negative of the voltage in the resistance of 10 [Ω], that is, the negative of the potential difference between node C and Earth. It is for this reason that there is a Probe Volt at node B and another at node C in Fig. 5.124. Thus, the actual requested voltages are shown in Figs. 5.134 and 5.135.

From Figs. 5.136 and 5.137, with the plotter cursor, it can be determined that:

$$e_{L_1}(\Delta t) = 71.415[V] \text{ and that } e_{L_2}(\Delta t) = 142.83[V].$$

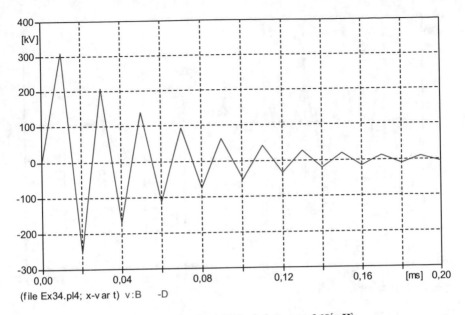

Fig. 5.128 Zoom at transient digital voltage in the inductance of 60[mH]

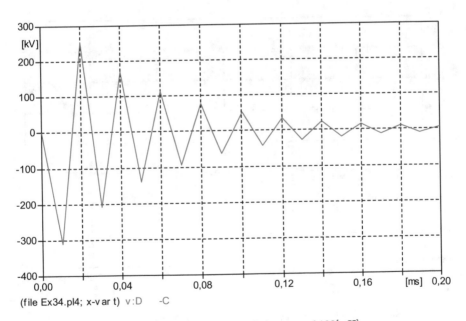

Fig. 5.129 Zoom at transient digital voltage in the inductance of 120[mH]

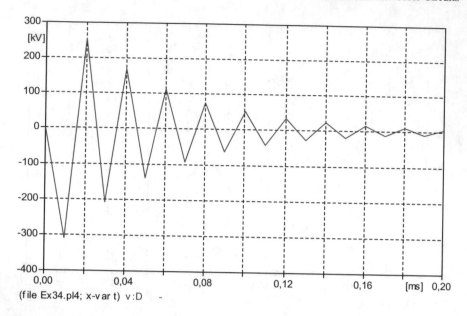

Fig. 5.130 Zoom at transient digital voltage in the inductance of $-40[\text{mH}]$

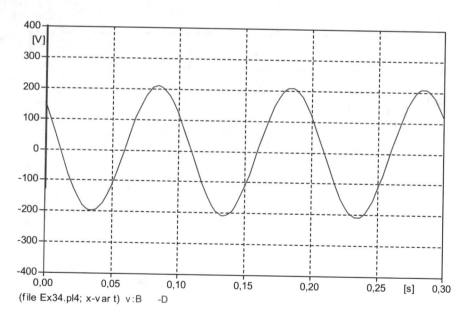

Fig. 5.131 Steady-state digital voltage in the inductance of $60[\text{mH}]$

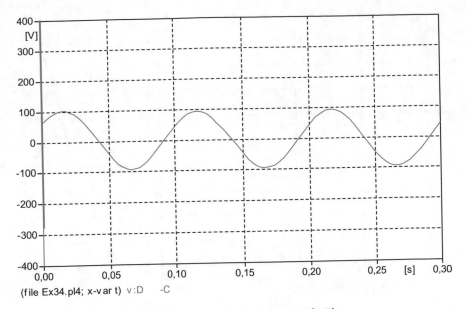

Fig. 5.132 Steady-state digital voltage in the inductance of $120[\mathbf{mH}]$

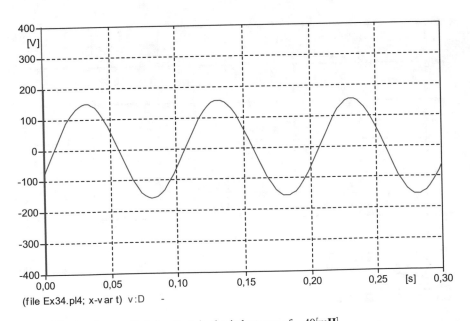

Fig. 5.133 Steady-state digital voltage in the inductance of $-40[\mathbf{mH}]$

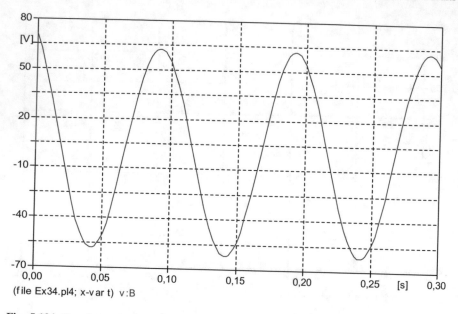

(file Ex34.pl4; x-var t) v:B

Fig. 5.134 Transient voltage $e_{L_1}(t)$ digital solution for circuit of Example 5.35

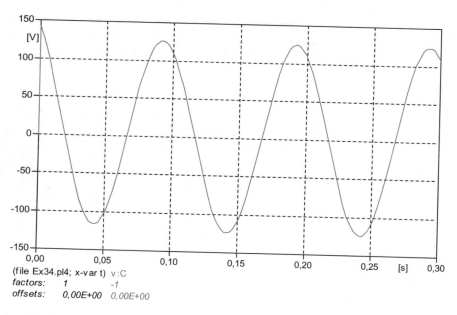

(file Ex34.pl4; x-var t) v:C
factors: 1   -1
offsets: 0,00E+00 0,00E+00

Fig. 5.135 Transient voltage $e_{L_2}(t)$ digital solution for circuit of Example 5.35

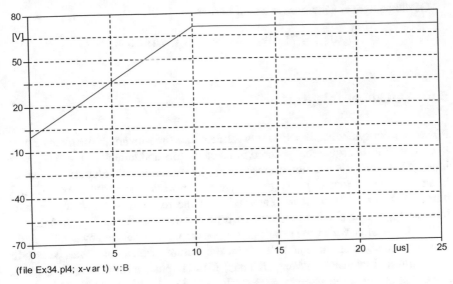

(file Ex34.pl4; x-var t) v:B

Fig. 5.136 Zoom at transientvoltage $e_{L_1}(t)$ digital solution for circuit of Example 5.35

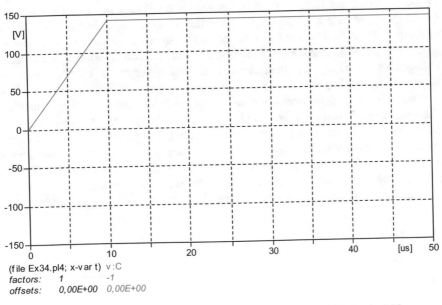

(file Ex34.pl4; x-var t) v:C
factors: 1   -1
offsets: 0,00E+00   0,00E+00

Fig. 5.137 Zoom at transientvoltage $e_{L_2}(t)$ digital solution for circuit of Example 5.35

Of the analytical solutions, as noted earlier,

$$e_{L_1}(0+) = 71.419[\text{V}] \text{ and } e_{L_2}(0+) = 142.854[\text{V}].$$

5.6 DuHamel Integrals

Figure 5.138a symbolically represents a linear and time-invariant (or fixed) circuit —CLI, with no energy initially stored in its capacitors and inductors at $t = 0_-$, with input (excitation) $x(t)$ and output (response) $y(t)$. Assuming an arbitrary and causal input $x(t)$, that is, that $x(t) = 0$ for $t < 0$, we want to determine the corresponding output $y(t)$, for $t \geq 0$, from the knowledge of the response $h(t)$, when a unitary impulse $U_0(t)$ is applied at the input, at time $t = 0$, as shown in Fig. 5.138b.

The **invariance with time** means that when the input is delayed by a value λ, the only consequence is an equal delay in the output, which, however, keeps its waveform and intensity unchanged. Thus, if the impulsive input is applied not at $t = 0$, but at $t = \lambda$, the same response will exist, but only from $t = \lambda$. This means that the response to input $U_0(t - \lambda)$ is simply $h(t - \lambda)$, as illustrated in Fig. 5.139a.

The **homogeneity property of linear systems** states that: if the input is multiplied by a constant factor, the output is simply multiplied by the same constant. So, if the input is not a unit impulse applied at time $t = \lambda$, but an impulse of intensity $x(\lambda)$, applied at time $t = \lambda$, where $x(\lambda)$ is the value of $x(t)$ for $t = \lambda$, that is, a constant, the answer will also be multiplied by $x(\lambda)$. In other words, if the input is changed to $x(\lambda)U_0(t - \lambda)$, then the corresponding output is $x(\lambda)U_0(t - \lambda)$, as shown in Fig. 5.139b.

The **principle of superposition of linear systems**, called the **superposition theorem** when it comes to electrical circuits, establishes that: the response of a circuit to several independent sources is the sum (superposition) of the individual responses of these sources. Thus, considering as input the sum of the infinite inputs $x(\lambda)U_0(t - \lambda)$, for all possible values of $\lambda (0 \leq \lambda < +\infty)$, the superposition theorem imposes that the output must be equal to the sum responses resulting from the use of λ, equally for all possible values. In other words: the input integral produces the

Fig. 5.138 Linear and time-invariant (or fixed) circuit—CLI

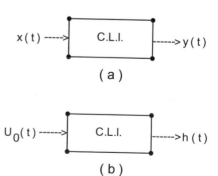

$x(t)$ ----> | C.L.I. | ---->$y(t)$

(a)

$U_0(t)$ ----> | C.L.I. | ---->$h(t)$

(b)

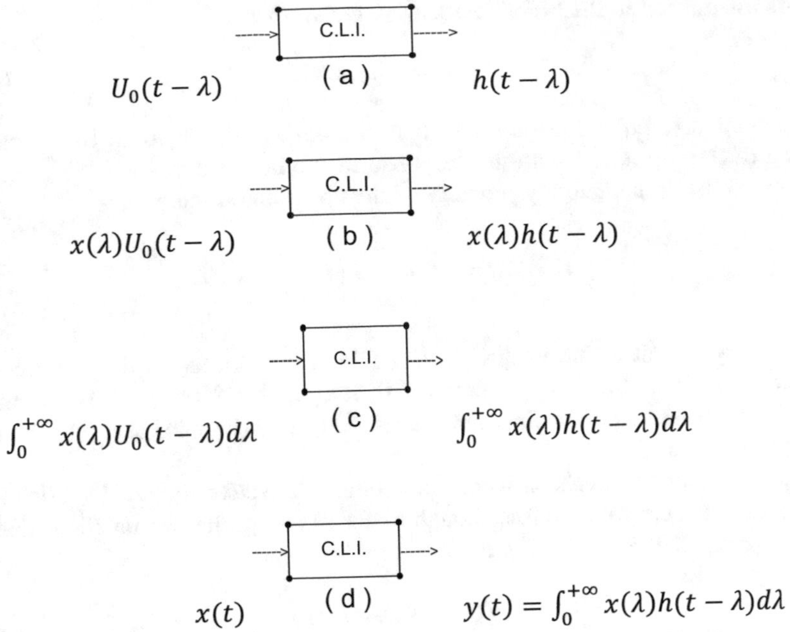

Fig. 5.139 a Invariance with time; b homogeneity property; c principle of superposition; d DuHamel, Carson, Superposition Integral or, simply, Convolution Integral

output integral, as shown in Fig. 5.139c. However, according to Eq. 2.112, adapted for causal signs,

$$\int\limits_{0}^{+\infty} x(\lambda)U_0(t - \lambda)d\lambda = x(t).$$

Therefore, the corresponding output is

$$y(t) = \int\limits_{0}^{+\infty} x(\lambda)h(t - \lambda)d\lambda, \tag{5.21}$$

as shown in Fig. 5.138d.

Equation 5.21 is called the DuHamel, Carson, Superposition Integral or, simply, Convolution Integral.

Considering, however, that $h(t) = 0$ for $t < 0$, then $h(t - \lambda) = 0$ for $\lambda > t$, and Eq. 5.21 can be written as

$$y(t) = \int\limits_{0}^{t} x(\lambda)h(t - \lambda)d\lambda, \text{ for all } t \geq 0. \tag{5.22}$$

As established in 2.12 (convolution integrals), then

$$y(t) = x(t)*h(t) \tag{5.23}$$

In summary, Eq. 5.23 establishes that: the **output $y(t)$ is equal to the convolution of the input $x(t)$** with the **response to the impulse $h(t)$**.

Due to the commutability property of the convolution integral,

$$y(t) = h(t)*x(t) = \int_{0_-}^{t} h(\lambda)x(t-\lambda)d\lambda. \tag{5.24}$$

The lower limit of the integral in Eq. 5.24 was taken as $\lambda = 0_-$, predicting a possible presence of impulse, at time $t = 0$, at $h(t)$; as in Example 5.16, where the impulse response also contains impulse. If there is no impulse at $t = 0$, the limit must be taken as $\lambda = 0$.

The property that deals with the convolution derivative, Eq. 2.116, establishes that to derive a convolution it is enough to derive any of the two functions that are being convoluted. So, from Eq. 5.23,

$$y'(t) = x'(t) * h(t). \tag{5.25}$$

Equation 5.25 shows **that when deriving the input from an invariant linear circuit, the output is also derived**.

Logically, the inverse is also true, that is: integrating the function h(t) in Eq. 5.25 yields $y(t)$. It turns out that the integral of $h(t)$ is $r(t)$, the response to the unitary step $U_{-1}(t)$. Therefore,

$$y(t) = x'(t)*r(t) = \int_{0_-}^{t} x'(\lambda)r(t-\lambda)d\lambda, for\, all\, t \geq 0. \tag{5.26}$$

In summary, Eq. 5.26 states that: **the output $y(t)$ is equal to the convolution of the derivative of the input, $x'(t)$, with the response to the step $r(t)$**.

The lower limit of the integral was taken as 0_- because it may happen that $x(t)$ presents finite discontinuity at $t = 0$, causing an impulse at $x'(\lambda)$.

Due to the property of commutativity,

$$y(t) = r(t)*x'(t) = \int_{0}^{t} r(\lambda)\frac{dx(t-\lambda)}{d(t-\lambda)}d\lambda. \tag{5.27}$$

In conclusion, it can then be said that Eq. 5.22 is a consequence of the fact that the sign $x(t)$ has been decomposed into impulses, and that Eq. 5.26 is a consequence of it having been decomposed in steps.

In general, Eqs. 5.24 and 5.27 are more difficult to apply.

Similarly, if the input signal $x(t)$ is decomposed into ramps, the output $y(t)$ will be given by

$$y(t) = x''(t)*a(t) = \int_{0_-}^{t} x''(\lambda)a(t-\lambda)d\lambda, \qquad (5.28)$$

where $a(t) = \int r(t)dt$ is the response to the unitary ramp $U_{-2}(t)$.

Equation 5.23, together with property 2.11 of Sect. 2.12 of Chap. 2, shows why the response of a circuit to a causal pulse of duration T_x lasts much longer than T_x

According to the aforementioned property, the duration of the output signal, T_y, is the sum of the duration of the input signal, T_x, with the duration of the impulse response, T_h:

$$T_y = T_x + T_h. \qquad (5.29)$$

Therefore, when the impulse response is an exponential signal with time constant T, the output signal lasts approximately $T_y = T_x + 5T$.

When $h(t)$ contains impulse, that is, when $h(t) = AU_0(t) + h_1(t)$, where $h_1(t)$ does not contain impulse, according to Eq. 5.23,

$$y(t) = x(t) * h(t) = x(t) * [AU_0(t) + h_1(t)] = x(t) * AU_0(t) + x(t) * h_1(t)$$

$$= A\int_{0_-}^{t_+} x(\lambda)U_0(t-\lambda)d\lambda + \int_{0}^{t} x(\lambda)h_1(t-\lambda)d\lambda$$

Therefore,

$$y(t) = Ax(t) + \int_{0}^{t} x(\lambda)h_1(t-\lambda)d\lambda. \qquad (5.30)$$

Equation 5.30 indicates that, in this special case, the input signal must appear as part of the output signal.

Example 5.36 If the impulse response of a linear and time-invariant circuit is given by $h(t) = (\epsilon^{-t} - \epsilon^{-3t})U_{-1}(t)$, determine the your answer $y(t)$ for the input $x(t) = (\epsilon^{-2t} - \epsilon^{-4t})U_{-1}(t)$.

Solution

$$x(\lambda) = \left(\epsilon^{-2\lambda} - \epsilon^{-4\lambda}\right)U_{-1}(\lambda)$$

$$h(t-\lambda) = \left(\epsilon^{-t+\lambda} - \epsilon^{-3t+3\lambda}\right)U_{-1}(t-\lambda)$$

By Eq. 5.20,

$$y(t) = x(t) * h(t) = \int_0^t \left(\epsilon^{-2\lambda} - \epsilon^{-4\lambda}\right)\left(\epsilon^{-t+\lambda} - \epsilon^{-3t+3\lambda}\right)d\lambda$$

$$= \epsilon^{-t}\int_0^t \left(\epsilon^{-\lambda} - \epsilon^{-3\lambda}\right)d\lambda - \epsilon^{-3t}\int_0^t \left(\epsilon^{\lambda} - \epsilon^{-\lambda}\right)d\lambda$$

$$= \epsilon^{-t}\left(\frac{\epsilon^{-\lambda}}{-1} + \frac{\epsilon^{-3\lambda}}{3}\right)\Big|_0^t - \epsilon^{-3t}\left(\epsilon^{\lambda} + \epsilon^{-\lambda}\right)\Big|_0^t$$

$$= \epsilon^{-t}\left(-\epsilon^{-t} + \frac{\epsilon^{-3t}}{3} + \frac{2}{3}\right) - \epsilon^{-3t}\left(\epsilon^{t} + \epsilon^{-t} - 2\right)$$

So,

$$y(t) = 2\left(\frac{\epsilon^{-t}}{3} - \epsilon^{-2t} + \epsilon^{-3t} - \frac{\epsilon^{-4t}}{3}\right)U_{-1}(t).$$

Input $x(t)$ lasts approximately 2.5 units of time; $h(t)$ lasts approximately 5 units of time. Therefore, $y(t)$ lasts approximately 7.5 units of time.

Example 5.37 Using the convolution integral, find the current $i(t)$ in the circuit of Fig. 5.140 when the input voltage is:

(a) $e(t) = \epsilon^{-t}U_{-1}(t)$,
(b) $e(t) = \sin t[U_{-1}(t) - U_{-1}(t - \pi)]$.

Solution
The impulse response of this circuit was determined in example 5.14, and it is worth:

$$h(t) = i(t) = \frac{1}{R}U_0(t) - \frac{1}{R^2C}\epsilon^{-t/RC}U_{-1}(t), \text{ for } e(t) = U_0(t).$$

Then,

$$h(t - \lambda) = \frac{1}{R}U_0(t - \lambda) - \frac{1}{R^2C}\epsilon^{-(t-\lambda)/RC}U_{-1}(t - \lambda).$$

Fig. 5.140 Electric circuit of
Example 5.37

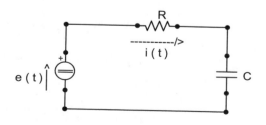

So,

$$i(t) = e(t) * h(t) = \int\limits_{0_-}^{t} e(\lambda)h(t - \lambda)d\lambda.$$

(a) For $e(t) = \epsilon^{-t}U_{-1}(t)$,

$$i(t) = \epsilon^{-t}U_{-1}(t) * \left[\frac{1}{R}U_0(t) - \frac{1}{R^2C}\epsilon^{-t/RC}U_{-1}(t)\right]$$

$$= \frac{1}{R}\int\limits_{0}^{t_+}\epsilon^{-\lambda}U_0(t-\lambda)d\lambda - \frac{1}{R^2C}\int\limits_{0}^{t}\epsilon^{-\lambda}\epsilon^{-(t-\lambda)/RC}d\lambda$$

$$= \frac{1}{R}\cdot\epsilon^{-t} - \frac{\epsilon^{-t/RC}}{R^2C}\int\limits_{0}^{t}\epsilon^{\frac{1-RC}{RC}\lambda}d\lambda = \frac{\epsilon^{-t}}{R} - \frac{\epsilon^{-t/RC}}{R^2C}\cdot\frac{RC}{1-RC}\epsilon^{\frac{1-RC}{RC}\lambda}\Big|_{0}^{t}$$

$$= \frac{\epsilon^{-t}}{R} - \frac{\epsilon^{-t/RC}}{R(1-RC)}\left(\epsilon^{\frac{1-RC}{RC}t} - 1\right) = \frac{\epsilon^{-t}}{R} - \frac{\epsilon^{-t} - \epsilon^{-t/RC}}{R(1-RC)}$$

$$= \frac{\epsilon^{-t/RC} - RC\epsilon^{-t}}{R(1-RC)}\cdot U_{-1}(t).$$

This result can be confirmed by directly solving the circuit by the usual classic method, for $e(t) = \epsilon^{-t}U_{-1}(t)$. From Example 5.14, or directly from Fig. 5.140, we have that:

$$i'(t) + \frac{1}{RC}i(t) = \frac{1}{R}e'(t).$$

If $e(t) = \epsilon^{-t}U_{-1}(t)$, $e'(t) = U_0(t) - \epsilon^{-t}U_{-1}(t)$. So,

$$i'(t) + \frac{1}{RC}i(t) = \frac{1}{R}U_0(t) - \frac{1}{R}\epsilon^{-t}U_{-1}(t) \doteq \frac{1}{R}U_0(t) - \frac{1}{R}U_{-1}(t) + \ldots, i(0_-) = 0.$$

For $t > 0$,

$$\left(D + \frac{1}{RC}\right)i(t) = -\frac{\epsilon^{-t}}{R}.$$

$$i_H(t) = k_1\epsilon^{-t/RC}.$$

$$i_P(t) = \frac{1}{D + \frac{1}{RC}}\left(-\frac{\epsilon^{-t}}{R}\right) = \frac{-\frac{\epsilon^{-t}}{R}}{-1 + \frac{1}{RC}} = \frac{C\epsilon^{-t}}{RC - 1}.$$

$$i(t) = k_1 \epsilon^{-t/RC} + \frac{C\epsilon^{-t}}{RC - 1}.$$

From the previous differential equation, we have that $i(0_+) = \frac{1}{R}$. So,

$$\frac{1}{R} = k_1 + \frac{C}{RC - 1}. \text{ Then, } k_1 = \frac{1}{R} - \frac{C}{RC - 1} = \frac{1}{R(1 - RC)}.$$

Therefore,

$$i(t) = \frac{\epsilon^{-t/RC}}{R(1 - RC)} - \frac{C\epsilon^{-t}}{1 - RC} = \frac{\epsilon^{-t/RC} - RC\epsilon^{-t}}{R(1 - RC)}.U_{-1}(t),$$

confirming the result obtained through the convolution integral.

(b) For $0 < t < \pi$, $e(t) = \sin t$ and

$$i(t) = \frac{1}{R}\int_0^{t_+} \sin\lambda U_0(t - \lambda)d\lambda - \frac{1}{R^2C}\int_0^t \sin\lambda\epsilon^{-(t-\lambda)/RC}d\lambda$$

$$= \frac{1}{R}\sin t - \frac{\epsilon^{-t/RC}}{R^2C}\int_0^t \epsilon^{\lambda/RC}\sin\lambda d\lambda$$

From $\int \epsilon^{ax}\sin bx dx = \frac{\epsilon^{ax}(a\sin bx - b\cos bx)}{a^2 + b^2}$, with $x = \lambda$, $a = 1/RC$ and $b = 1$, comes that:

$$i(t) = \frac{1}{R}\sin t - \frac{\epsilon^{-t/RC}}{R^2C}\cdot\frac{\epsilon^{\lambda/RC}\left(\frac{1}{RC}\sin\lambda - \cos\lambda\right)}{\left(\frac{1}{RC}\right)^2 + 1}\bigg|_0^t$$

$$= \frac{1}{R}\sin t - \frac{\epsilon^{-t/RC}}{R^2C}\cdot\left[\frac{\epsilon^{t/RC}\left(\frac{1}{RC}\sin t - \cos t\right)}{\left(\frac{1}{RC}\right)^2 + 1} - \frac{-1}{\left(\frac{1}{RC}\right)^2 + 1}\right]$$

$$= \frac{1}{R}\sin t + \frac{RC\cos t - \sin t}{R\left[1 + (RC)^2\right]} - \frac{C\epsilon^{-t/RC}}{1 + (RC)^2}$$

$$= -\frac{C\epsilon^{-t/RC}}{1 + (RC)^2} + \frac{C(\cos t + RC\sin t)}{1 + (RC)^2} = \frac{-C\epsilon^{-t/RC}}{1 + (RC)^2} + \frac{C\cos[t - arctg(RC)]}{\sqrt{1 + (RC)^2}}, \text{ for } 0 < t < \pi.$$

For $t > \pi$ the integral from 0 to t must be replaced by the integral from 0 to π, since $e(t) = 0$ for $t > \pi$.

$$i(t) = e(t) * h(t) = \int_0^\pi \sin \lambda U_0(t - \lambda)d\lambda - \frac{1}{R^2C}\int_0^\pi \sin \lambda \epsilon^{-(t-\lambda)/RC}d\lambda.$$

Since $U_0(t - \lambda)$ occurs only at $\lambda = t > \pi$, $\int_0^\pi \sin \lambda U_0(t - \lambda)d\lambda = 0$. So,

$$i(t) = -\frac{\epsilon^{-t/RC}}{R^2C}\int_0^\pi \epsilon^{\lambda/RC}\sin \lambda d\lambda = -\frac{\epsilon^{-t/RC}}{R^2C} \cdot \frac{\epsilon^{\lambda/RC}\left(\frac{1}{RC}\sin \lambda - \cos\lambda\right)}{\left(\frac{1}{RC}\right)^2 + 1}\Bigg|_0^\pi$$

$$= -\frac{\epsilon^{-t/RC}}{R^2C} \cdot \frac{\epsilon^{\lambda/RC}\left(RC\sin \lambda - R^2C^2\cos\lambda\right)}{1 + (RC)^2}\Bigg|_0^\pi = -\frac{\epsilon^{-t/RC}}{R^2C} \cdot \frac{R^2C^2\left(\epsilon^{\pi/RC} + 1\right)}{1 + (RC)^2}$$

$$= \frac{-C\left(\epsilon^{\pi/RC} + 1\right)}{1 + (RC)^2} \cdot \epsilon^{-t/RC}, \text{ for } t > \pi.$$

To confirm the results of this item b, using the classic method, the superposition theorem will be used, considering that:

$$e(t) = \sin t[U_{-1}(t) - U_{-1}(t - \pi)] = \sin t.U_{-1}(t) - \sin t.U_{-1}(t - \pi)$$

$$= \sin t.U_{-1}(t) + \sin(t - \pi).U_{-1}(t - \pi).$$

Therefore, the response $i(t)$ to $e(t) = \sin t.U_{-1}(t)$. will be obtained first. Recalling, from Example 5.14, that the differential equation of the circuit is

$$i'(t) + \frac{1}{RC}i(t) = \frac{1}{R}e'(t), \text{ then } e'(t) = cost.U_{-1}(t) \text{ and,}$$

$$i'(t) + \frac{1}{RC}i(t) = \frac{1}{R}cost.U_{-1}(t) = \frac{1}{R}U_{-1}(t) + \ldots, i(0_-) = 0.$$

$$i_H(t) = k_1\epsilon^{-t/RC}.$$

Using the phasor method to find the particular integral, it comes that:

$$\left(j1 + \frac{1}{RC}\right)\dot{I}_P = 1\angle 0°$$

$$\dot{I}_P = \frac{C}{1 + jRC}\angle 0° = \frac{C}{\sqrt{1 + (RC)^2}}\angle - arctg(RC)$$

$$i_P(t) = \frac{C}{\sqrt{1+(RC)^2}} \cos\left[t - tg^{-1}(RC)\right].$$

$$i(t) = k_1 \epsilon^{-t/RC} + \frac{C}{\sqrt{1+(RC)^2}} \cos\left[t - tg^{-1}(RC)\right].$$

Denominating $S = \sqrt{1+(RC)^2}$ and $\phi = tg^{-1}(RC)$, it becomes that

$$i(t) = k_1 \epsilon^{-t/RC} + \frac{C}{S} \cos(t - \phi).$$

From the previous differential equation, $i(0_+) = 0$.
So,

$0 = k_1 + \frac{C}{S} \cos \phi \therefore k_1 = -\frac{C}{S} \cos \phi$. Then,

$i(t) = -\frac{C}{S} \cos \phi \epsilon^{-t/RC} + \frac{C}{S} \cos(t - \phi)$. On the other hand,

$$tg\phi = \frac{\sin \phi}{\cos \phi} = \frac{RC}{1},$$

$$\frac{\sin^2 \phi}{\cos^2 \phi} = \frac{(RC)^2}{1}; \frac{\sin^2 \phi + \cos^2 \phi}{\cos^2 \phi} = \frac{(RC)^2 + 1}{1}$$

$$\cos \phi = \frac{1}{\sqrt{1+(RC)^2}}; \frac{\cos \phi}{S} = \frac{1}{1+(RC)^2}.$$

$$i(t) = \left[\frac{-C\epsilon^{-t/RC}}{1+(RC)^2} + \frac{C}{\sqrt{1+(RC)^2}} \cos(t - \phi)\right].U_{-1}(t). \qquad (5.31)$$

Therefore, the answer to $e(t) = \sin(t - \pi).U_{-1}(t - \pi)$ is

$$i(t-\pi) = \left[\frac{-C\epsilon^{\pi/RC}\epsilon^{-t/RC}}{1+(RC)^2} + \frac{C}{\sqrt{1+(RC)^2}}\cos(t-\pi-\phi)\right].U_{-1}(t-\pi)$$

$$= \left[\frac{-C\epsilon^{\pi/RC}\epsilon^{-t/RC}}{1+(RC)^2} - \frac{C}{\sqrt{1+(RC)^2}}\cos(t-\phi)\right].U_{-1}(t-\pi).$$

(5.32)

The answer sought is to overlap Eqs. 5.31 and 5.32, resulting in

$$i(t) = \begin{cases} \frac{-C\epsilon^{-t/RC}}{1+(RC)^2} + \frac{C}{\sqrt{1+(RC)^2}}\cos(t-\phi), & \text{for } 0 < t < \pi \\ \frac{-C\left(\epsilon^{\pi/RC}+1\right)}{1+(RC)^2}.\varepsilon^{-t/RC}, & \text{for } t > \pi, \end{cases}$$

thus confirming the results obtained through the convolution integral.

Example 5.38 In Fig. 5.141 we have what is called cascade circuits, since the output of circuit N_1 is the input of circuit N_2. If the connection of N_2 does not affect the output of N_1, they are called isolated. Otherwise, it is said that N_2 carries N_1. If $h_1(t)$ and $h_2(t)$ are the impulse responses of two isolated circuits connected in cascade, as in Fig. 5.141, find the impulse response of the combination. Also find the answer to the step of the cascading connection.

Solution
From the statement,
$\quad h_1(t) = z(t)$ when $x(t) = U_0(t)$ and $h_2(t) = y(t)$ when $z(t) = U_0(t)$.
$\quad$ The Convolution Theorem applied to the N_2 circuit states that $y(t) = z(t) * h_2(t)$. If $x(t) = U_0(t)$, then $z(t) = h_1(t)$. Therefore, $y(t) = h_1(t) * h_2(t)$. On the other hand, for the connection as a whole, $y(t) = h(t) * x(t) = h(t) * U_0(t) = h(t)$. So by comparison,

$$h(t) = h_1(t) * h_2(t). \tag{5.33}$$

$r_1(t) = z(t)$ when $x(t) = U_{-1}(t)$ and $r_2(t) = y(t)$ when $z(t) = U_{-1}(t)$.

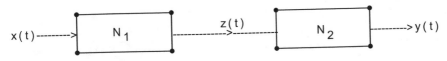

$x(t)$ -------> N_1 -------- $z(t)$ -------- N_2 -------->$y(t)$

Fig. 5.141 Cascade or latter circuit of Example 5.38

The DuHamel Integral applied to the N_2 circuit results in $y(t) = z'(t) * r_2(t)$. If $x(t) = U_{-1}(t)$, then $z(t) = r_1(t)$ and $z'(t) = r_1'(t)$. Therefore, $y(t) = r_1'(t) * r_2(t)$. However, $y(t) = x'(t) * r(t) = r(t) * x'(t) = r(t) * U_0(t) = r(t)$. By comparison,

$$r(t) = r_1'(t) * r_2(t). \qquad (5.34)$$

Example 5.39 If the impulse response of each of the circuits in Fig. 5.141 is $h_1(t) = h_2(t) = A\epsilon^{-\alpha t}$ for $t > 0$, determine the impulse response of the cascade combination.

Solution

$$h(t) = h_1(t) * h_2(t) = \int_0^t h_1(\lambda)h_2(t - \lambda)d\lambda = \int_0^t A\epsilon^{-\alpha\lambda}A\epsilon^{-\alpha(t-\lambda)}d\lambda$$

$$= A^2\epsilon^{-\alpha t}\int_0^t d\lambda = A^2 t\epsilon^{-\alpha t}.U_{-1}(t).$$

5.7   Proposed Problems

P.5-1 Find the circuit time constant in Fig. 5.142, for $t > 0$.

Answer: $T = 0.2[\text{s}]$.

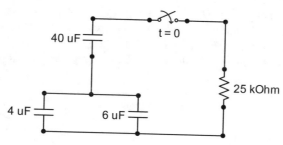

Fig. 5.142 Electric circuit of Problem P.5-1

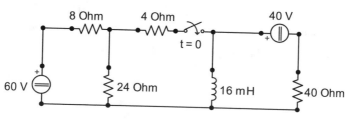

Fig. 5.143 Electric circuit of Problem P.5-2

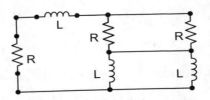

Fig. 5.144 Electric circuit of Problem P.5-3

P.5-2 Find the circuit time constant shown in Fig. 5.143, for $t < 0$ and for $t > 0$.

Answer: $T = 0.4[ms]$ for $t < 0$ and $T = 2[ms]$ for $t > 0$.

P.5-3 Find the circuit time constant in Fig. 5.144.

Answer: $T = L/R$.

P.5-4 The switch shown in the circuit of Fig. 5.145 has been open for a long time, before being closed at time $t = 0$.

(a) Determine

$i_1(0_-), e_1(0_-), i_C(0_-), e_C(0_-), i_2(0_-), e_2(0_-), i_k(0_-), e_k(0_-), i_3(0_-), e_3(0_-), i_L(0_-), e_L(0_-).$

(b) Determine

$i_1(0_+), e_1(0_+), i_C(0_+), e_C(0_+), i_2(0_+), e_2(0_+), i_k(0_+), e_k(0_+), i_3(0_+), e_3(0_+), i_L(0_+), e_L(0_+).$

(c) Determine

$i_1(\infty), e_1(\infty), i_C(\infty), e_C(\infty), i_2(\infty), e_2(\infty), i_k(\infty), e_k(\infty), i_3(\infty), e_3(\infty), i_L(\infty), e_L(\infty).$

Answer:

(a) $i_1(0_-) = 1[A], e_1(0_-) = 40[V], i_C(0_-) = 0[A], e_C(0_-) = 80[V], i_2(0_-) = 1[A],$

$e_2(0_-) = 50[V], i_k(0_-) = 0[A], e_k(0_-) = 30[V], i_3(0_-) = 1[A], e_3(0_-) = 30[V],$

$$i_L(0_-) = 1[A], e_L(0_-) = 0[V].$$

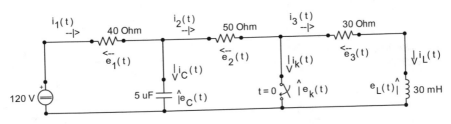

Fig. 5.145 Electric circuit of Problem P.5-4

(b) $i_1(0_+) = 1[A], e_1(0_+) = 40[V], i_C(0_+) = -0.6[A], e_C(0_+) = 80[V],$

$i_2(0_+) = 1.6[A], e_2(0_+) = 80[V], i_k(0_+) = 0.6[A], e_k(0_+) = 0[V],$

$i_3(0_+) = 1[A], e_3(0_+) = 30[V], i_L(0_+) = 1[A], e_L(0_+) = -30[V].$

(c) $i_1(\infty) = 4/3[A], e_1(\infty) = 160/3[V], i_C(\infty) = 0[A], e_C(\infty) = 200/3[V],$

$i_2(\infty) = 4/3[A], e_2(\infty) = 200/3[V], i_k(\infty) = 4/3[A], e_k(\infty) = 0[V],$

$i_3(\infty) = 0[A], e_3(\infty) = 0[V], i_L(\infty) = 0[A], e_L(\infty) = 0[V].$

P.5-5 A $10[\mu F]$ capacitor, with an initial charge of 200 $[\mu C]$, is connected in series with a 10 $[\Omega]$ resistor, by closing a switch at $t = 0$. (a) Establish the expression of the voltage $e_R(t)$ in the resistor for $t > 0$. (b) Calculate the time required for the transient voltage in the resistor to drop from 15 $[V]$ to 5 $[V]$.

Answer: (a) $e_R(t) = 20\epsilon^{-10,000t}[V]$, to $t > 0$ [s]. (b) 109.9 [μs].

P.5-6 The circuit capacitor in Fig. 5.146 has an initial charge of 600 $[\mu C]$. Get the expressions of $i(t)$, $e(t)$ and $q(t)$, for $t \geq 0$ [s].

Answer: $i(t) = -0.1\epsilon^{c-500t}[A]$, for $t > 0$[s].

$e(t) = 200 + 100\epsilon^{-500t}[V]$, for $t \geq 0$[s].

$q(t) = 400 + 200\epsilon^{-500t}[\mu C]$, for $t \geq 0$[s].

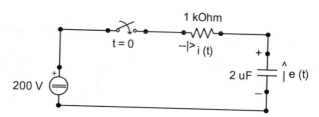

Fig. 5.146 Electric circuit of Problem P.5-6

Fig. 5.147 Electric circuit of
Problem P.5-7

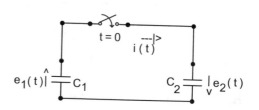

P.5-7 If capacitances $C_1 = 2[\mu F]$ and $C_2 = 1[\mu F]$, shown in Fig. 5.147, have voltages $e_1(0_-) = 150[V]$ and $e_2(0_-) = 50[V]$, find the voltages $e_1(0_+)$ and $e_2(0_+)$ and the current $i(t)$.

Answer: $e_1(0_+) = 250/3[V]$, $e_2(0_+) = -250/3[V]$,

$$i(t) = \frac{C_1 C_2}{C_1 + C_2}[e_1(0_-) + e_2 0_-)]U_0(t) = (400/3)U_0(t)[\mu A].$$

P.5-8 In the circuit shown in Fig. 5.148, the capacitor is de-energized at $t = 0_-[s]$. Find $i(t)$ and $e(t)$, for $t > 0[s]$.

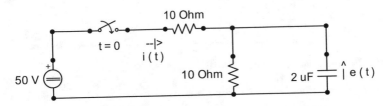

Fig. 5.148 Electric circuit of Problem P.5-8

Answer: $i(t) = 2.5(1 + \epsilon^{-t/10})[A]$, for t > $0[\mu s]$.

$$e(t) = 25\left(1 - \epsilon^{-t/10}\right)[V], \text{ for } t \geq 0[\mu s].$$

P.5-9 Capacitors C_1 and C_2 in the circuit of Fig. 5.149 are de-energized at $t = 0_-[s]$. If the switch is closed at $t = 0[s]$, establish the expressions of $i(t)$, $e_1(t)$ and $e_2(t)$, for $t > 0$.

Answer: $i(t) = \frac{E}{R}\epsilon^{-(C_1 + C_2)t/RC_1C_2} \cdot U_{-1}(t).$

$$e_1(t) = \frac{C_2 E}{C_1 + C_2}\left[1 - \epsilon^{-(C_1 + C_2)t/RC_1C_2}\right]U_{-1}(t).$$

Fig. 5.149 Electric circuit of Problem P.5-9

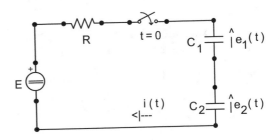

$$e_2(t) = \frac{C_1 E}{C_1 + C_2}\left[1 - \epsilon^{-(C_1 + C_2)t/RC_1C_2}\right]U_{-1}(t).$$

P.5-10 Determine $e_1(t)$ and $e_2(t)$ in the circuit of Fig. 5.150, for $t \geq 0$, knowing that $e_1(0_-) = E$ and $e_2(0_-) = E/3$.

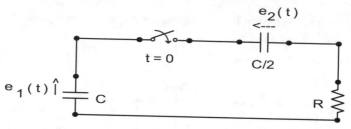

Fig. 5.150 Electric circuit of Problem P.5-10

Answer: $e_1(t) = \frac{E}{9}\left(7 + 2\epsilon^{-3t/RC}\right)$, for $t \geq 0$.

$$e_2(t) = \frac{E}{9}\left(7 - 4\epsilon^{-3t/RC}\right), \text{ for } t \geq 0.$$

P.5-11 The circuit shown in Fig. 5.151 was operating in a steady state when, at time $t = 0$, switch k was closed. Find the current $i_k(t)$ and the voltage $e_1(t)$, for $t > 0$, knowing that $e_2(0_-) = E/2$.

Fig. 5.151 Electric circuit of Problem P.5-11

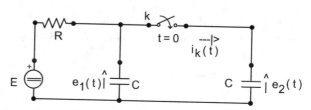

Answer: $i_k(t) = \frac{CE}{4}U_0(t) + \frac{E}{8R}\epsilon^{-t/2RC}U_{-1}(t)$.

$$e_1(t) = e_2(t) = E - \frac{E}{4}\epsilon^{-t/2RC}, \text{ for } t > 0.$$

P.5-12 The circuit in Fig. 5.152 was operating in a steady state when, instantly, switch k was opened at $t = 0$. Determine $e_k(t)$ and $e_C(t)$, for $t > 0$.

Answer: $e_k(t) = E\left(1 - \frac{1}{2}\epsilon^{-t/2RC}\right)U_{-1}(t)$,

$$e_C(t) = E\epsilon^{-t/2RC}, \text{ for } t \geq 0.$$

Fig. 5.152 Electric circuit of Problem P.5-12

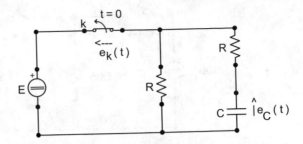

P.5-13 A simple *RC* timer (circuit that measures time, indirectly, through the voltage that the capacitor reaches) has a switch that, when closed, connects in series a constant source of E [V], a resistor of R [M Ω] and a C [µF], capacitor, discharged. Find the time between closing and opening the switch if the capacitor has its voltage changed from zero to E/k[V], $k > 1$, during this time interval.

Answer: $t = RC . \ln \frac{k}{k-1}$.

P.5-14 The circuit in Fig. 5.153 was operating in a steady state when switch k was opened at $t = 0$. Determine the voltage $e_k(t)$ at the switch terminals, for $t > 0$.

Answer: $e_k(t) = E\left(\frac{1}{2} + \frac{1}{6}\epsilon^{-2Rt/L}\right)U_{-1}(t)$.

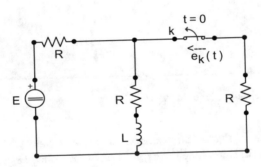

Fig. 5.153 Electric circuit of Problem P.5-14

P.5-15 The circuit in Fig. 5.154 operated in a steady state when the switch was closed at time $t = 0$. Find the expression of $i_0(t)$, for $t > 0$.

Answer: $i_0(t) = \left(\frac{20}{3} - \frac{5}{3}\epsilon^{-7.5t}\right)U_{-1}(t)[A]$.

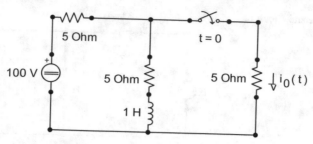

Fig. 5.154 Electric circuit of Problem P.5-15

P.5-16 The circuit shown in Fig. 5.155 was operating in steady state when, at time $t = 0$, switch k was opened. Find the voltage expression $e_k(t)$ at the terminals of switch k, for $t > 0$.

Answer: $e_k(t) = E\left(1 + \epsilon^{-Rt/L}\right)U_{-1}(t)[V]$.

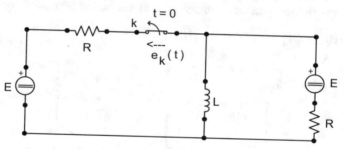

Fig. 5.155 Electric circuit of Problem P.5-16

P.5-17 The circuit in Fig. 5.156 operated in a steady state when, at time $t = 0$, switch k was opened. Determine the expression of the voltage $e_k(t)$ at its terminals, for $t \geq 0$.

Answer: $e_k(t) = \frac{2LE}{3R} U_0(t) + E\left(1 + \frac{1}{3}\epsilon^{-2Rt/L}\right)U_{-1}(t)$.

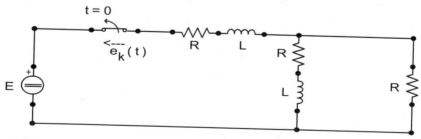

Fig. 5.156 Electric circuit of Problem P.5-17

P.5-18 The circuit shown in Fig. 5.157 was operating in a steady state when, at an instant considered as $t = 0$, switch k was closed. Determine the expression of the current $i_k(t)$ circulating through the switch, for $t > 0$. (Use the loop method).

Answer: $i_k(t) = \frac{E}{3R}\left(1 - \epsilon^{-Rt/L}\right)U_{-1}(t)$.

Fig. 5.157 Electric circuit of Problem P.5-18

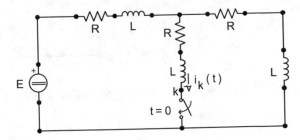

P.5-19 The circuit in Fig. 5.158 was operating in a steady state when switch k was opened at $t = 0$. Determine $e_k(t)$ and $i(t)$, for $t > 0$.

Answer: $e_k(t) = 2U_0(t) + \left(6 + \frac{8}{3}\epsilon^{-2t/3}\right)U_{-1}(t)[V]$.

$$i(t) = 3 + 2\epsilon^{-2t/3}[A], \text{ for } t \geq 0.$$

Fig. 5.158 Electric circuit of Problem P.5-19

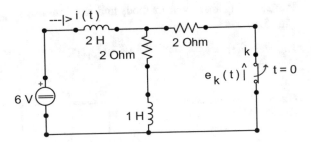

P.5-20 The switch in the circuit shown in Fig. 5.159 was closed for a long time before opening at $t = 0$. Determine $e_0(t)$ and $e_k(t)$, for $t > 0$.

Answer: $e_k(t) = 6RI\left(1 + \frac{1}{5}\epsilon^{-6Rt/L}\right)U_{-1}(t)[V]$.

$$e_0(t) = -\frac{12}{5}RI\epsilon^{-6Rt/L} \cdot U_{-1}(t)[V].$$

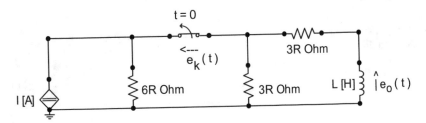

Fig. 5.159 Electric circuit of Problem P.5-20

P.5-21 Determine the step response, $r(t)$, of the circuit in Fig. 5.160.

Answer: $r(t) = \frac{1}{3}\left(1 - \epsilon^{-3Rt/2L}\right)U_{-1}(t)$.

Fig. 5.160 Electric circuit of
Problem P.5-21

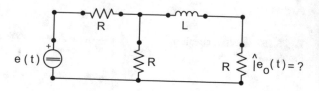

P.5-22 Find the step response, $r(t)$, of the circuit shown in Fig. 5.161.

Answer: $r(t) = \frac{1}{3}\epsilon^{-2t/3RC}.U_{-1}(t)$.

Fig. 5.161 Electric circuit of
Problem P.5-22

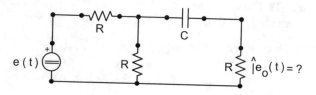

P.5-23 Using the nodal method, find the impulse response, $h(t)$, of the circuit in Fig. 5.162.

Answer: $h(t) = \frac{1}{(N+1)C}\epsilon^{-t/(N+1)RC}.U_{-1}(t)$.

Fig. 5.162 Electric circuit of
Problem P.5-23

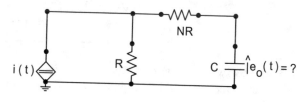

P.5-24 Find the impulse response, $h(t)$, of the circuit contained in Fig. 5.163. (Use the nodal method).

Answer: $h(t) = NRU_0(t) - \frac{N(N+1)R^2}{L}\epsilon^{-(N+1)Rt/L}.U_{-1}(t)$.

Fig. 5.163 Electric circuit of
Problem P.5-24

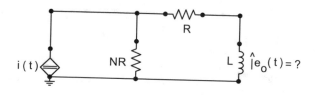

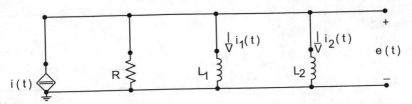

Fig. 5.164 Electric circuit of Problem P.5-25

P.5-25 Find the impulse responses $h(t) = e(t)$, $h_1(t) = i_1(t)$ and $h_2(t) = i_2(t)$ in the circuit in Fig. 5.164.

Answer: $h(t) = RU_0(t) - \frac{(L_1 + L_2)R^2}{L_1 L_2} \epsilon^{-R(L_1 + L_2)t/L_1 L_2} . U_{-1}(t).$

$$h_1(t) = \frac{R}{L_1} \epsilon^{-R(L_1 + L_2)t/L_1 L_2} . U_{-1}(t).$$

$$h_2(t) = \frac{R}{L_2} \epsilon^{-R(L_1 + L_2)t/L_1 L_2} . U_{-1}(t).$$

P.5-26 Find the impulse responses $h_1(t) = e_C(t)$ and $h_2(t) = i_C(t)$ in the circuit in Fig. 5.165.

Answer: $h_1(t) = \frac{1-B}{2C} \epsilon^{-(1+B)t/2RC} . U_{-1}(t).$

$$h_2(t) = \frac{1-B}{2} U_0(t) + \frac{B^2 - 1}{4RC} \epsilon^{-(1+B)t/2RC} . U_{-1}(t).$$

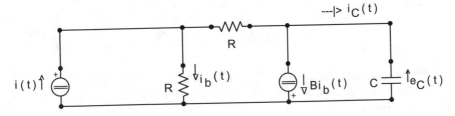

Fig. 5.165 Electric circuit of Problem P.5-26

P.5-27 Determine the voltage $e_0(t)$ in the circuit of Fig. 5.166, for $t \geq 0$, knowing that $i(t) = \begin{cases} 0 & for\, t < 0. \\ 2 & for\, 0 < t < 1. \\ -1 & for\, 1 < t < 2. \\ 0 & for\, t > 2. \end{cases}$ or $i(t) = 2U_{-1}(t) - 3U_{-1}(t-1) + U_{-1}(t-2).$

Fig. 5.166 Electric circuit of
Problem P.5-27

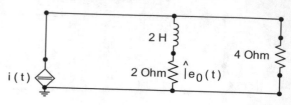

$$\text{Answer: } e_0(t) = \begin{cases} 0 & for \ t \leq 0. \\ \frac{4}{3}(1 - \epsilon^{-3t}) & for \ 0 \leq t \leq 1. \\ -\frac{2}{3} + (2\epsilon^3 - \frac{4}{3})\epsilon^{-3t} & for \ 1 \leq t \leq 2. \\ (2\epsilon^3 - \frac{2}{3}\epsilon^6 - \frac{4}{3})\epsilon^{-3t} & for \ t \geq 2. \end{cases}$$

P.5-28 Find $e_0(t)$, for $-\infty < t < \infty$, in the circuit shown in Fig. 5.167, knowing
that:

$$i(t) = \begin{cases} I & for \ t < 0. \\ \frac{I}{2} & for \ 0 < t < a. \\ I & for \ t > a. \end{cases}$$

Or,

$$i(t) = I - \frac{I}{2}U_{-1}(t) + \frac{I}{2}U_{-1}(t - a) = IU_{-1}(t + \infty) - \frac{I}{2}U_{-1}(t) + \frac{I}{2}U_{-1}(t - a).$$

$$\text{Answer: } e_0(t) = \begin{cases} \frac{RI}{2} & for \ t \leq 0. \\ (RI/4)[1 + e^{-(2Rt/L)}] & for \ 0 \leq t \leq a. \\ (RI/4)[2 + (1 - e^{(2Ra/L)})e^{(-2Rt/L)}] & for \ t \geq a. \end{cases}$$

Fig. 5.167 Electric circuit of
Problem P.5-28

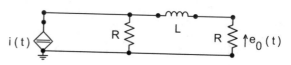

P.5-29 (a) Find the step response, $r(t)$, of the circuit in Figure P.5.168.
   (b) Find the answer $a(t)$, to the unitary ramp $U_{-2}(t)$, of the circuit in
Figure P.5.168, considering that: $a(t) = \int_{0_-}^{t} r(\lambda)d\lambda$.
   (c) Break down the voltage into singular functions

$$e(t) = \begin{cases} 0[V] & for \ t < 0. \\ -\frac{t}{2} + 2[V] & for \ 0 < t < 2[s]. \\ 0[V] & for \ t > 2[s]. \end{cases}$$

   (d) Find the voltage $e_0(t)$ in the circuit of Fig. 5.168 for the signal $e(t)$ of item c.

Fig. 5.168 Electric circuit of
Problem P.5-29

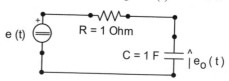

Answer: (a) $r(t) = (1 - \epsilon^{-t})U_{-1}(t)$.

(b) $a(t) = (t - 1 + \epsilon^{-t})U_{-1}(t)$.

(c) $e(t) = 2U_{-1}(t) - \frac{1}{2}U_{-2}(t) - U_{-1}(t-2) + \frac{1}{2}U_{-2}(t-2)$.

(d) $e_0(t) = \begin{cases} 0[V] & for\ t < 0 \\ -\frac{t}{2} + \frac{5}{2}(1 - \epsilon^{-t})[V] & for\ 0 \le t \le 2[s]. \\ (\frac{3}{2}\epsilon^2 - \frac{5}{2})\epsilon^{-t}[V] & for\ t \ge 2[s]. \end{cases}$

P.5-30 Calculate the voltage $e_0(t)$ in the circuit of Fig. 5.169 for any t.

$$e(t) = \begin{cases} 10[V] & for\ t \le 0[s]. \\ -5t + 10[V] & for\ 0 \le t \le 2[s]. \\ 0[V] & for\ t \ge 2[s]. \end{cases}$$

Answer: $e_0(t) = \begin{cases} \frac{10}{3}[V] & for\ t \le 0[s]. \\ -\frac{5}{3}t + \frac{40}{9} - \frac{10}{9}\epsilon^{-3t/2}[V] & for\ 0 \le t \le 2[s]. \\ \frac{10}{9}(\epsilon^3 - 1)\epsilon^{-3t/2}[V] & for\ t \ge 2[s]. \end{cases}$

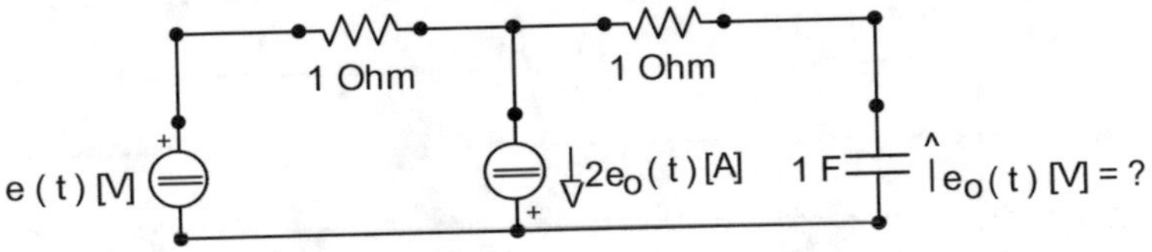

Fig. 5.169 Electric circuit of Problem P.5-30

P.5-31 If the voltage at a capacitance C, initially de-energized, is given by $e(t) = E\cos\omega t.U_{-1}(t)$, establish the expressions of current $i(t)$, power $p(t)$ and energy $W(t)$.

Answer: $i(t) = CEU_0(t) - \omega CE \sin \omega t.U_{-1}(t)$.

$$p(t) = \frac{CE^2}{2}U_0(t) - \frac{1}{2}\omega CE^2 \sin 2\omega t.U_{-1}(t).$$

$$W(t) = \frac{1}{2}CE^2 \cos^2 \omega t.U_{-1}(t).$$

P.5-32 Find the current and voltage expressions in the passive elements of the circuit in Fig. 5.170, knowing that the capacitor is initially de-energized. Then find the respective numerical solutions (Figs. 5.171 and 5.172).

$$e(t) = 180\cos 377t[V], \quad for\ t > 0[s], R = 10[\Omega], C = 1.000[\mu F].$$

Answer: $i(t) = [1.18\epsilon^{-100t} + 17.4\cos(377t + 14.86°)]U_{-1}(t)[A]$

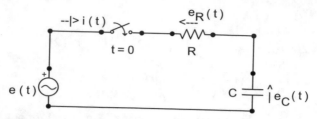

Fig. 5.170 Electric circuit of Problem P.5-32

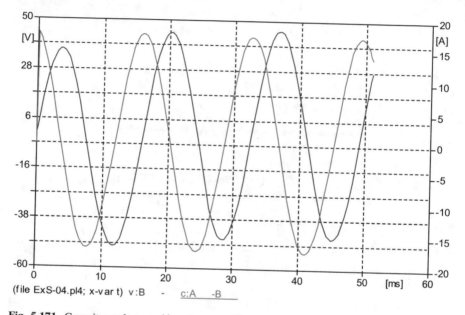

(file ExS-04.pl4; x-var t) v:B   -   c:A   -B

Fig. 5.171 Capacitor voltage $e_C(t)$ and current $i(t)$ of Problem P.5-32

$$e_R(t) = \left[11.8\epsilon^{-100t} + 174\cos(377t + 14.86°)\right]U_{-1}(t)[V]$$

$$e_C(t) = \left[-11.8\epsilon^{-100t} + 46.15\cos(377t - 75.14°)\right]U_{-1}(t)[V].$$

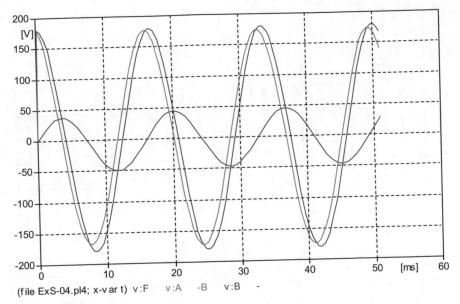

(file ExS-04.pl4; x-v ar t) v:F   v:A   -B   v:B   -

Fig. 5.172 Source $e(t)$, resistor $e_R(t)$ and capacitor $e_C(t)$ voltages of Problem P.5-32

P.5-33 The switch in the circuit of Fig. 5.173 is closed at time $t = 0$, applying the voltage $e(t) = Ecos(\omega t + \theta)$ in the series $R - C$ connection. The capacitor is initially energized with the voltage $e_C(0_-) = E_0$. Determine the current $i(t)$ that will circulate through the circuit and the voltages $e_R(t)$ and $e_C(t)$, for $t > 0$. Then perform the numerical solution for the same quantities mentioned (Figs. 5.174, 5.175 and 5.176).

   Datas:

$$E = 180[V], f = 600[Hz], \theta = 30°, E_0 = 60[V], R = 10[\Omega], C = 1,000[\mu F].$$

Answer: $i(t) = \left[-5.75\epsilon^{-100t} + 17.99\cos(1,200\pi t + 31.5°)\right]U_{-1}(t)[A]$

$$e_R(t) = \left[-57.5\epsilon^{-100t} + 17.99\cos(1,200\pi t + 31.5°)\right]U_{-1}(t)[V]$$

$$e_C(t) = \left[57.5\epsilon^{-100t} + 4.71\cos(1,200\pi t - 58.1°)\right][V], \text{ for } t \geq 0.$$

Fig. 5.173 Electric circuit of Problem P.5-33

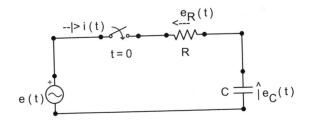

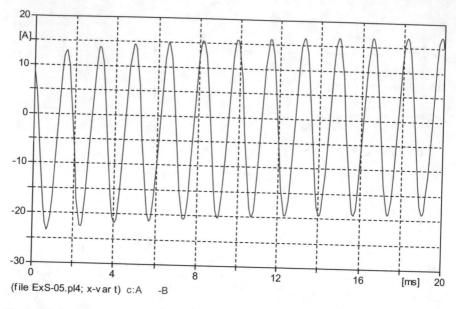

Fig. 5.174 Capacitor current $i(t)$ of Problem P.5-33

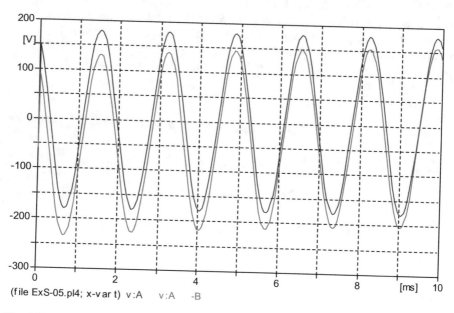

Fig. 5.175 Source voltage $e(t)$ and Voltage at resistor $e_R(t)$ of Problem P.5-33

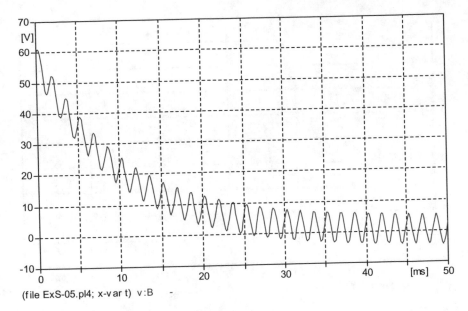

(file ExS-05.pl4; x-var t) v:B   -

Fig. 5.176 Capacitor voltage $e_C(t)$ of Problem P.5-33

P.5-34 The switch in the circuit of Fig. 5.177 is closed in an instant arbitrarily considered to be $t = 0$, causing $e(t) = E \sin(\omega t + \theta)$. Find the current $i(t)$ and the voltages $e_R(t)$ and $e_L(t)$, for $t > 0$. Then proceed to the numerical solution of the problem (Figs. 5.178 and 5.179).

Data: $E = 200[V], f = 180[Hz], \theta = 20°, R = 20[\Omega], L = 100[mH]$.

Answer: $i(t) = 1.51\epsilon^{-200t} - 1.74 \cos(360\pi t + 30°)[A], \, for \, t \geq 0$

$$e_R(t) = 30.2\epsilon^{-200t} - 34.8 \cos(360\pi t + 30°)[V], \, for \, t \geq 0$$

$$e_L(t) = -30.2\epsilon^{-200t} + 197 \cos(360\pi t - 60°)[V], \, for \, t > 0$$

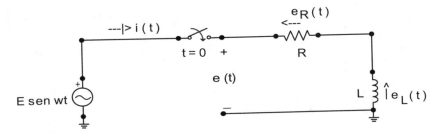

Fig. 5.177 Electric circuit of Problem P.5-34

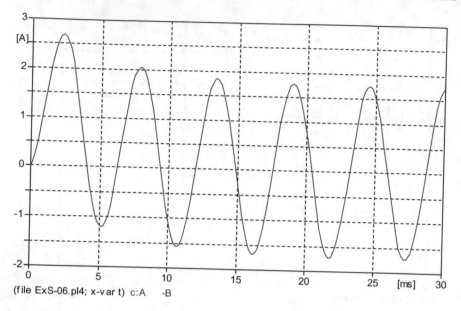

(file ExS-06.pl4; x-var t) c:A   -B

Fig. 5.178 Inductor current $i(t)$ of Problem P.5-34

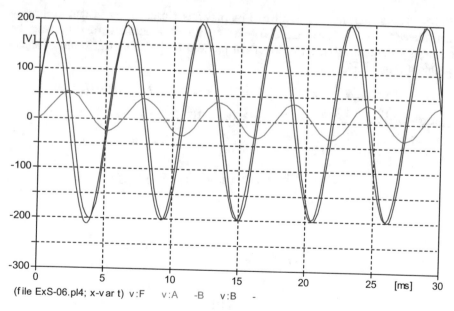

(file ExS-06.pl4; x-var t) v:F   v:A   -B   v:B   -

Fig. 5.179 Source, resistor, inductor voltages of Problem P.5-34

P.5-35 The circuit shown in Fig. 5.180 operated in steady state with the switch in position 1 when, at time $t = 0$, it was suddenly moved to position 2. Determine the voltage $e(t)$, for $t > 0$.

Answer: $e(t) = -10\epsilon^{-5t}$[V], for $t > 0$[s].

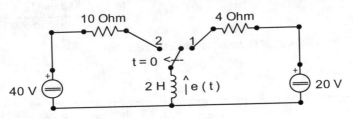

Fig. 5.180 Electric circuit of Problem P.5-35

P.5-36 In the circuit of Fig. 5.181, find the voltages $e_L(t)$ and $e_R(t)$, for $t > 0$. Using the ATPDraw program, find the numerical solution for $e_L(t)$ and $e_R(t)$, as well as for $i(t)$ and $i_T(t)$. Also plot the power and energy curves in the inductance (Figs. 5.182, 5.183, 5.184, 5.185, 5.186, 5.187, 5.188 and 5.189).

$$e(t) = 200 \cos^3 2000\pi t.U_{-1}(t)[\text{V}].$$

Answer:

$$e_L(t) = 3.08\epsilon^{-1,000t} + 118.51 \cos(2,000\pi t + 9.04°) + 39.94\cos(6,000\pi t + 3.04°)$$

$$e_R(t) = 1.23\epsilon^{-1,000t} + 119.05 \cos(2,000\pi t + 3.59°) + 39.96 \cos(6,000\pi t + 1.22°)$$

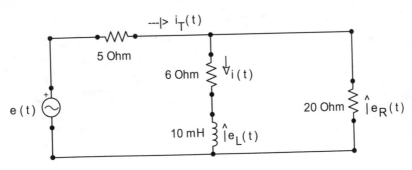

Fig. 5.181 Electric circuit of Problem P.5-36

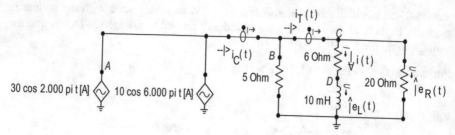

Fig. 5.182 Electric circuit of Problem P.5-36 for digital solution

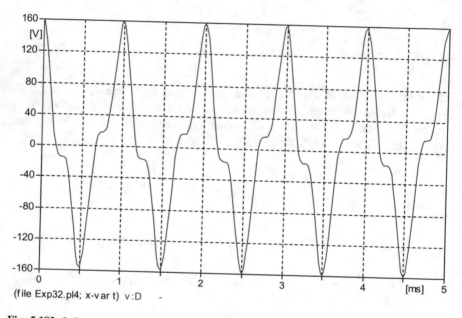

Fig. 5.183 Inductor voltage $e_L(t)$ of Problem P.5-36

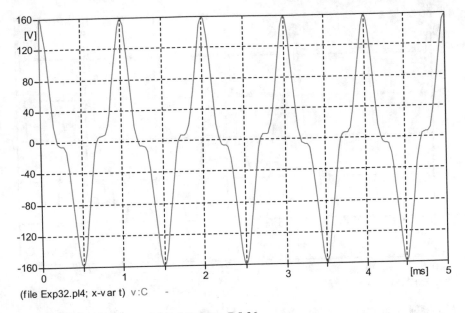

(file Exp32.pl4; x-var t) v:C   -

Fig. 5.184 Resistor voltage $e_R(t)$ of Problem P.5-36

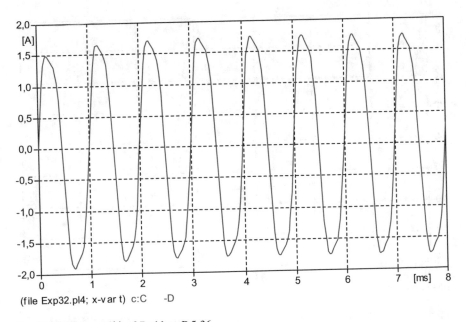

(file Exp32.pl4; x-var t) c:C   -D

Fig. 5.185 Current $i(t)$ of Problem P.5-36

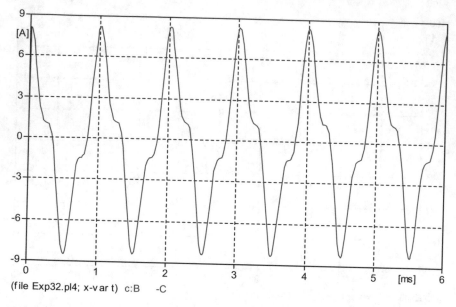

Fig. 5.186 Current $i_T(t)$, which flows through the voltage $e(t)$ source of Problem P.5-36

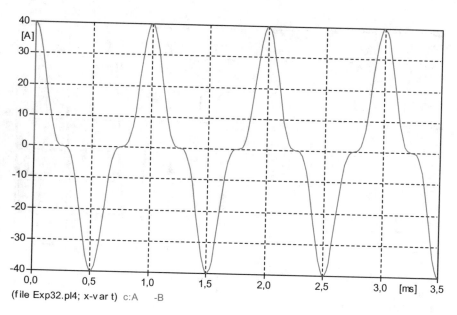

Fig. 5.187 Current $i_C(t)$, which flows through the voltage source of Problem P.5-36

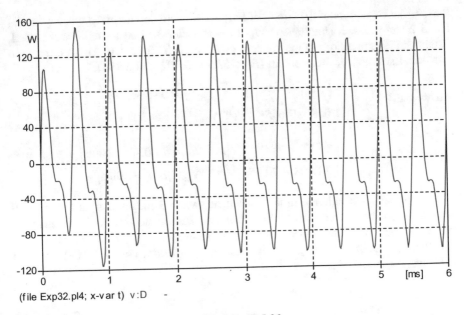

(file Exp32.pl4; x-var t) v:D   -

Fig. 5.188 Power $p_L(t)$ in the inductor of Problem P.5-36

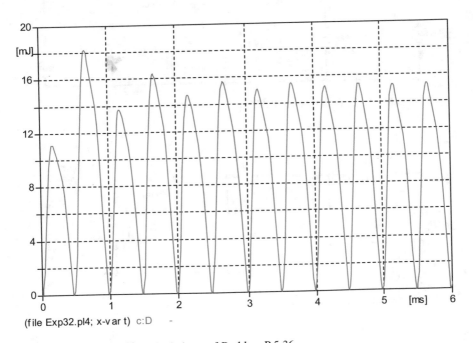

(file Exp32.pl4; x-var t) c:D   -

Fig. 5.189 Energy $W_L(t)$ in the inductor of Problem P.5-36

P.5-37 The capacitors of the circuit of Fig. 5.190 were de-energized when switches 1 and 2 were closed, simultaneously, at $t = 0$. After 30 [μs] switch 3 was closed. Establish the expressions of the voltages $e_1(t)$ and $e_2(t)$ for $t > 0$. Also, perform the numerical solution of the problem (Figs. 5.191, 5.192 and 5.193).

Answer: $e_1(t) = \begin{cases} 16\left(1 - \epsilon^{-10^6 t/20}\right)[V], & for\ 0 \leq t < 30[\mu s] \\ 11.008 - 7.655\epsilon^{-10^6 t/16}[V], & for\ t > 30[\mu s] \end{cases}$

$e_2(t) = \begin{cases} 6\left(1 - \epsilon^{-10^6 t/12}\right)[V], & for\ 0 \leq t < 30[\mu s] \\ 11.008 - 7.655\epsilon^{-10^6 t/16}[V], & for\ t > 30[\mu s] \end{cases}$

The numerical oscillation in the current of switch 3 means that, at time $t = 30$ [μs], an impulsive current passed through it, and it circulated clockwise through the mesh formed by the capacitances and switch 3, to make $e_1(30\mu s_+) = e_2(30\mu s_+) = 9.834$ [V], because $e_1(30\mu s_-) = 12.430[V]$ and $e_2(30\mu s_-) = 5.507[V]$.

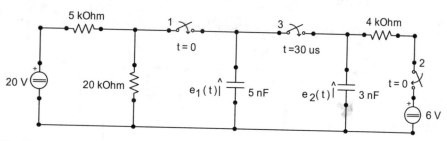

Fig. 5.190 Electric circuit of Problem P.5-37

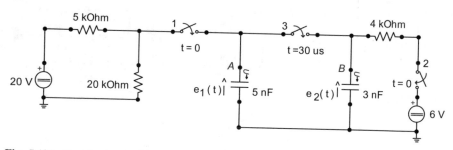

Fig. 5.191 Electric circuit of Problem P.5-37 for digital solution

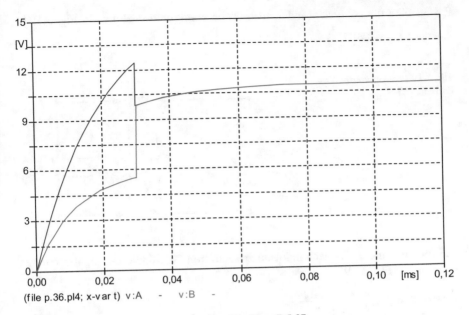

Fig. 5.192 Voltage $e_1(t)$ and Voltage $e_2(t)$ of Problem P.5-37

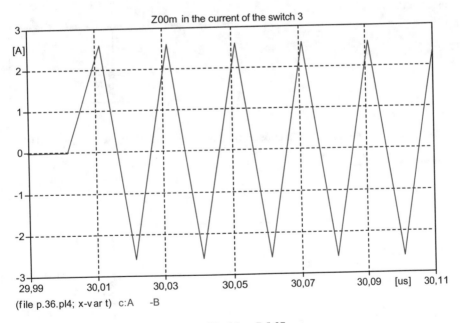

Fig. 5.193 Zoom at current in switch 3 of Problem P.5-37

P.5-38 Find the equivalent inductance of the configuration shown in Fig. 5.194.

Answer: $L_{eq} = 84[H]$

Fig. 5.194 Electric circuit of Problem P.5-38

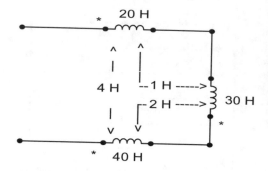

P.5-39 Find the equivalent inductance seen from terminals a-a' in the circuital structure of Fig. 5.195.

Answer: $L_{eq} = \frac{L_1 L_2 + (L_1 + L_2)M - 3M^2}{L_2 + M}$

Fig. 5.195 Electric circuit of Problem P.5-39

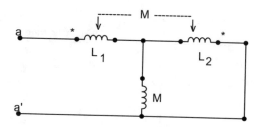

P.5-40 Find the equivalent inductance of the configuration shown in Fig. 5.196.

Answer: $L_{eq} = \frac{5}{2}L - \frac{2M^2}{L}$

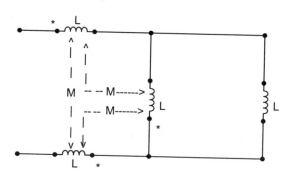

Fig. 5.196 Electric circuit of Problem P.5-40

P.5-41 Find the currents $i_1(t)$ and $i_2(t)$, for $t > 0$, in the circuit of Fig. 5.197, knowing that the coupling coefficient k is unitary and that:

$$e(t) = \begin{cases} 0 & for\ t \leq 0 \\ Et[V] & for\ 0 \leq t \leq 1[s] \\ E[V] & for\ t \geq 1[s]. \end{cases}$$

Answer:

$$i_1(t) = \begin{cases} 0[A] & for\ t \leq 0[s] \\ \frac{Et}{R_1} + \frac{ET_1}{R_1}\left(e^{\frac{-t}{T_1+T_2}} - 1\right)[A] & for\ 0 \leq t \leq 1[s] \\ \frac{E}{R_1} + \frac{ET_1}{R_1}\left(1 - e^{\frac{1}{T_1+T_2}}\right)e^{\frac{-t}{T_1+T_2}}[A] & for\ t \geq 1[s]. \end{cases}$$

$$i_2(t) = \begin{cases} 0[A] & for\ t \leq 0[s] \\ \sqrt{\frac{T_1 T_2}{R_1 R_2}}\left(e^{\frac{-t}{T_1+T_2}} - 1\right)E[A] & for\ 0 \leq t \leq 1[s] \\ \sqrt{\frac{T_1 T_2}{R_1 R_2}}\left(1 - e^{\frac{1}{T_1+T_2}}\right)Ee^{\frac{-t}{T_1+T_2}}[A] & for\ t \geq 1[s]. \end{cases}$$

where,

$$T_1 = \frac{L_1}{R_1}; T_2 = \frac{L_2}{R_2}.$$

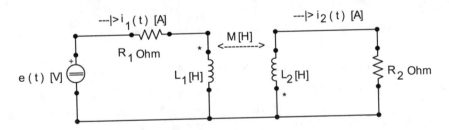

Fig. 5.197 Electric circuit of Problem P.5-41

P.5-42 The circuit shown in Fig. 5.198 is a cutter, which has the function of controlling the average current through resistor R. For this, the switch oscillates periodically, starting from $t = 0$, between positions A and B, at intervals equal to $T = L/R$ seconds. After a large number of cycles, the current $i(t)$ tends towards a periodic behavior, in which it oscillates between two levels: $I_1 = i_{min}$ and $I_2 = i_{max}$. Determine the general expression of the current $i(t)$, for $t > 0$, as well as the values of the current levels I_1 and I_2.

Answer:

$$i(t) = \frac{E}{R}\left\{\frac{1+(-1)^k}{2} - \left[\sum_{j=0}^{k}(-\epsilon)^j\right]\epsilon^{-t/T}\right\}, kT \le t \le (k+1)T, k = 0,1,2,\ldots$$

$$I_1 = \frac{1}{\epsilon+1}\cdot\frac{E}{R} = 0.269\frac{E}{R}.$$

$$I_2 = \frac{\epsilon}{\epsilon+1}\cdot\frac{E}{R} = 0.731\frac{E}{R}.$$

Fig. 5.198 Electric circuit of Problem P.5-42

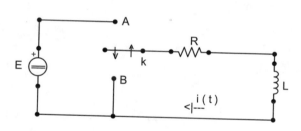

P.5-43 Solve Examples 5.18 and 5.19 using the convolution integral.

P.5-44 The response of a circuit to the unitary ramp $U_{-2}(t)$ is $a(t) = 1 - \epsilon^{-2t}$ for $t > 0$. If the circuit does not contain stored energy for $t < 0$, use the convolution integral to determine your response $y(t)$ to input $x(t) = \epsilon^{-t}U_{-1}(t)$.

Answer: $y(t) = (4\epsilon^{-2t} - 2\epsilon^{-t})U_{-1}(t)$.

P.5-45 Using the convolution integral, solve problems P.5-27 and P.5-28.

P.5-46 If the impulse response of each of the circuits in Fig. 5.141 is $h(t) = t\epsilon^{-t}U_{-1}(t)$, find the response to step $r(t)$ of cascading connection.

Answer: $r(t) = \frac{1}{48}[3 - (4t^3 + 6t^2 + 6t + 3)\epsilon^{-2t}]U_{-1}(t)$.

P.5-47 Two linear and time-invariant circuits, N_1 and N_2, are connected in cascade as in Fig. 5.141. If they are exchanged, show that the total connection response, $y(t)$, to any input $x(t)$ remains unchanged.

P.5-48 The impulse response of a fixed linear circuit is $h(t) = \epsilon^{-(t-1)}U_{-1}(t-1)$. Find the answer $y(t)$ to the input $x(t) = \epsilon^{-t}U_{-1}(t)$, by the four forms of the convolution integral.

Answer: $y(t) = (t-1)\epsilon^{-(t-1)}U_{-1}(t-1)$.

5.8   Conclusions

In this chapter it was presented the fundamental topic of transients in first order circuits. Extensive number of examples were provided with their analytical and digital solution using the software ATP—Alternative Transients Program, with its graphical interface ATPDraw. For didactic reasons, the study was organized as a large collection of various case studies, where the fundamentals of circuit analysis, mathematical time domain solution of ordinary differential equations with singular forcing functions, and physical principles were presented in details. Therefore, a good understanding of this chapter is an indispensable condition for the good follow-up of the following chapters.

References

1. AdkinsWA, Davidson MG (2012) Ordinary differential equations, undergraduate texts in mathematics. Springer
2. Boyce WE, Diprima RC (2002) Elementary differential equations and boundary value problems, 7th edn. LTC
3. Chicone C (2006) Ordinary differential equations with applications, 22nd edn. Springer
4. Close CM (1975) Linear circuits analysis. LTC
5. Doering CI, Lopes AO (2008) Ordinary differential equations, 3rd edn. IMPA
6. Irwin JD (1996) Basic engineering circuit analysis. Prentice Hall
7. Kreider DL, Kluler RG, Ostberg DR (2002) Differential equations. Edgard Blucher Ltda
8. Leithold L (1982) The calculus with analytical geometry, vol 1, 2nd edn. Harper & How of Brasil
9. Perko L (2001) Differential equations and dynamical systems, 3rd edn. Springer, New York
10. Stewart J (2010) Calculus, vol 1, 6th edn. Thomson
11. Zill DG, Cullen MR (2003) Differential equations, 3rd edn. Makron Books

Chapter 6
Transients in Circuits of Any Order

6.1 Introduction

In this chapter, circuits with any number of capacitances and inductances will be covered. This means that the differential equations can now be of any order. In general, the order of the differential equation for a variable of interest in a given circuit is at most equal to the sum of the number of capacitances and inductances it contains. Logically, before calculating this sum, it is necessary to reduce the capacitances that constitute series, parallel or series-parallel connections to their equivalent capacitances. The same goes for inductances that form series, parallel or series-parallel connections, which must be replaced by their equivalent inductances. Initially, circuits powered by independent sources, starting at $t = 0$, containing capacitors and inductors without energy initially stored at $t = 0_-$, that is, circuits initially de-energized, will be addressed. Then those with energy initially stored at $t = 0_-$ will be studied, with or without external sources. The latter will be solved, analytically, with the use of the Thévenin and Norton equivalents of capacitances and inductances initially energized. Typical examples of this second situation, that is, circuits containing capacitive elements with initial voltage at $t = 0_-$ and inductors with initial current at $t = 0_-$, the transients caused by switching. Whenever the examples are not exclusively literal, the respective numerical solutions will also be presented using the ATPDraw program.

In Chap. 5, both analytical and numerical procedures were developed to determine the transient response of first order circuits. As shown, they are so named because the differential equations [1–11] that describe them are first order. Such circuits have a single capacitance or a single inductance. Or they may contain several capacitances or several inductances, but they can nevertheless be reduced to a single equivalent capacitance or a single equivalent inductance, as they are series, parallel or series-parallel connections.

In this chapter, circuits with any number of capacitances and inductances will be covered. All analytical solutions will be performed in the Time Domain, by the also

© The Author(s), under exclusive license to Springer Nature Switzerland AG 2021
J. C. Goulart de Siqueira and B. D. Bonatto, *Introduction to Transients in Electrical Circuits*, Power Systems, https://doi.org/10.1007/978-3-030-68249-1_6

called Classic Method, without the use of the so-called Operational Methods, such as, for example, the solution of differential equations, in the Complex Frequency Domain, $s = \sigma + j\omega$, for Laplace transforms. The only exception will be the determination of the particular integral of differential equations with sinusoidal forcing functions, when Phasors (Simple Frequency Domain, with $s = j\omega$, $\omega = $ constant) will be used, corresponding to circuits excited by sinusoidal independent sources.

It is necessary to remember that the expression **"Transitional Response" refers to the Complete Circuit Response, that is, the sum of the Transient and Steady-state Regime Responses.**

6.2   Circuits Initially Deenergized

Example 6.1   Find the impulse response $h(t)$ of the circuit shown in Fig. 6.1.

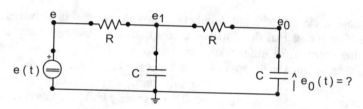

Fig. 6.1   Electric circuit of Example 6.1

Solution

The nodal equations of the circuit are:

Node e_1: $\frac{e_1 - e}{R} + C\frac{de_1}{dt} + \frac{e_1 - e_0}{R} = 0$

Node e_0: $\frac{e_0 - e_1}{R} + C\frac{de_0}{dt} = 0$

Introducing the differential operator $D = d/dt$ and simplifying, it comes that:

$$(RCD + 2)e_1 - e_0 = e$$
$$- e_1 + (RCD + 1) = 0$$

Solving this system of algebraic equations for e_0, it follows that:

$$\begin{vmatrix} RCD+2 & -1 \\ -1 & RCD+1 \end{vmatrix} e_0 = \begin{vmatrix} RCD+2 & e \\ -1 & 0 \end{vmatrix}$$

or,

$$R^2C^2 e_0''(t) + 3RC e_0'(t) + e_0(t) = e(t)$$

Considering now that $e(t) = U_0(t)$, then $e_0(t) = h(t)$ and the equation becomes:

$$R^2C^2h''(t) + 3RCh'(t) + h(t) = U_0(t), \text{ with } h(0_-) = h'(0_-) = 0. \qquad (6.1)$$

Reintroducing the operator D, and considering Eq. (6.1) for $t > 0$, we have:

$$(R^2C^2D^2 + 3RCD + 1)h(t) = 0.$$

The roots of the polynomial operator are:

$$r_{1,2} = \frac{-3 \pm \sqrt{5}}{2RC}$$

Therefore, $r_1 = -0.382/RC$ and $r_2 = -2.618/RC$.
So, the solution of Eq. (6.1), for $t > 0$ is:

$$h(t) = k_1\epsilon^{-0.382t/RC} + k_2\epsilon^{-2.618t/RC}$$
$$h'(t) = \frac{-0.382}{RC}k_1\epsilon^{-0.382t/RC} + \frac{-2.618}{RC}k_2\epsilon^{-2.618t/RC}$$

The balance of Eq. (6.1), at time $t = 0$, allows to write that:

$$h(t) \doteq 0.U_{-1}(t).$$

$$h'(t) \doteq C_0U_{-1}(t).$$

$$h''(t) \doteq C_0U_0(t) + C_1U_{-1}(t).$$

Substituting in Eq. (6.1), it turns out that

$$R^2C^2C_0U_0(t) + (R^2C^2C_1 + 3RCC_0)U_{-1}(t) \doteq U_0(t) + 0.U_{-1}(t).$$

Therefore,

$R^2C^2C_0 = 1$, and $C_0 = 1/R^2C^2$.
$R^2C^2C_1 + 3RCC_0 = 0$, or $R^2C^2C_1 + 3RC \cdot (1/R^2C^2) = 0$, and $C_1 = -3/R^3C^3$.

Since the initial condition at $t = 0_+$ of any function is always equal to the sum of the initial condition at $t = 0_-$ with the amplitude of the step function present in it,

$$h(0_+) = h(0_-) + 0 = 0.$$

$$h'(0_+) = h'(0_-) + C_0 = \frac{1}{R^2C^2}.$$

$$h''(0_+) = h''(0_-) + C_1 = \frac{-3}{R^3 C^3}.$$

So,

$$k_1 + k_2 = 0$$

$$\frac{-0.382}{RC} k_1 + \frac{-2.618}{RC} k_2 = \frac{1}{R^2 C^2}$$

Therefore, $k_1 = \frac{0.447}{RC}$ and $k_2 = -\frac{0.447}{RC}$ and

$$h(t) = \frac{0.447}{RC} \left(\epsilon^{-0.382t/RC} - \epsilon^{-2.618t/RC} \right) U_{-1}(t).$$

It is interesting to note that the circuit response contains two time constants:

$$T_1 = RC/0.382 = 2.618RC \text{ and } T_2 = RC/2.618 = 0.382RC.$$

Example 6.2 Find the step response $r(t)$ of the circuit in Fig. 6.1.

Solution

$$r(t) = \int_{-\infty}^{t} h(\lambda)d\lambda = \frac{0.447}{RC} \int_{0_+}^{t} \left(\epsilon^{-0.382\lambda/RC} - \epsilon^{-2.618\lambda/RC} \right) d\lambda = 0.447 \left(0.382\epsilon^{-2.618t/RC} - 2.618\epsilon^{-0.382t/RC} \right) \Big|_{0_+}^{t}$$

$$r(t) = \left(1 + 0.17\epsilon^{-2.618t/RC} - 1.17\epsilon^{-0.382t/RC} \right) U_{-1}(t).$$

$$(6.2)$$

Verification:

$$h(t) = \frac{d}{dt} r(t) = \frac{d}{dt} \left[\left(1 + 0.17\epsilon^{-2.618t/RC} - 1.17\epsilon^{-0.382t/RC} \right) U_{-1}(t) \right]$$

$$= \left(-\frac{0.45}{RC} \epsilon^{-2.618t/RC} + \frac{0.45}{RC} \epsilon^{-0.382t/RC} \right) U_{-1}(t)$$

$$+ \left(1 + 0.17\epsilon^{-2.618t/RC} - 1.17\epsilon^{-0.382t/RC} \right) \Big|_{t=0} \cdot U_0(t)$$

Therefore,

$$h(t) = \frac{0.45}{RC} \left(\epsilon^{-0.382t/RC} - \epsilon^{-2.618t/RC} \right) U_{-1}(t).$$

Example 6.3 Find the circuit response in Fig. 6.1 to the rectangular pulse defined by:

$$e(t) = \begin{cases} 0 & for \quad t<0 \\ E & for \quad 0<t<a \\ 0 & for \quad t>a. \end{cases}$$

Solution
The excitation of the circuit can be expressed by

$$e(t) = EU_{-1}(t) - EU_{-1}(t-a)$$

Since the circuit is linear and time-invariant, that is, with constant parameters, then the response to the rectangular pulse, according to the superposition theorem, is given by:

$$e_0(t) = Er(t) - Er(t-a)$$

Thus, using (Eq. 6.2),

$$e_0(t) = E\left(1+0.17\epsilon^{-2.618t/RC} - 1.17\epsilon^{-0.382t/RC}\right)U_{-1}(t)$$
$$- E\left[\left(1+0.17\epsilon^{-2.618(t-a)/RC} - 1.17\epsilon^{-0.382(t-a)/RC}\right)U_{-1}(t-a)\right]$$
$$= E\left(1+0.17\epsilon^{-2.618t/RC} - 1.17\epsilon^{-0.382t/RC}\right)U_{-1}(t)$$
$$- E\left(1+0.17\epsilon^{2.618a/RC}\epsilon^{-2.618t/RC} - 1.17\epsilon^{0.382a/RC}\epsilon^{-0.382t/RC}\right)U_{-1}(t-a).$$

Summarizing,

$$e_0(t) = \begin{cases} 0, & t \leq 0 \\ E\left(1+0.17\epsilon^{-2.618t/RC} - 1.17\epsilon^{-0.382t/RC}\right), & 0 \leq t \leq a \\ E\left[0.17\left(1-\epsilon^{2.618a/RC}\right)\epsilon^{-2.618t/RC} - 1.17\left(1-\epsilon^{0.382a/RC}\right)\epsilon^{-0.382t/RC}\right], & t \geq a. \end{cases}$$

Example 6.4 Find the digital solution of Example 6.1 for the following numerical values: $e(t) = 100U_{-1}(t)$ [V], $R = 10$ [Ω], $C = 1,000$ [μF].

Solution
The analytical solution, according to the homogeneity property of linear circuits, is $e_0(t) = 100r(t)$. So, by Eq. (6.2),

$$e_0(t) = 100\left(1+0.17\epsilon^{-2.618t/RC} - 1.17\epsilon^{-0.382t/RC}\right)U_{-1}(t).$$

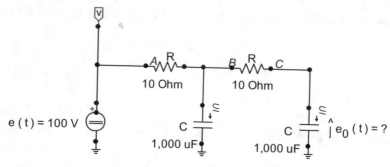

Fig. 6.2 Electric circuit of Example 6.4

By replacing the numerical values,

$$e_0(t) = \left(100 + 17\epsilon^{-261.8t} - 117\epsilon^{-38.2t}\right)U_{-1}(t)\,[\text{V}].$$

The time constants are: $T_1 = 26.16\,[\text{ms}]$ and $T_2 = 3.82\,[\text{ms}]$.

The circuit for the numerical solution by the ATPDraw program is shown in Fig. 6.2.

The voltage $e_0(t)$, on the right capacitance, together with the voltage on the left capacitance, are plotted in Fig. 6.3.

The currents circulating in the two capacitances are plotted in Fig. 6.4.

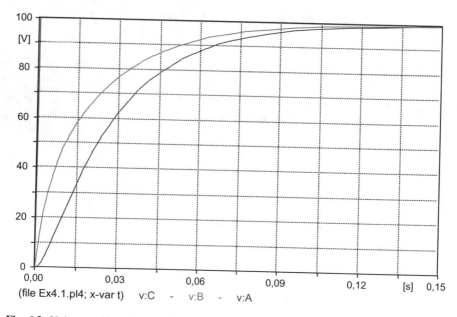

(file Ex4.1.pl4; x-var t)   v:C   -   v:B   -   v:A

Fig. 6.3 Voltage $e_0(t)$, voltage at left capacitance, source voltage $e(t)$

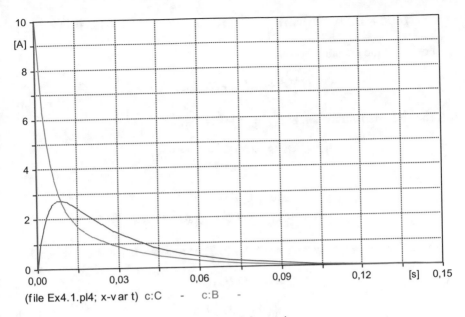

Fig. 6.4 Current at right capacitance, current at left capacitance

Example 6.5 Find voltage $e_0(t)$ and mesh currents $i_0(t)$ and $i_1(t)$ in the circuit of Fig. 6.5, for $t > 0$.

Fig. 6.5 Electric circuit of Example 6.5

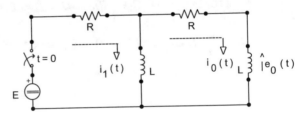

Solution

The circuit mesh equations are:

Mesh 1: $(LD + R)i_1(t) - LDi_0(t) = EU_{-1}(t)$

Mesh 0: $-LDi_1(t) + (2LD + R)i_0(t) = 0$.

Solving the system to $i_0(t)$, it comes that:

$$\begin{vmatrix} LD+R & -LD \\ -LD & 2LD+R \end{vmatrix} i_0(t) = \begin{vmatrix} LD+R & EU_{-1}(t) \\ -LD & 0 \end{vmatrix}$$

And the differential equation for $i_0(t)$ is:

$$(L^2D^2 + 3RLD + R^2)i_0(t) = LEU_0(t) \text{ with } i_0(0_-) = i_0'(0_-) = 0. \tag{6.3}$$

For $t > 0$ the equation becomes

$$(L^2D^2 + 3RLD + R^2)i_0(t) = 0$$

The characteristic roots are: $r_{1,2} = \frac{-3 \pm \sqrt{5}}{2} \cdot \frac{R}{L}$, So,

$$r_1 = -0.382R/L; r_2 = -2.618R/L.$$

Therefore,

$$i_0(t) = k_1 \epsilon^{-0.382Rt/L} + k_2 \epsilon^{-2.618Rt/L}$$
$$i_0'(t) = -0.382\frac{R}{L}k_1\epsilon^{-0.382Rt/L} - 2.618\frac{R}{L}k_2\epsilon^{-2.618Rt/L}$$

In view of the necessary balance of Eq. (6.3), at time $t = 0$, it can be written that:

$$i_0(t) \doteq 0.U_{-1}(t).$$

$$i_0'(t) \doteq C_0 U_{-1}(t).$$

$$i_0''(t) \doteq C_0 U_0(t) + C_1 U_{-1}(t).$$

Substituting in Eq. (6.3),

$$L^2 C_0 U_0(t) + (L^2 C_1 + 3RLC_0)U_{-1}(t) \doteq LEU_0(t) + 0.U_{-1}(t).$$

$$C_0 = \frac{E}{L}; L^2 C_1 + 3RLC_0 = 0, L^2 C_1 + 3RE = 0 \text{ and } C_1 = \frac{-3RE}{L^2}.$$

Therefore,

$$i_0(0_+) = i_0(0_-) + 0 = 0.$$

$$i_0'(0_+) = i_0'(0_-) + C_0 = \frac{E}{L}.$$

$$i_0''(0_+) = i_0''(0_-) + C_1 = \frac{-3RE}{L^2}.$$

Consequently,

$$i_0(0_+) = k_1 + k_2 = 0$$

$$i_0'(0_+) = -0.382\frac{R}{L}k_1 - 2.618\frac{R}{L}k_2 = \frac{E}{L},$$

where, $k_1 = 0.447E/R$ and $k_2 = -0.447E/R$

Therefore,

$$i_0(t) = 0.447 \frac{E}{R} \left(\epsilon^{-0.382Rt/L} - \epsilon^{-2.618Rt/L} \right) U_{-1}(t).$$

The voltage at the right inductance is:

$$e_0(t) = L \frac{d}{dt} i_0(t) = \left(1.17 \epsilon^{-2.618Rt/L} - 0.17 \epsilon^{-0.382Rt/L} \right) E \text{ for } t > 0.$$

Therefore, $e_0(0_+) = E$. Which is confirmed by $e_0(0_+) = L i_0'(0_+) = L \frac{E}{L} = E$. The solution of the mesh equations for $i_1(t)$ provides the differential equation

$$\left(L^2 D^2 + 3RLD + R^2 \right) i_1(t) = 2LEU_0(t) + REU_{-1}(t), \text{ with } i_1(0_-) = i_1'(0_-) = 0 \tag{6.4}$$

For $t > 0$, the Eq. (6.4) becomes

$$\left(L^2 D^2 + 3RLD + R^2 \right) i_1(t) = RE$$

$$i_{1_H}(t) = k_3 \epsilon^{-0.382Rt/L} + k_4 \epsilon^{-2.618Rt/L}$$

$$i_{1_P}(t) = \frac{1}{L^2 D^2 + 3RLD + R^2} \left(RE\epsilon^{0 \times t} \right) = \frac{E}{R}$$

So,

$$i_1(t) = \frac{E}{R} + k_3 \epsilon^{-0.382Rt/L} + k_4 \epsilon^{-2.618Rt/L}$$
$$i_1'(t) = -0.382 \frac{R}{L} k_3 \epsilon^{-0.382Rt/L} - 2.618 \frac{R}{L} k_4 \epsilon^{-2.618Rt/L}$$

The balance of the two members of Eq. (6.4) authorizes writing that:

$$i_1(t) \doteq 0.U_{-1}(t).$$

$$i_1'(t) \doteq D_0 U_{-1}(t).$$

$$i_1''(t) \doteq D_0 U_0(t) + D_1 U_{-1}(t).$$

Substituting in Eq. (6.4), it comes that

$$L^2 D_0 U_0(t) + \left(L^2 D_1 + 3RLD_0 \right) U_{-1}(t) \doteq 2LEU_0(t) + REU_{-1}(t).$$

$$D_0 = \frac{2E}{L} ; L^2 D_1 + 3RL \frac{2E}{L} = RE, D_1 = \frac{-5RE}{L^2}.$$

Therefore,

$$i_1(0_+) = i_1(0_-) + 0 = 0$$

$$i_1'(0_+) = i_1'(0_-) + D_0 = \frac{2E}{L}$$

$$i_1''(0_+) = i_1''(0_-) + D_1 = \frac{-5RE}{L^2}.$$

That way,

$$\frac{E}{R} + k_3 + k_4 = 0$$

$$-0.382\frac{R}{L}k_3 - 2.618\frac{R}{L}k_4 = \frac{2E}{L}$$

where,

$$k_3 = -0.276\frac{E}{R} \text{ and } k_4 = -0.724\frac{E}{R}.$$

Finally,

$$i_1(t) = \left(1 - 0.276\epsilon^{-0.382Rt/L} - 0.724\epsilon^{-2.618Rt/L}\right)\frac{E}{R}U_{-1}(t).$$

Example 6.6 Solve Example 6.5 numerically for $E = 100$ [V], $R = 10$ [Ω] and $L = 100$ [mH].

Solution
With the numerical values, the analytical solutions are:

$$e_0(t) = 117\epsilon^{-261.8t} - 17\epsilon^{-38.2t} \text{ [V]}, for\, t > 0 \text{ [s]}$$
$$i_0(t) = 4.47\left(\epsilon^{-38.2t} - \epsilon^{-261.8t}\right) \text{ [A], } for\, t \geq 0 \text{ [s]}$$
$$i_1(t) = 10\left(1 - 0.276\epsilon^{-38.2t} - 0.724\epsilon^{-261.8t}\right) \text{ [A], } for\, t \geq 0 \text{ [s]}.$$

The time constants are: $T_1 = 26.2$ [ms] and $T_2 = 3.8$ [ms].

The circuit for the corresponding numerical solution, using the ATPDraw program, is shown in Fig. 6.6.

Note that the simulation time $t_{max} = 150$ [ms] $\cong 5T_1$ was adopted.

The voltage $e_0(t)$, scaled on the vertical axis on the left, and the voltage from the external source, scaled on the vertical axis on the right, are shown in Fig. 6.7.

The currents $i_0(t)$ and $i_1(t)$ are recorded in Fig. 6.8.

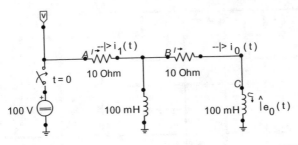

Fig. 6.6 Electric circuit of Example 6.6 for the numerical solution by the ATPDraw program

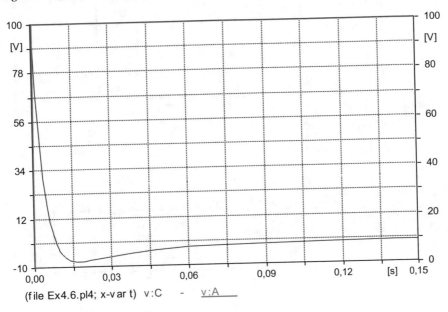

(file Ex4.6.pl4; x-v ar t) v:C - v:A

Fig. 6.7 Voltage $e_0(t)$, source voltage

Example 6.7 Find the initial conditions $e_0(0_+)$, $e_0'(0_+)$ and $e_0''(0_+)$ in the circuit of Fig. 6.9, knowing that it is totally de-energized at $t = 0_-$, and that the source voltage is $e(t) = E\cos(\omega t + \theta)U_{-1}(t)$.

Solution
If $e(t) = L\frac{d}{dt}i(t) = LDi(t)$, then $i(t) = \frac{1}{LD}e(t)$, and the nodal equations of the circuit are:

Node 1: $\frac{e_1 - e}{R} + \frac{1}{LD}e_1 + \frac{e_1 - e_0}{R} = 0$

Node 0: $\frac{e_0 - e_1}{R} + \frac{1}{LD}e_0 = 0$.

Multiplying both equations by RLD and factoring, it follows that:

$$(2LD + R)e_1 - LDe_0 = LDe$$
$$-LDe_1 + (LD + R)e_0 = 0.$$

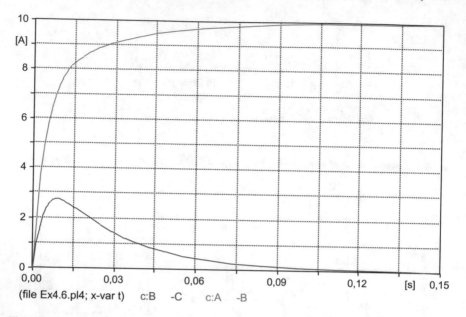

(file Ex4.6.pl4; x-var t)   c:B   -C   c:A   -B

Fig. 6.8 Current $i_0(t)$, Current $i_1(t)$

Fig. 6.9 Electric circuit of
Example 6.7

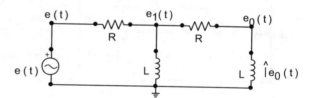

Therefore,

$$\begin{vmatrix} 2LD+R & -LD \\ -LD & LD+R \end{vmatrix} e_0(t) = \begin{vmatrix} 2LD+R & LDe(t) \\ -LD & 0 \end{vmatrix}$$

So, the differential equation for e_0 (t) is:

$$(L^2 D^2 + 3RLD + R^2) e_0(t) = L^2 e''(t)$$

Or,

$$e_0''(t) + \frac{3R}{L} e_0'(t) + \frac{R^2}{L^2} e_0(t) = e''(t).$$

$$e(t) = E \cos(\omega t + \theta) U_{-1}(t).$$

$$e'(t) = E \cos \theta U_0(t) - \omega E \sin(\omega t + \theta) U_{-1}(t).$$

$$e''(t) = E \cos \theta U_1(t) - \omega E \sin \theta U_0(t) - \omega^2 E \cos(\omega t + \theta) U_{-1}(t).$$

Using only the first term of the series expansion of singular functions of the last term of the expression of $e''(t)$, the differential equation in $e_0(t)$ becomes:

$$e_0''(t) + \frac{3R}{L} e_0'(t) + \frac{R^2}{L^2} e_0(t) \doteq E \cos \theta U_1(t) - \omega E \sin \theta U_0(t) - \omega^2 E \cos \theta U_{-1}(t).$$

The balance of the two members of this equation allows us to write that:

$$e_0(t) \doteq C_0 U_{-1}(t).$$

$$e_0'(t) \doteq C_0 U_0(t) + C_1 U_{-1}(t).$$

$$e_0''(t) \doteq C_0 U_1(t) + C_1 U_0(t) + C_2 U_{-1}(t).$$

Therefore

$$C_0 U_1(t) + \left(C_1 + \frac{3R}{L} C_0 \right) U_0(t) + \left(C_2 + \frac{3R}{L} C_1 + \frac{R^2}{L^2} C_0 \right) U_{-1}(t)$$
$$\doteq E \cos \theta U_1(t) - \omega E \sin \theta U_0(t) - \omega^2 E \cos \theta U_{-1}(t).$$

So,

$$C_0 = E \cos \theta.$$

$$C_1 + \frac{3R}{L} E \cos \theta = -\omega E \sin \theta; \; C_1 = -\omega E \sin \theta - \frac{3RE}{L} \cos \theta.$$

$$C_2 + \frac{3R}{L} \left(-\omega E \sin \theta - \frac{3RE}{L} \cos \theta \right) + \frac{R^2}{L^2} E \cos \theta = -\omega^2 E \cos \theta;$$

$$C_2 = \frac{3\omega RE}{L} \sin \theta + \frac{8R^2 - \omega^2 L^2}{L^2} E \cos \theta.$$

Since all initial conditions are null at $t = 0_-$, finally,

$$e_0(0_+) = E \cos \theta.$$

$$e_0'(0_+) = -\omega E \sin \theta - \frac{3RE}{L} \cos \theta.$$

$$e_0''(0_+) = \frac{3\omega RE}{L} \sin \theta + \frac{8R^2 - \omega^2 L^2}{L^2} E \cos \theta.$$

For $\theta = 0$, that is, $e(t) = E \cos \omega t U_{-1}(t)$,

$$e_0(0_+) = E.$$

$$e_0'(0_+) = -\frac{3RE}{L}.$$

and

$$e_0''(0_+) = \frac{8R^2 - \omega^2 L^2}{L^2} E.$$

Example 6.8 Find the impulse response $h(t)$ of the circuit in Fig. 6.9.

Solution

From Example 6.7, the differential equation at $e_0(t)$ is:

$$e_0''(t) + \frac{3R}{L} e_0'(t) + \frac{R^2}{L^2} e_0(t) = e''(t).$$

When $e(t) = U_0(t)$, $e_0(t) = h(t)$ and the differential equation becomes:

$$h''(t) + \frac{3R}{L} h'(t) + \frac{R^2}{L^2} h(t) = U_2(t), \text{ with } h(0_-) = h'(0_-) = 0. \qquad (6.5)$$

For $t > 0$,

$$\left(D^2 + \frac{3R}{L} D + \frac{R^2}{L^2} \right) h(t) = 0$$

The characteristic roots are:$r_1 = -0.382R/L$ and $r_2 = -2.618R/L$.
So,

$$h(t) = k_1 \epsilon^{-0.382Rt/L} + k_2 \epsilon^{-2.618Rt/L}$$

$$h'(t) = -0.382 \frac{R}{L} k_1 \epsilon^{-0.382Rt/L} - 2.618 \frac{R}{L} k_2 \epsilon^{-2.618Rt/L}$$

The balance of Eq. (6.5), at the time $t = 0$, allows to write that:

$$h(t) \doteq C_0 U_0(t) + C_1 U_{-1}(t).$$

$$h'(t) \doteq C_0 U_1(t) + C_1 U_0(t) + C_2 U_{-1}(t).$$

$$h''(t) \doteq C_0 U_2(t) + C_1 U_1(t) + C_2 U_0(t) + C_3 U_{-1}(t).$$

Substituting in Eq. (6.5), it comes that:

$$C_0 U_2(t) + \left(C_1 + \tfrac{3RC_0}{L}\right) U_1(t) + \left(C_2 + \tfrac{3RC_1}{L} + \tfrac{R^2 C_0}{L^2}\right) U_0(t)$$
$$+ \left(C_3 + \tfrac{3RC_2}{L} + \tfrac{R^2 C_1}{L^2}\right) U_{-1}(t) \doteq U_2(t) + 0.U_1(t) + 0.U_0(t) + 0.U_{-1}(t).$$

$$C_0 = 1$$

$$C_1 + \frac{3RC_0}{L} = 0; \; C_1 = -\frac{3R}{L}$$

$$C_2 + \frac{3RC_1}{L} + \frac{R^2 C_0}{L^2} = 0; \; C_2 = \frac{8R^2}{L^2}.$$

Therefore, $h(0_+) = C_1 = -\tfrac{3R}{L}$ and $h'(0_+) = C_2 = \tfrac{8R^2}{L^2}$. So,

$$-\frac{3R}{L} = k_1 + k_2$$

$$\frac{8R^2}{L^2} = -0.382 \frac{R}{L} k_1 - 2.618 \frac{R}{L} k_2.$$

Solving this system of algebraic equations results that

$$k_1 = 0.065 \; R/L \text{ and } k_2 = -3.065 \; R/L$$

Finally,

$$h(t) = U_0(t) + \left(0.065 \frac{R}{L} \epsilon^{-\frac{0.382Rt}{L}} - 3.065 \frac{R}{L} \epsilon^{-\frac{2.618Rt}{L}}\right) U_{-1}(t).$$

Example 6.9 Determine the initial conditions $h(0_+)$, $h'(0_+)$ and $h''(0_+)$ in the circuit shown in Fig. 6.10.

Solution
The circuit is the same as in Fig. 6.11. Just replace the part to the left of the capacitance with the Thévenin equivalent to confirm.

The loop (mesh) equations are:
Loop 1: $i_1 + \tfrac{1}{4D}(i_1 - i_0) = e$
Loop 0: $\tfrac{1}{4D}(i_0 - i_1) + 2Di_0 + i_0 = 0$
Or,

$$(4D + 1)i_1 - i_0 = 4De$$
$$-i_1 + (8D^2 + 4D + 1)i_0 = 0$$

Solving the system of equations for $i_0(t)$, the differential equation results

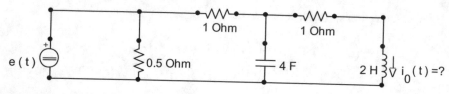

Fig. 6.10 Electric circuit of Example 6.9

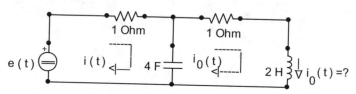

Fig. 6.11 Equivalent electric circuit of Example 6.9

$$(8D^2 + 6D + 2)i_0(t) = e(t)$$

For $e(t) = U_0(t)$ [V], $i_0(t) = h(t)$ [A] and the differential equation becomes:

$$8h''(t) + 6h'(t) + 2h(t) = U_0(t), \text{ with } h(0_-) = h'(0_-) = h''(0_-) = 0.$$

So,

$$h(t) \doteq 0.U_{-1}(t).$$

$$h'(t) \doteq C_0 U_{-1}(t).$$

$$h''(t) \doteq C_0 U_0(t) + C_1 U_{-1}(t).$$

Thus,

$$8C_0 U_0(t) + (8C_1 + 6C_0)U_{-1}(t) \doteq U_0(t) + 0.U_{-1}(t).$$

$$C_0 = 1/8$$

$$8C_1 + 6C_0 = 0; C_1 = -3/32.$$

Finally,

$$h(0_+) = h(0_-) + 0 = 0 \text{ [A]}$$

$$h'(0_+) = h'(0_-) + C_0 = 1/8 \text{ [A/s]}$$

$$h''(0_+) = h''(0_-) + C_1 = -3/32 \text{[A/s}^2\text{]}.$$

Example 6.10 The circuit shown in Fig. 6.12 is completely de-energized at $t = 0_-$. Find $e_0(0_+)$, $e'_0(0_+)$ and $e''_0(0_+)$, for $e(t) = E\epsilon^{-\alpha t}\cos(\omega t + \theta)U_{-1}(t)$.

Fig. 6.12 Electric circuit of Example 6.10

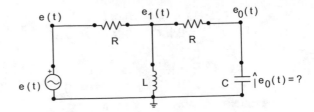

Solution

$$e(t) = E\epsilon^{-\alpha t}\cos(\omega t + \theta)U_{-1}(t).$$

$$e'(t) = -\alpha E\epsilon^{-\alpha t}\cos(\omega t + \theta)U_{-1}(t) - \omega E\epsilon^{-\alpha t}\sin(\omega t + \theta)U_{-1}(t) + E\cos\theta U_0(t)$$

$$= E\cos\theta U_0(t) - E\epsilon^{-\alpha t}[\alpha\cos(\omega t + \theta) + \omega\sin(\omega t + \theta)]U_{-1}(t)$$

$$= E\cos\theta U_0(t) - E(\alpha\cos\theta + \omega\sin\theta)U_{-1}(t) + \cdots$$

The two nodal equations of the circuit are:

Node 1: $\frac{e_1 - e}{R} + \frac{1}{LD}e_1 + \frac{e_1 - e_0}{R} = 0$

Node 0: $\frac{e_0 - e_1}{R} + CDe_0 = 0.$

Or,

$$(2LD + R)e_1 - LDe_0 = LDe$$
$$-e_1 + (RCD + 1)e_0 = 0$$

Solving for $e_0(t)$,

$$e_0''(t) + \frac{L + R^2C}{2RLC}e_0'(t) + \frac{1}{2LC}e_0(t) = \frac{1}{2RC}e'(t) = \frac{E\cos\theta}{2RC}U_0(t) -$$

$$\frac{E}{2RC}(\alpha\cos\theta + \omega\sin\theta)U_{-1}(t) + \cdots, \text{ with } e_0(0_-) = e_0'(0_-) = e_0''(0_-) = 0$$

$$e_0(t) \doteq 0.U_{-1}(t).$$

$$e_0'(t) \doteq C_0 U_{-1}(t).$$

$$e_0''(t) \doteq C_0 U_0(t) + C_1 U_{-1}(t).$$

Substituting in the previous differential equation,

$$C_0 U_0(t) + \left(C_1 + \frac{L + R^2C}{2RLC}C_0\right)U_{-1}(t) \doteq \frac{E\cos\theta}{2RC}U_0(t)$$

$$- \frac{E}{2RC}(\alpha\cos\theta + \omega\sin\theta)U_{-1}(t).$$

$$C_0 = \frac{E\cos\theta}{2RC}$$

$$C_1 + \frac{L+R^2C}{2RLC} \cdot \frac{E\cos\theta}{2RC} = -\frac{E}{2RC}(\alpha\cos\theta + \omega\sin\theta)\, e$$

$$C_1 = -\frac{E}{2RC}(\alpha\cos\theta + \omega\sin\theta) - \frac{E(R^2C+L)\cos\theta}{4R^2C^2L}$$

Therefore,

$$e_0(0_+) = 0$$

$$e_0'(0_+) = \frac{E\cos\theta}{2RC}$$

$$e_0''(0_+) = -\frac{E}{2RC}(\alpha\cos\theta + \omega\sin\theta) - \frac{E(R^2C+L)\cos\theta}{4R^2C^2L}.$$

Example 6.11 In the circuit of Fig. 6.13:

(a) Find the analytical expressions of de $i(t)$, $e_L(t)$ and $e_C(t)$), for $t > 0$, knowing that $e_C(0_-) = 0$ and that $E = 100\,[\mathrm{V}]$, $L = 100\,[\mathrm{mH}]$ and $C = 1,000\,[\mu\mathrm{F}]$.
(b) Obtain the expressions of power and energy in the three elements of the circuit.
(c) Simulate the item using the ATPDraw program.

Solution

(a) The circuit equation for the circuit is:

$$L\frac{d}{dt}i(t) + \frac{1}{C}\int i(t)dt = EU_{-1}(t)$$

Or,

$$i''(t) + \frac{1}{LC}i(t) = \frac{E}{L}U_0(t), \ \text{with } i(0_-) = i'(0_-) = 0. \tag{6.6}$$

Defining $\omega_n = 1/\sqrt{LC}$, called **the natural angular frequency of the circuit,** and considering the differential equation for $t > 0$, it results that

$$\left(D^2 + \omega_n^2\right)i(t) = 0$$

The characteristic roots are $r_{1,2} = \pm j\omega_n$ and

Fig. 6.13 Electric circuit of Example 6.11

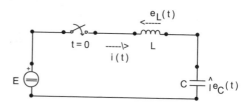

$$i(t) = k_1 \cos \omega_n t + k_2 \sin \omega_n t$$
$$i'(t) = -\omega_n k_1 \sin \omega_n t + \omega_n k_2 \cos \omega_n t.$$

By inspecting Eq. (6.6), it can be concluded that $i(0_+) = 0$ and $i'(0_+) = E/L$. So,

$$k_1 = 0 \text{ and } k_2 = \frac{E}{\omega_n L} = E\sqrt{\frac{C}{L}}.$$

Therefore

$$i(t) = E\sqrt{\frac{C}{L}}. \sin \omega_n t.U_{-1}(t).$$

$$e_L(t) = L\frac{d}{dt}i(t) = E \cos \omega_n t.U_{-1}(t).$$

$$e_C(t) = EU_{-1}(t) - e_L(t) = E(1 - \cos \omega_n t)U_{-1}(t).$$

Introducing the numerical values, it comes that:

$$\omega_n = \frac{1}{\sqrt{0.1 \times 0.001}} = 100 \text{ [rad/s]}$$

But, $\omega_n = 2\pi f_n = \frac{2\pi}{T_n} \therefore T_n = \frac{2\pi}{\omega_n} = 0.063 \text{ [s]} = 63 \text{ [ms]}$

$$i(t) = 10 \sin 100t U_{-1}(t) \text{ [A]}$$

$$e_L(t) = 100 \cos 100t U_{-1}(t) \text{ [V]}$$

$$e_C(t) = 100(1 - \cos 100t)U_{-1}(t)\text{[V]}.$$

(b) The power at the inductance is

$$p_L(t) = e_L(t).i(t) = E \cos \omega_n t.U_{-1}(t).E\sqrt{\frac{C}{L}} \sin \omega_n t.U_{-1}(t)$$

$$p_L(t) = \frac{E^2}{2}\sqrt{\frac{C}{L}} \sin 2\omega_n t.U_{-1}(t).$$

In capacitance it is

$$p_C(t) = e_C(t) \cdot i(t) = (E - E\cos\omega_n t)U_{-1}(t) \cdot E\sqrt{\frac{C}{L}}\sin\omega_n t \cdot U_{-1}(t)$$

$$= E^2\sqrt{\frac{C}{L}}\sin\omega_n t \cdot U_{-1}(t) - \frac{E^2}{2}\sqrt{\frac{C}{L}}2\sin\omega_n t \cdot \cos\omega_n t \cdot U_{-1}(t)$$

$$p_C(t) = E^2\sqrt{\frac{C}{L}}\sin\omega_n t \cdot U_{-1}(t) - \frac{E^2}{2}\sqrt{\frac{C}{L}}\sin 2\omega_n t \cdot U_{-1}(t).$$

Therefore,

$$p_L(t) + p_C(t) = E^2\sqrt{\frac{C}{L}}\sin\omega_n t . U_{-1}(t).$$

The power of the voltage source is

$$p_F(t) = -EU_{-1}(t).i(t) = -E^2\sqrt{\frac{C}{L}}\sin\omega_n t . U_{-1}(t).$$

It is seen, then, that: $p_F(t) + p_L(t) + p_C(t) = 0$, as it could not be otherwise. The energy in the inductance is

$$W_L(t) = \int_{-\infty}^{t} p_L(\lambda)d\lambda = \frac{E^2}{2}\sqrt{\frac{C}{L}}\int_{0_+}^{t}\sin 2\omega_n\lambda d\lambda = \frac{E^2}{2}\sqrt{\frac{C}{L}} \cdot \frac{-\cos 2\omega_n\lambda}{2\omega_n}\bigg|_{0_+}^{t}$$

$$= \frac{CE^2}{4}(-\cos 2\omega_n t + 1) = \frac{CE^2}{2}\left(\frac{1 - \cos 2\omega_n t}{2}\right)$$

$$W_L(t) = \frac{CE^2}{2}\sin^2\omega_n t . U_{-1}(t).$$

Alternatively,

$$W_L(t) = \frac{L}{2}[i(t)]^2 = \frac{L}{2}.E^2.\frac{C}{L}.\sin^2\omega_n t.[U_{-1}(t)]^2 = \frac{CE^2}{2}\sin^2\omega_n t . U_{-1}(t).$$

The energy in the capacitance is

$$W_C(t) = \int_{-\infty}^{t} p_C(\lambda)d\lambda = E^2 \sqrt{\frac{C}{L}} \left[\int_{0_+}^{t} \sin \omega_n \lambda \, d\lambda - \frac{1}{2} \int_{0_+}^{t} \sin 2\omega_n \lambda \, d\lambda \right]$$

$$= E^2 \sqrt{\frac{C}{L}} \left[\frac{-\cos \omega_n \lambda}{\omega_n} \Big|_{0_+}^{t} - \frac{1}{2} \cdot \frac{1 - \cos 2\omega_n \lambda}{2\omega_n} \Big|_{0_+}^{t} \right]$$

$$= CE^2(-\cos \omega_n t + 1) + \frac{CE^2}{4}(\cos 2\omega_n t - 1)$$

$$= CE^2(1 - \cos \omega_n t) - \frac{CE^2}{2} \cdot \left(\frac{1 - \cos 2\omega_n t}{2} \right)$$

$$= CE^2(1 - \cos \omega_n t) - \frac{CE^2}{2} \sin^2 \omega_n t$$

$$= CE^2(1 - \cos \omega_n t) - \frac{CE^2}{2}(1 - \cos^2 \omega_n t)$$

$$= \frac{CE^2}{2}(2 - 2\cos \omega_n t - 1 + \cos^2 \omega_n t)$$

$$= \frac{CE^2}{2}(1 - 2\cos \omega_n t + \cos^2 \omega_n t)$$

$$W_C(t) = \frac{CE^2}{2}(1 - \cos \omega_n t)^2 \cdot U_{-1}(t)$$

Alternatively,

$$W_C(t) = \frac{C}{2}[e_C(t)]^2 = \frac{CE^2}{2}(1 - \cos \omega_n t)^2 \cdot U_{-1}(t).$$

So,

$$W_L(t) + W_C(t) = CE^2(1 - \cos \omega_n t) \cdot U_{-1}(t).$$

The energy at the voltage source is

$$W_F(t) = \int_{-\infty}^{t} p_F(\lambda)d\lambda = -E^2 \sqrt{\frac{C}{L}} \int_{0_+}^{t} \sin \omega_n \lambda \, d\lambda = E^2 \sqrt{\frac{C}{L}} \frac{\cos \omega_n \lambda}{\omega_n} \Big|_{0_+}^{t}$$

$$W_F(t) = CE^2(\cos \omega_n t - 1) \cdot U_{-1}(t).$$

Therefore, $W_F(t) + W_L(t) + W_C(t) = 0$, as it should to be.

Entering the numerical values, it results that:

$$p_L(t) = 500 \sin 200t . U_{-1}(t) \, [\text{W}]$$
$$p_C(t) = (1,000 \sin 100t - 500 \sin 200t).U_{-1}(t)$$
$$= 1,000 \sin 100t(1 - \cos 100t).U_{-1}(t) \, [\text{W}].$$
$$p_F(t) = -1,000 \sin 100t . U_{-1}(t) \, [\text{W}].$$
$$W_L(t) = 5 \sin^2 100t . U_{-1}(t) \, [\text{J}].$$
$$W_C(t) = 5(1 - \cos 100t)^2 . U_{-1}(t) \, [\text{J}].$$
$$W_F(t) = 10(\cos 100t - 1).U_{-1}(t) \, [\text{J}].$$

(c) Nyquist sampling theorem

The time domain Nyquist sampling theorem states that:
"If the Fourier transform, $X(f)$, of a signal $x(t)$ is null beyond a certain frequency f_c, called the cutoff frequency, that is, if $X(f) = 0$ for $|f| > f_c$, then the continuous function $x(t)$ can only be determined from the knowledge of its sampled values, provided that the sampling frequency $f_a = 1/\Delta t$, where Δt is the sampling interval, satisfies the condition $f_a > 2f_c$".

As $\mathcal{F}\left[E \cos \frac{2\pi}{T} t\right] = X(f) = \frac{E}{2} U_0(f + f_0) + \frac{E}{2} U_0(f - f_0)$, where $f_0 = f_c = \frac{1}{T}$, then, according to the theorem,

$$f_a = \frac{1}{\Delta t} > \frac{2}{T}, \text{ that is }, \Delta t < \frac{T}{2}.$$

In other words, what the theorem is saying, for the sinusoidal case, is that **if the function is not sampled with Δt less than its semiperiod, then it will not be possible to recover it, in continuous time, from the samples collected, that is, from the discrete time values.**

For this Example 6.11, therefore, it is necessary that $\Delta t \ll T_n/2 = 31.5$ [ms], considering the discretization of capacitance and inductance through the approximation of integrals by the trapezoidal method, adopted in the ATPDraw program. **A good practical value is something like $\Delta t = T_n/100$.** Thus, the sampling interval $\Delta t = 31.5.10^{-5}$ [s], will be considered for the numerical solution, or better yet, $\Delta t = 300$ [μs]. As the responses contain only steady state components, it will be considered that $t_{max} \cong 4.5T_n = 300$[ms]. And the circuit for the numerical solution by the ATPDraw program is shown in Fig. 6.14.

Fig. 6.14 Electric circuit of Example 6.11 for the numerical solution by the ATPDraw program

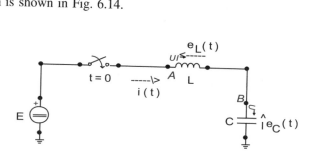

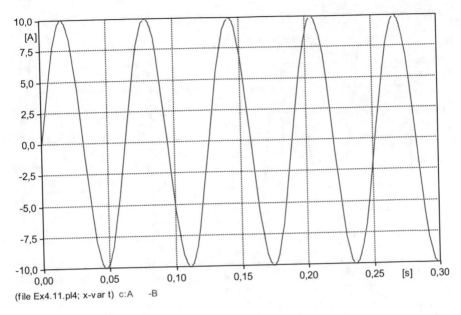

(file Ex4.11.pl4; x-var t) c:A   -B

Fig. 6.15 Current $i(t)$

Figures 6.15, 6.16 and 6.17 plot the curves of $i(t)$, $e_L(t)$ and $e_C(t)$. It is inter-esting to note that the voltage in the capacitance is never negative, and fluctuates between zero and two hundred volts.

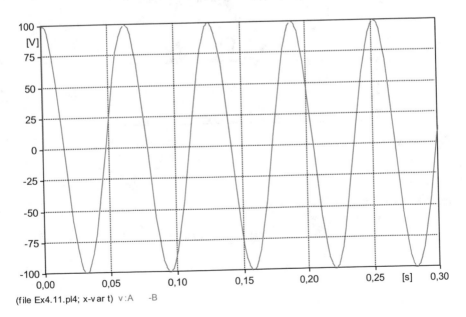

(file Ex4.11.pl4; x-var t) v:A   -B

Fig. 6.16 Voltage $e_L(t)$

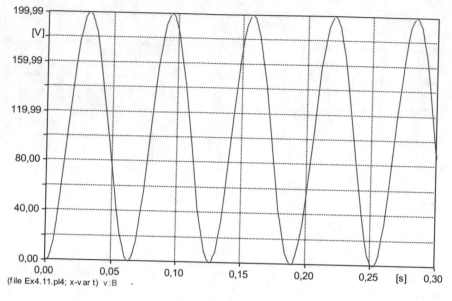

Fig. 6.17 Voltage $e_C(t)$

Example 6.12 Considering the circuit shown in Fig. 6.18:
(a) Find the current and voltage analytical expressions in all passive elements of
   the circuit, for $t > 0$, knowing that the capacitor is initially de-energized at
   $t = 0_-$ and that $E = 10\,[\mathrm{V}]$, $R = 10\,[\Omega]$, $L = 100\,[\mathrm{mH}]$ and $C = 1,000\,[\mu\mathrm{F}]$.
(b) Perform the numerical solution of the circuit.

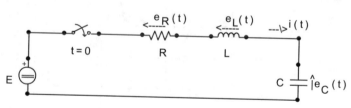

Fig. 6.18 Electric circuit of Example 6.12

Solution
(a) The mesh equation for the circuit is:

$$Ri(t) + L\frac{d}{dt}i(t) + \frac{1}{C}\int i(t)dt = EU_{-1}(t)$$

or,

$$i''(t) + \frac{R}{L}i'(t) + \frac{1}{LC}i(t) = \frac{E}{L}U_0(t), \text{ with } i(0_-) = i'(0_-) = 0. \qquad (6.7)$$

For $0+ \leq t \leq \infty$,

$$\left(D^2 + \frac{R}{L}D + \frac{1}{LC}\right)i(t) = 0$$

Substituting the numerical values results that

$$(D^2 + 100D + 10,000)i(t) = 0.$$

The characteristic roots are $r_{1,2} = -50 \pm j86.603$ and,

$$i(t) = \epsilon^{-50t}(k_1 \cos 86.603t + k_2 \sin 86.603t)$$
$$i'(t) = -50\epsilon^{-50t}(k_1 \cos 86.603t + k_2 \sin 86.603t)$$
$$+ 86.603\epsilon^{-50t}(-k_1 \sin 86.603t + k_2 \cos 86.603t)$$

From Eq. (6.7), by inspection,

$$i(0_+) = 0[\text{A}] \text{ and } i'(0_+) = E/L = 100[\text{V/s}]. \text{ Then,}$$

$$k_1 = 0; 86.603k_2 = 100 \text{ and } k_2 = 1.155,$$

Therefore,

$$i(t) = 1.155\epsilon^{-50t} \sin 86.603t.U_{-1}(t)[\text{A}].$$

The time constant is $T = 1/50 = 20\,[\text{ms}]; 5T = 100\,[\text{ms}]$.

$$\omega_n = \frac{2\pi}{T_n}; T_n = \frac{2\pi}{86.603} = 0.073\,[\text{s}] = 73\,[\text{ms}].$$

$$e_R(t) = Ri(t) = 10i(t)$$
$$e_R(t) = 11.55\epsilon^{-50t} \sin 86.603t.U_{-1}(t)[\text{V}].$$

The voltage at the inductance is

$$e_L(t) = L\frac{d}{dt}i(t) = 0.1\frac{d}{dt}\left[1.155\epsilon^{-50t}\sin 86.603t.U_{-1}(t)\right]$$
$$= \epsilon^{-50t}(10.0026\cos 86.603t - 5.7750\sin 86.603t)U_{-1}(t).$$

Or, according to the trigonometric identity

$$A\cos\omega t + B\sin\omega t = \sqrt{A^2+B^2}.\cos\left(\omega t - arctg\frac{B}{A}\right),$$

$$e_L(t) = 11.55\epsilon^{-50t}\cos(86.603t+30°)U_{-1}(t)[\text{V}].$$

$$e_C(t) = \frac{1}{C}\int_0^t i(\lambda)d\lambda = 1,000\int_0^t 1.155\epsilon^{-50\lambda}\sin 86.603\lambda d\lambda$$

$$= \left[10.0026 - \epsilon^{-50t}(10.0026\cos 86.603t + 5.7750\sin 86.603t)\right]U_{-1}(t)[\text{V}].$$

Or, using the same trigonometric identity as before,

$$e_C(t) = \left[10.0026 - 11.55\epsilon^{-50t}\cos(86.603t - 30°)\right]U_{-1}(t)[\text{V}].$$

(b) The circuit for the numerical solution by ATPDraw is shown in Fig. 6.19. $\Delta t = 1$ [μs] was adopted. As $5T = 100$ [ms] $= 0.1$ [s], the requested quantities were plotted up to $t_{max} = 0.2$ [s] $= 200$ [ms]. The current $i(t)$ is shown in Fig. 6.20 and the voltages in Fig. 6.21.

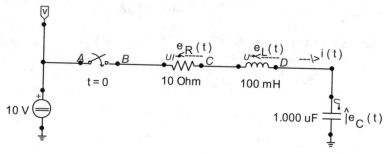

Fig. 6.19 Electric circuit of Example 6.12 for the numerical solution by the ATPDraw program

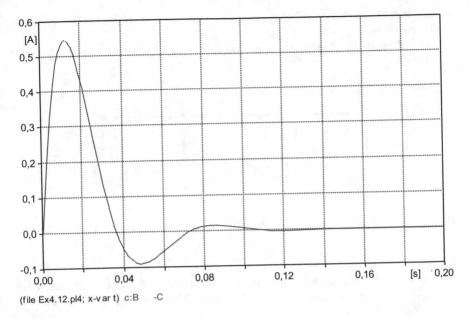

(file Ex4.12.pl4; x-var t) c:B   -C

Fig. 6.20 Current $i(t)$

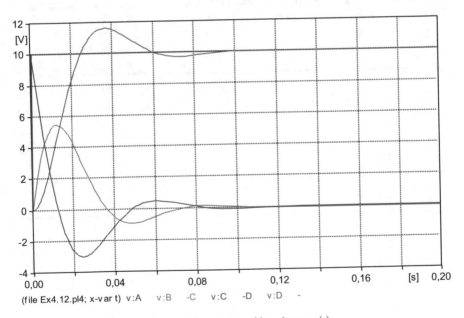

(file Ex4.12.pl4; x-var t) v:A   v:B   -C   v:C   -D   v:D   -

Fig. 6.21 Supply voltage, voltage $e_R(t)$, voltage $e_L(t)$, voltage $e_C(t)$

Example 6.13 Find, numerically, the voltage and current in the passive elements of the circuit in Fig. 6.22.

Data: $I = 10\,[\text{A}]$, $R = 10\,[\Omega]$, $L = 100\,[\text{mH}]$, $C = 1,000\,[\mu\text{F}]$, $\Delta t = 1\,[\mu\text{s}]$ and $t_{\max} = 0.2\,[\text{s}]$.

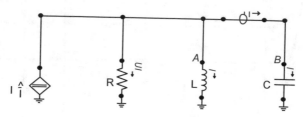

Fig. 6.22 Electric circuit of Example 6.13 for the numerical solution by the ATPDraw program

Solution

The voltage in the four parallel elements and the current in the resistor are shown in Fig. 6.23. Figure 6.24 shows the currents at inductance and capacitance, each with its vertical scale.

The Current Probe was incorporated into the circuit of Fig. 6.22 to allow the introduction of node B and, thus, facilitate the distinction between the current in the inductance and the current in the capacitance.

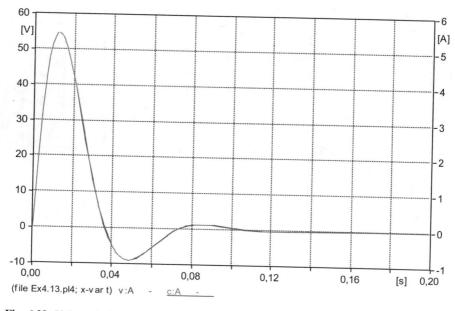

(file Ex4.13.pl4; x-var t) v:A   -   c:A   -

Fig. 6.23 Voltage in the elements, current in the resistance

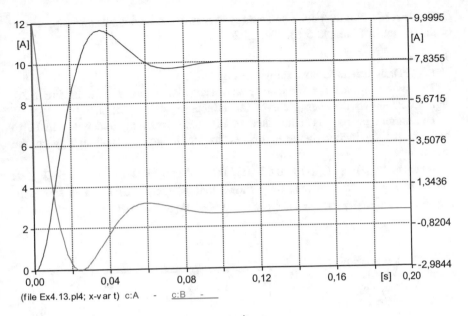

Fig. 6.24 Current at inductance, current at capacitance

The analytical solution, using the nodal method, produces the following results:

$$e(t) = 115.5\epsilon^{-50t}\sin 86.603t U_{-1}(t)\,[\text{V}]$$
$$i_R(t) = 11.55\epsilon^{-50t}\sin 86.603t U_{-1}(t)\,[\text{A}]$$
$$i_L(t) = \left[10 - 11.55\epsilon^{-50t}\cos\left(86.603t - 30°\right)\right]U_{-1}(t)\,[\text{A}]$$
$$i_C(t) = 11.55\epsilon^{-50t}\cos\left(86.603t + 30°\right)U_{-1}(t)\,[\text{A}].$$

Of these expressions we have that:

$$e(0_+) = 0\,[\text{V}], e(\infty) = 0\,[\text{V}]$$

$$i_R(0_+) = 0\,[\text{A}], i_R(\infty) = 0\,[\text{A}]$$

$$i_L(0_+) = 0\,[\text{A}]\,(open\ circuit), i_L(\infty) = 10\,[\text{A}]\,(short\ circuit)$$

$$i_C(0_+) = 10\,[\text{A}]\,(short\ circuit), i_C(\infty) = 0\,[\text{A}]\,(open\ circuit).$$

$$T = 1/50 = 20\,[\text{ms}]; 5T = 100\,[\text{ms}].$$

$$T_n = \frac{2\pi}{86.603} = 0.073\,[\text{s}] = 73\,[\text{ms}].$$

Example 6.14 Plot the power and energy curves in the inductance and capacitance of the circuit in Example 6.13, Fig. 6.22.

Solution

The ATPDraw circuit is now shown in Fig. 6.25.

The power curves for inductance and capacitance are shown in Fig. 6.26. Figure 6.27 shows the energy curves for inductance and capacitance.

The current probe was maintained in Fig. 6.25, retaining nodes A and B, to facilitate the distinction between energy in the inductance and energy in the capacitance.

Whereas $W_L(t) = \frac{1}{2}Li_L^2(t) = 0.05i_L^2(t)$, then $W_L(\infty) = 0.05i_L^2(\infty) = 5$ [J], as shown in Fig. 6.27. Dual reasoning shows that $W_C(\infty) = 0$ [J], also shown in Fig. 6.27, that is, $W_C(\infty) = \frac{1}{2}Ce_C^2(\infty) = 0.0005e_C^2(\infty) = 0$ [J].

Fig. 6.25 Electric circuit of Example 6.14 for the numerical solution by the ATPDraw program

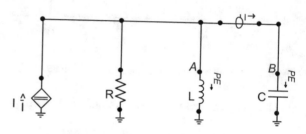

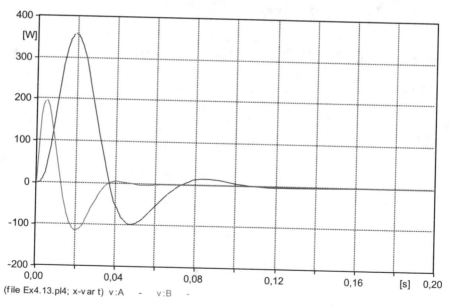

Fig. 6.26 Inductance power, capacitance power

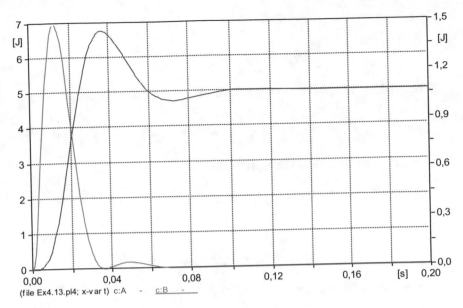

(file Ex4.13.pl4; x-var t) c:A   -   c:B   -

Fig. 6.27 Energy in inductance, energy in capacitance

Example 6.15 For the circuit in Fig. 6.28:

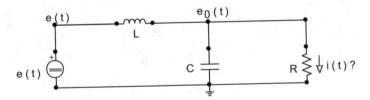

Fig. 6.28 Electric circuit of Example 6.15

(a) Find the impulse response $h(t)$ of the circuit, for $R = 5$ [Ω], $L = 100$ [mH] and $C = 1,000$ [μF].

(b) Obtain, numerically, using the ATPDraw program, an approximation of the response to the impulse $h(t)$.

(a) The nodal circuit equation is:

$$\frac{1}{LD}(e_0 - e) + CDe_0 + \frac{1}{R}e_0 = 0$$

Or,

$$\frac{1}{L}(e_0 - e) + CD^2e_0 + \frac{1}{R}De_0 = 0$$

As $e_0(t) = Ri(t)$, it follows that:

$$\frac{R}{L}i(t) + RCD^2 i(t) + Di(t) = \frac{1}{L}e(t)$$

When $e(t) = U_0(t)$, $i(t) = h(t)$ and the differential equation becomes

$$h''(t) + \frac{1}{RC}h'(t) + \frac{1}{LC}h(t) = \frac{1}{RLC}U_0(t), \text{ with } h(0_-) = h'(0_-) = 0.$$

Introducing the numerical values, it comes that:

$$h''(t) + 200h'(t) + 10,000h(t) = 2,000U_0(t), \text{ with } h(0_-) = h'(0_-) = 0. \quad (6.8)$$

For $t > 0$,

$$\left(D^2 + 200D + 10,000\right)h(t) = 0$$

The characteristic roots are: $r_{1,2} = -100$. Like this,

$$h(t) = (k_1 + k_2 t)\epsilon^{-100t}$$
$$h'(t) = k_2\epsilon^{-100t} - 100(k_1 + k_2 t)\epsilon^{-100t}$$

From Eq. (6.8),

$$h(t) \doteq 0.U_{-1}(t).$$

$$h'(t) \doteq C_0 U_{-1}(t).$$

$$h''(t) \doteq C_0 U_0(t) + C_1 U_{-1}(t).$$

Substituting in Eq. (6.8), it is

$$C_0 U_0(t) + (C_1 + 200C_0)U_{-1}(t) \doteq 2,000U_0(t) + 0.U_{-1}(t).$$

Therefore,

$$h(0_+) = 0 = k_1$$

$$h'(0_+) = C_0 = 2,000 = k_2$$

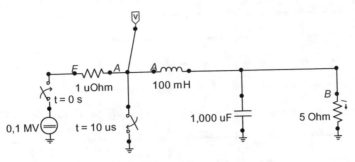

Fig. 6.29 Equivalent electric circuit of Example 6.15 for the numerical solution by the ATPDraw program

Solution

Finally,

$$h(t) = 2,000te^{-100t}U_{-1}(t) \, [\text{A}]$$

$h(t)$ goes through a maximum value of 7.36[A], exactly when $t = T = 1/100 = 10\,[\text{ms}]$. And the impulse response lasts approximately $5T = 50\,[\text{ms}]$.

(b) The circuit constructed in ATPDraw to obtain an approximation of the impulse response is shown in Fig. 6.29.

To approximate the unitary impulse to the voltage source $e(t)$), a base rectangular pulse (duration) $10\,[\mu\text{s}] = 10^{-5}\,[\text{s}]$ and amplitude of $0.1\,[\text{MV}] = 10^5\,[\text{V}]$ which logically results in a unitary area pulse, that is, $1\,[\text{V.s}]$. Therefore, a step Δt equal to one hundredth of the rectangular pulse duration was adopted, that is, $\Delta t = 0.1\,[\mu\text{s}] = 10^{-7}\,[\text{s}]$. As the circuit time constant is $T = 10\,[\text{ms}]$, a simulation time $t_{max} = 60\,[\text{ms}]$. was taken. **To prevent the two switches from having a common terminal, and the program not to "run", the resistance of $1\,[\mu\Omega]$ shown between nodes F and A in Fig.** 6.29 **was introduced.** It should be noted, however, that the assembly with the two switches and the resistance of $1\,[\mu\Omega]$ in Fig. 6.29 was not necessary, since ATPDraw allows to directly establish the mentioned rectangular font, just by simply assigning the following values for the voltage source: Amplitude Volt **Volt** $1E5$, **Tstart s** 0, **Tstop s** $1E - 5$.

Figure 6.30 shows the rectangular pulse of the voltage source $e(t)$, captured by the voltage probe in Fig. 6.29. Figure 6.31 shows a zoom of this same source, in the vicinity of $t = 0\,[\text{s}]$, on the right. And in Fig. 6.32, the approximation "$h(t)$" of the response to impulse $h(t)$ of the circuit in Fig. 6.28 is recorded.

With the aid of the ATPDraw plotter cursor, and applying an appropriate zoom, it is possible to determine, from the curve that was plotted, and that Fig. 6.32 reproduces, that for $t = 10.009\,[\text{ms}]$, "h" is worth 7.3576 [A], a result perfectly consistent with the maximum value of 7.3576 [A] obtained from the analytical solution for $t = 10\,[\text{ms}]$. The same coherence can be seen for all values of t between $t = 0\,[\text{ms}]$ and $t = t_{max} = 60\,[\text{ms}]$.

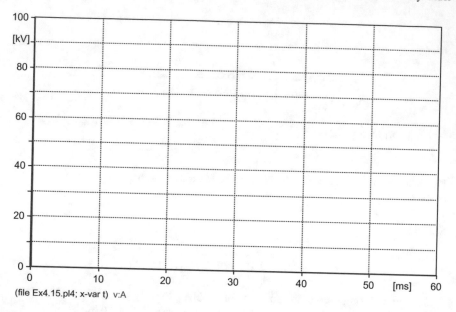

Fig. 6.30 Rectangular pulse of the voltage source $e(t)$

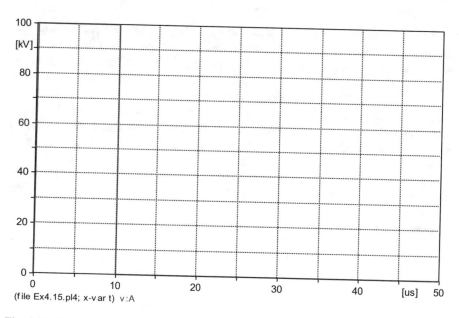

Fig. 6.31 Zoom on the rectangular pulse of the voltage source $e(t)$

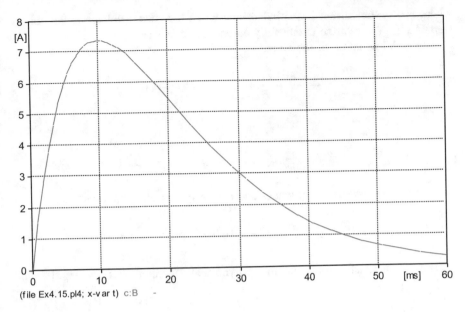

(file Ex4.15.pl4; x-v ar t) c:B   -

Fig. 6.32 Response to the impulse "$h(t)$"

Example 6.16 Considering the circuit in Fig. 6.33:
(a) Determine the current i(t) in the inductance of the circuit, for all t, knowing that
   $R = 1\,[k\Omega]$, $L = 4\,[mH]$, $C = 3\,[nF]$ and that the voltage of the source is:

$$e(t) = \begin{cases} -6\,[V], \ for \ t<0\,[s] \\ +6\,[V], \ for \ t>0\,[s] \end{cases}$$

(b) Perform the numerical solution of the problem.

Fig. 6.33 Electric circuit of
Example 6.16

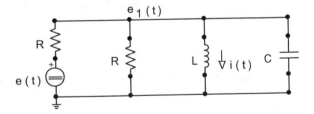

Solution
(a) The source voltage can also be expressed by $e(t) = -6 + 12U_{-1}(t)\,[V]$.

   Like $-6 = -6U_{-1}(t+\infty)$, then $e(t) = -6U_{-1}(t+\infty) + 12U_{-1}(t)\,[V]$.
   Therefore, by the superposition theorem, the current in the inductance is given by

$$i(t) = -6r(t+\infty) + 12r(t), \tag{6.9}$$

where $r(t)$ is the current in the inductance when the source voltage is the unitary step $U_{-1}(t)$ [V], with the circuit fully de-energized at $t = 0_{-}$.

The nodal equation of the circuit is:

$$\frac{e_1 - e}{R} + \frac{e_1}{R} + \frac{1}{LD}e_1 + CDe_1 = 0$$

Or,

$$e_1''(t) + \frac{2}{RC}e_1'(t) + \frac{1}{LC}e_1(t) = \frac{1}{RC}e'(t)$$

Substituting the numerical values and taking $e(t) = U_{-1}(t)$, comes that

$$e_1''(t) + \frac{2}{3}10^6 e_1'(t) + \frac{1}{12}10^{12}e_1(t) = \frac{1}{3}10^6 U_0(t); e_1(0_-) = e_1'(0_-) = 0 \quad (6.10)$$

The characteristic roots of the polynomial operator are: $r_1 = -\frac{1}{6}10^6$ and $r_2 = -\frac{1}{2}10^6$

$$e_1(t) = k_1 \epsilon^{-10^6 t/6} + k_2 \epsilon^{-10^6 t/2}$$

$$e_1'(t) = -\tfrac{10^6}{6}k_1\epsilon^{-10^6 t/6} - \tfrac{10^6}{2}k_2\epsilon^{-10^6 t/2}$$

By inspecting Eq. (6.10), $e_1(0_+) = 0$ [V] and $e_1'(0_+) = \frac{1}{3}10^6$ [V/s]. Therefore,

$$k_1 + k_2 = 0$$

$$-\tfrac{10^6}{6}k_1 - \tfrac{10^6}{2}k_2 = \tfrac{1}{3}10^6$$

From this system of equations it follows that $k_1 = 1$ and $k_2 = -1$. So,

$$e_1(t) = \left(\epsilon^{-10^6 t/6} - \epsilon^{-10^6 t/2}\right)U_{-1}(t) \text{ [V]}.$$

Thus, the step response (current) in the inductance is

$$r(t) = \frac{1}{L}\int_{0_+}^{t} e_1(\lambda)d\lambda = 250 \int_{0_+}^{t} \left(\epsilon^{-\frac{10^6\lambda}{6}} - \epsilon^{-\frac{10^6\lambda}{2}}\right)d\lambda$$

Following the integration process, it results that:

$$r(t) = \left(1 + 0.5\epsilon^{-10^6 t/2} - 1.5\epsilon^{-10^6 t/6}\right)U_{-1}(t) \text{ [mA]}$$

Therefore,

$$r(t+\infty) = \left[1 + 0.5\epsilon^{-\frac{10^6(t+\infty)}{2}} - 1.5\epsilon^{-\frac{10^6(t+\infty)}{6}}\right]U_{-1}(t+\infty) = 1\,[\mathrm{mA}]$$

So, according to Eq. (6.9), the current in the inductance, for all t, is

$$i(t) = -6 + \left(12 + 6\epsilon^{-10^6 t/2} - 18\epsilon^{-10^6 t/6}\right)U_{-1}(t)\,[\mathrm{mA}]$$

Or,

$$i(t) = \begin{cases} -6\,[\mathrm{mA}] & for\ t \le 0\,[\mathrm{s}] \\ 6 + 6\epsilon^{-10^6 t/2} - 18\epsilon^{-\frac{10^6 t}{6}}\,[\mathrm{mA}] & for\ t \ge 0\,[\mathrm{s}]. \end{cases}$$

And the time constants are: $T_1 = 2\,[\mu\mathrm{s}]$ and $T_2 = 6[\mu\mathrm{s}]$.

(b) For the numerical solution, it was necessary to use current sources, since all attempts at arrangements with voltage sources and switches failed, that is, the program "did not run". In addition, as the ATP does not work with negative time values, a situation present in the case of the source of $-6\,[\mathrm{V}]$, which acts from $-\infty$ to $t = 0_-$, it became mandatory to consider that the source of $-6\,[\mathrm{V}]$ excites the circuit, continuously, from $t = 0_+$, and not from $-\infty$. And, then, to consider that the source of $+12\,[\mathrm{V}]$ starts to act on the circuit only from the moment when the behavior of transient regime, due to the first source, of $-6\,[\mathrm{V}]$, has already extinguished, that is, that $t > 5T_2 = 30\,[\mu\mathrm{s}]$. Therefore, it was considered that the source of $+12\,[\mathrm{V}]$ enters the circuit by exciting only from $t = 50\,[\mu\mathrm{s}]$. To introduce the current sources, the Norton equivalent of the circuit part of Fig. 6.33, to the left of the inductance, was used. And the result is the circuit shown in Fig. 6.34.

In Fig. 6.35 the current $i(t)$ is plotted on the inductance.

Figures 6.36 and 6.37 show, respectively, the total current injected by the current sources and the voltage $e_1(t)$ in the three elements, $R - L - C$, connected in parallel. It is necessary to remember that in Figs. 6.35 to 6.37 the time $t = 0.00\,[\mathrm{ms}]$ corresponds, in fact, to $t \to -\infty$; and that the time $t = 50\,[\mu\mathrm{s}] = 0.05\,[\mathrm{ms}]$ corresponds to $t = 0\,[\mathrm{s}]$, when the voltage source went from $-6\,[\mathrm{V}]$ para $+6\,[\mathrm{V}]$. The step $\Delta t = T_1/100 = 2.10^{-8}\,[\mathrm{s}] = 20\,[\mathrm{ns}]$ was adopted.

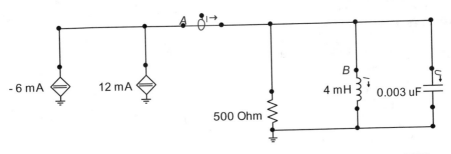

Fig. 6.34 Equivalent electric circuit of Example 6.16 for the numerical solution by the ATPDraw program

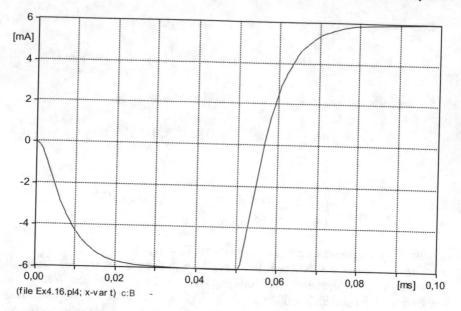

(file Ex4.16.pl4; x-var t) c:B   -

Fig. 6.35 Current at inductance $i(t)$

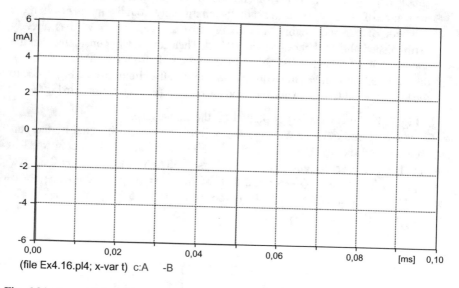

(file Ex4.16.pl4; x-var t) c:A   -B

Fig. 6.36 Current injected by sources

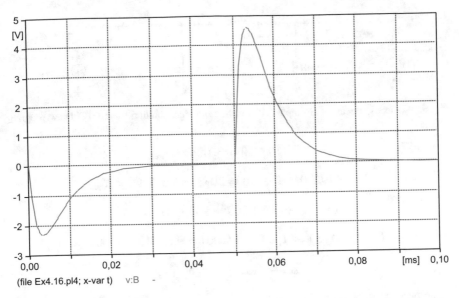

Fig. 6.37 Voltage $e_1(t)$ in R, L and C

Example 6.17 Determine, numerically, the voltage and current in the capacitance and inductance of 400 [mH], as well as the current in the resistances of the circuit of Fig. 6.38,   for   the   voltage   sources $e_1(t) = 120\cos(100t + 90°)[V] = -120 \sin 100t$ [V] and $e_2(t) = 80\cos 100t$ [V].

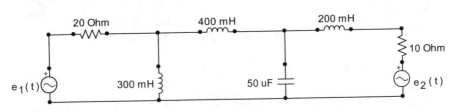

Fig. 6.38 Electric circuit of Example 6.17

Solution

The corresponding circuit for the numerical solution is shown in Fig. 6.39.

In order to establish an adequate value for the step Δt and the simulation time t_{max}, hypothetical time constants, maximum and minimum, will be calculated, assuming serial and parallel connections of passive elements, of the same type, of the given circuit. Hypothetical natural periods of oscillation, maximum and minimum, will also be calculated with the elements of the given circuit. To know:

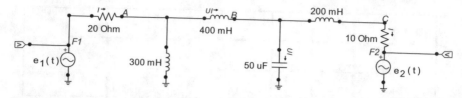

Fig. 6.39 Electric circuit of Example 6.17 for the numerical solution by the ATPDraw program

$$R_S = 10 + 20 = 30\ [\Omega] = R_{max}$$

$$R_P = (10.20)/(10+20) = 20/3 = 6.667\ [\Omega] = R_{min}$$

$$L_S = 300 + 400 + 200 = 900\ [\text{mH}] = L_{max}$$

$$L_P = 1/(1/300 + 1/400 + 1/200) = 92.308\ [\text{mH}] = L_{min}$$

$$C_{max} = C_{min} = C = 50\ [\mu F]$$

Time constants of type $T_C = RC$:

$$T_{C_{max}} = R_{max}.C_{max} = 30.50.10^{-6} = 0.0015\ [\text{s}] = 1.5\ [\text{ms}]$$

$$T_{C_{min}} = R_{min}.C_{min} = \frac{20}{3}.50.10^{-6} = 333\ [\mu s]$$

Time constants of type $T_L = L/R$:

$$T_{L_{max}} = L_{max}/R_{min} = 900.10^{-3}/(20/3) = 0.135\ [\text{s}] = 135\ [\text{ms}]$$
$$T_{L_{min}} = L_{min}/R_{max} = 92.308.10^{-3}/30 = 3.077\ [\text{ms}]$$

Natural periods of oscillation $T_n = 2\pi\sqrt{LC}$:

$$T_{n_{max}} = 2\pi\sqrt{L_{max}.C_{max}} = 2\pi\sqrt{900.10^{-3}.50.10^{-6}} = 0.042\ [\text{s}] = 42\ [\text{ms}]$$
$$T_{n_{min}} = 2\pi\sqrt{L_{mim}.C_{min}} = 2\pi\sqrt{92.308.10^{-3}.50.10^{-6}} = 0.0135\ [\text{s}] = 13.5\ [\text{ms}]$$

Comparing the three maximum and minimum times obtained, it can be concluded that

$$T_{min} = 333[\mu s]\ \text{and}\ T_{max} = 135[\text{ms}]$$

This means that: whatever the number of time constants that the given circuit presents, none of them will stop being located in the range between 333 [μs] and 135 [ms].

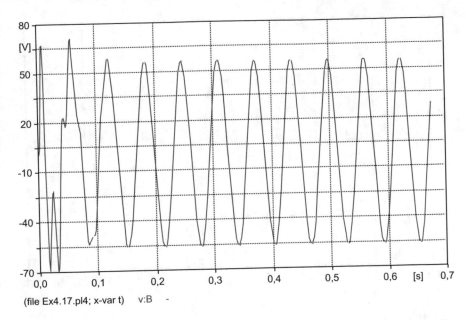

Fig. 6.40 Voltage at capacitance

Taking, for example, $\Delta t = T_{min}/100 = 3.33\,[\mu s]$ and $t_{max} = 5T_{max} = 675[\mathrm{ms}]$, then the number of points that the program should calculate is:

$$N_P = t_{max}/\Delta t + 1 = 675.10^{-3}/3.33.10^{-6} + 1 \cong 202,704\,\text{dots}.$$

The angular frequency (pulsation) of external voltage sources is $\omega = 100\,[\mathrm{rad/s}] = 2\pi/T\,[\mathrm{rad/s}]$. Thus, $T \cong 63\,[\mathrm{ms}]$ and $f = 15.91549\,[\mathrm{Hz}]$.

The conclusion is that it is reasonable to adopt the default value that ATP uses for step Δt, that is, $\Delta t = 1\,[\mu s]$.

The program was run, however, initially with the values obtained previously. Figure 6.40 shows the capacitance voltage.

As it turns out, the simulation time can perfectly be cut in half. In Fig. 6.41, the voltage at the capacitance is recorded again, but now with $t_{max} = 0.35\,[\mathrm{s}]$.

Figure 6.42 shows a zoom in on the capacitance voltage.

The current at the capacitance is shown in Fig. 6.43.

Figure 6.44 shows the voltage and current at 400 [mH]. inductance. And in Fig. 6.45, the voltages in the resistors.

Finally, in Fig. 6.46 the voltages from the two external sources are recorded.

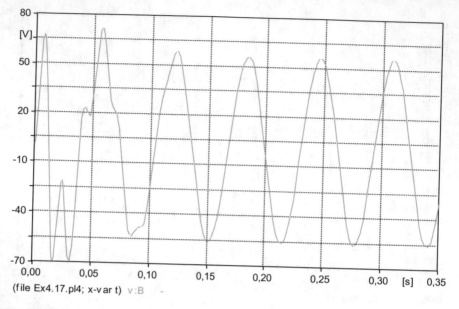

Fig. 6.41 Zoom at the capacitance voltage

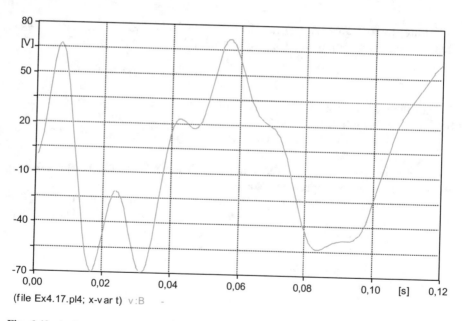

Fig. 6.42 A closer zoom at the capacitance voltage

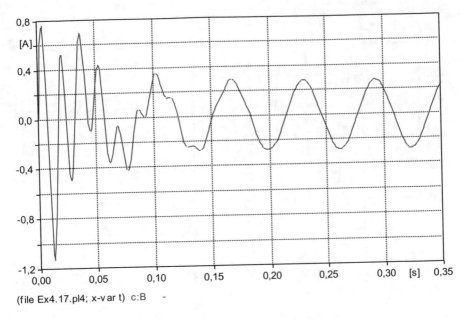

(file Ex4.17.pl4; x-var t) c:B   -

Fig. 6.43 Current at capacitance

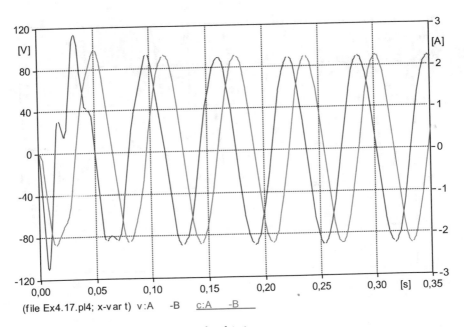

(file Ex4.17.pl4; x-var t) v:A   -B   c:A   -B

Fig. 6.44 Voltage and current at the 400 [mH] inductance

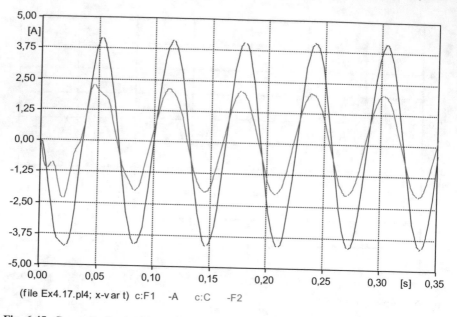

Fig. 6.45 Current in the resistors of 20 [Ω] and 10 [Ω]

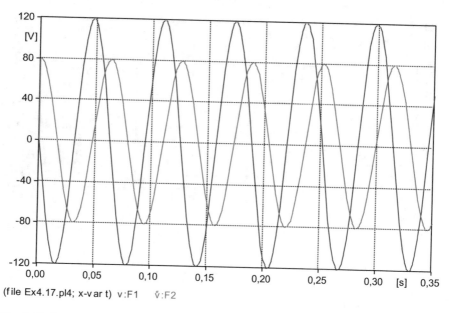

Fig. 6.46 Source voltage $e_1(t)$ and $e_2(t)$

6.3 Thévenin and Norton Equivalents for Initially Energized Capacitances and Inductances

Figure 6.47a shows a capacitance C with an initial voltage $e(0_-)$. Voltage $e(t)$ and current $i(t)$ are of interest only for $t \geq 0$. The voltage-current relationship for this passive circuit element is:

$$e(t) = \frac{1}{C} \int^t i(t)dt = \frac{1}{C} \int_{-\infty}^t i(\lambda)d\lambda = \frac{1}{C} \int_{-\infty}^{0_-} i(\lambda)d\lambda + \frac{1}{C} \int_{0_-}^t i(\lambda)d\lambda$$

$$= \frac{1}{C} q(0_-) + \frac{1}{C} \int_{0_-}^t i(\lambda)d\lambda$$

Since $q(t) = Ce(t), \frac{q(0_-)}{C} = e(0_-)$ and, then,

$$e(t) = e(0_-) + \frac{1}{C} \int_{0_-}^t i(\lambda)d\lambda \tag{6.11}$$

As the equivalents sought must be valid only for $t \geq 0$ specifically: $t > 0$ for $e(t)$ and $t \geq 0$ for $i(t)$ then the term $e(0_-)$ of Eq. 6.11 can be multiplied by the step function $U_{-1}(t)$, without creating any inconsistency, since the function $U_{-1}(t)$ is equal to 1 for $t > 0$. Thus, Eq. (6.11) can be rewritten as:

$$e(t) = e(0_-) . U_{-1}(t) + \frac{1}{C} \int_{0_-}^t i(\lambda)d\lambda \tag{6.12}$$

The Eq. (6.12) makes it possible to construct the Thévenin equivalent shown in Fig. 6.47b, in which the voltage source $e(0_-)U_{-1}(t)$ is connected in series with capacitance C, now initially de-energized. From Fig. 6.47b one can then obtain the

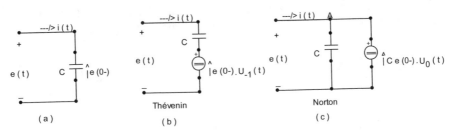

(a)

Thévenin

(b)

Norton

(c)

Fig. 6.47 Electric circuit with **a** a capacitance C with an initial voltage $e(0_-)$; **b** Thévenin equivalent with capacitance C, now initially de-energized; **c** Norton equivalent with capacitance C, now initially de-energized

Norton equivalent shown in Fig. 6.47c, where the impulsive current source $Ce(0_-)U_0(t)$ is in parallel with capacitance C, initially de-energized.

Equating node A in Fig. 6.47c, it turns out that

$$i(t) = C\frac{d}{dt}e(t) - Ce(0_-)U_0(t) \tag{6.13}$$

Taking $t = 0_+$ in Eq. (6.12) we get that

$$e(0_+) = e(0_-) + \frac{1}{C}\int_{0_-}^{0_+} i(\lambda)d\lambda \tag{6.14}$$

In the event that $i(t)$ does not contain impulse function at $t = 0$, then $\int_{0_-}^{0_+} i(\lambda)d\lambda = 0$ and $e(0_+) = e(0_-)$, confirming the Continuity Theorem of Chap. 5, that is, that the voltage in the capacitance does not suffer discontinuity. On the other hand, if there is an impulse in current $i(t)$ at $t = 0$, then $\int_{0_-}^{0_+} i(\lambda)d\lambda \neq 0$ and $(0_+) \neq e(0_-)$, causing discontinuity in the capacitance voltage at $t = 0$, as mentioned in Corollary 1 of Chap. 5. It is always necessary to keep in mind that capacitance C, both in the Thévenin equivalent and in the Norton equivalent, is always initially de-energized.

Equation (6.12) can also be written as

$$e(t) = e(0_-).U_{-1}(t) + \frac{1}{C}\int_{0_-}^{0_+} i(\lambda)d\lambda + \frac{1}{C}\int_{0_+}^{t} i(\lambda)d\lambda \tag{6.15}$$

Therefore, if there is no impulse in current $i(t)$, at $t = 0$, (Eq. 6.15) becomes

$$e(t) = e(0_-).U_{-1}(t) + \frac{1}{C}\int_{0_+}^{t} i(\lambda)d\lambda \tag{6.16}$$

Figure 6.48a shows an inductance L with an initial current $i(0_-)$. The current-voltage relationship in this passive circuit element is given by:

$$i(t) = \frac{1}{L}\int e(t)dt = \frac{1}{L}\int_{-\infty}^{t} e(\lambda)d\lambda = \frac{1}{L}\int_{-\infty}^{0_-} e(\lambda)d\lambda + \frac{1}{L}\int_{0_-}^{t} e(\lambda)d\lambda$$

$$= \frac{1}{L}\emptyset(0_-) + \frac{1}{L}\int_{0_-}^{t} e(\lambda)d\lambda = i(0_-) + \frac{1}{L}\int_{0_-}^{t} e(\lambda)d\lambda \tag{6.17}$$

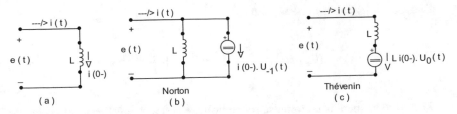

Fig. 6.48 Electric circuit with **a** an inductance L with an initial current $i(0_-)$; **b** Norton equivalent with inductance L, now initially de-energized; **c** Thévenin equivalent with inductance L, now initially de-energized

As the search equivalents must be valid only for $t \geq 0$, specifically: $t > 0$ for $i(t)$ and $t \geq 0$ for $e(t)$, the term $i(0_-)$ in Eq. (6.17) can be multiplied by the unitary step $U_{-1}(t)$, since $U_{-1}(t) = 1$ for $t > 0$. Therefore, nothing prevents (Eq. 6.17) from being rewritten as

$$i(t) = i(0_-).U_{-1}(t) + \frac{1}{L}\int_{0_-}^{t} e(\lambda)d\lambda \qquad (6.18)$$

The Eq. (6.18) depicts the Norton equivalent shown in Fig. 6.48b, in which a current source $i(0_-).U_{-1}(t)$ is connected in parallel with the inductance L, now initially de-energized. From Fig. 6.48b we can now obtain the Thévenin equivalent shown in Fig. 6.48c, in which the impulsive voltage source $Li(0_-)U_0(t)$ is in series with the inductance L, initially de-energized .

From Fig. 6.48c it follows that

$$e(t) = L\frac{d}{dt}i(t) - Li(0_-)U_0(t) \qquad (6.19)$$

Taking $t = 0_+$ in Eq. (6.18) results that

$$i(0_+) = i(0_-) + \frac{1}{L}\int_{0_-}^{0_+} e(\lambda)d\lambda \qquad (6.20)$$

In the event that there is no impulse in the voltage $e(t)$ at $t = 0$, $\int_{0_-}^{0_+} e(\lambda)d\lambda = 0$ and $i(0_+) = i(0_-)$, reaffirming the Continuity Theorem of Chap. 5, that is, that the current in the inductance does not suffer discontinuity. On the other hand, however, with an impulse in the voltage $e(t)$ at $t = 0$, $\int_{0_-}^{0_+} e(\lambda)d\lambda \neq 0$ and $i(0_+) \neq i(0_-)$, creating discontinuity in the inductance current, corroborating what was said in Corollary 2 of Chap. 5.

Equation 6.18 can also be written as

$$i(t) = i(0_-) \cdot U_{-1}(t) + \frac{1}{L} \int_{0_-}^{0_+} e(\lambda)d\lambda + \frac{1}{L} \int_{0_+}^{t} e(\lambda)d\lambda \tag{6.21}$$

Therefore, if there is no impulse in the voltage $e(t)$ at $t = 0$ Eq. (6.21) becomes

$$i(t) = i(0_-) \cdot U_{-1}(t) + \frac{1}{L} \int_{0_+}^{t} e(\lambda)d\lambda \tag{6.22}$$

It is interesting to note that Fig. 6.48a–c are dual, respectively, of Fig. 6.47a–c. Likewise, Eq. (6.18) is dual from Eq. (6.12); (6.19) is dual from (6.13); (6.21) is dual from (6.15) etc.

Finally, note that the reference arrows from the sources of the Thévenin and Norton equivalents always have the same sense (polarity or direction) as those of the respective initial conditions.

6.4 Circuits Initially Energized

Example 6.18 A capacitor of 2 [μF], with an initial charge of 200 [μC], is connected in series with a resistor of 10 [Ω], by closing a switch at $t = 0$. as shown in Fig. 6.49. Find the expression of the transient current that will circulate in the circuit for $t > 0$.

Solution

The capacitance voltage at 0_- is $e(0_-) = q(0_-)/C = 200.10^{-6}/2.10^{-6} = 100$ [V].

Replacing, then, the capacitance initially energized by its Thévenin equivalent, according to Fig. 6.47b, results in the circuit shown in Fig. 6.50.

For the closed switch, the circuit loop equation is given by

$$Ri(t) + \frac{1}{C} \int_{0_+}^{t} i(\lambda)d\lambda = e(0_-)U_{-1}(t)$$

Fig. 6.49 Electric circuit of Example 6.18

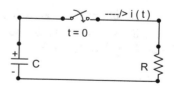

Fig. 6.50 Equivalent electric
circuit of Example 6.18

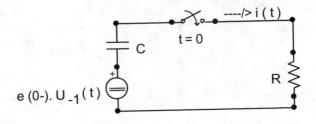

or,

$$i'(t) + \frac{1}{RC}i(t) = \frac{e(0_-)}{R}U_0(t)$$

Substituting the numerical values, it comes that:

$$i'(t) + 50,000i(t) = 10U_0(t), \quad with \ i(0_-) = 0.$$

$$i(t) = k\,\epsilon^{-50,000t}$$

$$i(0_+) = 10 = k$$

$$i(t) = 10\epsilon^{-50,000t}.U_{-1}(t)\,[\text{A}]$$

Example 6.19 Determine the voltage expressions $e_1(t)$ and $e_2(t)$, for $t > 0$, in the circuit shown in Fig. 6.51, knowing that $e_1(0_-) = E_1$ and $e_2(0_-) = E_2$. (This is the same Example 5.9).

Solution
Replacing the capacitors initially energized with their Thévenin equivalents results in the circuit in Fig. 6.52.

The nodal equations, for $t > 0$, are:

Node e_1 : $CD[e_1(t) - E_1U_{-1}(t)] + \frac{e_1(t)-e_2(t)}{R} = 0$

Node e_2 : $CD[e_2(t) - E_2U_{-1}(t)] + \frac{e_2(t)-e_1(t)}{R} = 0$

Fig. 6.51 Electric circuit of
Example 6.19

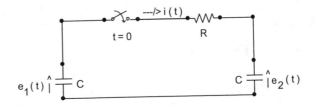

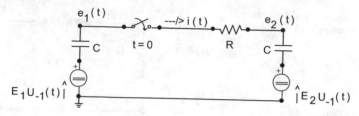

Fig. 6.52 Equivalent electric circuit of Example 6.19

Or,

$$(RCD+1)e_1(t) - e_2(t) = RCE_1U_0(t)$$
$$-e_1(t) + (RCD+1)e_2(t) = RCE_2U_0(t)$$

Solving the system for $e_1(t)$, it comes that:

$$e_1'(t) + \frac{2}{RC}e_1(t) = E_1U_0(t) + \frac{E_1+E_2}{RC}U_{-1}(t), \text{ with } e_1(0_-) = 0. \qquad (6.23)$$

(Note that $e_1(0_-) = 0$, because $e_1(0_-) = E_1$ has already been used in the Thévenin equivalent).

For $t > 0$,

$$\left(D + \frac{2}{RC}\right)e_1(t) = \frac{E_1+E_2}{RC}$$

$$e_{1_H}(t) = ke^{-2t/RC}$$

$$e_{1_P}(t) = \frac{\frac{E_1+E_2}{RC}}{\frac{2}{RC}} = \frac{E_1+E_2}{2}$$

$$e_1(t) = e_{1_P}(t) + e_{1_H}(t) = \frac{E_1+E_2}{2} + ke^{-2t/RC}$$

By inspecting Eq. (6.23), it is found that $e_1(0_+) = E_1$. So, $\frac{E_1+E_2}{2} + k = E_1$ and $k = \frac{E_1-E_2}{2}$

Finally,

$$e_1(t) = \frac{E_1+E_2}{2} + \frac{E_1-E_2}{2}e^{-2t/RC}, \text{ for } t \geq 0.$$

Solving the system of nodal equations for $e_2(t)$, it comes that

$$e_2'(t) + \frac{2}{RC}e_2(t) = E_2U_0(t) + \frac{E_1+E_2}{RC}U_{-1}(t), \text{ with } e_2(0_-) = 0. \qquad (6.24)$$

Therefore,

$$e_2(t) = \frac{E_1 + E_2}{2} + k_1 e^{-2t/RC}$$

Inspecting Eq. (6.24), it is concluded that $e_2(0_+) = E_2$. So, $\frac{E_1+E_2}{2} + k_1 = E_2$
and

$$k_1 = \frac{E_2 - E_1}{2}$$

Finally,

$$e_2(t) = \frac{E_1 + E_2}{2} + \frac{E_2 - E_1}{2} e^{-2t/RC}, \text{ for } t \geq 0.$$

The current is given by $i(t) = \frac{e_{1(t)} - e_2(t)}{R} = \frac{E_1 - E_2}{R} e^{-2t/RC}.U_{-1}(t).$

Example 6.20 (a) Find the expression of the voltage $e_1(t)$ in the circuit of Fig. 6.53, for $t > 0$, knowing that $e_1(0_-) = E$ and $e_2(0_-) = 0$. (b) Find, numerically, the voltages $e_1(t)$ and $e_2(t)$, for $t > 0$, where $E = 100$ [V], $R = 10[\Omega]$ and $C = 100$ [µF].

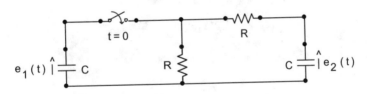

Fig. 6.53 Electric circuit of Example 6.20

Solution

(a) Replacing the capacitance initially energized with its corresponding Thévenin equivalent, the circuit shown in Fig. 6.54 results, for $t > 0$.

The nodal equations of the circuit in Fig. 6.54 are

$$CD[e_1(t) - EU_{-1}(t)] + \frac{e_1(t)}{R} + \frac{e_1(t) - e_2(t)}{R} = 0$$

$$\frac{e_2(t) - e_1(t)}{R} + CDe_2(t) = 0$$

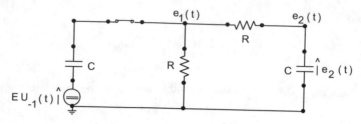

Fig. 6.54 Equivalent electric circuit of Example 6.20

Or,

$$(RCD+2)e_1(t) - e_2(t) = RCEU_0(t)$$
$$-e_1(t) + (RCD+1)e_2(t) = 0$$

Solving this system for $e_1(t)$, the differential equation results

$$\left(D^2 + \frac{3}{T}D + \frac{1}{T^2}\right)e_1(t) = EU_1(t) + \frac{E}{T}U_0(t), \text{ with } e_1(0_-) = e_1'(0_-) = 0. \quad (6.25)$$

where $T \triangleq RC$.

The characteristic roots are: $r_1 = -0.382/T$ and $r_2 = -2.618/T$. Soon,

$$e_1(t) = k_1\epsilon^{-\frac{0.382t}{T}} + k_2\epsilon^{-\frac{2.618t}{T}}$$
$$e_1'(t) = -\frac{0.382}{T}k_1\epsilon^{-\frac{0.382t}{T}} - \frac{2.618}{T}k_2\epsilon^{-\frac{2.618t}{T}}$$

Observing Eq. (6.25), it can be written that:

$$e_1(t) \doteq C_0 U_{-1}(t).$$

$$e_1'(t) \doteq C_0 U_0(t) + C_1 U_{-1}(t).$$

$$e_1''(t) = C_0 U_1(t) + C_1 U_0(t) + C_2 U_{-1}(t).$$

Substituting in Eq. (6.25), it comes that

$$C_0 U_1(t) + \left(C_1 + \frac{3}{T}C_0\right)U_0(t) + \left(C_2 + \frac{3}{T}C_1 + \frac{1}{T^2}C_0\right)U_{-1}(t) \doteq EU_1(t) + \frac{E}{T}U_0(t) + 0.U_{-1}(t).$$

So, $C_0 = E = e_1(0_+)$; $C_1 = -2E/T$.
Then,

$$k_1 + k_2 = E$$
$$-0.382k_1 - 2.618k_2 = -2E.$$

From this system of equations, $k_1 = 0.276E$ and $k_2 = 0.724E$ are obtained.

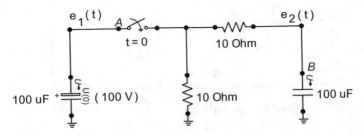

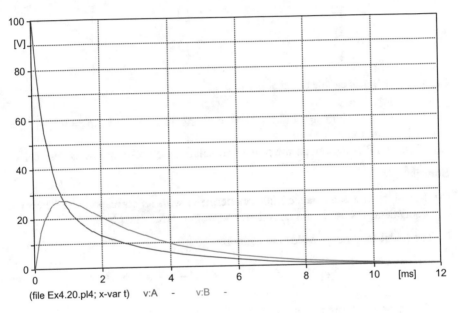

Fig. 6.55 Electric circuit of Example 6.20 for the numerical solution by the ATPDraw program

Fig. 6.56 Voltage $e_1(t)$, Voltage $e_2(t)$

Finally,

$$e_1(t) = E\left(0.276\epsilon^{-\frac{0.382t}{T}} + 0.724\epsilon^{-\frac{2.618t}{T}}\right), \text{ for } t \geq 0, \text{ where } T = RC.$$

(b) The circuit for the numerical solution is shown in Fig. 6.55, and the voltages, in Fig. 6.56.

Example 6.21 The switch in the circuit of Fig. 6.57 is closed at time $t = 0$, with capacitance C with voltage $e_C(0_-) = E$.

(a) Obtain the general analytical expressions for $e(t)$, $i(t)$, $i_1(t)$ and $i_2(t)$, for $t > 0$.
(b) Establish the analytical expressions of the voltage and currents of item a, for $k = 0$, $k = 1$ and $k = 2$.

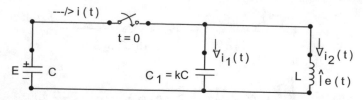

Fig. 6.57 Electric circuit of Example 6.21

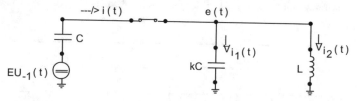

Fig. 6.58 Equivalent electric circuit of Example 6.21

(c) Obtain the analytical solutions corresponding to item b, for $E = 90\,[\text{V}]$, $L = 10\,[\text{mH}]$ and $C = 100\,[\mu\text{F}]$.

(d) Using ATPDraw, obtain the numerical solutions corresponding to item c).

Solution

(a) Replacing the capacitance initially energized with its corresponding Thévenin equivalent, results in the circuit shown in Fig. 6.58, for $t > 0$.

The nodal method provides the equation

$$CD[e(t) - EU_{-1}(t)] + kCDe(t) + \frac{1}{LD}e(t) = 0$$

Simplifying and regrouping, it comes that

$$e''(t) + \frac{1}{(k+1)LC}e(t) = \frac{E}{k+1}U_1(t), \text{ with } e(0_-) = e'(0_-) = 0. \qquad (6.26)$$

Defining $\omega_0 = \frac{1}{\sqrt{LC}}$, results

$$\left(D^2 + \frac{\omega_0^2}{k+1}\right)e(t) = 0, \text{ for } t > 0$$

The characteristic roots are

$$r_1 = j\frac{\omega_0}{\sqrt{k+1}}; r_2 = -j\frac{\omega_0}{\sqrt{k+1}}. \text{ So,}$$

$$e(t) = k_1 \cos \frac{\omega_0}{\sqrt{k+1}} t + k_2 \sin \frac{\omega_0}{\sqrt{k+1}} t \qquad (6.27)$$

$$e'(t) = -\frac{k_1\omega_0}{\sqrt{k+1}} \sin \frac{\omega_0}{\sqrt{k+1}} t + \frac{k_2\omega_0}{\sqrt{k+1}} \cos \frac{\omega_0}{\sqrt{k+1}} t \qquad (6.28)$$

Equation 6.57 allows us to write that

$$e(t) \doteq C_0 U_{-1}(t).$$

$$e'(t) \doteq C_0 U_0(t) + C_1 U_{-1}(t).$$

$$e''(t) \doteq C_0 U_1(t) + C_1 U_0(t) + C_2 U_{-1}(t).$$

Substituting in Eq. (6.26), it results

$$C_0 U_1(t) + C_1 U_0(t) + \left(C_2 + \frac{C_0\omega_0^2}{k+1}\right) U_{-1}(t) \doteq \frac{E}{k+1} U_1(t) + 0.U_0(t) + 0.U_{-1}(t).$$

Therefore, $C_0 = \frac{E}{k+1}$; $C_1 = 0$. So, $e(0_+) = \frac{E}{k+1}$ and $e'(0_+) = 0$.
From Eqs. (6.27) and (6.28), for $t = 0_+$, $\frac{E}{k+1} = k_1, 0 = \frac{k_2\omega_0}{\sqrt{k+1}}$ and $k_2 = 0$
Substituting k_1 and k_2 in Eq. (6.27), it turns out that

$$e(t) = \frac{E}{k+1} \cos \frac{\omega_0}{\sqrt{k+1}} t . U_{-1}(t) \qquad (6.29)$$

$$i(t) = C\frac{d}{dt}[EU_{-1}(t) - e(t)] = C\frac{d}{dt}\left[EU_{-1}(t) - \frac{E}{k+1}\cos\frac{\omega_0}{\sqrt{k+1}} t \cdot U_{-1}(t)\right]$$

$$= CEU_0(t) - \frac{CE}{k+1}\left[-\frac{\omega_0}{\sqrt{k+1}}\sin\frac{\omega_0}{\sqrt{k+1}} t . U_{-1}(t) + U_0(t)\right]$$

$$= CE\left[1 - \frac{1}{k+1}\right]U_0(t) + \frac{\omega_0 CE}{\sqrt{(k+1)^3}}\sin\frac{\omega_0}{\sqrt{k+1}} t . U_{-1}(t)$$

$$= \frac{kCE}{k+1}U_0(t) + \frac{\omega_0 CE}{\sqrt{(k+1)^3}}\sin\frac{\omega_0}{\sqrt{k+1}} t . U_{-1}(t)$$

$$i(t) = \frac{kCE}{k+1}U_0(t) + \frac{1}{\sqrt{(k+1)^3}}\sqrt{\frac{C}{L}}E\sin\frac{\omega_0}{\sqrt{k+1}} t . U_{-1}(t). \qquad (6.30)$$

$$i_1(t) = kC\frac{d}{dt}e(t) = kC\frac{d}{dt}\left[\frac{E}{k+1}\cos\frac{\omega_0}{\sqrt{k+1}}t\cdot U_{-1}(t)\right]$$

$$= \frac{kCE}{k+1}\left[-\frac{\omega_0}{\sqrt{k+1}}\sin\frac{\omega_0}{\sqrt{k+1}}t\cdot U_{-1}(t) + U_0(t)\right]$$

$$= \frac{kCE}{k+1}U_0(t) - \frac{k\omega_0 CE}{\sqrt{(k+1)^3}}\sin\frac{\omega_0}{\sqrt{k+1}}t\cdot U_{-1}(t)$$

$$i_1(t) = \frac{kCE}{k+1}U_0(t) - \frac{k}{\sqrt{(k+1)^3}}\sqrt{\frac{C}{L}}E\sin\frac{\omega_0}{\sqrt{k+1}}t.U_{-1}(t). \tag{6.31}$$

$$i_2(t) = i(t) - i_1(t) = \frac{(k+1)\omega_0 CE}{\sqrt{(k+1)^3}}\sin\frac{\omega_0}{\sqrt{k+1}}t\cdot U_{-1}(t)$$

$$= \frac{\omega_0 CE}{\sqrt{k+1}}\sin\frac{\omega_0}{\sqrt{k+1}}t.U_{-1}(t)$$

$$i_2(t) = \frac{1}{\sqrt{k+1}}\sqrt{\frac{C}{L}}E\sin\frac{\omega_0}{\sqrt{k+1}}t.U_{-1}(t) \tag{6.32}$$

(b) For $k = 0$, that is, $C_1 = 0$ (absence of capacitance C_1, that is, capacitance C_1 replaced by open circuit):

$$e(t) = E\cos\omega_0 t.U_{-1}(t)$$

$$i(t) = \sqrt{\frac{C}{L}}E\sin\omega_0 t.U_{-1}(t)$$

$$i_1(t) = 0$$

$$i_2(t) = \sqrt{\frac{C}{L}}E\sin\omega_0 t.U_{-1}(t) = i(t).$$

For $k = 1$, that is, $C_1 = C$,

$$e(t) = \frac{E}{2}\cos\frac{\omega_0}{\sqrt{2}}t.U_{-1}(t)$$

$$i(t) = \frac{CE}{2}U_0(t) + \frac{1}{\sqrt{3}}\sqrt{\frac{C}{L}}E\sin\frac{\omega_0}{\sqrt{2}}t.U_{-1}(t)$$

$$i_1(t) = \frac{CE}{2}U_0(t) - \frac{1}{\sqrt{8}}\sqrt{\frac{C}{L}}E\sin\frac{\omega_0}{\sqrt{2}}t.U_{-1}(t)$$

$$i_2(t) = \frac{1}{\sqrt{2}}\sqrt{\frac{C}{L}}E\sin\frac{\omega_0}{\sqrt{2}}t \cdot U_{-1}(t).$$

For $k = 2$, that is, $C_1 = 2C$,

$$e(t) = \frac{E}{3}\cos\frac{\omega_0}{\sqrt{3}}t.U_{-1}(t)$$

$$i(t) = \frac{2}{3}CEU_0(t) + \frac{1}{\sqrt{27}}\sqrt{\frac{C}{L}}E\sin\frac{\omega_0}{\sqrt{3}}t.U_{-1}(t)$$

$$i_1(t) = \frac{2}{3}CEU_0(t) - \frac{2}{\sqrt{27}}\sqrt{\frac{C}{L}}E\sin\frac{\omega_0}{\sqrt{3}}t.U_{-1}(t)$$

$$i_2(t) = \frac{1}{\sqrt{3}}\sqrt{\frac{C}{L}}E\sin\frac{\omega_0}{\sqrt{3}}t.U_{-1}(t).$$

(c) For $E = 90$ [V], $C = 100$ [μF], $L = 10$ [mH]; $\sqrt{\frac{C}{L}} = 0.1$ [s]   and
$\omega_0 == 1,000$ [rad/s].

For $k = 0$, $(C_1 = 0)$:

$$e(t) = 90\cos 1,000t.U_{-1}(t) \text{ [V]} \quad f_0 = 159.15 \text{ [Hz] and } T = 6.28 \text{ [ms]}$$

$$i(t) = 9\sin 1,000t.U_{-1}(t) \text{ [A]}$$

$$i_1(t) = 0 \text{ [A]}$$

$$i_2(t) = 9\sin 1,000t.U_{-1}(t) \text{ [A]}.$$

For $k = 1$, $(C_1 = C)$:

$$e(t) = 45\cos 707.11t.U_{-1}(t)\text{[V]} \quad f_0 = 112.54 \text{ [Hz] and } T = 8.89 \text{ [ms]}$$

$$i(t) = 4.5.10^{-3}U_0(t) + 3.18\sin 707.11t.U_{-1}(t) \text{ [A]}$$
$$i_1(t) = 4.5.10^{-3}U_0(t) - 3.18\sin 707.11t.U_{-1}(t) \text{ [A]}$$
$$i_2(t) = 6.36\sin 707.11t.U_{-1}(t) \text{ [A]}.$$

For $k = 2$, $(C_1 = 2C)$:

$$e(t) = 30\cos 577.35t.U_{-1}(t)[V] \quad f_0 = 91.89 \,[Hz] \text{ and } T = 10.88 \,[ms]$$

$$i(t) = 6.10^{-3}U_0(t) + 1.732\sin 577.35t.U_{-1}(t) \,[A]$$
$$i_1(t) = 6.10^{-3}U_0(t) - 3.464\sin 577.35t.U_{-1}(t) \,[A]$$
$$i_2(t) = 5.196\sin 577.35t.U_{-1}(t) \,[A].$$

(d)   As the analytical solution has already shown that there is an impulse in currents $i(t)$ and $i_1(t)$, **due to the left loop formed only by capacitances**, when entering the numerical value of capacitance C_1, **the factor k_s must be used**, or that is, introduce a small resistance $R_s = k_s.\Delta t/2C_1$ in series with C_1. The corresponding circuit for the numerical solution, then, is the one shown in Fig. 6.59.

(d-1)   **Numerical solution for $k = 0$.** In this case, $C_1 = 0$ and the period is $T = 6.28 \,[ms]$.

The program did not run with the simple assignment of zero value for capacitance C_1. Therefore, a small value (compared to capacitance C) was introduced for this capacitance C_1, that is, it was considered that $C_1 = C/1000 = 0.1 \,[\mu F]$. Even so, it only ran after the factor $k_s = 0.1$, that is, $R_s = 0.1.10^{-6}/(2.10^{-7}) = 0.5 \,[\Omega]$. Figure 6.60 shows the voltage $e(t)$.

Figure 6.61 shows the current $i(t)$, which is almost equal to $i_2(t)$, since $i_1(t) \cong 0$. (Capacitance C_1 is approximately an open circuit).

The currents $i_1(t)$ and $i_2(t)$ are recorded in Fig. 6.62.

Figure 6.63 shows the result of several zooms (horizontal and vertical) in the current $i_1(t)$, in which the attenuating effect of the resistance R_s in numerical oscillations can be seen.

(d-2)   **Numerical solution for k = 1.**

In this case, $C_1 = C = 100 \,[\mu F]$, and the period is $T = 8.89 \,[ms]$.
Figure 6.64 records voltage $e(t)$ and current $i(t)$.

As the scale of current $i(t)$—on the right, is very wide (determined by the amplitudes of the initial oscillations, and which will be quickly attenuated due to the introduction of the factor $k_s = 0.1$), it is necessary to manually modify the scale of current $i(t)$, which is plotted separately in Fig. 6.65.

Fig. 6.59 Electric circuit of Example 6.21 for the numerical solution by the ATPDraw program

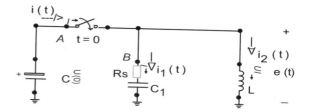

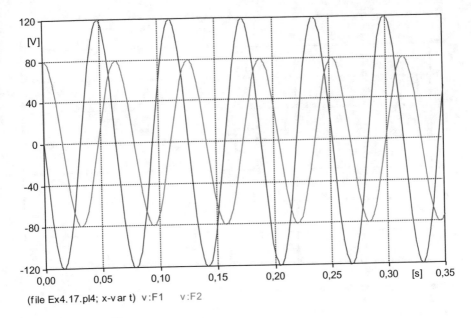

(file Ex4.17.pl4; x-var t) v:F1   v:F2

Fig. 6.60 Voltage $e(t)$

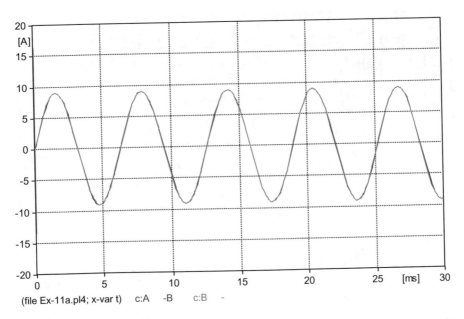

(file Ex-11a.pl4; x-var t)   c:A   -B   c:B   -

Fig. 6.61 Current $i(t) \cong i_2(t)$

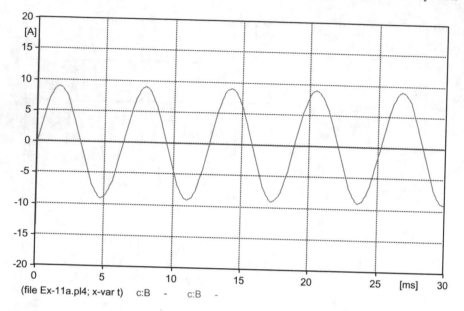

(file Ex-11a.pl4; x-var t)   c:B   -   c:B   -

Fig. 6.62 Currents $i_1(t)$ and $i_2(t)$

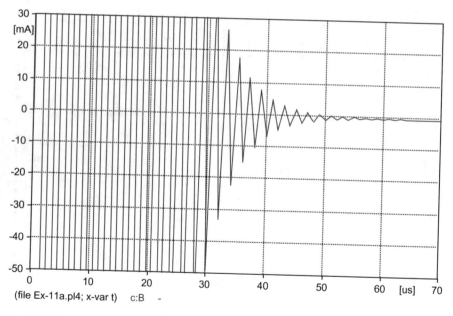

(file Ex-11a.pl4; x-var t)   c:B   -

Fig. 6.63 Current zooms $i_1(t)$

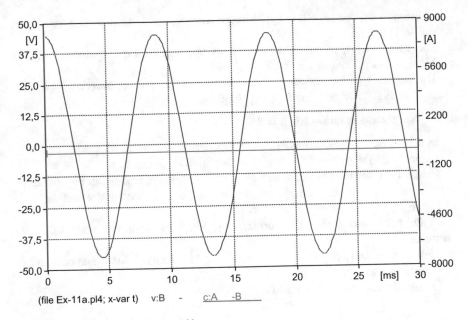

(file Ex-11a.pl4; x-var t) v:B - c:A -B

Fig. 6.64 Voltage $e(t)$ and current $i(t)$

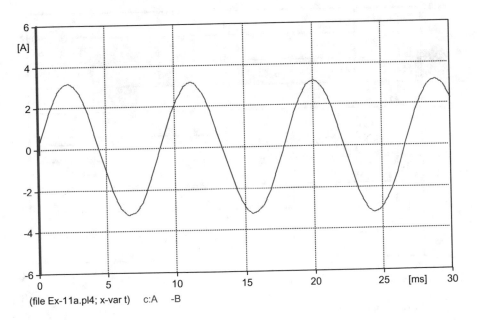

(file Ex-11a.pl4; x-var t) c:A -B

Fig. 6.65 Current $i(t)$

A zoom in the current curve $i(t)$, without moving the vertical scale manually, is shown in Fig. 6.66.

The current $i_2(t)$, in the inductance, does not present any impulse /oscillation problem and is plotted in Fig. 6.67.

The current $i_1(t)$ has the same "phenomena" as $i(t)$ and is plotted, with a scale change, in Fig. 6.68; and zoom in Fig. 6.69.

(d-3) Numerical solution for k = 2

In this case, $C_1 = 2C = 200\,[\mu F]$ and the period is $T = 10.88\,[ms]$.

Figure 6.70 shows the voltage $e(t)$ in the three elements in parallel.

Figure 6.71 shows the current $i(t)$, on a scale of kiloampers; and in Fig. 6.72, the same current $i(t)$, but in zoom, justifying the ratio of $\pm 12\,[kA]$ of the vertical scale of Fig. 6.71.

Figure 6.73 also shows the current $i(t)$, but now with the scale manually changed.

Figures 6.74 and 6.75 show the current $i_1(t)$, respectively with manually modified scale and zoom.

Finally, in Fig. 6.76 the current $i_2(t)$ is plotted.

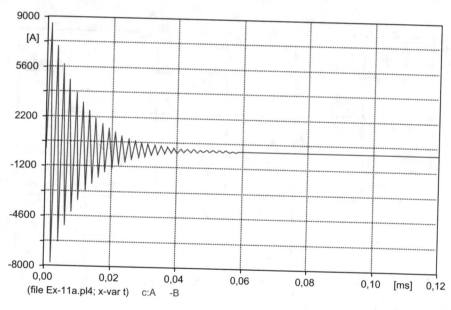

Fig. 6.66 Zoom in current $i(t)$

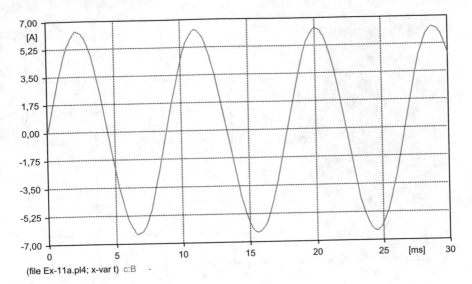

(file Ex-11a.pl4; x-var t) c:B   -

Fig. 6.67 Current $i_2(t)$

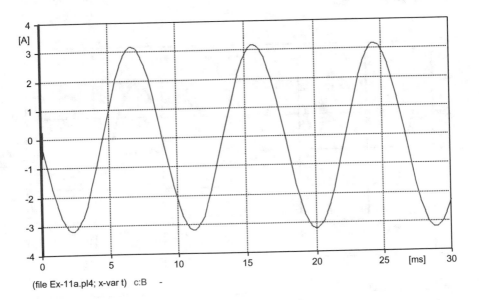

(file Ex-11a.pl4; x-var t) c:B   -

Fig. 6.68 Current $i_1(t)$

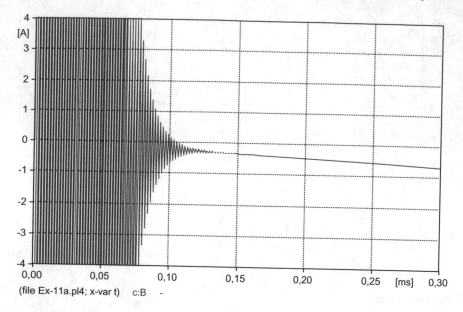

Fig. 6.69 Zoom in current $i_1(t)$

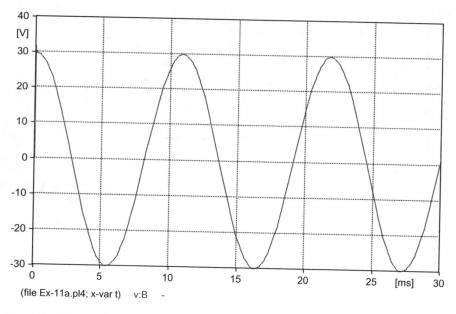

Fig. 6.70 Voltage $e(t)$

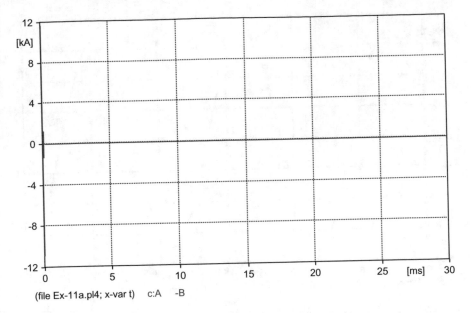

(file Ex-11a.pl4; x-var t)   c:A   -B

Fig. 6.71 Current $i(t)$

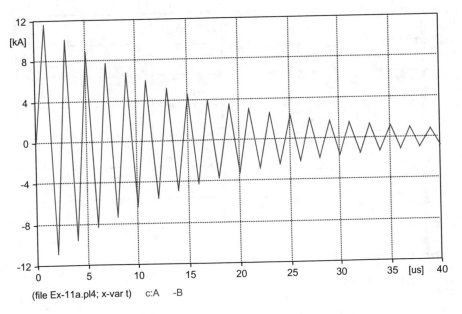

(file Ex-11a.pl4; x-var t)   c:A   -B

Fig. 6.72 Zoom at current $i(t)$

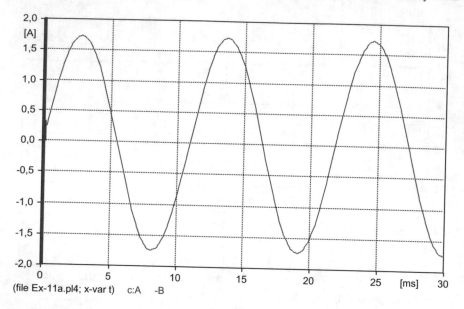

(file Ex-11a.pl4; x-var t)   c:A   -B

Fig. 6.73 Current $i(t)$ with manually altered scale

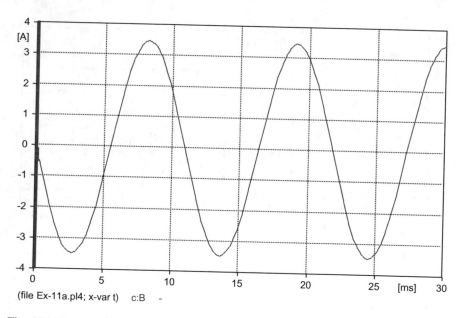

(file Ex-11a.pl4; x-var t)   c:B   -

Fig. 6.74 Current $i_1(t)$ with manually altered scale

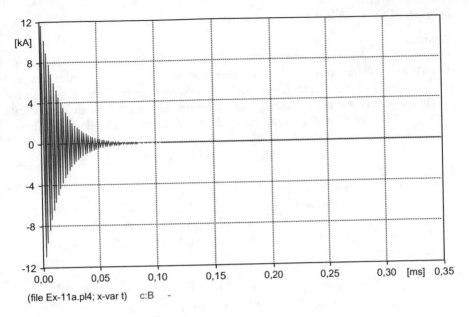

Fig. 6.75 Zoom at current $i_1(t)$

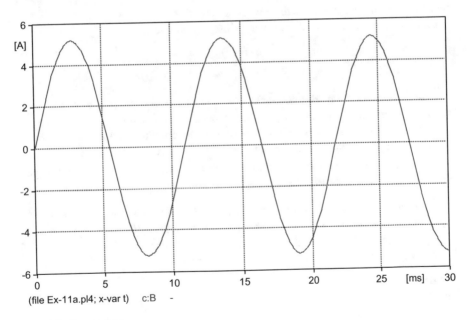

Fig. 6.76 Current $i_2(t)$

Example 6.22 Study the periodicity of the function $f(t) = A \cos \omega_1 t + B \cos \omega_2 t$.

Solution

A **rational number** is any real number that can be represented by the quotient of two whole numbers. So, k is a rational number when it is possible to write it in the form $k = m/n$, where m is an integer and n is also an integer, different from zero.

Therefore, any real number that cannot be written as a fraction of an integer and an integer (and non-zero) is an **irrational number**. Examples of irrational numbers, among many others, are the following:

$\pi, \sqrt{2},$ the golden number: $\frac{1+\sqrt{5}}{2} = 1.6180339...,$ the radian: $180/\pi = 57.2957795...$

A function $f(t)$ is periodic of period T, when $f(t+T) = f(t)$.

By iterating this equation, we also have that $f(t + nT) = f(t)$, for any integer n, that is: any integer multiple of the period of a periodic function is also its period.

Therefore, for the given trigonometric function to be periodic of period T, it is necessary that:

$$f(t+T) = A \cos \omega_1(t+T) + B \cos \omega_2(t+T) = A \cos(\omega_1 t + \omega_1 T)$$
$$+ B \cos(\omega_2 t + \omega_2 T) = A \cos \omega_1 t + B \cos \omega_2 t = f(t).$$

Considering, on the other hand, that $\cos(\theta + 2k\pi) = \cos \theta$, for any integer k, then the preceding equation will only be satisfied if $\omega_1 T = 2m\pi$ and $\omega_2 T = 2n\pi$, for any integers m and n. So,

$$\frac{\omega_1}{\omega_2} = \frac{m}{n}.$$

Conclusion: For the trigonometric function $f(t) = A \cos \omega_1 t + B \cos \omega_2 t$ to be periodic of period T, the ratio ω_1/ω_2 must be a rational number.

If the function is $f(t) = A \cos \omega_1 t + B \sin \omega_2 t$ the conclusion does not change, as $\sin \omega_2 t = \cos(\omega_2 t - 90°)$, and the lag does not interfere in the period.

Therefore, the trigonometric function $f(t) = A \cos \omega t + B \cos \sqrt{2}\omega t$ is not periodic, because $\omega_1/\omega_2 = 1/\sqrt{2}$ is an irrational number.

Example 6.23 The switch in the circuit of Fig. 6.77 is closed at time $t = 0$. If the capacitor to the left of the switch has an initial voltage $e_1(0_-) = 100 \,[\text{V}]$ and the right part of the circuit is initially de-energized, determine, numerically, $e_1(t)$, $e_4(t)$, $i_2(t)$ and $i_3(t)$, with the switch closed. Data: $C = 10 \,[\mu\text{F}]$ and $L = 10 \,[\text{mH}]$.

Solution

The analytical solution obtained for $e_1(t)$ is given by

$$e_1(t) = 27.6 \cos 0.618\omega_0 t + 72.4 \cos 1.618\omega_0 t \,[\text{V}], \; for \; t \geq 0,$$

where $\omega_0 = 1/\sqrt{LC}$.

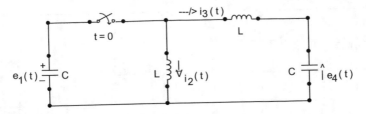

Fig. 6.77 Electric circuit of Example 6.23

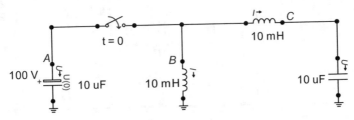

Fig. 6.78 Electric circuit of Example 6.23 for the numerical solution by the ATPDraw program

Since $k = 1.618\omega_0/0.618\omega_0 = 2.618122977\ldots$ is an irrational number, so, according to example 6.22, $e_1(t)$ is not a periodic voltage. (Ditto for the other requested signs). With the numerical values,

$$e_1(t) = 27.6\cos 1{,}954.29t + 72.4\cos 5{,}116.57t\;[\text{V}],\;for\,t \geq 0.$$

The periods are $T_1 = 3.215\,[\text{ms}]$ and $T_2 = 1.228\,[\text{ms}]$, corresponding to the frequencies $f_1 = 311.041991\,[\text{Hz}]$ and $f_2 = 814.332248\,[\text{Hz}]$.

The circuit for the numerical solution by ATPDraw is shown in Fig. 6.78. The simulation was performed with $\Delta t = 1\,[\mu s]$.

Figure 6.79 shows the voltage $e_1(t)$. And Fig. 6.80, a zoom at the same voltage $e_1(t)$. It is interesting to observe, in Fig. 6.80, that $t_{max} = 6\,[\text{ms}] > T_1 = 3.215\,[\text{ms}]$, the largest individual period of the analytical solution of $e_1(t)$, and, however, there is not even any periodicity in the signal.

Figure 6.81 records the voltage $e_4(t)$. And Fig. 6.82 shows a zoom at the same voltage.

The current $i_2(t)$ is plotted in Fig. 6.83, and shown in zoom in Fig. 6.84.

In Fig. 6.85 the current $i_3(t)$ is recorded; and in Fig. 6.86, the same zoom current.

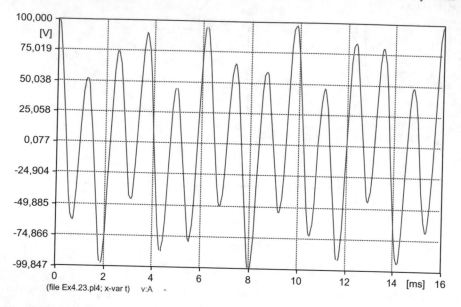

Fig. 6.79   Voltage $e_1(t)$

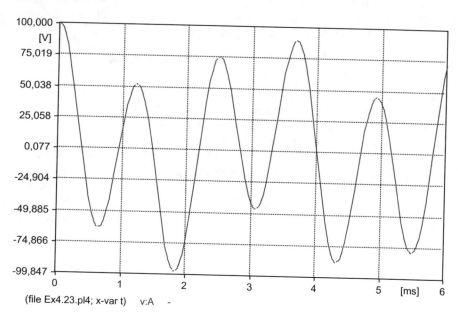

Fig. 6.80   Zoom in voltage $e_1(t)$

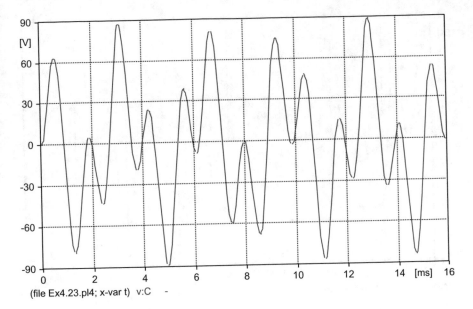

Fig. 6.81 Voltage $e_4(t)$

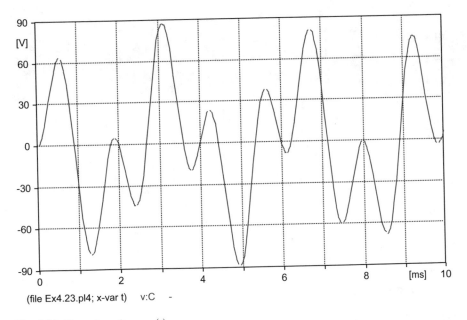

Fig. 6.82 Zoom in voltage $e_4(t)$

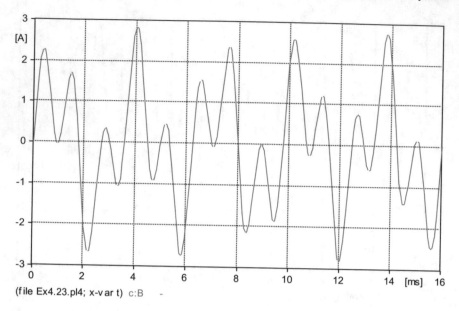

(file Ex4.23.pl4; x-var t) c:B   -

Fig. 6.83 Current $i_2(t)$

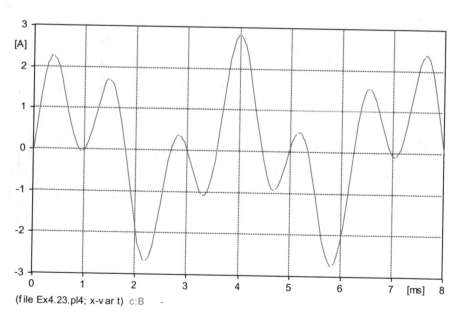

(file Ex4.23.pl4; x-var t) c:B   -

Fig. 6.84 Zoom in current $i_2(t)$

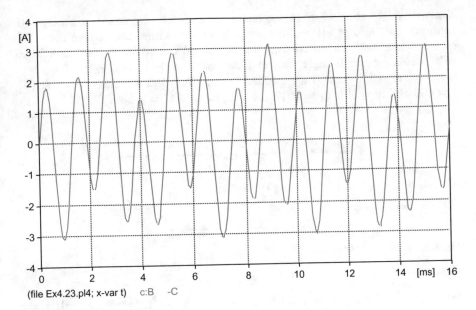

(file Ex4.23.pl4; x-var t)   c:B   -C

Fig. 6.85 Current $i_3(t)$

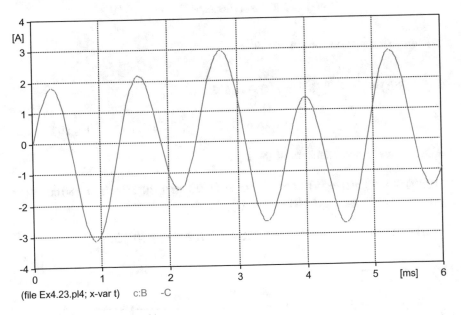

(file Ex4.23.pl4; x-var t)   c:B   -C

Fig. 6.86 Zoom in current $i_3(t)$

Example 6.24 The switch in the circuit of Fig. 6.87 is closed at time $t = 0$, applying the voltage $e(t) = E\cos(\omega t + \theta)$ in the series R-C connection. The capacitor is initially energized with voltage E_0. (a) Determine the current $i(t)$ that will circulate through the circuit and the voltages $e_R(t)$ and $e_C(t)$, for $t > 0$. (b) Perform the numerical solution using ATPDraw. Data: $E = 180\,[\mathrm{V}]$, $f = 600\,[\mathrm{Hz}]$, $\theta = 30°$, $E_0 = 60\,[\mathrm{V}]$, $R = 10\,[\Omega]$ and $C = 1,000\,[\mu\mathrm{F}]$.

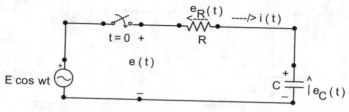

Fig. 6.87 Electric circuit of Example 6.24

Solution

(a) Replacing the capacitor initially energized with its Thévenin equivalent, the circuit shown in Fig. 6.88 results, for $t > 0$.

$$e(t) = 180\cos(1,200\pi t + 30°)U_{-1}(t)\,[\mathrm{V}]$$

The mesh equation for the circuit is

$$Ri(t) + \frac{1}{C}\int_{0_+}^{t} i(\lambda)d\lambda = e(t) - E_0 U_{-1}(t)$$

$$= 180\cos(1,200\pi t + 30°)U_{-1}(t) - 60U_{-1}(t)$$

Deriving the two members of the equation, there is

$$10i'(t) + 1,000i(t) = -180 \times 1,200\pi\sin(1,200\pi t + 30°)U_{-1}(t) + 180\cos 30°$$
$$\cdot U_0(t) - 60U_0(t)$$

$$i'(t) + 100i(t) = 9.59U_0(t) - 67,858.4\sin(1,200\pi t + 30°)U_{-1}(t), i(0_-) = 0.$$
$$(6.33)$$

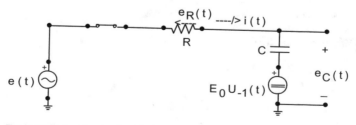

Fig. 6.88 Equivalent electric circuit of Example 6.24

For, $t > 0$,

$$(D+100)i(t) = 67,858.4\cos(1,200\pi t + 120°)$$

$$i_H(t) = k_1 \epsilon^{-100t}$$

$$(100+j1,200\pi)\dot{I}_P = 67,858.4e^{j120°}$$

$$\dot{I}_P = 17.99e^{j31.5°}$$

$$i_P(t) = 17.99\cos(1,200\pi t + 31.5°)$$

$$i(t) = k_1 \epsilon^{-100t} + 17.99\cos(1,200\pi t + 31.5°)$$

From Eq. (6.33), $i(0_+) = 9.59$. Where $k_1 + 15.34 = 9.59$ and $k_1 = -5.75$. So,

$$i(t) = -5.75\epsilon^{-100t} + 17.99\cos(1,200\pi t + 31.5°)U_{-1}(t) \, [A]$$

$$e_R(t) = Ri(t)$$

$$e_R(t) = -57.5\epsilon^{-100t} + 179.9\cos(1,200\pi t + 31.5°)U_{-1}(t) \, [V]$$

$$e_C(t) = e(t) - e_R(t)$$

$$e_C(t) = 57.5\epsilon^{-100t} + 180\cos(1,200\pi t + 30°) - 179.9\cos(1,200\pi t + 31.5°)$$

Considering $z(t) = 180\cos(1,200\pi t + 30°) - 179.9\cos(1,200\pi t + 31.5°)$, then the corresponding phasor is:

$$\dot{Z} = 180e^{j30°} - 179.9e^{j31.5°} = 155.88 + j90.00 - 153.39 - j94.00 = 4.71\epsilon^{-j58.1°}$$

Therefore, $z(t) = 4.71\cos(1,200\pi t - 58.1°)$ and,

$$e_C(t) = 57.5\epsilon^{-100t} + 4.71\cos(1,200\pi t - 58.1°)[2V], \text{for } t \geq 0 \, [s].$$

The time constant is $T = 1/100 = 10$ [ms]. And the period of sinusoidal oscillations is worth $T_n = 1/600 = 1.67$ [ms]. Therefore, $5T = 50$ [ms] and $\Delta t = T_n/1000 = 1.7$ [μs].

(b) The circuit built for the numerical solution by ATPDraw is shown in Fig. 6.89. It was considered that $\Delta t = 1$ [μs].

The current $i(t)$ circulating in the circuit, for $t > 0$ [s], is recorded in Fig. 6.90. Figure 6.91 shows a zoom in current $i(t)$, close to $t = 0$ [s], on the right. And in Fig. 6.92 only the end of the transient regime and the beginning of the permanent regime of current $i(t)$ are shown, that is, the interval 50 [ms] $\leq t \leq$ 60 [ms].

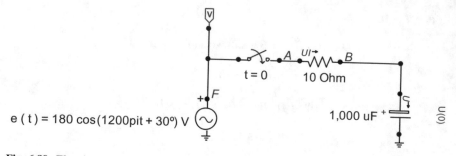

Fig. 6.89 Electric circuit of Example 6.24 for the numerical solution by the ATPDraw program

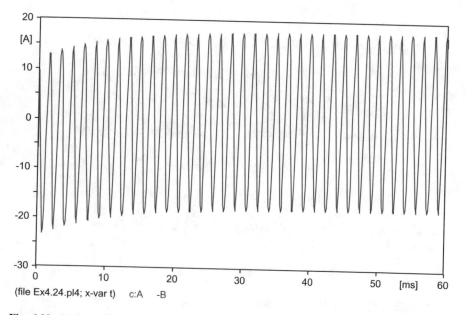

(file Ex4.24.pl4; x-var t)   c:A   -B

Fig. 6.90 Current $i(t)$

Figures 6.93 and 6.94 show the voltage $e_R(t)$, respectively, in the normal pattern and in zoom.

In Fig. 6.95 the voltage $e_C(t)$, is recorded, and in Fig. 6.96 a zoom is shown at the same voltage.

Finally, in Fig. 6.97, the steady state component of the voltage $e_C(t)$, is recorded, in the range $90\,[\text{ms}] \leq t \leq 100\,[\text{ms}]$.

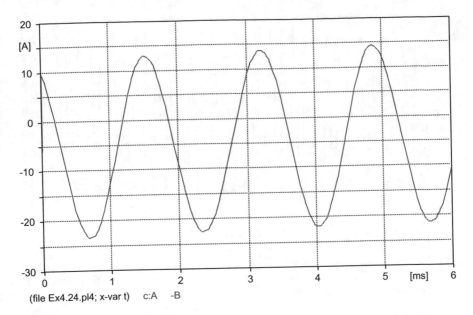

Fig. 6.91 Zoom at current $i(t)$

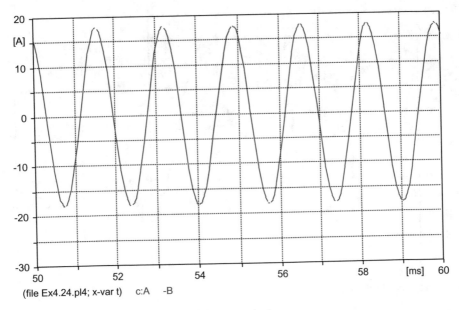

Fig. 6.92 Current $i(t)$ in the range $50\,[\text{ms}] \leq t \leq 60\,[\text{ms}]$

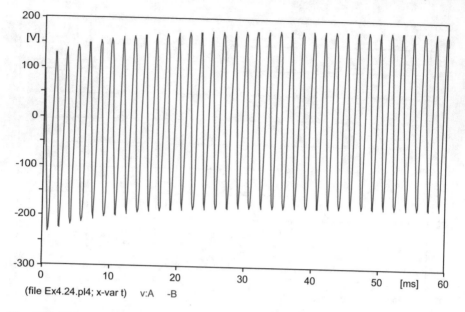

Fig. 6.93 Voltage $e_R(t)$

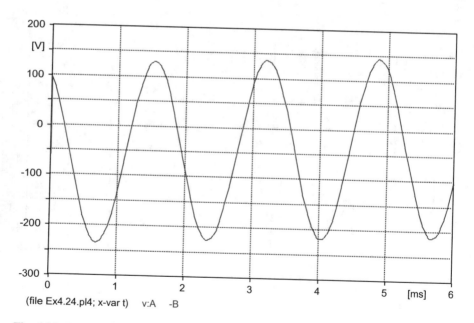

Fig. 6.94 Zoom in voltage $e_R(t)$

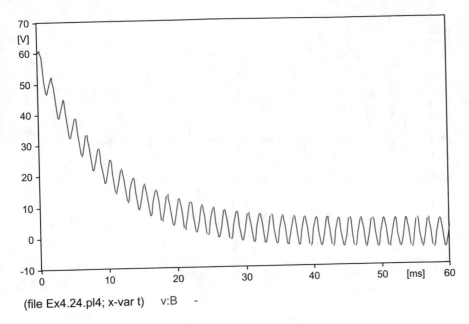

(file Ex4.24.pl4; x-var t)   v:B   -

Fig. 6.95 Voltage $e_C(t)$

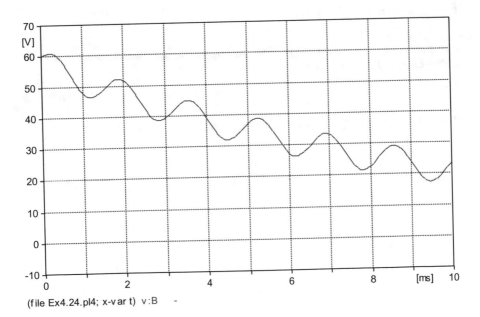

(file Ex4.24.pl4; x-var t) v:B   -

Fig. 6.96 Zoom in voltage $e_C(t)$

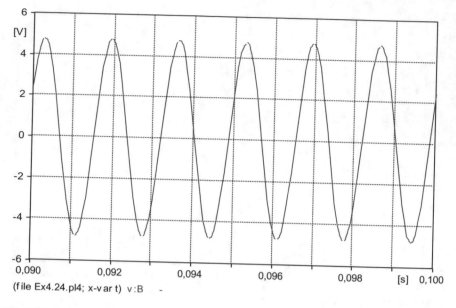

Fig. 6.97 Voltage $e_C(t)$ in the range $90\,[\text{ms}] \leq t \leq 100\,[\text{ms}]$

6.5   Switching Transients

Example 6.25   The circuit in Fig. 6.98 was operating in a steady state when switch k was closed at time $t = 0$.

(a) Determine the current and voltage in each passive element of the circuit, for $t \geq 0$.
(b) Perform an energy balance on the capacitances, in the transition from $t = 0_-$ to $t = 0_+$.
(c) Perform the numerical solution using ATPDraw.

Data : $E = 100\,[\text{V}]$, $R = 10\,[\Omega]$, $C_1 = C_2 = 50\,[\mu\text{F}]$, $e_2(0_-) = E/2 = 50\,[\text{V}]$.

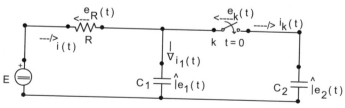

Fig. 6.98 Electric circuit of Example 6.25

Solution

In steady state, with the switch k still open, $e_1(0_-) = E = 100$ [V],

$$q_1(0_-) = C_1 e_1(0_-) = C_1 E = 5 \text{ [mC]}$$

$$q_2(0_-) = C_2 e_2(0_-) = C_2 \frac{E}{2} = 2.5 \text{ [mC]}$$

$$q_{total}(0_-) = q_1(0_-) + q_2(0_-) = 7.5 \text{ [mC]}$$

Immediately after closing the switch,

$$q_1(0_+) = C_1 e_1(0_+)$$

$$q_2(0_+) = C_2 e_2(0_+) = C_2 e_1(0_+)$$

$$q_{total}(0_+) = q_1(0_+) + q_2(0_+) = (C_1 + C_2) e_1(0_+) = 10^{-4} e_1(0_+)$$

By the Load Conservation Principle, $q_{total}(0_+) = q_{total}(0_-)$. So,

$10^{-4} e_1(0_+) = 7.5 \cdot 10^{-3}$ and $e_1(0_+) = e_2(0_+) = 75$ [V]. So,
$$q_1(0_+) = q_2(0_+) = 3.75 \text{ [mC]}$$

$$q_{total}(0_+) = 7.5 \text{ [mC]}$$

Load variations in capacitances are:

$$\Delta q_1 = q_1(0_+) - q_1(0_-) = -1.25 \text{ [mC]} \text{(loss of electrical charge)}$$

$$\Delta q_2 = q_2(0_+) - q_2(0_-) = +1.25 \text{ [mC]} \text{(electric charge gain)}.$$

(a) Replacing the capacitances initially energized with their Thévenin equivalents,
the circuit in Fig. 6.99 results, for $t > 0$.

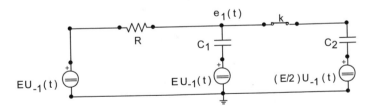

Fig. 6.99 Equivalent electric circuit of Example 6.25

The nodal circuit equation is

$$\frac{e_1(t) - EU_{-1}(t)}{R} + C_1 \frac{d}{dt}[e_1(t) - EU_{-1}(t)] + C_2 \frac{d}{dt}\left[e_1(t) - \frac{E}{2}U_{-1}(t)\right] = 0$$

Simply put, and taking into account that $C_2 = C_1$, the differential equation results

$$e_1'(t) + \frac{1}{2RC_1}e_1(t) = \frac{3E}{4}U_0(t) + \frac{E}{2RC_1}U_{-1}(t), \text{ with } e_1(0_-) = 0$$
$$e_1'(t) + 1,000e_1(t) = 75U_0(t) + 100,000U_{-1}(t), \text{ with } e_1(0_-) = 0. \tag{6.34}$$

For $t > 0$,

$$(D + 1,000)e_1(t) = 100,000$$

$$e_{1_H}(t) = k_1\epsilon^{-1,000t}$$

$$e_{1_P}(t) = 100,000/1,000 = 100$$

$$e_1(t) = 100 + k_1\epsilon^{-1,000t}$$

From Eq. (6.34), $e_1(0_+) = 75$. Hence $75 = 100 + k_1$ and $k_1 = -25$. So,

$$e_1(t) = 100 - 25\epsilon^{-1,000t} \text{ [V]}, \textit{ for } t > 0.$$

Or, for all t,

$$e_1(t) = 100U_{-1}(-t) + \left(100 - 25\epsilon^{-1,000t}\right)U_{-1}(t)[\mathbf{V}].$$

So, also for all t,

$$e_2(t) = 50U_{-1}(-t) + \left(100 - 25\epsilon^{-1,000t}\right)U_{-1}(t)[\mathbf{V}].$$

The resistance voltage, for $t > 0$, is

$$e_R(t) = EU_{-1}(t) - e_1(t)$$

$$e_R(t) = 25\epsilon^{-1,000t}.U_{-1}(t)[V].$$

$$i_R(t) = \frac{1}{R}e_R(t)$$

$$i_R(t) = 2.5\epsilon^{-1,000t}.U_{-1}(t)[A].$$

The current at capacitance C_1, obtained from Fig. 6.98, is

$$i_1(t) = C_1 \frac{d}{dt} e_1(t) = 50.10^{-6} \frac{d}{dt} \left[100 U_{-1}(-t) + \left(100 - 25\epsilon^{-1,000t} \right) U_{-1}(t) \right]$$
$$= 5.10^{-3} \cdot U_0(-t) \cdot (-1) + 1.25\epsilon^{-1,000t} \cdot U_{-1}(t) + 3.75.10^{-3} \cdot U_0(t)$$

Since the impulse function is even, $U_0(-t) = U_0(t)$ and

$$i_1(t) = -1.25.10^{-3} U_0(t) + 1.25\epsilon^{-1,000t}.U_{-1}(t) \, [A].$$

The impulsive current $1.25.10^{-3} U_0(t)$ [A], circulating upwards at capacitance C_1, is responsible for removing the electric charge $\Delta q_1 = 1.25$ [mC] of capacitance C_1, and that is transferred to capacitance C_2, at $t = 0$ [s].

The current at capacitance C_2, which is the same at switch k, obtained from Fig. 6.98, is

$$i_k(t) = C_2 \frac{d}{dt} e_2(t) = 50.10^{-6} \frac{d}{dt} \left[50 U_{-1}(-t) + \left(100 - 25\epsilon^{-1,000t} \right) U_{-1}(t) \right]$$
$$= -2.5.10^{-3} \cdot U_0(t) + 1.25\epsilon^{-1,000t} \cdot U_{-1}(t) + 3.75.U_0(t).$$

Finally,

$$i_k(t) = +1.25.10^{-3} U_0(t) + 1.25\epsilon^{-1,000t}.U_{-1}(t) \, [A].$$

Note that the same impulsive current that went up through capacitance C_1, at $t = 0$, closed the path down through capacitance C_2. And that $i_k(t) + i_1(t) = i_R(t) = i(t)$.

(b) The Energy Conservation Law states that energy cannot be destroyed or created, only transformed from one type to another. Therefore, the energy cannot just disappear! That is, $W_{total}(0_+)$ must be equal to $W_{total}(0_-)$.

$$W_1(0_-) = \frac{1}{2} C_1 e_1^2(0_-) = 0.25 \, [J]$$

$$W_2(0_-) = \frac{1}{2} C_2 e_2^2(0_-) = 0.0625 \, [J]$$

$$W_{total}(0_-) = W_1(0_-) + W_2(0_-) = 0.3125 \, [J].$$

$$W_1(0_+) = \frac{1}{2} C_1 e_1^2(0_+) = 0.140625 \, [J]$$

$$W_2(0_+) = \frac{1}{2} C_2 e_2^2(0_+) = 0.140625 \, [J]$$

$$W_{total}(0_+) = W_1(0_+) + W_2(0_+) = 0.281250 \, [J].$$

So,

$$\Delta W = W_{total}(0_-) - W_{total}(0_+) = 0.312500 - 0.281250 = 0.031250 \, [\text{J}] \neq 0?!$$

How to explain the disappearance of ΔW energy?
The voltage at the terminals of switch k, in view of Fig. 6.98, is

$$e_k(0_-) = e_1(0_-) - e_2(0_-) = 100 - 50 = 50 \, [\text{V}]$$

$$e_k(0_+) = e_1(0_+) - e_2(0_+) = 75 - 75 = 0 \, [\text{V}]$$

Or, yet,

$$e_k(t) = \begin{cases} 50 \, [\text{V}] \; for \; t < 0 \, [\text{s}] \\ 0 \, [\text{V}] \; for \; t > 0 \, [\text{s}] \end{cases}$$

Alternatively, for all t,

$$e_k(t) = 50 \, U_{-1}(-t) = 25 + 25 \, U_{-1}(-t) - 25 \, U_{-1}(t) = 25 - 25.sgn \; t,$$

where $sgnt = \frac{|t|}{t} = \begin{cases} -1 \; for \; t < 0 \, [\text{s}] \\ +1 \; for \; t > 0 \, [\text{s}] \end{cases}$, called the **sign function of t**.
The power in switch k is

$$p_k(t) = e_k(t).i_k(t)$$

For $t < 0 \, [\text{s}]$ the switch k is open and $e_k \neq 0$, but $i_k = 0$. So, $p_k = 0$.
For $t > 0 \, [\text{s}]$, the switch k is closed and $e_k = 0$, but $i_k \neq 0$. So, $p_k = 0$.
Therefore,

$$p_k(t) = 0 \; for \; t \neq 0.$$

And how much is $p_k(t)$ for $t = 0 \, [\text{s}]$?
Remembering that when $I(t)$ is an odd function,

$\int_{-a}^{+a} I(t)dt = 0$, then $\int_{-\Delta t}^{+\Delta t} I(\lambda)d\lambda = 0$.
Whereas

$$\lim_{\Delta t \to 0} (+\Delta t) = 0_+$$

$$\lim_{\Delta t \to 0} (-\Delta t) = 0_-$$

So,

$$\lim_{\Delta t \to 0} \int_{-\Delta t}^{+\Delta t} I(\lambda)d\lambda = \int_{0_-}^{0_+} I(t)dt = 0$$

So, the power in switch k **for $t = 0$**, is

$$p_k(t) = e_k(t).i_k(t) = (25 - 25.sgnt).1.25.10^{-3}.U_0(t) \text{ [W]}$$

Therefore, the energy in switch k, at time $t = 0_+$, is worth:

$$W_k(0_+) = \int_{0_-}^{0_+} p_k(t)dt = 0.031250 \int_{0_-}^{0_+} U_0(t)dt - 0.031250 \int_{0_-}^{0_+} sgnt.U_0(t)dt$$

The last integral has an odd integrating, as it is the product of the odd function $sgn\ t$ by the even function $U_0(t)$, and that is null, due to $\int_{0_-}^{0_+} I(t)dt = 0$.

Therefore,

$$W_k(0_+) = \Delta W = 0.031250 \text{ [J]}, \text{ since } \int_{0_-}^{0_+} U_0(t)dt = 1.$$

Alternatively, you can proceed as follows:

$$p_k(t) = e_k(t) \cdot i_k(t) = 50U_{-1}(-t) \cdot [+1.25 \cdot 10^{-3}U_0(t) + 1.25\epsilon^{-1,000t} \cdot U_{-1}(t)]$$
$$= 62.5.10^{-3} \cdot U_{-1}(-t) \cdot U_0(t) + 0 = 62.5.10^{-3} \cdot [1 - U_{-1}(t)].U_0(t)$$
$$= 62.5 \cdot 10^{-3} \cdot U_0(t) - 62.5 \cdot 10^{-3} \cdot U_{-1}(t) \cdot U_0(t)$$

But, according to Eq. (2.67),

$$U_{-1}(t).U_0(t) = \frac{1}{2}U_0(t)$$

So,

$$p_k(t) = 62.5.10^{-3}\left(1 - \frac{1}{2}\right)U_0(t) = 31.25.10^{-3}.U_0(t) = 0.03125U_0(t) \text{ [W] and}$$

$$W_k(0_+) = \int_{0_-}^{0_+} p_k(t)dt = 0.03125 \int_{0_-}^{0_+} U_0(t)dt = 0.03125 \text{ [J]}.$$

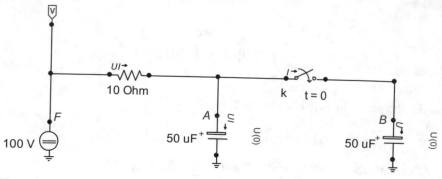

Fig. 6.100 Electric circuit of Example 6.25 for the numerical solution by the ATPDraw program

Thus, the energy that apparently disappeared, as shown, was consumed in the process of closing switch k. And this suggests that in the statement of the present example, instead of saying that switch k was closed instantly at time $t = 0$, it would be better to say that the switch was closed in the interval between $t = 0_-$ (start of switching) and $t = 0_+$ (end of switching). And that within this interval, that is, at $t = 0$ the circulated impulsive current was $1.25 \cdot 10^{-3} U_0(t)$ [A], which equalized the voltages at capacitances C_1 and C_2 in $t = 0_+$ [s]. **Remember, however, that this is a very idealized example!**

(c) For the numerical solution by ATPDraw, the circuit of Fig. 6.100 was built. As the circuit time constant is $T = 1/1000 = 1$ [ms], the step $\Delta t = 1$ [μs] $= T/1000$ and $t_{max} = 5T = 5$ [ms] was adopted.

In Fig. 6.101 the voltages $e_1(t)$, $e_2(t)$ and $e_R(t)$ are plotted, as well as the source voltage, E, for $t > 0$ [s].

The plotted curves are absolutely in accordance with the analytical solutions obtained for the referred voltages.

It is necessary to point out that the vertical scale of the voltages was changed from -100 [V], $+100$ [V] to 0 [V], $+100$ [V], manually.

The current in the resistor, $i_R(t) = i(t)$, is shown in Fig. 6.102. The vertical scale has also been changed manually.

Knowing that the analytical solutions of $i_1(t)$ and $i_k(t)$ contain impulses, which have infinite amplitudes, it is logical not to expect the ATPDraw program to be able to plot them. This is evidenced by the attempts to plot $i_1(t)$ and $i_k(t)$ shown, respectively, in Figs. 6.103 and 6.104, which contain Numerical Oscillations.

After successive zooms on the "graphs" of $i_1(t)$ and $i_k(t)$ the results shown in Figs. 6.105 and 6.106 are obtained, in which the numerical oscillation is evident.

As already highlighted in Example 5.20, although the program is unable to plot the currents, because of the impulses contained therein, resulting from the discontinuities in the capacitance voltages, it nevertheless manages to supply the loads carried by these impulsive currents. In Fig. 6.105, for example, the base triangle $\Delta t = 1$ [μs] and height $-2,500$ [A] has an area of -1.25 [mC], exactly the load

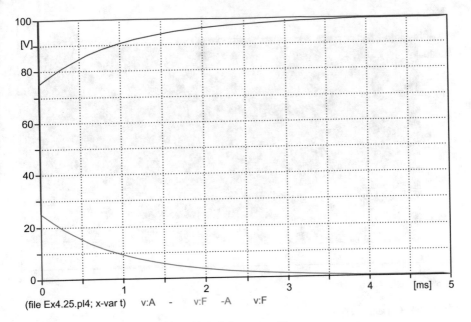

Fig. 6.101 Voltages $e_1(t) = e_2(t)$, $e_R(t)$ and the source E

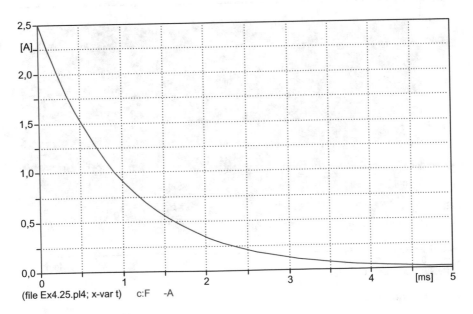

Fig. 6.102 Current $i_R(t)$

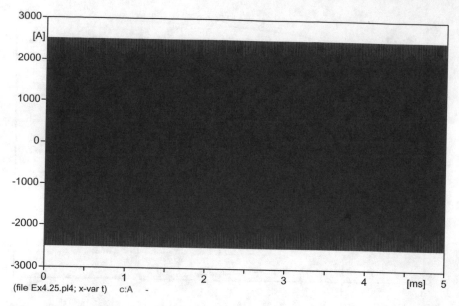

Fig. 6.103 Current $i_1(t)$

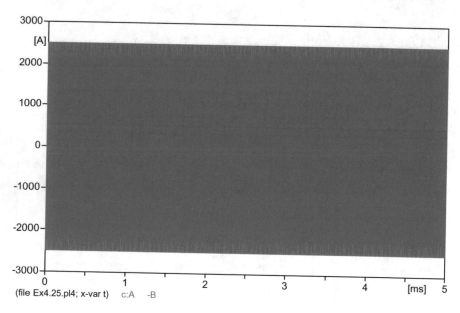

Fig. 6.104 Current $i_k(t)$

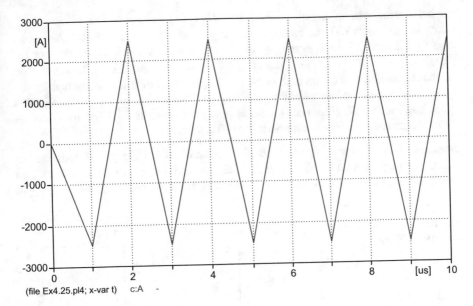

Fig. 6.105 Zoom in current $i_1(t)$

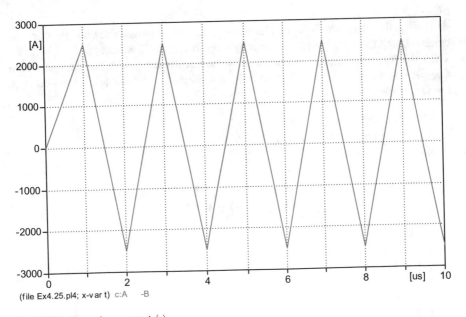

Fig. 6.106 Zoom in current $i_k(t)$

carried by the current impulse $-1.25.10^{-3}U_0(t)$ [A] present in the current $i_1(t)$. Likewise, in Fig. 6.106, the base triangle $\Delta t = 1$ [μs] and height $+2,500$ [A] has an area of 1.25 [mC], which is exactly the load carried, at $t = 0$ [s], by the current impulse $+1.25.10^{-3}U_0(t)$ [A] present in $i_k(t)$. We must not forget, however, that all this happened due to the high degree of idealization of the loop formed by the two capacitances and the switch k. The simple introduction of a resistance in this loop, however small, eliminates the impulses in the currents $i_1(t)$ and $i_k(t)$, and all its consequences, mainly the numerical oscillations.

Example 6.26 Knowing that the switch k, in the circuit of Fig. 6.107, remained in position 1 long enough for the capacitor to charge and that it goes from position 1 to position 2 at time $t = 0$, determine the voltage $e_C(t)$ at the capacitor terminals to $t \geq 0$.

Fig. 6.107 Electric circuit of Example 6.26

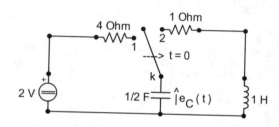

Solution

At $t = 0_-$ the capacitor behaves like an open circuit, because the voltage source is constant, and the steady state has already been reached. Therefore, $e_C(0_-) = 2$ [V]. Then replacing the capacitance with its Norton equivalent, with the current source $Ce_C(0_-)U_0(t)$ [A] $= U_0(t)$ [A], results in the circuit of null initial conditions at $t = 0_-$, shown in Fig. 6.108.

The nodal equations of the circuit are:
Node C: $\frac{1}{2}De_C(t) + e_C(t) - e_L(t) = U_0(t)$
Node L: $e_L(t) - e_C(t) + \frac{1}{D}e_L(t) = 0$

$$\begin{cases} (D+2)e_C(t) - 2e_L(t) = 2U_0(t) \\ -De_C(t) + (D+1)e_L(t) = 0 \end{cases}$$

Solving for $e_C(t)$, comes that

$$e_C''(t) + e_C'(t) + 2e_c(t) = 2U_1(t) + 2U_0(t), \text{ with } e_C(0_-) = e_C'(0_-) = 0. \quad (6.35)$$

Fig. 6.108 Equivalent electric circuit of Example 6.26 for the numerical solution by ATPDraw program

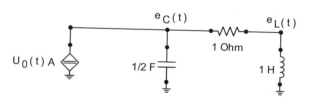

For t > 0,

$$\left(D^2 + D + 2\right)e_C(t) = 0$$

$$r_{1,2} = -\frac{1}{2} \pm j\frac{\sqrt{7}}{2}$$

$$e_C(t) = \epsilon^{-t/2}\left(k_1 \cos\frac{\sqrt{7}}{2}t + k_2 \sin\frac{\sqrt{7}}{2}t\right)$$

$$e_C'(t) = -\frac{1}{2}\epsilon^{-t/2}\left(k_1 \cos\frac{\sqrt{7}}{2}t + k_2 \sin\frac{\sqrt{7}}{2}t\right) + \epsilon^{-t/2}\left(-\frac{\sqrt{7}}{2}k_1 \sin\frac{\sqrt{7}}{2}t + \frac{\sqrt{7}}{2}k_2 \cos\frac{\sqrt{7}}{2}t\right)$$

From Eq. (6.35),

$$e_C(t) \doteq C_1 U_{-1}(t).$$

$$e_C'(t) \doteq C_1 U_0(t) + C_2 U_{-1}(t).$$

$$e_C''(t) \doteq C_1 U_1(t) + C_2 U_0(t) + C_3 U_{-1}(t)$$

Substituting in Eq. (6.35),

$$C_1 U_1(t) + (C_2 + C_1)U_0(t) + (C_3 + C_2 + 2C_1)U_{-1}(t) \doteq 2U_1(t) + 2U_0(t) + 0.U_{-1}(t).$$

Therefore, $C_1 = 2 = e_C(0_+)$; $C_2 + C_1 = 2$. Therefore, $C_2 = 0 = e_C'(0_+)$.
Then,

$$2 = k_1$$

$$0 = -\frac{1}{2}k_1 + \frac{\sqrt{7}}{2}k_2 \text{ and } k_2 = \frac{2}{\sqrt{7}} = \frac{2\sqrt{7}}{7}$$

Finally,

$$e_C(t) = 2\epsilon^{-t/2}\left(\cos\frac{\sqrt{7}}{2}t + \frac{\sqrt{7}}{7}\sin\frac{\sqrt{7}}{2}t\right), \text{ for } t \geq 0.$$

Example 6.27 The circuit in Fig. 6.109 was operating in a steady-state when, at an arbitrary time taken as $t = 0$, switch k was opened.

(a) Determine the expression of the voltage $e_k(t)$ at the terminals of the same, for $t \geq 0$. Also find the voltage and current in the inductance.
(b) Perform the numerical solution by ATPDraw, for $E = 100$ [V], $R = 10$ [Ω], $L = 100$ [mH] and $C = 1,000$ [μF].

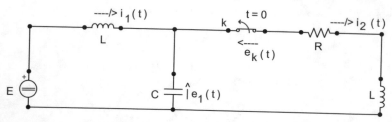

Fig. 6.109 Electric circuit of Example 6.27

Solution

(a) Being the constant voltage source, in the steady state the inductances behave as short circuits and the capacitance as an open circuit. Soon,

$$e_1(0_-) = E \text{ and } i_1(0_-) = i_2(0_-) = E/R.$$

Replacing the elements initially energized with their Thévenin equivalents, the circuit of null initial conditions appears at $t = 0_-$ shown in Fig. 6.110.
The nodal circuit equation is

$$\frac{1}{L} \int \left[e_1(t) - \frac{LE}{R} U_0(t) - E U_{-1}(t) \right] dt + C \frac{d}{dt} \left[e_1(t) - E U_{-1}(t) \right] = 0$$

$$e_1''(t) + \frac{1}{LC} e_1(t) = E U_1(t) + \frac{E}{RC} U_0(t) + \frac{E}{LC} U_{-1}(t)$$

Defining $\omega_0 = 1/\sqrt{LC}$,

$$e_1''(t) + \omega_0^2 e_1(t) = E U_1(t) + \frac{E}{RC} U_0(t) + \omega_0^2 E U_{-1}(t); e_1(0_-) = e_1'(0_-) = 0 \quad (6.36)$$

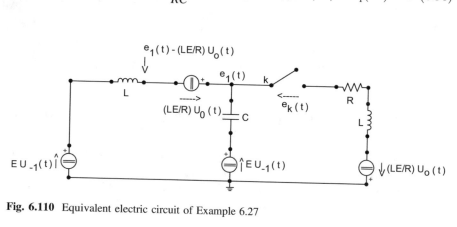

Fig. 6.110 Equivalent electric circuit of Example 6.27

For $t > 0$,

$$\left(D^2 + \omega_0^2\right)e_1(t) = \omega_0^2 E$$

$$e_{1_H}(t) = k_1 \cos \omega_0 t + k_2 \sin \omega_0 t$$

$$e_{1_P}(t) = \frac{1}{D^2 + \omega_0^2}\left(\omega_0^2 E \epsilon^{0.t}\right) = E$$

$$e_1(t) = E + k_1 \cos \omega_0 t + k_2 \sin \omega_0 t$$

$$e_1'(t) = -\omega_0 k_1 \sin \omega_0 t + \omega_0 k_2 \cos \omega_0 t$$

In view of the differential Eq. (6.36), it can be written that:

$$e_1(t) \doteq C_0 U_{-1}(t).$$

$$e_1'(t) \doteq C_0 U_0(t) + C_1 U_{-1}(t).$$

$$e_1''(t) \doteq C_0 U_1(t) + C_1 U_0(t) + C_2 U_{-1}(t).$$

Substituting in the differential Eq. (6.36), it turns out that

$$C_0 U_1(t) + C_1 U_0(t) + \left(C_2 + \omega_0^2 C_0\right)U_{-1}(t) \doteq EU_1(t) + \frac{E}{RC}U_0(t) + \omega_0^2 EU_{-1}(t).$$

$$C_0 = e_1(0_+) = E = E + k_1; k_1 = 0.$$

$$C_1 = e_1'(0_+) = \frac{E}{RC} = \omega_0 k_2; k_2 = \frac{E}{R}\sqrt{\frac{L}{C}}$$

Therefore,

$$e_1(t) = E + \frac{E}{R}\sqrt{\frac{L}{C}} \sin \omega_0 t, \ for \ t \geq 0.$$

By the Kirchhoff Voltage Law, applied to the right of the circuit,

$$e_k(t) - e_1(t) - \frac{LE}{R}U_0(t) = 0$$

Finally,

$$e_k(t) = \frac{LE}{R}U_0(t) + E\left(1 + \frac{1}{R}\sqrt{\frac{L}{C}}\sin \omega_0 t\right)U_{-1}(t).$$

From Fig. 6.109, the voltage at the inductance, positive at its left terminal, is worth:

$$e_L(t) = E - e_1(t) = E - E - \frac{E}{R}\sqrt{\frac{L}{C}}\sin\omega_0 t = -\frac{E}{R}\sqrt{\frac{L}{C}}\sin\omega_0 t$$

As $e_L(t) = 0$ for $t < 0$, as the inductance is short-circuited,

$$e_L(t) = -\frac{E}{R}\sqrt{\frac{L}{C}}\sin\omega_0 t . U_{-1}(t).$$

With the switch k open, the current in the inductance is the same as the capacitance and can be obtained from the voltage $e_1(t)$, as follows:

$$i_L(t) = C\frac{d}{dt}e_1(t) = C\frac{d}{dt}\left(E + \frac{E}{R}\sqrt{\frac{L}{C}}\sin\omega_0 t\right) = C\frac{E}{R}\sqrt{\frac{L}{C}}\omega_0\cos\omega_0 t$$
$$i_L(t) = \frac{E}{R}\cos\omega_0 t, \text{ for } t \geq 0.$$

(b) For the numerical values given, the analytical solutions are:

$$e_1(t) = 100 + 100\sin 100t \text{ [V]}, \text{ for } t \geq 0$$

$$e_k(t) = U_0(t) + 100(1 + \sin 100t)U_{-1}(t)[\text{V}]$$

$$e_L(t) = -100\sin 100t . U_{-1}(t)[\text{V}]$$

$$i_L(t) = 10\cos 100t \text{ [A]}, \text{ for } t \geq 0.$$

Figure 6.111 shows the circuit assembled for the numerical solution, using the ATPDraw program. As the models of passive energy storage elements already include the initial conditions, namely: $e_1(0_-) = E = 100\,[\text{V}]$ in capacitance and $i_1(0_-) = i_2(0_-) = E/R = 10\,[\text{A}]$ in both inductances, **the external constant voltage source was considered to be acting from $t = 0$.** For switch k, the values $T - op = 0\,[\text{s}]$ and $T - cl = 1.000\,[\text{s}]$ were adopted. $\Delta t = 1\,[\mu\text{s}]$ was adopted.

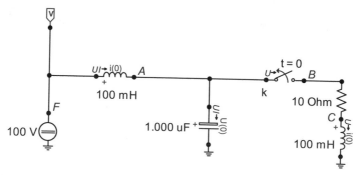

Fig. 6.111 Electric circuit of Example 6.27 for the numerical solution by the ATPDraw program

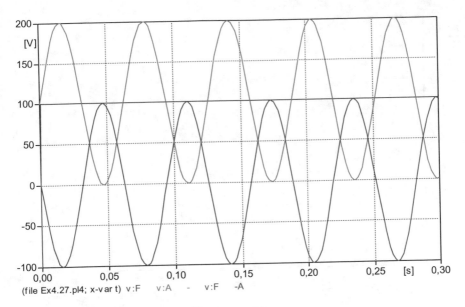

Fig. 6.112 Source voltage, voltage $e_1(t)$ voltage $e_L(t)$

Figure 6.112 shows the voltage of the external source, $E = 100\,[\mathrm{V}]$, the voltage at capacitance $e_1(t)$ and the voltage at the left inductance $e_L(t)$. It can be seen that all the results are consistent with the analytical solutions obtained previously.

Figure 6.113 "shows" the voltage in switch k, $e_k(t)$, as well as that of the right inductance, $e_{L_D}(t)$.

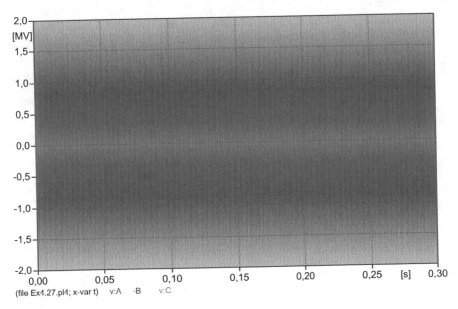

Fig. 6.113 Voltage $e_k(t)$ and voltage $e_{L_D}(t)$

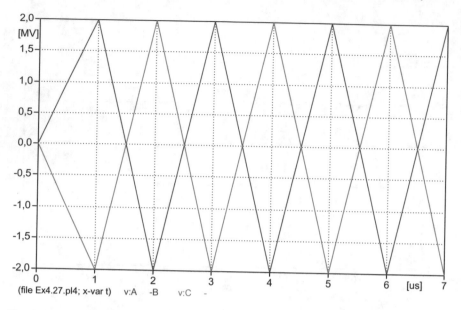

Fig. 6.114 Zoom in voltage $e_k(t)$ and voltage $e_{L_D}(t)$

Figure 6.113 reaffirms the total impossibility of the program working with the impulse function, whose amplitude tends to infinity, and which is present in $e_k(t)$ and $e_{L_D}(t)$. Hence the numerical oscillations that "colored" Fig. 6.113, and which are shown in zoom in Fig. 6.114. In the case of the right inductance, the current varies from $E/R = 10\,[\text{A}]$ to zero, when opening switch k at $t = 0$. Therefore, we have that $i_2(t) = 10U_{-1}(-t)\,[\text{A}]$. Thus, the voltage $e_{L_D}(t)$, opposite to the current $i_2(t)$, is worth $e_{L_D}(t) = Li_2'(t) = 0.1.10U_0(-t).(-1) = -U_0(t)\,[\text{V}]$. The area of this voltage impulse, $-1\,[\text{Wb}]$, represents the flow that was associated with the current $i_2(t)$, and which was zeroed together with the current at $t = 0$. It is equal to the area of the base triangle $1\,[\mu s] = \Delta t$ and height $-2\,[\text{MV}]$ of the green curve in Fig. 6.114. As the voltage at the right inductance has a downward reference arrow, that is why it appears positively in the analytical expression of the voltage $e_k(t)$.

The current at the left inductance is shown in Fig. 6.115.

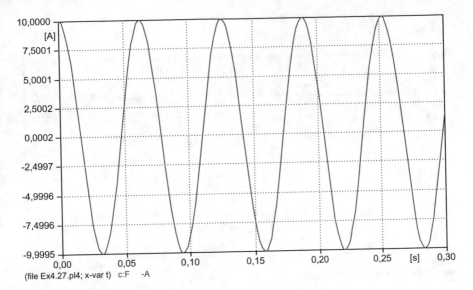

Fig. 6.115 Current $i_L(t)$

Example 6.28 In the circuit shown in Fig. 6.116, the voltage source is applied by closing the switch k_1, at time $t = 0$. At time $t = 3$ [ms] the switch k_2 is then closed. Numerically, determine all voltages and currents in the passive elements of the circuit, for $0 \leq t \leq 9$ [ms]. Numerical data: $E = 90$ [V], $R_1 = 10$ [Ω], $R_2 = 5$ [Ω], $C = 100$ [μF].

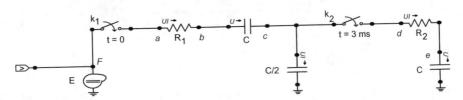

Fig. 6.116 Electric circuit of Example 6.28 for the numerical solution by the ATPDraw program

Solution

In Fig. 6.117 the voltages in R_1, C and $C/2$, and the source voltage, $E = 90$ [V] are plotted. In Fig. 6.118, the voltages in R_2 and the capacitance C on the right.

$$e_{R_1}(0_+) = 90\,[\text{V}]; \ e_C(0_+) = 0\,[\text{V}]; \ e_{c/2}(0_+) = 0\,[\text{V}]$$

$$e_{R_1}(3ms) = 0\,[\text{V}]; \ e_C(3ms) = 30\,[\text{V}]; \ e_{c/2}(3ms) = 60\,[\text{V}]$$

$$e_{R_1}(9ms) = 0\,[\text{V}]; \ e_C(9ms) = 54\,[\text{V}]; \ e_{c/2}(9ms) = 36\,[\text{V}]$$

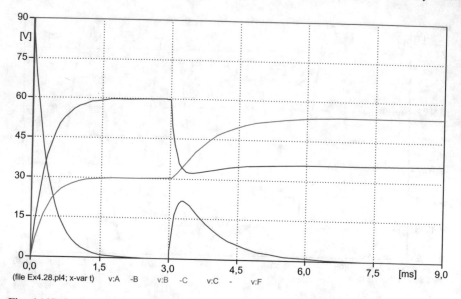

Fig. 6.117 Voltage in R_1, voltage in C, voltage in $C/2$, source voltage

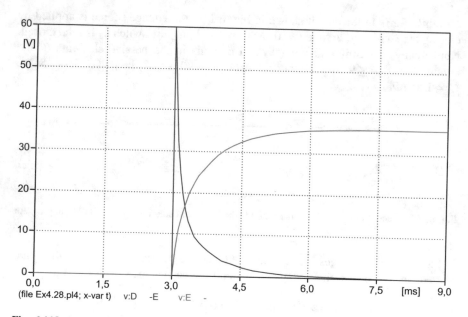

Fig. 6.118 Voltage in R_2, voltage in C

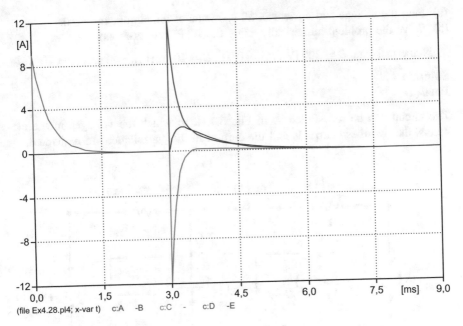

Fig. 6.119 Current in R_1-C Current in C_2 Current in R_2-C

$$e_{R_2}(3ms_+) = 60\,[\text{V}];\ e_C(3ms) = 0\,[\text{V}]$$

$$e_{R_2}(9ms) = 0\,[\text{V}];\ e_C(9ms) = 36\,[\text{V}]$$

Figure 6.119 shows currents in branches $R_1 - C$, $C/2$ and $R_2 - C$.

$$i_{R_1}(0_+) = 9\,[\text{A}] = i_C(0_+);\ i_{C/2}(0_+) = 9\,[\text{A}]$$

$$i_{R_1}(3ms_+) = 0\,[\text{A}] = i_C(3ms_+);\ i_{C/2}(3ms_+) = -12\,[\text{A}];\ i_{R_2}(3ms_+) = 12\,[\text{A}]$$

Example 6.29 The *Ch* switch in the circuit of Fig. 6.120 is closed at time $t = 0$, with the energy storage elements fully de-energized at $t = 0_-$.

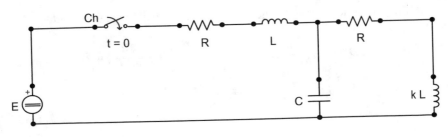

Fig. 6.120 Electric circuit of Example 6.29

(a) Find the voltage and current in all elements of the circuit, for $t \geq 0$.
(b) Solve the problem numerically using the ATPDraw program.

Numerical data: $E = 200\,[\text{V}], R = 10\,[\Omega], L = 10\,[\text{mH}], k = 4$ and $C = 100\,[\mu\text{F}]$.

Solution
Part (a)

The circuit can be redesigned as in Fig. 6.121, in which the voltages in the elements, the two mesh currents and the current in the capacitance are registered.

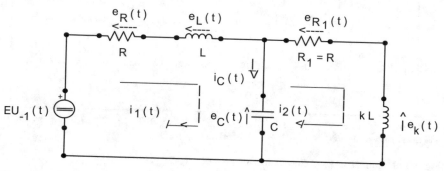

Fig. 6.121 Equivalent electric circuit of Example 6.29

The mesh equations are:

$$Ri_1 + LDi_1 + \frac{1}{CD}(i_1 - i_2) = EU_{-1}(t)$$

$$kLDi_2 + Ri_2 + \frac{1}{CD}(i_2 - i_1) = 0$$

Or,

$$\left(LCD^2 + RCD + 1\right)i_1 - i_2 = CDEU_{-1}(t)$$

$$-i_1 + \left(kLCD^2 + RCD + 1\right)i_2 = 0.$$

Solving the system for $i_1(t)$, the third order differential equation results

$$\left\{ D^3 + \frac{(1+k)R}{kL}D^2 + \left[\frac{1+k}{kLC} + \frac{R^2}{kL^2}\right]D + \frac{2R}{kL^2C} \right\} i_1(t)$$
$$= \left(\frac{1}{L}D^2 + \frac{R}{kL^2}D + \frac{1}{kL^2C}\right)EU_{-1}(t)$$

By substituting the numerical values, the equation becomes

$$i_1'''(t) + 1.25.10^3 i_1''(t) + 1.5.10^6 i_1'(t) + 5.10^8 i_1(t)$$
$$= 2.10^4 U_1(t) + 5.10^6 U_0(t) + 5.10^9 U_{-1}(t),$$

$$\text{with } i_1(0_-) = i_1'(0_-) = i_1''(0_-) = 0.$$

For $t > 0$, it results:

$$(D^3 + 1,250D^2 + 1,500,000D + 500,000,000)i_1(t) = 5,000,000,000$$

The particular integral is

$$i_{1p}(t) = \frac{5,000,000,000}{500,000,000} = 10$$

Factoring the polynomial operator $F(D)$, we have that:

$$F(D) = D + 1,250).D + 1,500,000).D + 500,000,000$$

To find an interval containing at least one root of $F(D) = 0$, that is, two points, a and b, so that $F(a)$ and $F(b)$ have opposite signs, Table 6.1 was constructed:
Since $F(-400) = +36,000,000$ and $F(-500) = -62,500,000$, there is at least one real root of $F(D) = 0$ between -400 e -500. Thus, an initial estimate for it is $D_0 = -440$. So, by Newton-Raphson, it comes that

$$D_{i+1} = D_i - \frac{F(D_i)}{F'(D_i)} = D_i - \frac{D_i^3 + 1,250D_i^2 + 1,500,000D_i + 500,000,000}{3D_i^2 + 2,500D_i + 1,500,000}$$

$$= \frac{2D_i^3 + 1,250D_i^2 - 500,000,000}{3D_i^2 + 2,500D_i + 1,500,000}, i = 0,1,2,\ldots$$

Expressed in factored form, it comes that

$$D_{i+1} = \frac{2)D_i + 1,250)D_iD_i - 500,000,000}{3)D_i + 2,500)D_i + 1,500,000}$$

Considering, then, that $D_0 = -440$, the following sequence of values is obtained:

$$D_1 = \frac{-428,368,000}{980,800} = -436.7537$$

$$D_2 = \frac{-428,182,608}{980,377.1334} = -436.7530$$

$$D_3 = \frac{-428,182,571.1}{980,377.0490} = -436.7530.$$

Table 6.1 Values for D and $F(D)$

D	0	-100	-200	-300	-400	-500
$F(D)$	500,000,000	361,500,000	242,000,000	135,500,000	36,000,000	$-62,500,000$

So, with four exact decimal places, $r_1 = -436,7530$. Performing polynomial deflation results in the equation

$$D^2 + 813.2470D + 1,144,811.933 = 0,$$

whose roots are $D_2 = -406.6235 + j989.6814$ and $D_3 = -406.6235 - j989.6814$. Therefore, the complementary solution is

$$i_{1_H}(t) = k_1 \epsilon^{-436.7530t} + \epsilon^{-406.6235t}(k_2 \cos 989.6814t + k_3 \sin 989.6814t)$$

Defining $r_2 = -406.6235$ and $\omega_n = 989.6814$, then the general solution is

$$i_1(t) = 10 + k_1 \epsilon^{r_1 t} + \epsilon^{r_2 t}(k_2 \cos \omega_n t + k_3 \sin \omega_n t)$$
$$i'_1(t) = r_1 k_1 \epsilon^{r_1 t} + r_2 \epsilon^{r_2 t}(k_2 \cos \omega_n t + k_3 \sin \omega_n t) + \epsilon^{r_2 t}(-\omega_n k_2 \sin \omega_n t + \omega_n k_3 \cos \omega_n t)$$
$$i''_1(t) = r_1^2 k_1 \epsilon^{r_1 t} + r_2^2 \epsilon^{r_2 t}(k_2 \cos \omega_n t + k_3 \sin \omega_n t) + r_2 \epsilon^{r_2 t}(-\omega_n k_2 \sin \omega_n t + \omega_n k_3 \cos \omega_n t)$$
$$+ r_2 \epsilon^{r_2 t}(-\omega_n k_2 \sin \omega_n t + \omega_n k_3 \cos \omega_n t) + \epsilon^{r_2 t}(-\omega_n^2 k_2 \cos \omega_n t - \omega_n^2 k_3 \sin \omega_n t).$$

Therefore,

$$i_1(0_+) = 10 + k_1 + k_2$$
$$i'_1(0_+) = r_1 k_1 + r_2 k_2 + \omega_n k_3$$
$$i''_1(0_+) = r_1^2 k_1 + (r_2^2 - \omega_n^2)k_2 + 2r_2 \omega_n k_3$$

Substituting the values of r_1, r_2 and ω_n, results:

$$10 + k_1 + k_2 = i_1(0_+)$$

$$-436.7530k_1 - 406.6235k_2 + 989.6814k_3 = i'_1(0_+)$$

$$190,753.1830k_1 - 814,126.6027k_2 - 804,855.4295k_3 = i''_1(0_+)$$

From the differential equation of $i_1(t)$, we can write that

$$i_1(t) \doteq 0.U_{-1}(t).$$

$$i'_1(t) \doteq C_0 U_{-1}(t).$$

$$i''_1(t) \doteq C_0 U_0(t) + C_1 U_{-1}(t).$$

$$i'''_1(t) \doteq C_0 U_1(t) + C_1 U_0(t) + C_2 U_{-1}(t).$$

Substituting in the differential equation of $i_1(t)$, comes that

$$C_0 U_1(t) + \left(C_1 + 1.25.10^3 C_0\right) U_0(t) + \left(C_2 + 1.25.10^3 C_1 + 1.5.10^6 C_0\right) U_{-1}(t)$$
$$\dot{=} 2.10^4 U_1(t) + 5.10^6 U_0(t) + 5.10^9 U_{-1}(t).$$

Therefore,

$$C_0 = 20,000$$

$$C_1 = -20,000,000$$

The initial conditions at $t = 0_+$ are:

$$i_1(0_+) = i_1(0_-) + 0 = 0\,[\text{A}]$$

$$i_1'(0_+) = i_1'(0_-) + C_0 = 20,000\,[A/s]$$

$$i_1''(0_+) = i_1''(0_-) + C_1 = -20,000,000[A/s^2].$$

Then, the system of algebraic equations in k_1, k_2 and k_3 is:

$$k_1 + k_2 = -10$$

$$-436.7530k_1 - 406.6235k_2 + 989.6814k_3 = 20,000$$

$$190,753.1830k_1 - 814,26.6027k_2 - 804,855.4295k_3 = -20,000,000$$

And the solution of this system of equations provides that:

$$k_1 = 4.8928; k_2 = -14.8928; k_3 = 16.2490.$$

Therefore,

$$i_1(t) = 10 + 4.8928\epsilon^{-436.7530t}$$
$$+ \epsilon^{-406.6235t}(-14.8928\cos 989.6814t + 16.2490\sin 989.6814t)$$

Or,

$$i_1(t) = 10 + 4.8928\epsilon^{-436.7530t}$$
$$+ 22.0414\epsilon^{-406.6235t}\cos\left(989.6814t - 132.5064^\circ\right)[\text{A}], \textbf{ for } \mathbf{t \geq 0}.$$

The solution of the system of differential equations for $i_2(t)$ results in

$$\left\{ D^3 + \frac{(1+k)R}{kL} D^2 + \left[\frac{1+k}{kLC} + \frac{R^2}{kL^2}\right] D + \frac{2R}{kL^2C} \right\} i_2(t) = \frac{E}{kL^2C} U_{-1}(t).$$

By substituting the numerical values, the equation becomes

$$i_2'''(t) + 1.25.10^3 i_2''(t) + 1.5.10^6 i_2'(t) + 5.10^8 i_2(t) = 5.10^9 U_{-1}(t),$$

$$\text{with } i_2(0_-) = i_2'(0_-) = i_2''(0_-) = 0.$$

For $t > 0$, results

$$\left(D^3 + 1,250D^2 + 1,500,000D + 500,000,000\right)i_2(t) = 5,000,000,000$$

So, the particular solution is

$$i_{2P}(t) = \frac{5.10^9}{5.10^8} = 10$$

And the complementary solution,

$$i_{2H}(t) = k_4 \epsilon^{r_1 t} + \epsilon^{r_2 t}(k_5 \cos \omega_n t + k_6 \sin \omega_n t)$$

Thus,

$$i_2(t) = 10 + k_4 \epsilon^{r_1 t} + \epsilon^{r_2 t}(k_5 \cos \omega_n t + k_6 \sin \omega_n t)$$

$$i_2'(t) = r_1 k_4 \epsilon^{r_1 t} + r_2 \epsilon^{r_2 t}(k_5 \cos \omega_n t + k_6 \sin \omega_n t) + \epsilon^{r_2 t}(-\omega_n k_5 \sin \omega_n t + \omega_n k_6 \cos \omega_n t)$$

$$i_2''(t) = r_1^2 k_4 \epsilon^{r_1 t} + r_2^2 \epsilon^{r_2 t}(k_5 \cos \omega_n t + k_6 \sin \omega_n t) + r_2 \epsilon^{r_2 t}(-\omega_n k_5 \sin \omega_n t + \omega_n k_6 \cos \omega_n t)$$

$$+ r_2 \epsilon^{r_2 t}(-\omega_n k_5 \sin \omega_n t + \omega_n k_6 \cos \omega_n t) + \epsilon^{r_2 t}(-\omega_n^2 k_5 \cos \omega_n t - \omega_n^2 k_6 \sin \omega_n t).$$

So,

$$i_2(0_+) = 10 + k_4 + k_5$$

$$i_2'(0_+) = r_1 k_4 + r_2 k_5 + \omega_n k_6$$

$$i_2''(0_+) = r_1^2 k_4 + \left(r_2^2 - \omega_n^2\right)k_5 + 2r_2 \omega_n k_6.$$

The analysis of the differential equation of $i_2(t)$ shows that there will only be a step, $U_{-1}(t)$, in $i_2'''(t)$. So, $i_2(0_+) = i_2'(0_+) = i_2''(0_+) = 0$. Then, replacing the numerical values in the preceding expressions, it comes that:

$$10 + k_4 + k_5 = 0$$

$$-436.7530\,k_4 - 406.6235 k_5 + 989.6814 k_6 = 0$$

$$190,753.1830 k_4 - 814,126.6027 k_5 - 804,855.4295 k_6 = 0$$

Or,

$$k_4 + k_5 = -10$$

$$-k_4 - 0.9310k_5 + 2.2660k_6 = 0$$

$$k_4 - 4.2680k_5 - 4.2194k_6 = 0$$

The solution to this system of algebraic equations is

$$k_4 = -11.6773; k_5 = 1.6773; k_6 = -4.4642.$$

Therefore,

$$i_2(t) = 10 - 11.6773\epsilon^{-436.7530t} + \epsilon^{-406.6235t}(1.6773\cos 989.6814t \\ - 4.4642\sin 989.6814t)$$

Or,

$$i_2(t) = 10 - 11.6773\epsilon^{-436.7530t} + 4.7689\epsilon^{-406.6235t}\cos(989.6814t + 69.4077°) \, [\text{A}], p/t \geq 0.$$

$$i_C(t) = i_1(t) - i_2(t)$$

$$i_C(t) = 16.5701\epsilon^{-436.7530t} + 26.5256\epsilon^{-406.6235t}\cos(989.6814t - 128.6590°) \, [\text{A}], p/t \geq 0.$$

$$e_R(t) = Ri_1(t) = 10i_1(t).$$

$$e_R(t) = 100 + 48.928\epsilon^{-436.7530t} + 220.414\epsilon^{-406.6235t}\cos(989.6814t - 132.5064°) \, [\text{V}], t \geq 0.$$

$$e_L(t) = Li_1'(t) = 0.01i_1'(t).$$

$$e_L(t) = -21.3695\epsilon^{-436.7530t} \\ - 235.8338\epsilon^{-406.6235t}\cos(989.6814t + 159.8296°) \, [\text{V}], p/t > 0.$$

$$e_C(t) = 200 - [e_R(t) + e_L(t)].$$

$$e_C(t) = 100 - 27.5585\epsilon^{-436.7530t} \\ - 254.3430\epsilon^{-406.6235t}\cos(989.6814t - 73.4517°) \, [\text{V}], para \, t \geq 0.$$

$$e_{R_1}(t) = Ri_2(t) = 10i_2(t).$$

$$e_{R_1}(t) = 100 - 116.773\epsilon^{-436.7530t} + 47.689\epsilon^{-406.6235t}\cos(989.6814t + 69.4077°)\,[\text{V}],\,para\,t \ge 0.$$

$$e_k(t) = e_C(t) - e_{R_1}(t).$$

$$e_k(t) = 89.2145\epsilon^{-436.7530t}$$
$$- 218.2352\epsilon^{-406.6235t}\cos(989.6814t - 65.8701°)\,[\text{V}],\,para\,t \ge 0.$$

Part (b)

The responses have two time constants:

$$T_1 = \frac{1}{436.7530} = 2.3\,[\text{ms}] \text{ and } T_2 = \frac{1}{406.6235} = 2.5\,[\text{ms}].$$

The angular frequency is $\omega_n = 989.6814\,[\text{rad/s}] = 2\pi f_n = 2\pi/T_n$. Therefore, $f_n = 157.51\,[\text{Hz}]$, $T_n = 6.35\,[\text{ms}]$.

Thus, the transitional regime components last approximately $5T_2 = 12.5\,[\text{ms}]$.

The circuit built for the numerical solution, using the ATPDraw program, is shown in Fig. 6.122.

Considering $\Delta t \cong T_1/10^4\,[\text{s}]$, it was adopted that $\Delta t = 10^{-7}\,[\text{s}]$ and $t_{max} = 20\,[\text{ms}]$.

In Fig. 6.123, the mesh currents $i_1(t)$ and $i_2(t)$ are plotted, and also the current in the capacitance, $i_C(t)$.

As in the permanent regime of direct current (constant source), the inductances behave as short circuits and the capacitance, as an open circuit, logically, $i_C(\infty) = 0\,[\text{A}]$ and $i_1(\infty) = i_2(\infty) = E/2R = 200/20 = 10\,[\text{A}]$.

The source, resistance R, inductance L and capacitance C voltages are shown in Fig. 6.124.

Finally, Fig. 6.125 plots the voltage curves $e_{R1}(t)$ e $e_k(t)$, the latter being the voltage at the inductance kL.

This Example 6.29 **fully illustrates the stupidity of solving analytically transient problems in circuits of a higher order than the second.** Thankfully,

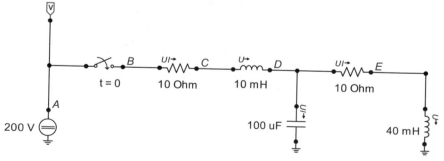

Fig. 6.122 Electric circuit of Example 6.29 for the numerical solution by the ATPDraw program

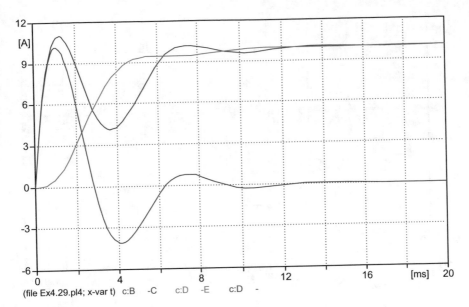

Fig. 6.123 Current $i_1(t)$, Current $i_2(t)$, Current $i_C(t)$

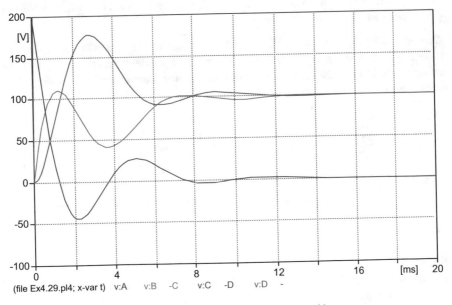

Fig. 6.124 Source voltage voltage $e_R(t)$ Voltage $e_L(t)$ Volltage $e_C(t)$

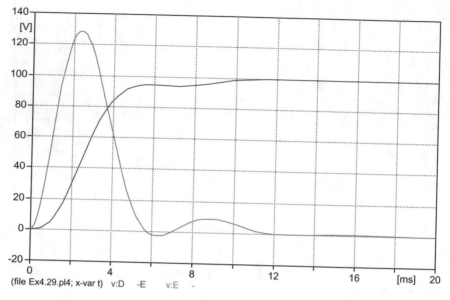

(file Ex4.29.pl4; x-var t)   v:D   -E   v:E   -

Fig. 6.125 Voltage $e_{R_1}(t)$ Voltage $e_k(t)$

there are several programs for the digital solution, such as ATPDraw, of transient phenomena in higher order circuits.

Example 6.30 The circuit in Fig. 6.126 was operating in a steady state when, at time $t = 0$, switch Ch was opened. Find the voltages $e_C(t)$, $e_k(t)$, $e_{Ch}(t)$ and the current $i_2(t)$ in the right mesh, for $t \geq 0$. Also, perform the numerical solution. The numerical data is the same as in Example 6.29.

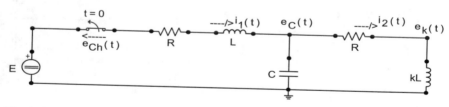

Fig. 6.126 Electric circuit of Example 6.30

Solution

In DC steady-state, in $t = 0_-$, the circuit can be visualized as shown in Fig. 6.127, with the inductors in short-circuit and the capacitor in open circuit.

Therefore, $i_1(0_-) = i_2(0_-) = E/2R$ and $e_C(0_-) = Ri_2(0_-) = E/2$. Then introducing the Thévenin equivalents in the inductances and capacitance, the circuit for $t \geq 0$ results in the one shown in Fig. 6.128.

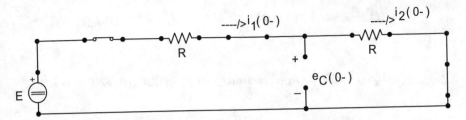

Fig. 6.127 Electric circuit of Example 6.30 for DC steady-state solution

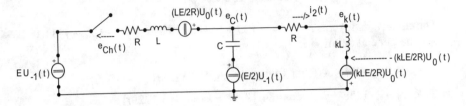

Fig. 6.128 Equivalent electric circuit of Example 6.30

The nodal equations of the circuit in Fig. 6.128 are:

$$CD\left[e_C(t) - \frac{E}{2}U_{-1}(t)\right] + \frac{e_C(t) - e_k(t)}{R} = 0$$

$$\frac{e_k(t) - e_C(t)}{R} + \frac{1}{kL}D^{-1}\left[e_k(t) + \frac{kLE}{2R}U_0(t)\right] = 0$$

Or,

$$(RCD + 1)e_C(t) - e_k(t) = \frac{RCE}{2}DU_{-1}(t)$$

$$-kLDe_C(t) + (kLD + R)e_k(t) = -\frac{kLE}{2}U_0(t).$$

Solving this system for $e_C(t)$ results in the differential equation

$$\left(D^2 + \frac{R}{kL}D + \frac{1}{kLC}\right)e_C(t) = \frac{E}{2}U_1(t) + \left(\frac{R}{kL} - \frac{1}{RC}\right)\frac{E}{2}U_0(t).$$

Introducing the numerical values, comes that

$$\left(D^2 + 250D + 250{,}000\right)e_C(t) = 100U_1(t) - 75{,}000U_0(t)$$

Or,

$$e_C''(t) + 250e_C'(t) + 250{,}000e_C(t) = 100U_1(t) - 75{,}000U_0(t); e_C(0_-) = e_C'(0_-)$$
$$= 0.$$

For $t > 0$,

$$\left(D^2 + 250D + 250{,}000\right)e_C(t) = 0$$

The characteristic roots are $r_1 = -125 + j484.1229$ and $r_2 = -125 - j484.1229$. So, the general solution is

$$e_C(t) = \epsilon^{-125t}(k_1 \cos 484.1229t + k_2 \sin 484.1229t)$$
$$e_C'(t) = -125\epsilon^{-125t}(k_1 \cos 484.1229t + k_2 \sin 484.1229t)$$
$$+ \epsilon^{-125t}(-484.1229k_1 \sin 484.1229t + 484.1229k_2 \cos 484.1229t)$$

From the original differential equation, it can be written that

$$e_C(t) \doteq C_0 U_{-1}(t).$$

$$e_C'(t) \doteq C_0 U_0(t) + C_1 U_{-1}(t).$$

$$e_C''(t) \doteq C_0 U_1(t) + C_1 U_0(t) + C_2 U_{-1}(t).$$

Substituting in the same original differential equation,

$$C_0 U_1(t) + (C_1 + 250C_0)U_0(t) + (C_2 + 250C_1 + 250{,}000C_0)U_{-1}(t)$$
$$\doteq 100U_1(t) - 75{,}000U_0(t) + 0.U_{-1}(t).$$

Therefore,

$$C_0 = 100 = e_C(0_+) = k_1$$

$$C_1 = -75{,}000 - 250C_0 = -100{,}000 = e_C'(0_+) = -125k_1 + 484.1229k_2$$

So, $k_1 = 100\,[\text{V}]$ and $k_2 = -180.7392\,[\text{V}]$, and the solution is

$$e_C(t) = \epsilon^{-125t}(100 \cos 484.1229t - 180.7392 \sin 484.7392t)\,[\text{V}]$$

Or,

$$e_C(t) = 206.5591\epsilon^{-125t}\cos(484.1229t + 61.0450°)\,[\text{V}],\ \textit{for}\ t \geq 0.$$

The solution of the system of nodal equations for $e_k(t)$ generates the differential equation

$$\left(D^2 + \frac{R}{kL}D + \frac{1}{kLC}\right)e_k(t) = \frac{E}{2RC}U_0(t)$$

Substituting the numerical values, comes that

$$e_k''(t) + 250e_k'(t) + 250,000e_k(t) = 100,000U_0(t),\ \textit{with}\ e_k(0_-) = e_k'(0_-) = 0.$$

So,

$$e_k(t) = \epsilon^{-125t}(k_3\cos 484.1229t + k_4\sin 484.1229t)$$
$$e_k'(t) = -125\epsilon^{-125t}(k_3\cos 484.1229t + k_4\sin 484.1229t)$$
$$+ \epsilon^{-125t}(-484.1229k_3\sin 484.1229t + 484.1229k_4\cos 484.1229t)$$

Analyzing the two members of the differential equation of $e_k(t)$, one can write that

$$e_k(0_+) = 0\,[\text{V}] = k_3$$

$$e_k'(0_+) = 100,000 = -125k_3 + 484.1229k_4$$

Therefore, $k_3 = 0\,[\text{V}]$ and $k_4 = 206.5591\,[\text{V}]$. And the solution for $e_k(t)$ is

$$e_k(t) = 206.5591\epsilon^{-125t}\sin 484.1229t\,[\text{V}],\ \textit{for}\ t \geq 0.$$

From Fig. 6.128, according to Kirchhoff's voltage law, applied to the left loop,

$$e_C(t) - \frac{LE}{2R}U_0(t) + e_{Ch}(t) - E = 0.$$

So,

$$e_{Ch}(t) = \frac{LE}{2R}U_0(t) - e_C(t) + E$$

Therefore,

$$e_{Ch}(t) = 0.1U_0(t) + \left[200 - 206.5591\epsilon^{-125t}\cos(484.1229t + 61.0450°)\right]U_{-1}(t)\,[\text{V}].$$

The current $i_2(t)$, in the right mesh, is

$$i_2(t) = \frac{e_C(t) - e_k(t)}{R}. \text{ So,}$$

$$i_2(t) = 40\epsilon^{-125t}\cos(484.1229t + 75.5225°)\,[\text{A}], \text{ for } t \geq 0$$

For the numerical solution by ATPDraw, the circuit shown in Fig. 6.129 was built.

Figure 6.130 shows the source, capacitance and inductance voltages on the right. The current circulating in the right mesh, $i_2(t)$, is recorded in Fig. 6.131.

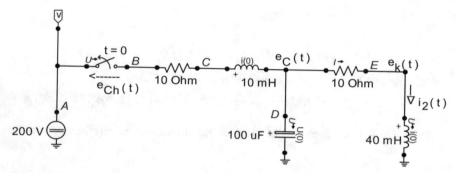

Fig. 6.129 Electric circuit of Example 6.30 for the numerical solution by the ATPDraw program

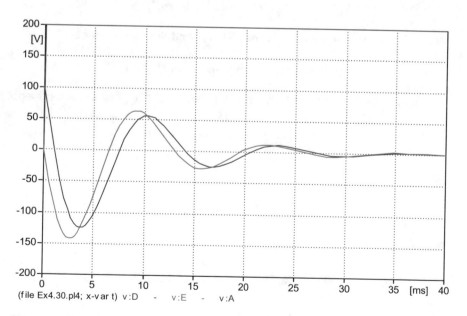

Fig. 6.130 Source voltage, Voltage $e_C(t)$, Voltage $e_k(t)$

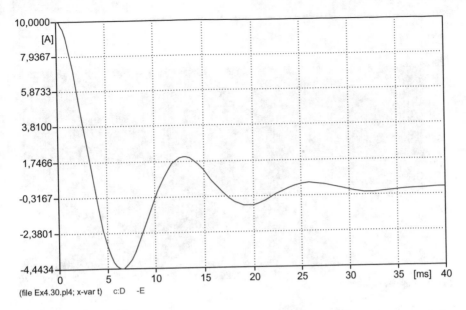

Fig. 6.131 Current $i_2(t)$

As it could not be otherwise, the voltage at the open switch terminals presents a numerical oscillation, caused by the voltage impulse $0.1U_0(t)$ [V], as determined in the analytical solution of $e_{Ch}(t)$. This voltage impulse is generated at the 10 [mH] inductance terminals, as a result of the sudden current drop from 10 [A] to zero, with the opening of the switch in series with it at $t = 0$. And the area of this impulse, 0.1 [Wb], represents the flow that was associated with the 10 [A] current of the 10 [mH] inductance. The numerical oscillation in $e_{Ch}(t)$ is explained in Fig. 6.132. Figure 6.133 shows the result of several zooms in $e_{Ch}(t)$. Once again, it can be confirmed that the area of the first triangle, base 1 [μs] $= \Delta t$ and height 200 [kV] is worth 0.1 [Wb], exactly the area of the impulse present in the analytical solution of $e_{Ch}(t)$.

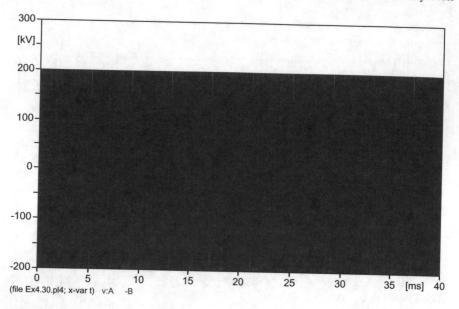

Fig. 6.132 Voltage in the open switch $e_{Ch}(t)$

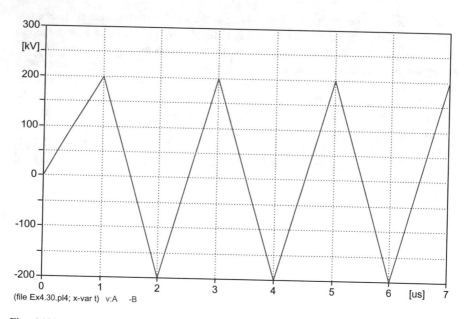

Fig. 6.133 Zoom at voltage in the open switch $e_{Ch}(t)$

Example 6.31 The switch Ch in the circuit of Fig. 6.134 is closed at time $t = 0$ [s] and opened at time $t = 30$ [ms]. Numerically, find the voltages $e_C(t)$, $e_k(t)$ and $e_{Ch}(t)$, as well as the currents $i_1(t)$ in L and $i_2(t)$ in kL, in the range $0 \leq t \leq 80$[ms]. Data: $E = 200$ [V], $R = 10$ [Ω], $L = 10$ [mH], $k = 4$ and $C = 100$ [μF].

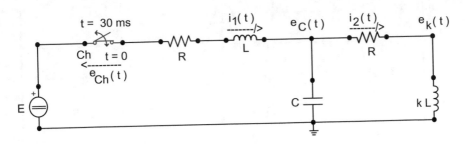

Fig. 6.134 Electric circuit of Example 6.31

Solution
The circuit built for the solution by ATPDraw is shown in Fig. 6.135.

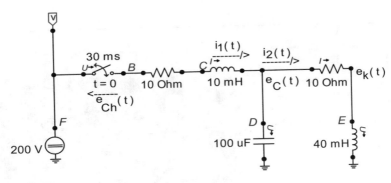

Fig. 6.135 Electric circuit of Example 6.31 for the numerical solution by the ATPDraw program

The Fig. 6.136 shows the source, capacitor and right inductor voltages.

As seen in Example 6.30, the voltage at the switch terminals, $e_{Ch}(t)$, must present a numerical oscillation, from the opening of the switch at $t = 30$ [ms], due to the sudden drop in the current at inductance L which is in series with the $Ch-$ switch, it instantly changes from 10 [A] to zero, generating a impulse in $e_{Ch}(t)$. And Fig. 6.137 confirms this prediction.

Figure 6.138 shows the result of successive zooms in Fig. 6.137.

Finally, Fig. 6.139 records the currents $i_1(t)$ and $i_2(t)$.

As is easy to see, this Example 6.31 is a kind of "junction" of Examples 6.29 and 6.30. A most relevant fact is that: **any switch that will be opened in ATPDraw**

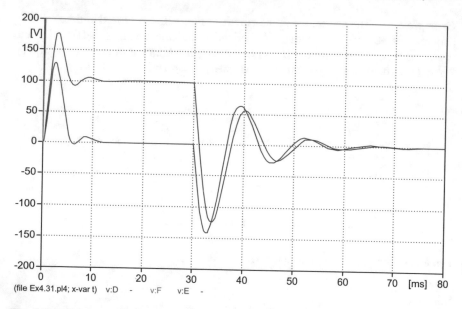

Fig. 6.136 Voltage E, Voltage $e_C(t)$, Voltage $e_k(t)$

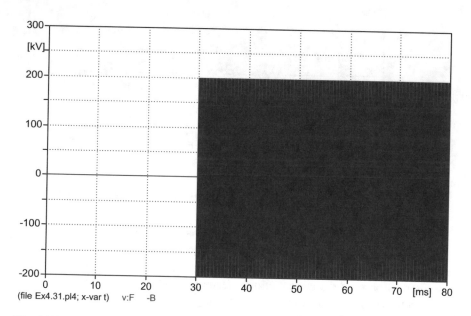

Fig. 6.137 Voltage $e_{Ch}(t)$

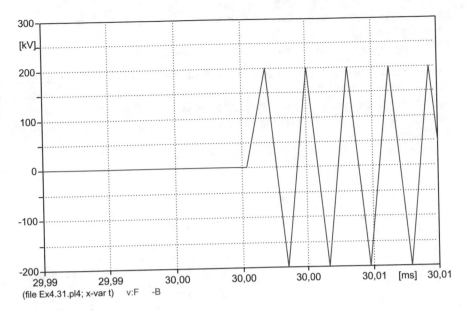

Fig. 6.138 Zoom in voltage $e_{Ch}(t)$

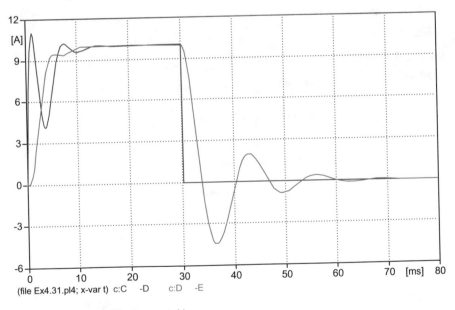

Fig. 6.139 Current $i_1(t)$, Current $i_2(t)$

must do so when the current through it is passing zero. This restriction is implicit in **Imar Amps 0**, from **Component: TSWITCH**. As in this Example 6.31 the steady-state current $i_1(t)$, passing through the switch Ch, is constant and is worth 10 [A], immediately before $t = 30$ [ms], **the switch will not open without**

first changing Imar Amps for a value greater than $10\,[A]$. Therefore, it was necessary to change Imar to a value greater than $10\,[A]$. In the present case, **Imar** $=$ $12\,[A]$, was used to force the switch to open at $t = 30\,[\text{ms}]$.

Example 6.32 Switch 2 in the circuit of Fig. 6.140 has been closed for a long time, shorting capacitance C. At time $t = 0\,[\text{s}]$ switch 1 is closed. After $40[\text{ms}]$, switch 2 is opened. (a) Determine the analytical expressions of voltage and current in all elements of the circuit. (b) Perform the numerical solution of the problem using ATPDraw.

  Data: $E = 100\,[\text{V}]$, $R = 5\,[\Omega]$, $L = 20\,[\text{mH}]$, $C = 50\,[\mu\text{F}]$.

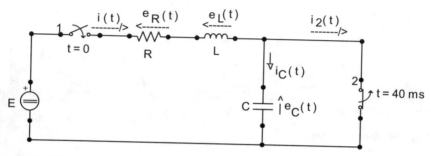

Fig. 6.140 Electric circuit of Example 6.32

Solution

(a) For $0 \le t < 40\,[\text{ms}]$ (switch 2 closed).

  The equation of the external loop of the circuit is

$$Li'(t) + Ri(t) = EU_{-1}(t)$$

Therefore,

$$i'(t) + 250i(t) = 5,000U_{-1}(t), \ \text{with } i(0_-) = 0.$$

$$i_H(t) = k_1\epsilon^{-250t}$$

$$i_P(t) = \frac{5,000\epsilon^{0.t}}{D + 250} = 20$$

$$i(t) = 20 + k_1\epsilon^{-250t}$$

As the current differential equation shows that there is no step in $i(t)$,

$$i(0_+) = i(0_-) = 0. \ \text{So}, \ k_1 = -20 \ \text{and},$$

$$i(t) = \left(20 - 20\epsilon^{-250t}\right)\,[\text{A}], \ for \ 0 \le t < 40\,[\text{ms}].$$

So,

$$e_R(t) = Ri(t) = 5i(t) = \left(100 - 100\epsilon^{-250t}\right) [V], \; for \; 0 \le t < 40 \; [ms].$$

$$e_L(t) = Li'(t) = 20.10^{-3} \frac{d}{dt} \left[(20 - 20\epsilon^{-250t}) U_{-1}(t) \right]$$

$$e_L(t) = 100\epsilon^{-250t} [V], \; for \; 0 < t < 40 [ms].$$

The time constant is worth $T = 1/250 = 4$ [ms] and $5T = 20$ [ms]. Therefore, before $t = 40$ [ms] the circuit is already operating in a steady state and, $i(40 \; ms_-) = 20$ [A].
Consequently,

$$e_R(40 \; ms_-) = 100 \; [V].$$
$$e_L(40 \; ms_-) = 0 \; [V].$$

For $t > 40$ [ms] (switch 2 open).
Considering, **temporarily**, the instant of opening of switch 2, $t = 40$ [ms], as being a **new** $t = 0$ [s], that is, **taking a new origin of the time axis over** 40 [ms], so that $i(40ms_-)$ become $i(0_-) = 20$ [A], the circuit of Fig. 6.141 can be constructed, in which the initially energized inductance is represented by its Thévenin equivalent, with voltage $Li(0_-)U_0(t) = 20.10^{-3}.20.U_0(t) = 0.4U_0(t)$ [V].
The mesh equation for the circuit in Fig. 6.141 is

$$5i(t) + 20.10^{-3} i'(t) + \frac{1}{50.10^{-6}} \int i(t)dt = 0.4U_0(t) + 100U_{-1}(t)$$

Or,

$$i''(t) + 250i'(t) + 10^6 i(t) = 20U_1(t) + 5,000U_0(t), \; with \; i(0_-) = i'(0_-) = 0.$$

For $t > 0$,

$$i''(t) + 250i'(t) + 10^6 i(t) = 0$$

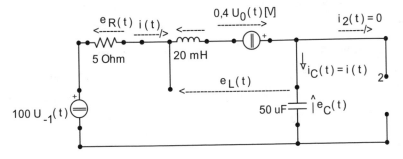

Fig. 6.141 Equivalent electric circuit of Example 6.32

And the roots of the polynomial operator $D^2 + 250D + 10^6 = 0$ are $r_{1,2} = -125 \pm j992.1567$. So,

$$i(t) = \epsilon^{-125t}(k_2 \cos 992.1567t + k_3 \sin 992.1567t)$$
$$i'(t) = -125\epsilon^{-125t}(k_2 \cos 992.1567t + k_3 \sin 992.1567t)$$
$$+ 992.1567\epsilon^{-125t}(-k_2 \sin 992.1567t + k_3 \cos 992.1567t).$$

From the differential equation,

$$i(t) \doteq C_0 U_{-1}(t).$$

$$i'(t) \doteq C_0 U_0(t) + C_1 U_{-1}(t).$$

$$i''(t) \doteq C_0 U_1(t) + C_1 U_0(t) + C_2 U_{-1}(t).$$

Substituting in the differential equation results

$$C_0 U_1(t) + (C_1 + 250C_0)U_0(t) + (C_2 + 250C_1 + 10^6 C_0)U_{-1}(t) \doteq$$

$$20U_1(t) + 5,000U_0(t) + 0.U_{-1}(t).$$

So, $C_0 = 20$; $C_1 + 250.20 = 5,000$, where $C_1 = 0$. And,

$$i(0_+) = 20 = k_2$$

$$i'(0_+) = 0 = -125k_2 + 992.1567k_3$$

Therefore, $k_2 = 20$ and $k_3 = 2.5198$. So,

$$i(t) = \epsilon^{-125t}(20 \cos 992.1567t + 2.5198 \sin 992.1567t)$$

Or,

$$i(t) = 20.1581\epsilon^{-125t} \cos(992.1567t - 7.1809°) \, [A], \, for \; "t \geq 0".$$

This expression of current $i(t)$, however, refers to the new origin, taken, temporarily, at the instant equal to 40 [ms]. Therefore, it is necessary to refer to the first origin, which is the time $t = 0$ of the closing of switch 1. For this, **it is necessary to replace t with $t - 40$ [ms] $= t - 40.10^{-3}$ [s]**. So,

$$i(t) = 20.1581\epsilon^{-125(t-40.10^{-3})} \cos\left[992.1567\left(t - 40.10^{-3}\right) - 7.1809°\right]$$
$$= 2,991.7273\epsilon^{-125t} \cos(992.1567t - 39.6863\text{rad} - 7.1809°)$$
$$= 2,991.7173\epsilon^{-125t} \cos(992.1567t - 2,273.8575° - 7.1809°)$$
$$= 2,991.7173\epsilon^{-125t} \cos(992.1567t - 2,281.0384°).$$

Finally, removing the number of entire circumferences from the phase angle, it follows that:

$$i(t) = 2,991.7273\epsilon^{-125t}\cos(992.1567t - 121.0384°)\,[\text{A}], \; for\; t \geq 40[\text{ms}].$$

The new circuit time constant is $T_1 = 1/125 = 8\,[\text{ms}]$, and $5T_1 = 40\,[\text{ms}]$. Where $\omega_n = \frac{2\pi}{T_n} = 992.1567\,[\text{rad/s}]$, $T_n = 6.33\,[\text{ms}]$. The resistance voltage is

$$e_R(t) = Ri(t) = 5i(t)$$

$$e_R(t) = 14,958.6365\epsilon^{-125t}\cos(992.1567t - 121.0384°)\,[\text{V}], para\; t \geq 40[\text{ms}].$$

The voltage at the inductance is

$$e_L(t) = Li'(t) = 20.10^{-3}\frac{d}{dt}\left[2,991.7273\epsilon^{-125t}\cos(992.1567t - 121.0384°)\right]$$

$$e_L(t) = 59,834.5435\epsilon^{-125t}\cos(992.1567t - 23.8576°)\,[\text{V}], for\; t \geq 40[\text{ms}].$$

And the voltage across the capacitance is

$$e_C(t) = 100 - e_R(t) - e_L(t)$$
$$e_C(t) = 100 + 59,834.5321\epsilon^{-125t}\cos(992.1567t + 141.7809°)[\text{V}], \; for\; t \geq 40[\text{ms}].$$

(b) The circuit built for the numerical solution is registered in Fig. 6.142.

Figure 6.143 shows the voltage curves in the resistance, $e_R(t)$, and in the inductance, $e_L(t)$.

Figure 6.144 shows the voltage at the capacitance, $e_C(t)$, scaled on the left vertical axis, and the current at the capacitance, $i_C(t)$, scaled on the right vertical axis. Figure 6.145 shows the currents in branch R-L, $i(t)$, and in switch 2, $i_2(t)$.

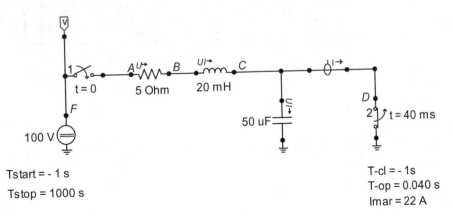

Fig. 6.142 Electric circuit of Example 6.32 for the numerical solution by the ATPDraw program

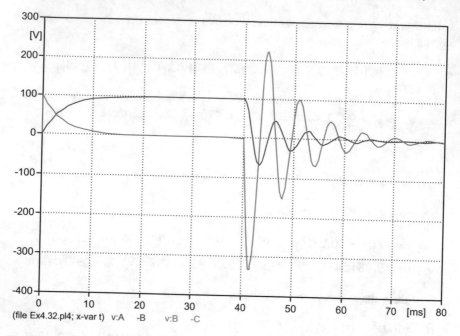

Fig. 6.143 Voltage $e_R(t)$, Voltage $e_L(t)$

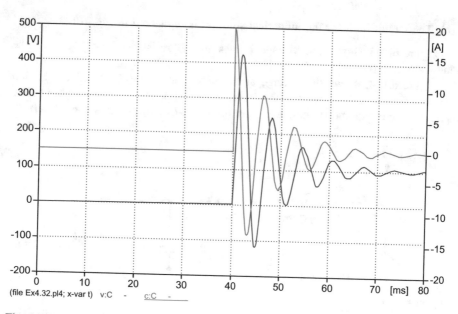

Fig. 6.144 Voltage $e_C(t)$, Voltage $i_C(t)$

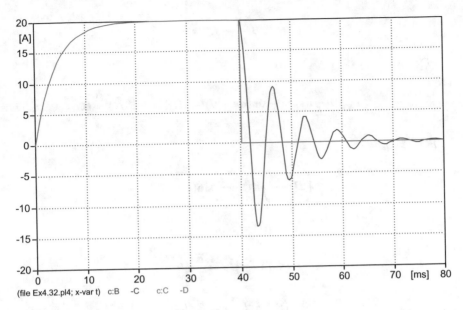

(file Ex4.32.pl4; x-var t)   c:B   -C   c:C   -D

Fig. 6.145 Current $i(t)$, Current $i_2(t)$

Example 6.33 Switch k in the circuit of Fig. 6.146 is closed at the instant considered, arbitrarily, as $t = 0$.

Fig. 6.146 Electric circuit of
Example 6.33

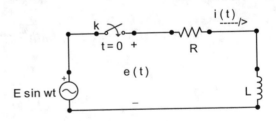

(a) Find the analytical expression of current $i(t)$, for $t > 0$.
(b) Establish the best and worst conditions for closing the switch, considering, respectively, the absence and presence of the transitional regime component in $i(t)$.
(c) Perform the digital solution of the problem.

   Numerical data: $E = 112.7\,[\text{kV}]$, $f = 60\,[\text{Hz}]$, $R = 16[\Omega]$, $L = 600\,[\text{mH}]$.

Solution

(a) The voltage applied to the R-L series circuit is $e(t) = E\sin(\omega t + \theta)U_{-1}(t)$, where the initial phase θ is associated with the closing time of switch k, that is, the amplitude initial sinusoidal source at time $t = 0$. Therefore,

$$L\frac{di}{dt} + Ri = e$$

Or,

$$i'(t) + \frac{R}{L}i(t) = \frac{E}{L}\sin(\omega t + \theta)U_{-1}(t), \text{ with } i(0_-) = 0. \tag{6.37}$$

For $t > 0$,

$$\left(D + \frac{R}{L}\right)i(t) = \frac{E}{L}\sin(\omega t + \theta)$$

$$i_H(t) = K_1\epsilon^{-Rt/L}$$

$$\left(j\omega + \frac{R}{L}\right)\dot{I}_P = \frac{E}{L}\angle\theta$$

$$(R + j\omega L)\dot{I}_P = E\angle\theta$$

$$\dot{I}_P = \frac{E\angle\theta}{R + j\omega L} = \frac{E\angle\theta}{\sqrt{R^2 + (\omega L)^2}\angle tg^{-1}(\omega L/R)}$$

Defining:

$$Z = R + j\omega L; |Z| = \sqrt{R^2 + (\omega L)^2}; \angle Z = \varphi = tg^{-1}(\omega L/R),$$

So,

$$\dot{I}_P = \frac{E\angle\theta}{|Z|\angle\varphi} = \frac{E}{|Z|}\angle(\theta - \varphi).$$

Therefore,

$$i_P(t) = \frac{E}{|Z|}\sin(\omega t + \theta - \varphi)$$

So,

$$i(t) = K_1\epsilon^{-Rt/L} + \frac{E}{|Z|}\sin(\omega t + \theta - \varphi)$$

Equation (6.37) can be rewritten as:

$$i'(t) + \frac{R}{L}i(t) = \frac{E}{L}\sin\theta U_{-1}(t) + \frac{\omega E}{L}\cos\theta U_{-2}(t) + \cdots$$

So, $i'(0_+) = \frac{E}{L}\sin\theta$ and $i(0_+) = 0$.
Therefore,

$$i(0_+) = K_1 + \frac{E}{|Z|}\sin(\theta - \varphi) = 0 \text{ and } k_1 = -\frac{E}{|Z|}\sin(\theta - \varphi)$$

Finally,

$$i(t) = \frac{E}{|Z|}\left[\sin(\omega t + \theta - \varphi) - \sin(\theta - \varphi)\epsilon^{-Rt/L}\right]U_{-1}(t).$$

By replacing the numerical values,

$$i(t) = 497.00\left[\sin(376.99t + \theta - 85.954°) - \sin(\theta - 85.954°)\epsilon^{-26.667t}\right]U_{-1}(t) \text{ [A]}.$$

(b) So that there is no transient component (better situation) in current $i(t)$, then
$\theta = 85.954°$. So,

$$i(t) = 497.00\sin 376.99t.U_{-1}(t)[\text{A}] \text{ and,} \tag{6.38}$$

$$e(t) = 112.700\sin(376.99t + 85.954°)U_{-1}(t)[\text{V}]. \tag{6.39}$$

This current reaches the peak (maximum) value, 497 [A], at
$t = \frac{\pi/2}{2\pi.60} = 4.167$ [ms]. On the other hand,

$$i'(t) = 497 \times 376.99\cos 376.99t.U_{-1}(t) = 187,364.03\cos 376.99t.U_{-1}(t) \text{ [A/s]}.$$

Then,

$$i'(0_+) = 187,364.03 \text{ [A/s]}.$$

Also, from the analytical solution of Eq. (6.37),

$$i'(0_+) = \frac{E}{L}\sin\theta = \frac{112.700}{0.6}\sin 85.954° = 187,365.20 \text{ [A/s]}.$$

In order for a transient component to occur in current $i(t)$, and of maximum
amplitude (worst situation), at the time of closing of switch k, that is, at $t = 0$, then
$\sin(\theta - 85.954°) = +1$. Thus, $\theta - 85.954° = 90°$ and $\theta = 175.954°$. Therefore,

$$i(t) = 497.00\left[\sin(376.99t + 90°) - \epsilon^{-26.667t}\right]U_{-1}(t)\,[\text{A}].$$

Or,

$$i(t) = 497.00\left(\cos 376.99t - \epsilon^{-26.667t}\right)U_{-1}(t)[\text{A}] \text{ and,} \qquad (6.40)$$

$$e(t) = 112,700\sin(376.99t + 175.954°)\,[\text{V}]. \qquad (6.41)$$

The time constant is $T = 1/26.667 = 37.5\,[\text{ms}]$.

This current will be maximum when $\cos 376.99t = -1$, that is, $376.99t = \pi$. So, $t = \frac{\pi}{2\pi.60} = 8.333\,[\text{ms}]$ and,

$$i_{max} = 497.00\left(-1 - \epsilon^{-26.667 \times 8.333.10^{-3}}\right) = -894.969\,[\text{A}].$$

$$i'(t) = 497\left(-376.99\sin 376.99t + 26.667\epsilon^{-26.667t}\right)$$
$$- 187,364.030\sin 376.99t + 13,253.499\epsilon^{-26.667t}[\text{A/s}].$$

So,

$$i'(0_+) = 13,253.499\,[\text{A/s}].$$

Also,

$$i'(0_+) = \frac{E}{L}\sin\theta = \frac{112,700}{0.6}\sin 175.954° = 13,253.022\,[\text{A/s}]. \text{ This value corre-}$$
sponds to a slope $\alpha = arctg\,13,253.022 = 90.00°$ in $i(t)$, for $t = 0_+$.

(c) The circuit built in ATPDraw for the requested numerical solution, in the case of the absence of a transitional regime component, is shown in Fig. 6.147.

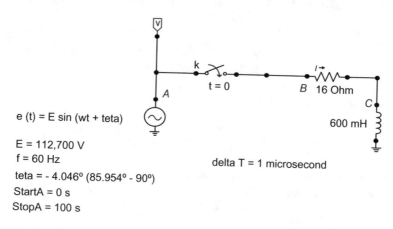

e (t) = E sin (wt + teta)

E = 112,700 V
f = 60 Hz
teta = - 4.046° (85.954° - 90°)
StartA = 0 s
StopA = 100 s

Fig. 6.147 Electric circuit of Example 6.33 for the numerical solution by the ATPDraw program

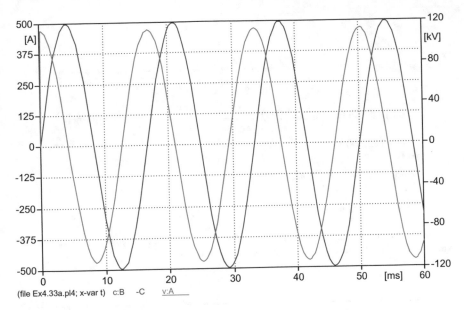

Fig. 6.148 Current $i(t)$, Source Voltage $e(t)$

Figure 6.148 shows the waveforms of the current $i(t)$ and the voltage of the source $e(t)$, in the case where there is no transient regime component in the current (best condition).

The ATPDraw circuit for the most unfavorable condition, when switch k is closed at $t = 0$, but with $\theta = 175.954°$ in the initial phase of the voltage source $e(t)$, is recorded in Fig. 6.149.

The current $i(t)$, with the worst transient regime, is shown in Fig. 6.150.

As the ATPDraw program operates with the source in cosine form, it is necessary to change the expression

$$e(t) = 112,700\sin(376.99t + 175.954°)$$

for its equivalent

$$e(t) = 112,700\cos(376.99t + 175.954° - 90°)$$
$$= 112,700\cos(376.99t + 85.954°)$$

Figure 6.150 suggests that $i'(0_+) < 0$, according to the geometric interpretation of the derivative concept, in a frontal inconsistency with the analytically obtained value of $+13,253.022\,[\text{A/s}]$, calculated from $i'(0_+) = \frac{E}{L}\sin\theta$. However, the result of successive zooms in the vicinity of $t = 0$, on the right, recorded in Fig. 6.151, shows that the current $i(t)$, for a short time (approximately $37\,[\text{ms}]$), is positive. Furthermore, Fig. 6.152 shows that:

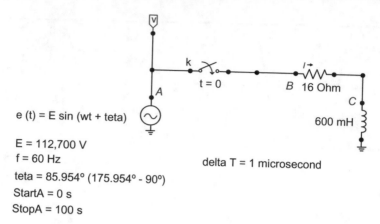

Fig. 6.149 Electric circuit of Example 6.33 for the numerical solution by the ATPDraw program, when switch *k* is closed at *t* = **0**, in the worst condition

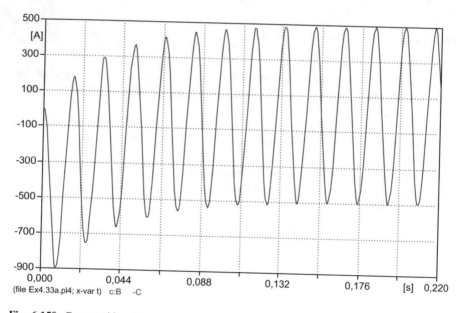

Fig. 6.150 Current $i(t)$, with the most intense transient phenomenon

$i = 6.5911\,[\text{mA}]$ for $t = 1\,[\mu s]$ so that $tg\alpha = 6.5911.10^{-3}/10^{-6} = 6,591.1$. That is, that $\alpha = 89.99\,[°]$. Therefore, there is no inconsistency.

Equations 6.38 and 6.39 indicate that when switch *k* is closed, in an instant arbitrarily considered to be $t = 0\,[\text{s}]$, generating no transient regime component in current $i(t)$, the current and voltage are worth, respectively, $0\,[\text{A}]$ and $112,419.121\,[\text{V}]$. It happens, however, that *the voltage source was already*

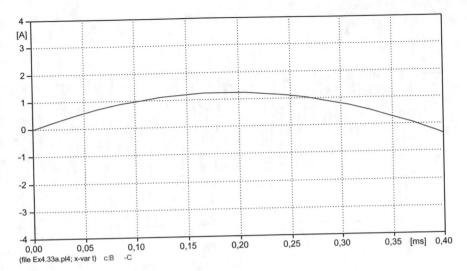

Fig. 6.151 Zoom at current $i(t)$, with the most intense transient phenomenon

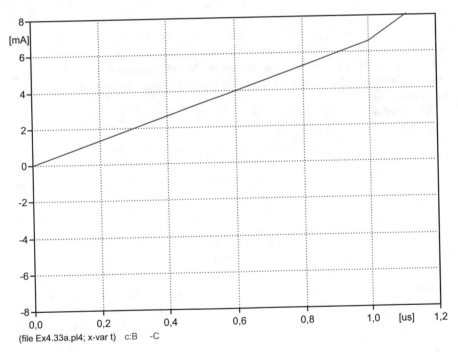

Fig. 6.152 Zoom at current $i(t)$, with the most intense transient phenomenon

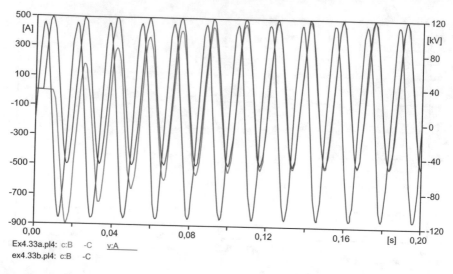

Ex4.33a.pl4: c:B   -C   v:A
ex4.33b.pl4: c:B   -C

Fig. 6.153 Current $i(t)$ for   switch k closing   at $t = 3.979$ ms, current $i(t)$ for switch k closing at $t = 8.146$ ms, voltage $e(t)$ of the inusoidal source

operating sinusoidally, from its own time $t = 0$ [s], *before the closing of switch k.* This means that the question can be asked in another way; namely: **at what moment, counting the time from the beginning of the operation of the voltage source, should the switch be closed so as not to produce a transient component in current $i(t)$?**

Considering,   then,   that   the   voltage   source   is $e(t) = E \sin \omega t = 112,700 \sin 376.99t$, that is

$112,419.121 = 112,700 \sin 85.954° = 112,700 \sin(1.5 \text{rad})$.   Then, $1.5 = 376.99t$ and $t = 3.979$ [ms].

For a similar reasoning, to obtain a maximum intensity transitional regime, according to Eq. (6.41), we have to

$112,700 \sin 175.954° = 7,951.813 = 112,700 \sin 3.071 = 112,700 \sin 376.99t$.   So. $3.071 = 376.99t$ and $t = 8.146$ [ms].

In summary: counting the time from the operation of the sinusoidal voltage source, the switch must be closed at time $t = 3.979$ [ms] so as not to produce a transient component in current $i(t)$. Or closed at time $t = 8.146$ [ms], to produce a transient regime of maximum intensity in current $i(t)$. Logically, the maximum values of the current, in both cases, that is, 497.00 [A] and -894.97 [A], now occur at $t = 4.167 + 3.979 = 8.146$ [ms] and $t = 8.333 + 8.146 = 16.479$ [ms]. The currents, with this new interpretation, together with the source voltage, are shown in Fig. 6.153. And, for a better view, zoom in Fig. 6.154.

The analytical expressions of Eqs. (6.38) and (6.40), *referred to the zero of the voltage source,* are obtained by replacing t, respectively, by $t - 3.979.10^{-3}$ and $t - 8.146.10^{-3}$. So,

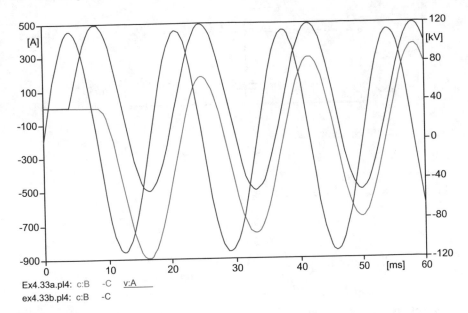

Ex4.33a.pl4: c:B   -C   v:A
ex4.33b.pl4: c:B   -C

Fig. 6.154 Zoom in Fig. 6.153

$$i(t) = 497.00 \sin[376.99(t - 3.979.10^{-3})]U_{-1}(t - 3.979.10^{-3})$$
$$= 497.00 \sin(376.99t - 1.500043)U_{-1}(t - 3.979.10^{-3})[A].$$

So,

$$i(t) = 497 \sin(376.99t - 85.95°)U_{-1}(t - 3.979.10^{-3})\,[A]. \qquad (6.42)$$

And,

$$i(t) = 497.00\left[\cos 376.99(t - 8.146.10^{-3}) - \epsilon^{-26.667(t-8.146.10^{-3})}\right]U_{-1}(t - 8.146.10^{-3})$$
$$= 497.00\left[\cos 376.99(t - 3.070961) - 1.243\epsilon^{-26.667t}\right]U_{-1}(t - 8.146.10^{-3})$$

Or,

$$i(t) = 497\left[\cos(376.99t - 175.95°) - 1.243\epsilon^{-26.667t}\right]U_{-1}(t - 8.146.10^{-3})[A] \qquad (6.43)$$

Equations 6.42 and 6.43 are the analytical expressions of the current curves plotted in Figs. 6.153 and 6.154.

Example 6.34 Find the analytical expression of the voltage $e_0(t)$ in the inductance L of the circuit shown in Fig. 6.155, for $t > 0$, considering that the source voltage is $e(t) = 1,000 \sin 377t.U_{-1}(t)\,[V]$, $R = 10\,[\Omega]$, $L = 10\,[mH]$, $C = 100\,[\mu F]$.

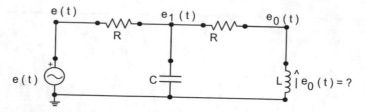

Fig. 6.155 Electric circuit of Example 6.34

Solution

The nodal equations of the circuit are

$$\frac{e_1 - e}{R} + C\frac{d}{dt}e_1 + \frac{e_1 - e_0}{R} = 0$$

$$\frac{e_0 - e_1}{R} + \frac{1}{L}\int e_0(t)dt = 0$$

Simply put, it comes that:

$$(RCD + 2)e_1 - e_0 = e$$

$$-LDe_1 + (LD + R)e_0 = 0$$

Solving the system for $e_0(t)$, results

$$e_0''(t) + \left(\frac{R}{L} + \frac{1}{RC}\right)e_0'(t) + \frac{2}{LC}e_0(t) = \frac{1}{RC}e'(t)$$

By entering the numerical values, the equation becomes

$$e_0''(t) + 2.10^3 e_0'(t) + 2.10^6 e_0(t) = 377.10^6 \cos 377t.U_{-1}(t); e_0(0_-) = e_0'(0_-) = 0.$$

The characteristic roots, obtained from $F(D) = D^2 + 2.10^3 D + 2.10^6 = 0$, are: $r_1 = -1,000 + j1,000$ and $r_2 = -1,000 - j1,000$. Therefore, the homogeneous solution is:

$$e_{0_H}(t) = \epsilon^{-1,000t}(k_1 \cos 1,000t + k_2 \sin 1,000t)$$

The particular solution is obtained using the phasor method, with D replaced by $j377$:

$$(-142,129 + j754,000 + 2,000,000)\dot{E}_{0P} = 377 \times 10^6 \angle 0^{\circ}$$

So,

$\dot{E}_{0P} = 188.03\angle - 22.1^{\circ}$. So,

$$e_{0P}(t) = 188.03\cos(377t - 22.1^{\circ})$$

Therefore, the general solution is:

$$e_0(t) = 188.03\cos(377t - 22.1^{\circ}) + \epsilon^{-1,000t}(k_1 \cos 1,000t + k_2 \sin 1,000t)$$
$$e_0'(t) = -70,887.31\sin(377t - 22.1^{\circ}) - 1,000\epsilon^{-1,000t}(k_1 \cos 1,000t + k_2 \sin 1,000t)$$
$$+ \epsilon^{-1,000t}(-1,000k_1 \sin 1,000t + 1,000k_2 \cos 1,000t)$$

In $t = 0_+$,

$$e_0(0_+) = 174.22 + k_1$$
$$e_0'(0_+) = 26,669.53 - 1,000k_1 + 1,000k_2$$

On the other hand, the differential equation of $e_0(t)$ can be rewritten as:

$$e_0''(t) + 2.10^3 e_0'(t) + 2.10^6 e_0(t) = 377.10^6.U_{-1}(t) + \ldots$$

Therefore, only $e_0''(t)$ contains a step. Thus,

$$e_0(0_+) = 0 = 174.22 + k_1$$
$$e_0'(0_+) = 0 = 26,669.53 - 1,000k_1 + 1,000k_2$$

So,

$$k_1 = -174.22 \text{ and } k_2 = -200.89$$

Finally,

$$e_0(t) = -\epsilon^{-1,000t}(174.22\cos 1,000t + 200.89\sin 1,000t) + 188.03\cos(377t - 22.1^{\circ}) \, [\text{V}], t \geq 0 \, [\text{s}]$$

Or,

$$e_0(t) = -265.91\epsilon^{-1,000t}\cos(1,000t - 49.1^{\circ}) + 188.03\cos(377t - 22.1^{\circ}) \, [\text{V}], t \geq 0 \, [\text{s}]$$

Example 6.35 The circuit shown in Fig. 6.156 was operating steadily when switch k was closed at time $t = 0$. Knowing that at the moment of switching the voltage of the source was maximum, determine the expression of the current $i_k(t)$, for $t > 0$. Consider that $\omega_0 = 1/\sqrt{LC} = 2\omega$.

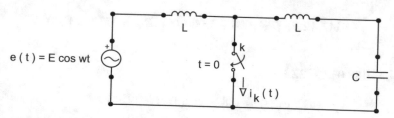

Fig. 6.156 Electric circuit of Example 6.35

Solution
AC Steady-state solution for $<0(-\infty<t\leq0_-)$:

The circuit in the single frequency domain, for the sinusoidal steady state solution, is shown in Fig. 6.157.

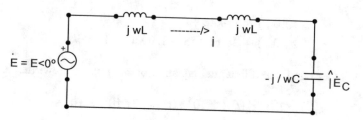

Fig. 6.157 Electric circuit of Example 6.35 for AC steady-state solution

The phasor current is $\dot{I}=\frac{\dot{E}}{Z}=\frac{E\angle0°}{j(2\omega L-1/\omega C)}=\frac{\frac{E\omega}{L}\angle0°}{j\left(2\omega^2-\omega_0^2\right)}=\frac{\frac{E\omega}{L}\angle0°}{-j2\omega^2}=\frac{E}{2\omega L}\angle90°$

So, the sinusoidal steady state current is

$$i(t)=\frac{E}{2\omega L}\cos(\omega t+90°)=-\frac{E}{2\omega L}\sin\omega t.$$

On the other hand,

$$\dot{E}_C=-j\frac{1}{\omega C}\cdot\dot{I}=\frac{1}{\omega C}\angle-90°\cdot\frac{E}{2\omega L}\angle90°=\frac{E}{2\omega^2LC}\angle0°=2E\angle0°$$

Therefore,

$$e_C(t)=2E\cos\omega t.$$

As at the moment of closing of switch k the voltage source $e(t)$ is going through its maximum value, so $E\cos\omega t=E$. Therefore, $\omega t=0$, that is, $t=0$; zero this is the same closing time of switch k. Consequently, $i(0_-)=0$ and $e_C(0_-)=2E$.

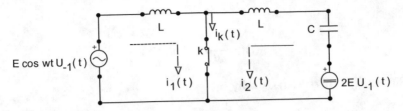

Fig. 6.158 Equivalent electric circuit of Example 6.35

Solution for $t > 0 (0_+ \leq t < \infty)$:

Since the initial current in the inductances is zero, there is neither the Thévenin nor the Norton equivalent for them. Then, replacing the capacitance initially energized with its Thévenin equivalent, results in the circuit shown in Fig. 6.158.

For the left mesh,

$E \cos \omega t . U_{-1}(t) = L \frac{d}{dt} i_1(t), \text{ with } i_1(0_-) = 0 = i_1(0_+).$ So,

$$i_1(t) = \frac{E}{L} \int_{0_+}^{t} \cos \omega \lambda d\lambda = \frac{E \sin \omega \lambda}{L} \frac{1}{\omega} \Big|_{0_+}^{t} = \frac{E}{\omega L} \sin \omega t, t \geq 0.$$

For the right mesh,

$$L i_2'(t) + \frac{1}{C} \int_{0_+}^{t} i_2(\lambda) d\lambda = 2E U_{-1}(t)$$

$$i_2''(t) + 4\omega^2 i_2(t) = \frac{2E}{L} U_0(t), \text{ with } i_2(0_-) = i_2'(0_-) = 0.$$

For $t > 0$,

$$(D^2 + 4\omega^2) i_2(t) = 0.$$

The characteristic roots are $r_1 = j2\omega$ and $r_2 = -j2\omega$. So,

$$i_2(t) = k_1 \cos 2\omega t + k_2 \sin 2\omega t$$
$$i_2'(t) = -2\omega k_1 \sin 2\omega t + 2\omega k_2 \cos 2\omega t.$$

From the differential equation of $i_2(t)$ and the preceding expressions, it follows that:

$$i_2(0_+) = 0 = k_1$$
$$i_2'(0_+) = \frac{2E}{L} = 2\omega k_2; k_2 = \frac{E}{\omega L}$$

So,

$$i_2(t) = \frac{E}{\omega L}\sin 2\omega t, t \geq 0.$$

Since $i_k(t) = i_1(t) + i_2(t)$, finally

$$i_k(t) = \frac{E}{\omega L}(\sin \omega t + \sin 2\omega t).U_{-1}(t).$$

Note that since the circuit is not dissipative, that is, without resistances, the solution of the homogeneous equation in $i_2(t)$, which is the free answer, does not have a transitory character. And that the current in the switch, $i_k(t)$, is periodic, because $\omega_2/\omega_1 = 2\omega/\omega = 2/1$, a rational number.

Example 6.36 The circuit shown in Fig. 6.159 operated in a steady state when switch k was opened at $t = 0$.

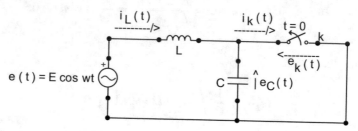

Fig. 6.159 Electric circuit of Example 6.36

(a) If at the time of switching $i_k = 0$, with $i'_k > 0$, determine the voltage $ek(t) = 0$ at the open switch terminals, for $t > 0$. Also determine the current in the left loop, with the switch open.
(b) Perform the numerical solution using ATPDraw.

Data: $E = 1,000\,[\text{V}], f = 60\,[\text{Hz}], L = 10\,[\text{mH}]$ e $C = 312.7\,[\mu\text{F}]$.

Solution

(a) For $t < 0$, in steady state,

$$e(t) = E\cos \omega t$$

$$e_C(t) = 0$$

$$i_k(t) = i_L(t) = \frac{E}{\omega L}\cos(\omega t - 90°) = \frac{E}{\omega L}\sin \omega t.$$

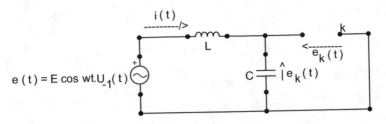

Fig. 6.160 Equivalent electric circuit of Example 6.36

This current $i_k(t)$ meets the condition of being $i_k = 0$, with $i'_k > 0$, at time $t = 0$ that is, at the exact moment of opening the switch k. In other words: **the time $t = 0$ of the opening of the switch k coincides with the time $t = 0$ of the voltage source** $e(t)$, therefore, there is no need to consider two different times $t = 0$.

So,

$$i_L(0_-) = 0 \ and \ e_C(0_-) = 0.$$

This means that, for $t > 0$, there is no Thévenin or Norton equivalent, neither for inductance nor for capacitance. So that the corresponding circuit for $t > 0$ is the one shown in Fig. 6.160.

The left mesh equation is

$$L\frac{d}{dt}i(t) + \frac{1}{C}\int_{0_+}^{t} i(\lambda)d\lambda = e(t).$$

Considering that $i(t) = C\frac{d}{dt}e_k(t)$, $i'(t) = C\frac{d^2}{dt^2}e_k(t)$, the preceding equation can be rewritten as

$$LCe''_k(t) + \frac{1}{C}\int_{0_+}^{t} C\frac{d}{d\lambda}e_k(\lambda)d\lambda = e(t)$$

$$LCe''_k(t) + \int_{0_+}^{t} de_k(\lambda) = e(t)$$

$$LCe''_k(t) + e_k(t) = e(t)$$

Defining $\omega_0 = 1/\sqrt{LC}$, $\omega_0^2 = 1/LC$, results that

$$(D^2 + \omega_0^2)e_k(t) = \omega_0^2 E \cos \omega t.U_{-1}(t) = \omega_0^2 E.U_{-1}(t) + \ldots; e_k(0_-) = e'_k(0_-) = 0.$$

For $t > 0$,

$$\left(D^2 + \omega_0^2\right)e_k(t) = \omega_0^2 E \cos \omega t.$$

The homogeneous solution is:

$$e_{k_H}(t) = k_1 \cos \omega_0 t + k_2 \sin \omega_0 t$$

The particular integral comes from

$$\left(-\omega^2 + \omega_0^2\right)\dot{E}_{k_P} = \omega_0^2 E \angle 0°$$

$$\dot{E}_{k_P} = \frac{\omega_0^2 E \angle 0°}{\omega_0^2 - \omega^2}$$

$$e_{k_P}(t) = \frac{\omega_0^2 E}{\omega_0^2 - \omega^2} \cos \omega t.$$

So, the general solution is

$$e_k(t) = \frac{\omega_0^2 E}{\omega_0^2 - \omega^2} \cos \omega t + k_1 \cos \omega_0 t + k_2 \sin \omega_0 t$$

$$e_k'(t) = \frac{-\omega \omega_0^2 E}{\omega_0^2 - \omega^2} \sin \omega t - \omega_0 k_1 \sin \omega_0 t + \omega_0 k_2 \cos \omega_0 t.$$

From the differential equation of $e_k(t)$, it follows that:

$$e_k(0_+) = 0 \text{ and } e_k'(0_+) = 0.$$

Then,

$$\frac{\omega_0^2 E}{\omega_0^2 - \omega^2} + k_1 = 0$$

$$\omega_0 k_2 = 0$$

So, $k_1 = -\frac{\omega_0^2 E}{\omega_0^2 - \omega^2}$ and $k_2 = 0$.

Finally,

$$e_k(t) = \frac{\omega_0^2 E}{\omega_0^2 - \omega^2} (\cos \omega t - \cos \omega_0 t).U_{-1}(t).$$

As $i(t) = C\frac{d}{dt}e_k(t)$,

$$i(t) = \frac{C\omega_0^2 E}{\omega_0^2 - \omega^2}(-\omega \sin \omega t + \omega_0 \sin \omega_0 t), \; for \; t \geq 0.$$

Substituting the numerical values, comes that

$$e_k(t) = 179.99(\cos 376.99t - \cos 565.50t).U_{-1}(t)[V]. \tag{6.44}$$

$$i(t) = -21.22 \sin 376.99t + 31.83 \sin 565.50t[A], \; for \; t \geq 0 \tag{6.45}$$

The periods are: $T = 1/60 = 16.67[ms]$ and $T_0 = 2\pi/\omega_0 = 11.11[ms]$.

The steady-state current $i_k(t)$, passing through the switch before $t = 0$, has a maximum value of $E/\omega L = 26.53[A]$.

(b) The circuit for the numerical solution is shown in Fig. 6.161.

Figure 6.162 shows the curves of the source voltage, $e(t)$, and the voltage at the open switch terminals, $e_k(t)$.

In Fig. 6.163, the current in the left loop, $i(t)$.

A detailed observation of Figs. 6.162 and 6.163 allows us to conclude that the switch was not opened at $t = 0$, but at $t = T/2 = 8.33[ms]$. This is because the ATPDraw program is pre-programmed to only open switch k when the current through it passes through zero for the first time, after $t = 0$, due to *Component* : *TSWITCH*, *Imar* = 0. Taking, however, *Imar* = 30, which is greater than 26.53[A], the maximum value of the steady-state current that passes through the switch, so it is opened exactly at $t = 0$, as shown in Figs. 6.164 and 6.165.

Logically, the analytical expressions obtained in Eqs. (6.44) and (6.45) refer to Figs. 6.164 and 6.165. For example, for $t = 0.02[s]$, Eq. (6.45) provides that $i(0.02) = -50.451[A]$. And, from Fig. 6.165, with the aid of the cursor, it is determined that $i(0.02) = -50.447[A]$. Also, from Eq. (6.44), $e_k(4.305.10^{-3}) = 127.45[V]$, while from Fig. 6.164, $e_k(4.305.10^{-3}) = 127.46[V]$.

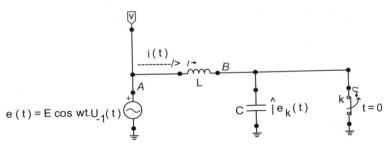

Fig. 6.161 Electric circuit of Example 6.36 for the numerical solution by the ATPDraw program

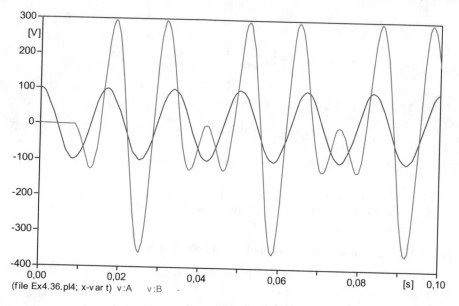

Fig. 6.162 Source voltage $e(t)$ and Voltage in switch $e_k(t)$

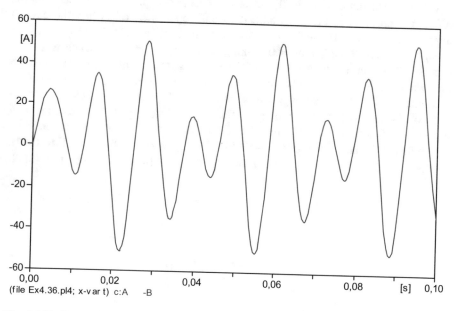

Fig. 6.163 Current in the left loop $i(t)$

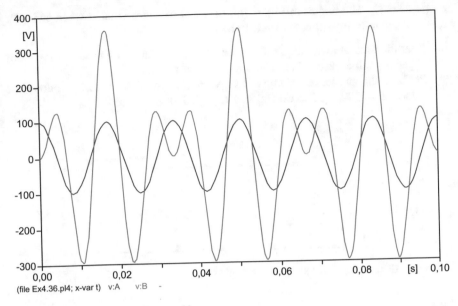

Fig. 6.164 Voltage $e(t)$, Voltage $e_k(t)$

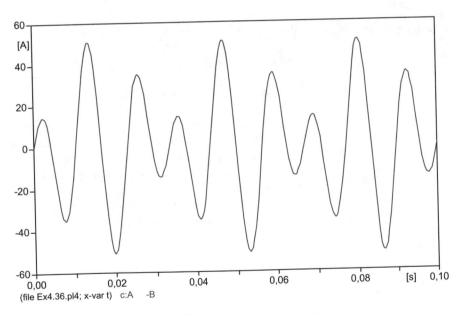

Fig. 6.165 Current in the left loop $i(t)$

Example 6.37   The circuit in Fig. 6.166 operated in a steady state when switch k was opened at an instant considered to be $t = 0$.

(a) If, at the time of switching, the source voltage was at its minimum (maximum negative) value, determine $e_1(t)$ and $e_2(t)$ for $t > 0$.
(b) Perform the numerical solution by ATPDraw for the following numerical values: $E = 200[V]$, $f = 60[Hz]$ and $C = 1,000[\mu F]$.

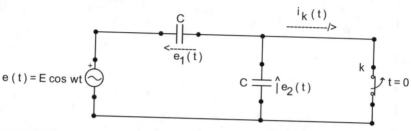

Fig. 6.166   Electric circuit of Example 6.37

Solution

(a) In the AC steady state, with the switch k closed, that is, for $-\infty < t \leq 0_-$, we have that:

$e_2(t) = 0$ and, therefore, that $e_2(0_-) = 0$.

$$e_1(t) = e(t) = E\cos\omega t.$$

$$i_k(t) = \omega CE\cos(\omega t + 90°) = -\omega CE\sin\omega t.$$

In the present case, the time $t = 0$ of the voltage source $e(t)$, when it is maximum, does not coincide with the time $t = 0$ of the opening of switch k, as this is opened when the voltage of the source is at a minimum. Therefore, it is necessary to correct the analytical expression of the voltage source, resulting, then, that $e(t) = -E\cos\omega t$, *in relation to the time $t = 0$ of the opening of switch k*. Thus, $e_1(0_-) = -E$ and $i_k(0_-) = 0$. Therefore, for $t > 0$, the circuit shown in Fig. 6.167 results, in which the polarity of the voltage sources was inverted, so as not to use negative values in their analytical expressions.

The equation for node $e_2(t)$ is

$$CDe_2(t) + CD[e_2(t) - EU_{-1}(t) + e(t)] = 0$$

$$CDe_2(t) + CD[e_2(t) - EU_{-1}(t) + E\cos\omega t.U_{-1}(t)] = 0$$

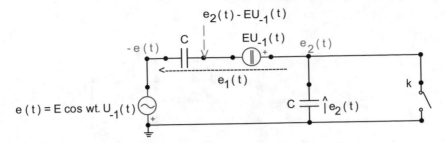

Fig. 6.167 Electric circuit of Example 6.37 for $t > 0$ solution

$$2CDe_2(t) - CEU_0(t) - \omega CE \sin \omega t.U_{-1}(t) + CEU_0(t) = 0$$

Therefore,

$$\frac{d}{dt}e_2(t) = \frac{\omega E}{2} \sin \omega t.U_{-1}(t)$$

$$e_2(t) = \frac{\omega E}{2} \int_{0_+}^{t} \sin \omega \lambda d\lambda = \frac{\omega E}{2} \cdot \frac{-\cos \omega \lambda}{\omega}\bigg|_{0_+}^{t}$$

$$e_2(t) = \frac{E}{2}(1 - \cos \omega t).U_{-1}(t).$$

The voltage law applied to the circuit mesh in Fig. 6.167 provides the equation

$$e_2(t) + e_1(t) + e(t) = 0.$$

$$e_1(t) = -e(t) - e_2(t) = -E \cos \omega t - \frac{E}{2} + \frac{E}{2} \cos \omega t$$

$$e_1(t) = -\frac{E}{2}(1 + \cos \omega t), \text{ for } t \geq 0.$$

Obviously, the expressions of $e_1(t)$ and $e_2(t)$ are referred to at $t = 0$ of the opening of the switch k. To refer to the $t = 0$ original time of the voltage source, it is necessary to introduce a delay of $\pi[\text{rad}]$, (180°), in the cosine ωt arguments; which is equivalent to replacing t with $(t - \pi/\omega)[s]$. So,

$$e_1(t) = -\frac{E}{2}[1 + \cos(\omega t - \pi)], \text{ for } t - \pi/\omega \geq 0.$$

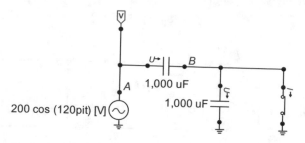

Fig. 6.168 Electric circuit of Example 6.37 for the numerical solution by the ATPDraw program

Or,

$$e_1(t) = \frac{E}{2}(\cos \omega t - 1), \; \text{for } t \geq \pi/\omega. \tag{6.46}$$

$$e_2(t) = \frac{E}{2}[1 - \cos(\omega t - \pi)].U_{-1}(t - \pi/\omega).$$

$$e_2(t) = \frac{E}{2}(1 + \cos \omega t).U_{-1}(t - \pi/\omega). \tag{6.47}$$

For the numerical values given,

$$e_1(t) = 100(\cos 376.99t - 1)[\text{V}], \; \text{for } t \geq 8.33 \, [\text{ms}]. \tag{6.48}$$

$$e_2(t) = 100(1 + \cos 376.99t).U_{-1}(t - 0.00833) \, [\text{V}]. \tag{6.49}$$

The origin correction that was made is only necessary so that it is possible to compare the values plotted in item b) with Eqs. (6.48) and (6.49). Logically, with the origin correction, it remains evident that the switch k is then opened at $t = \pi/\omega$, when the sinusoidal voltage source is passing through the minimum value: $E = -200[\text{V}]$.

(b) The circuit for the numerical solution is shown in Fig. 6.168, and the results, in Fig. 6.169.

Note that switch k opened at exactly time $t = 8.33[\text{ms}]$, when the source voltage $e(t)$ passed the minimum value (-200) and the current through the switch, $i_k(t)$, passed by zero. And that the curves of $e_1(t)$ and $e_2(t)$ are consistent with Eqs. (6.48) and (6.49), that is, consistent with the respective analytical expressions.

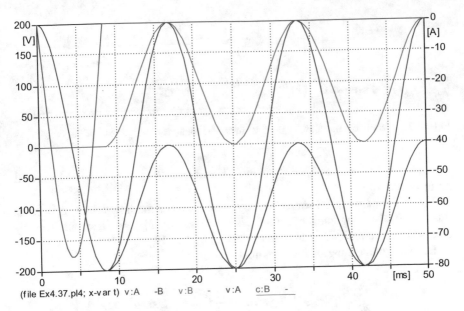

(file Ex4.37.pl4; x-v ar t) v :A   -B   v :B   -   v :A   c:B   -

Fig. 6.169 $e_1(t)$, $e_2(t)$, $e(t)$, $i_k(t)$

Example 6.38 (a) The circuit shown in Fig. 6.170 was operating steadily when switch k was opened at time $t = 0$. If the current $i_k(t)$ circulating through the switch was maximum at the moment of switching, find the expression of the voltage $e_k(t)$ in its terminals for $t > 0$. Correct the expression of $e_k(t)$ so that it refers to the origin of the voltage source $e(t)$. (b) Perform the numerical solution using the ATPDraw program for the following numerical values: $E = 400[\text{V}]$, $f = 60[\text{Hz}]$, $L = 6[\text{mH}]$ and $C = 70[\mu\text{F}]$.

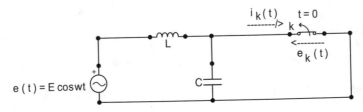

Fig. 6.170 Electric circuit of Example 6.38

Solution

(a) For $t < 0$, with switch k closed, the steady-state current flowing through the switch is

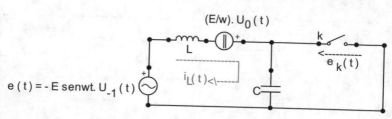

Fig. 6.171 Electric circuit of Example 6.38 for $t > 0$ solution

$i_k(t) = \frac{E}{\omega L} \cos(\omega t - 90^\circ) = \frac{E}{\omega L} \sin \omega t$. Therefore, $i_k(0_-) = \frac{E}{\omega L} = i_L(0_-) = $ maximum current in switch k.

For this condition, the source voltage, corrected for the new origin, that is, the time $t = 0$ of the opening of switch, is expressed by $e(t) = -E \sin \omega t$. And the corresponding circuit, for $t > 0$, is the one shown in Fig. 6.171.

Equating the left mesh, it comes that

$$L \frac{d}{dt} i_L(t) + \frac{1}{C} \int_{0_+}^{t} i_L(\lambda) d\lambda = \frac{E}{\omega} U_0(t) - E \sin \omega t . U_{-1}(t)$$

Where,

$$i_L''(t) + \omega_0^2 i_L(t) = \frac{E}{\omega L} U_1(t) - \frac{\omega E}{L} \cos \omega t . U_{-1}(t) = \frac{E}{\omega L} U_1(t) - \frac{\omega E}{L} U_{-1}(t) + \ldots,$$

with $i_L(0_-) = i_L'(0_-) = 0$ and $\omega_0^2 = 1/LC$.

For $t > 0$,

$$\left(D^2 + \omega_0^2\right) i_L(t) = -\frac{\omega E}{L} \cos \omega t$$

So, the homogeneous solution is

$$i_{L_H}(t) = k_1 \cos \omega_0 t + k_2 \sin \omega_0 t.$$

For the particular integral, using the phasor method, we have that:

$$\left(-\omega^2 + \omega_0^2\right) \dot{I}_{L_P} = -\frac{\omega E}{L} \angle 0^\circ \text{ and } \dot{I}_{L_P} = \frac{-\frac{\omega E}{L} \angle 0^\circ}{\omega_0^2 - \omega^2}$$

$$i_{L_P}(t) = \frac{\omega E}{L\left(\omega^2 - \omega_0^2\right)} \cos \omega t.$$

And the general solution is

$$i_L(t) = \frac{\omega E}{L(\omega^2 - \omega_0^2)} \cos \omega t + k_1 \cos \omega_0 t + k_2 \sin \omega_0 t.$$

$$i'_L(t) = -\frac{\omega^2 E}{L(\omega^2 - \omega_0^2)} \sin \omega t - k_1 \omega_0 \sin \omega_0 t + k_2 \omega_0 \cos \omega_0 t.$$

Examining the differential equation we can write that:

$$i_L(t) \doteq C_0 U_{-1}(t).$$

$$i'_L(t) \doteq C_0 U_0(t) + C_1 U_{-1}(t).$$

$$i''_L(t) \doteq C_0 U_1(t) + C_1 U_0(t) + C_2 U_{-1}(t).$$

Substituting in the differential equation, it turns out that

$$C_0 U_1(t) + C_1 U_0(t) + (C_2 + \omega_0^2 C_0) U_{-1}(t) \doteq \frac{E}{\omega L} U_1(t) + 0.U_0(t) - \frac{\omega E}{L} U_{-1}(t).$$

So,

$$C_0 = \frac{E}{\omega L}; C_1 = 0.$$

Therefore,

$$i_L(0_+) = \frac{E}{\omega L} = \frac{\omega E}{L(\omega^2 - \omega_0^2)} + k_1$$

$$i'_L(0_+) = 0 = k_2 \omega_0.$$

Then,

$$k_1 = \frac{-\omega_0^2 E}{\omega L(\omega^2 - \omega_0^2)}; k_2 = 0.$$

$$i_L(t) = \frac{\omega E}{L(\omega^2 - \omega_0^2)} \cos \omega t - \frac{\omega_0^2 E}{\omega L(\omega^2 - \omega_0^2)} \cos \omega_0 t$$

or,

$$i_L(t) = \frac{E}{\omega L(\omega^2 - \omega_0^2)} (\omega^2 \cos \omega t - \omega_0^2 \cos \omega_0 t), \text{ for } t \geq 0. \tag{6.50}$$

Therefore, the voltage at the terminals of the open switch k is

$$e_k(t) = \frac{1}{C} \int_{0_+}^{t} i_L(\lambda) d\lambda = \frac{E}{\omega L C (\omega^2 - \omega_0^2)} \int_{0_+}^{t} (\omega^2 \cos \omega\lambda - \omega_0^2 \cos \omega_0\lambda) d\lambda$$

$$= \frac{\omega_0^2 E}{\omega (\omega^2 - \omega_0^2)} (\omega \sin \omega\lambda - \omega_0 \sin \omega_0\lambda) \Big|_{0_+}^{t} = \frac{\omega_0^2 E}{\omega (\omega^2 - \omega_0^2)} (\omega \sin \omega t - \omega_0 \sin \omega_0 t) \cdot U_{-1}(t)$$

Finally,

$$e_k(t) = \frac{\omega_0^2 E}{(\omega^2 - \omega_0^2)} \left(\sin \omega t - \frac{\omega_0}{\omega} \sin \omega_0 t \right) . U_{-1}(t). \tag{6.51}$$

In Eqs. (6.50) and (6.51), $t = 0$ refers to the opening time of switch k. To refer to the original time $t = 0$, from the voltage source $e(t)$, it is necessary to introduce a delay of $\pi/2[\text{rad}]$, (90°), in the arguments in ωt of the trigonometric functions, that is, replace $\omega t[\text{rad}]$ by $(\omega t - \pi/2)[\text{rad}]$. This is equivalent to replacing $t[\text{s}]$ with $(t - \pi/2\omega)[\text{s}]$. (Remember that ATPDraw works with the variable t on the horizontal axis and not with ωt). Thus, the expressions of current $i_L(t)$ and voltage $e_k(t)$ become:

$$i_L(t) = \frac{E}{\omega L (\omega^2 - \omega_0^2)} [\omega^2 \cos \omega(t - \pi/2\omega) - \omega_0^2 \cos \omega_0(t - \pi/2\omega)]; (t - \pi/2\omega) \geq 0.$$

Or,

$$i_L(t) = \frac{E}{\omega L (\omega^2 - \omega_0^2)} [\omega^2 \sin \omega t - \omega_0^2 \cos \omega_0(t - \pi/2\omega)], \text{ for } t \geq \pi/2\omega. \tag{6.52}$$

$$e_k(t) = \frac{\omega_0^2 E}{\omega^2 - \omega_0^2} \left[\sin \omega(t - \pi/2\omega) - \frac{\omega_0}{\omega} \sin \omega_0(t - \pi/2\omega) \right] . U_{-1}(t - \pi/2\omega)$$

Or,

$$e_k(t) = \frac{\omega_0^2 E}{\omega^2 - \omega_0^2} \left[-\cos \omega t - \frac{\omega_0}{\omega} \sin \omega_0(t - \pi/2\omega) \right] . U_{-1}(t - \pi/2\omega). \tag{6.53}$$

Obviously, now, Eqs. (6.52) and (6.53) relate to the switching at $t = \pi/2\omega[\text{s}]$, when the current $i_k(t)$ is passing its maximum value: $E/\omega L$.

Entering the numerical values, the Eqs. (6.52) and (6.53) become:

$$i_L(t) = -11.226 \sin 376.99t + 188.065 \sin[1,543.03(t - 4.167.10^{-3})] [\text{A}] \text{ for } t \geq 4.167 \text{ [ms]}. \tag{6.54}$$

$$e_k(t) = (425.392 \cos 376.99t + 1,741.142 \sin[1,543.03(t - 4.167.10^{-3})]) [\text{V}], \text{ for } t \geq 4.167 \text{ [ms]}. \tag{6.55}$$

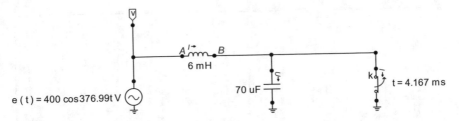

Fig. 6.172 Electric circuit of Example 6.38 for the numerical solution by the ATPDraw program

(b) The circuit for the numerical solution is shown in Fig. 6.172.

Na Figura 6.173 estão plotadas as curvas da tensão da fonte, $e(t)$, e da tensão nos terminais da chave, $e_k(t)$. E na Figura 6.174, as curvas de $i_k(t)$ and $i_L(t)$.

The curves of $i_L(t)$ and $e_k(t)$, recorded in the previous figures, are in accordance with Eqs. (6.54) and (6.55). For example, for $t = 10[\text{ms}]$ the Eq. (6.54) provides that the current i_L is worth $-164.79[\text{A}]$. On the other hand, with the cursor, we obtain, from Fig. 6.174, that $i_L = -164.74[\text{A}]$. Also, from Eq. (6.55), for $t = 15[\text{ms}]$ we have that $e_k = -1,128.15[\text{V}]$. And, from Fig. 6.173, we have that $e_k = --1,127.50[\text{V}]$.

Since this is not a book of electromagnetic transients in Power Systems, obviously, all the examples presented so far, without exception, refer to Electrical Circuits. Were it a book on power systems, the examples would be like that of the next problem.

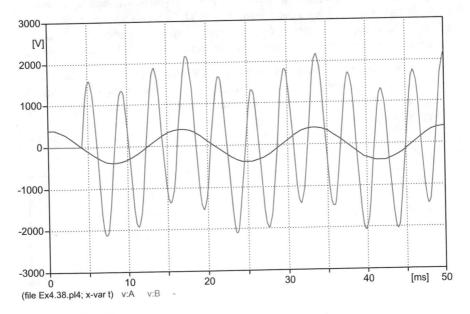

Fig. 6.173 $e(t)$, $e_k(t)$

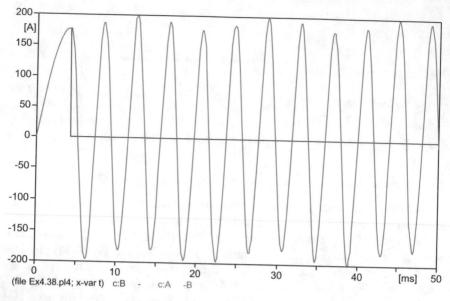

Fig. 6.174 $i_k(t)$, $i_L(t)$

Example 6.39 An industrial electrical system is supplied by a 138 [kV] three-phase network. As the set is symmetrical and balanced, and also aiming at simplifications, only a single-phase analysis is made. The single-phase equivalent supply system has a resistance of 1 [Ω] in series with an inductance of 10 [mH]. The equivalent industry load can be represented by a resistance of 200 [Ω] in series with an inductance of 1,200 [mH]. A capacitor bank for power factor correction purposes is used in the industry, as shown in the single-line diagram shown in Fig. 6.175. It is requested:

(a) Supply current and load energization, $i(t)$, with indication of initial value, maximum asymmetric value and instant of occurrence, maximum symmetrical value and value in the instant immediately before the closing of switch k_2, that is, $i(250.10^{-3}_{-})$.

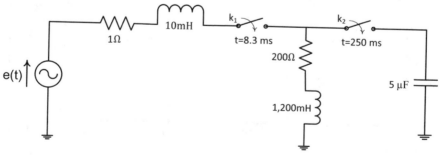

Fig. 6.175 Electric circuit of Example 6.39

(b) Voltage in the resistance, $e_R(t)$, and in the load inductance, $e_L(t)$, as well as the voltage in the industrial load itself, $e_A(t)$, explaining initial values, maximum asymmetric values—and occurrence moments, maximum symmetrical values and values in the instant immediately before the closing of switch k_2, that is, before the energization of the capacitor bank.

(c) Load power factor before correction: $fp_1 = \cos \varphi_1$.

(d) Instantaneous power in the resistance, $p_R(t)$, and in the load inductance, $p_L(t)$, with indication of the initial values, maximum asymmetric values—and instant of occurrence, and maximum symmetrical values.

(e) Power triangle in the load, before energizing the capacitor bank.

(f) Supply current of the system, $i(t)$, after insertion of the capacitor bank, highlighting its "initial value" at $t = 250[\text{ms}]$, maximum asymmetric value—with instant of occurrence, and maximum value of AC steady-state, that is, maximum symmetrical value.

(g) Voltage, $e_C(t) = e_A(t)$, and energizing current of the capacitor bank, $i_C(t)$, with its "initial" values, maximum asymmetric values and instantaneous occurrences, as well as their AC steady-state values .

(h) Current that supplies the industrial load, $i_F(t)$, after energizing the capacitor bank, with the exposure of the "initial" value at $t = 250[\text{ms}]$, the maximum asymmetric value, with the time of occurrence, and the AC steady-state value.

(i) New power factor: $fp_2 = \cos \varphi_2$.

(j) Instantaneous power curve in the load resistance and values of the elements of the new power triangle

(k) New power triangle.

Considering the three-phase voltage source connected to the star, then the effective voltage in the phase is $138/\sqrt{3}$. And the maximum voltage in the phase is $138\sqrt{2}/\sqrt{3}$. Therefore,

$$e(t) = E_{\max} \sin \omega t = \frac{138}{\sqrt{3}} . \sqrt{2} . \cos\left(120\pi t - \frac{\pi}{2}\right) = 112,677 \cos(377 - 90°)[\text{kV}].$$

Solution

The corresponding ATPDraw circuit is shown in Fig. 6.176.

1. **Switch k_1 closed.**

   (a) The circuit time constant is

   $$T = 1,210.10^{-3}/201 = 6.02 \ [\text{ms}]; 5T = 30.1 \ [\text{ms}]. \text{ So,}$$

   $$t_{\max} \gg 8.3 + 30.1 = 38.4 \ [\text{ms}] = 0.0384 \ [\text{s}].$$

   The oscillation period of the source is $T_F = 1/f = 1/60 = 0.01667 \ [\text{s}] \cong 16.7 \ [\text{ms}]$.

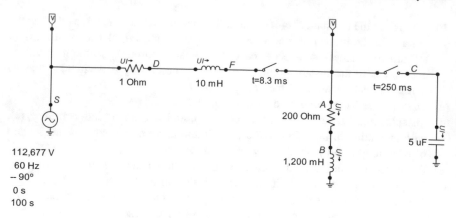

Fig. 6.176 Electric circuit of Example 6.39 for the numerical solution by the ATPDraw program

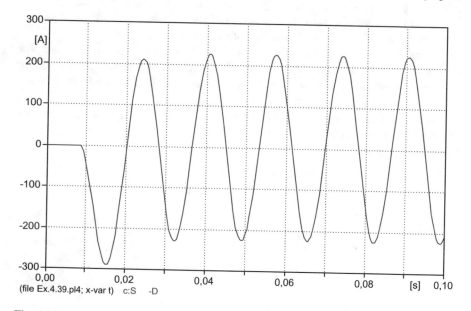

Fig. 6.177 Load energizing current

The energizing current of the industrial load is shown in Fig. 6.177 for $t_{max} = 100\,[\text{ms}] = 0.1\,[\text{s}]$.

The initial value of the charge energizing current is $i(8.3.10_+^{-3}) = 0\,[\text{A}]$, since $i(8.3010.10^{-3}) = i(8.3.10^{-3} + \Delta t) = 5.6749.10^{-4} = 0.0006[\text{A}]$. On the other hand, the maximum asymmetric current value is $-290.01[\text{A}]$, occurring at time $t = 15.225[\text{ms}] = 0.015225[\text{s}]$. These results were obtained with the cursor over the zoom applied at the beginning of Fig. 6.177, which is recorded in Fig. 6.178.

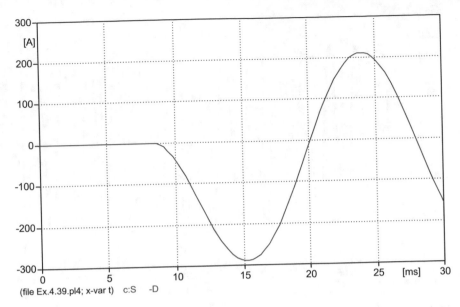

(file Ex.4.39.pl4; x-var t)   c:S   -D

Fig. 6.178 Zoom at the start of the load energizing current

Figure 6.179 again shows the energizing current of the industrial load, but now for $t_{max} = 250[\text{ms}] = 0.25[\text{s}]$.

Figure 6.180 shows a zoom at the end of Fig. 6.179. It can be obtained, with the aid of the cursor, that the maximum symmetrical value of the load energization current, that is, its maximum steady state value is $226.04[\text{A}]$ and that $i(250.10^{-3}) = -206.85[\text{A}]$.

(b) The voltage curves in the resistance, $e_R(t)$, and in the inductance, $e_L(t)$, of the load, are shown in Fig. 6.181.

Figure 6.182 shows a zoom at the beginning of Fig. 6.181. Apparently, the initial values are null at $t = 8.3_+[\text{ms}]$. However, as the voltage range varies from $-120[\text{kV}]$ to $+120[\text{kV}]$, new zooms were performed, including manually changing the voltage scale values. And the result is shown in Fig. 6.183.

Finally, Fig. 6.183 shows that $e_R(8.301.10^{-3}) = e_R(8.3.10^{-3} + \Delta t) = 0.1135[\text{V}]$ and that $e_L(8.301.10^{-3}) = e_L(8.3.10^{-3} + \Delta t) = 1,362[\text{V}]$. Thus, it can be established that $e_R(8.3.10_+^{-3}) = 0[\text{V}]$ and $e_L(8.3.10_+^{-3}) \cong 1,400[\text{V}] = 1.4 [\text{kV}]$.. The only possibility of being $e_R(8.3.10_+^{-3}) \neq 0[\text{V}]$ would be the presence of discontinuity in current $i(t)$. But this would imply an impulse in the inductance voltage, $e_L(t)$, which Fig. 6.182 proves there is not. However, there is discontinuity in $e_L(t)$.

Figure 6.182 shows that the maximum asymmetric value of the resistance voltage is $-58,002[\text{V}] = -58.002 [\text{kV}]$. occurring at $t = 0.015226[\text{s}] = 15.226[\text{ms}]$. On the other hand, from the same Fig. 6.182, it is also obtained that

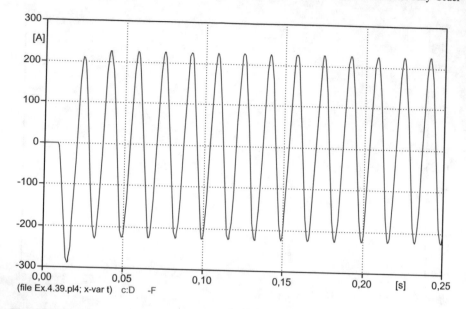

Fig. 6.179 Load energizing current, but now for $t_{max} = 250$ ms $= 0.25$ [s]

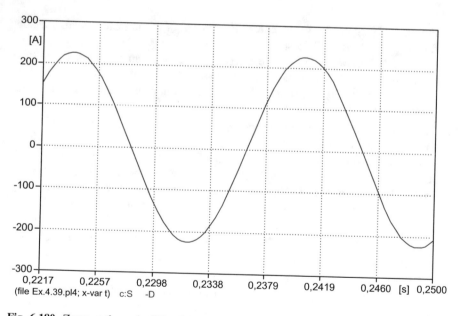

Fig. 6.180 Zoom at the end of Fig. 6.179

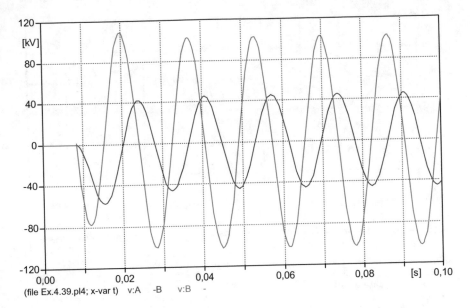

Fig. 6.181 Voltages $e_R(t)$ and $e_L(t)$

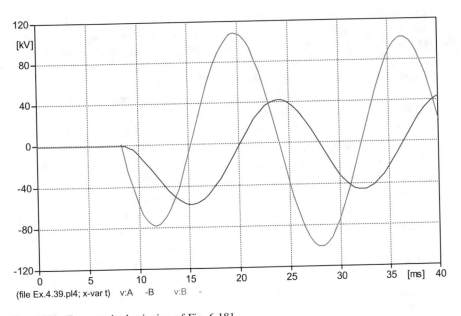

Fig. 6.182 Zoom at the beginning of Fig. 6.181

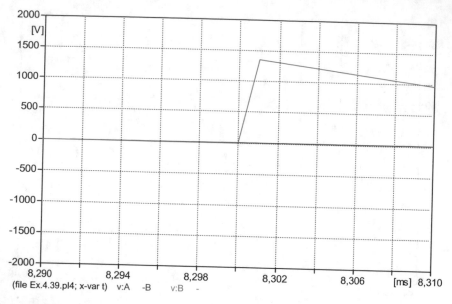

Fig. 6.183 Result of new zooms in Fig. 6.181

the maximum asymmetric value of the voltage at the inductance is $108,500 [V] = 108.5 [kV]$, occurring at $t = 0.019661 [s] = 19.661 [ms]$. Figure 6.184 shows the same curves for $e_R(t)$ and $e_L(t)$, but now for $t_{max} = 250 [ms] = 0.25 [s]$.

Figure 6.185 shows a zoom at the end of Fig. 6.184.

From Fig. 6.185 it can be extracted that the maximum symmetrical values of the voltages $e_R(t)$ and $e_L(t)$ are, respectively, $45,208 [V] = 45.208 [kV]$ and $102,260 [V] = 102.26 [kV]$. Also, that $e_R(250.10_-^{-3}) = -41,370 [V] = -41.37 [kV]$ and $e_L(250.10_-^{-3}) = 41,233 [V] = 41.233 [kV]$.

Figure 6.186 shows the voltage in the industrial load, $e_A(t)$, for $t_{max} = 250 [ms] = 0.25 [s]$, that is, before energizing the capacitor bank.

A zoom at the beginning of Fig. 6.186 is recorded in Fig. 6.187.

Applying successive zooms in Fig. 6.187, in addition to manually reducing the values of the voltage scale, results the graph in Fig. 6.188.

Whereas $e_A(t) = e_R(t) + e_L(t)$, then $e_A(8.3.10_+^{-3}) = e_R(8.3.10_+^{-3}) + e_L(8.3.10_+^{-3}) = 0 + 1,400 = 1,400 [V]$, which is in accordance with Fig. 6.188. The maximum asymmetric value of $e_A(t)$ is obtained from Fig. 6.187 and is worth $-111,870 [V] = -111.87 [kV]$, occurring at $t = 12.504 [ms]$.

By zooming at the end of Fig. 6.186, the waveform shown in Fig. 6.189 is obtained. It can be obtained that the maximum symmetrical value of the voltage in the load is $111.81 [kV]$ and that $e_A(250.10_-^{-3}) = -136.76 [V]$. Note that $e_R(250.10_-^{-3}) + e_L(250.10_-^{-3}) = -41,370 + 41,233 = -137 [V]$.

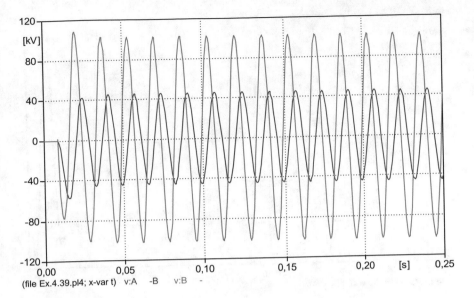

Fig. 6.184 Voltages $e_R(t)$ e $e_L(t)$

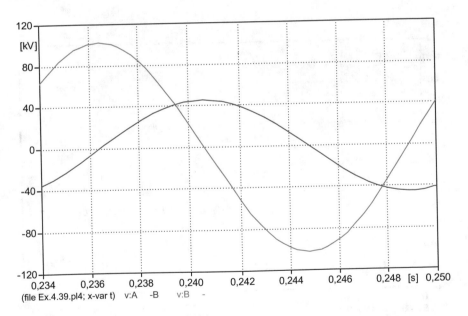

Fig. 6.185 Zoom at the end of Fig. 6.184

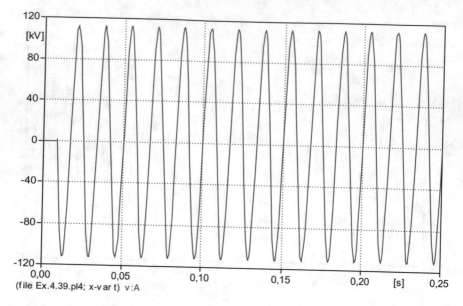

Fig. 6.186 Industrial load voltage $e_A(t)$

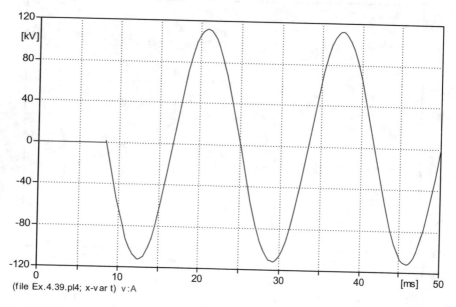

Fig. 6.187 Zoom at the beginning of Voltage $e_A(t)$

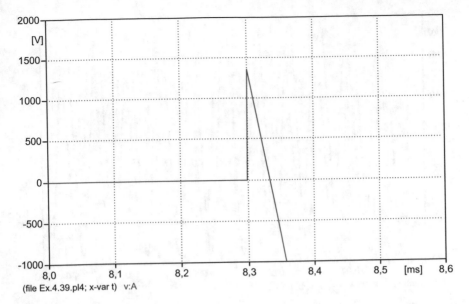

(file Ex.4.39.pl4; x-var t)   v:A

Fig. 6.188 Result of new zooms in Fig. 6.187

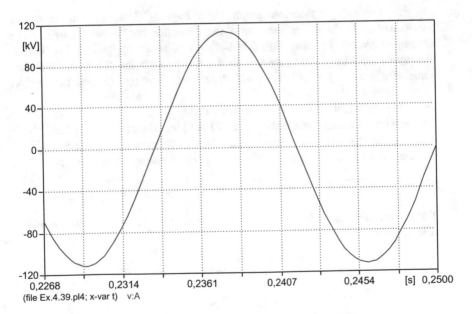

(file Ex.4.39.pl4; x-var t)   v:A

Fig. 6.189 Zoom at the end of Fig. 6.186

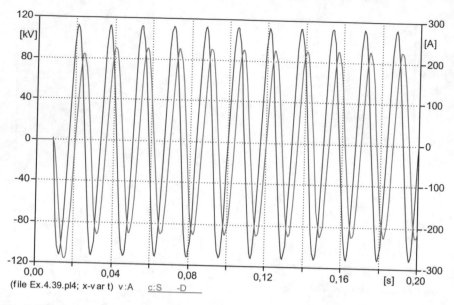

Fig. 6.190 Voltage and current in industrial load

(c)   Fig. 6.190 records, in the same graph, the voltage $e_A(t)$ and the current $i(t)$ in the industrial load, before energizing the capacitor bank, that is, before the closing of switch k_2. And Fig. 6.191 shows a zoom applied in Fig. 6.190, already in the steady state phase, with the two waveforms oscillating with the same frequency $f = 60[\text{Hz}]$ and, therefore, having the same period $T_F = 1/f = 1/60 = 0.01667$ [s].

From Fig. 6.191, the following can be extracted:

The maximum steady state voltage,111.81 [kV], occurs at $t_1 = 0.087505$[s].

The maximum steady state current, 226.04[A], occurs at $t_2 = 0.090564$ [s].

So, the current delay in relation to the voltage, measured in seconds is:

$$t_2 - t_1 = 0.003059 \,[\text{s}].$$

To convert this delay from seconds to decimal degrees, just use the following rule of three:

$$T_F = 0.016667 \,[s] \rightarrow 360[°]$$
$$0.003059 \,[s] \rightarrow \varphi_1[°].$$

$$\varphi_1 = \frac{0.003059 \times 360}{0.016667} = 66.0731[°]$$

Therefore, the load power factor is

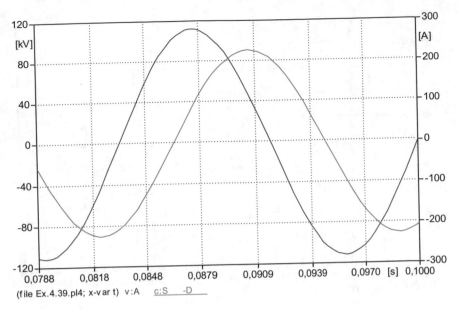

(file Ex.4.39.pl4; x-v ar t)   v:A   c:S   -D

Fig. 6.191 Zoom at the Fig. 6.190

$$f_{p_1} = \cos \varphi_1 = \cos 66.0731° = 0.4056_{IND}.$$

An analytical solution, using the industrial load impedance, is:
$$Z = R + j\omega L = 200 + j120\pi.1.2 = 200 + j452.3893 = 494.6272\angle 66.1499° \, [\Omega].$$
So,
$$\varphi_1 = 66.1499°; f_{p_1} = \cos \varphi_1 = \cos 66.1499° = 0.4043_{IND}.$$

(d)   Fig. 6.192 shows the instantaneous power curve, $p_R(t)$, in the load resistance,
for $t_{max} = 0.12 \, [s] = 120 \, [ms]$.

Figure 6.193 shows a zoom at the beginning of the instantaneous power curve
$p_R(t)$. As it shows, and also considering that the initial current in the resistance
is $i(8.3.10_+^{-3}) = 0 \, [A]$, $p_R(8.3.10_+^{-3}) = 0 \, [W]$.

Also from Fig. 6.193 it is extracted that the maximum asymmetric value of $p_R(t)$
is $1.6821.10^7 \, [W] = 16.821 \, [MW]$, occurring at time $t = 15.225 \, [ms]$.

Figure 6.194 shows a zoom already in the steady state phase of the power $p_R(t)$.
From Fig. 6.194 it is possible to determine that the maximum steady state power
in the load resistance is $p_{R_{max}} = 1.0219.10^7 \, [W] = 10.219 \, [MW]$.

Figure 6.195 records the instantaneous power curve, $p_L(t)$, at the load
inductance.

Figure 6.196 shows a zoom at the beginning of the instantaneous power curve
$p_L(t)$. From it we get that the initial value is $p_L(8.301.10^{-3}) = 0.7729 \, [W]$. And
also that the maximum asymmetric value is $-16.531 \, [MW]$, occurring at
$t = 17.619 \, [ms]$.

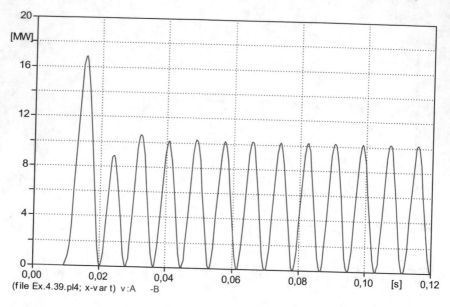

Fig. 6.192 Instantaneous power, $p_R(t)$, in the load resistance

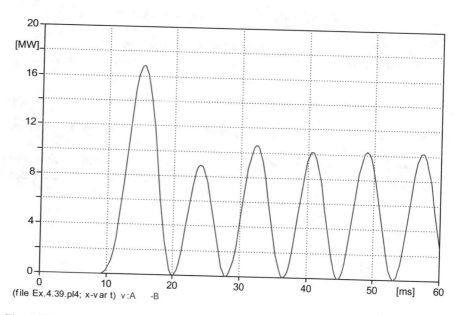

Fig. 6.193 Zoom in the transient phase of the instantaneous power $p_R(t)$

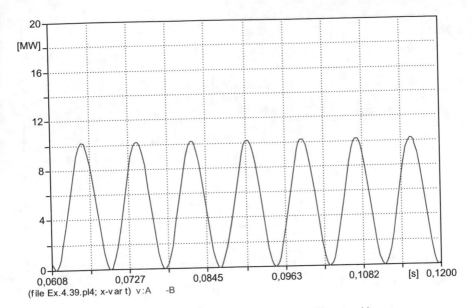

Fig. 6.194 Zoom in the steady-state phase of the instantaneous power $p_R(t)$

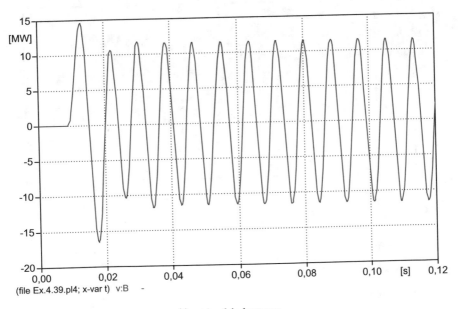

Fig. 6.195 Instantaneous power, $p_L(t)$, at load inductance

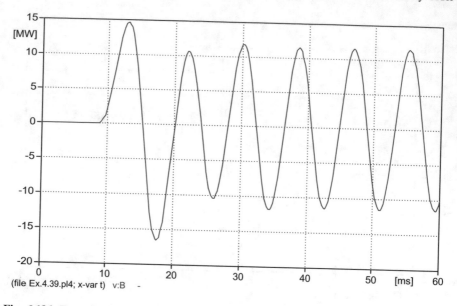

Fig. 6.196 Zoom in the transient phase of the instantaneous power $p_L(t)$

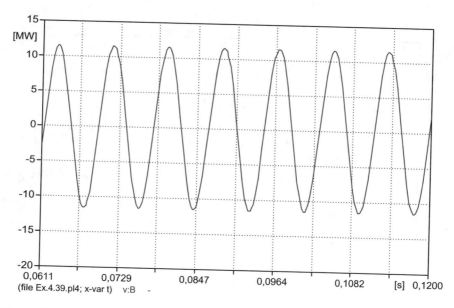

Fig. 6.197 Zoom in the steady-state phase of the instantaneous power $p_L(t)$

Figure 6.197 shows a zoom already in the steady state phase of the power $p_L(t)$. Figure 6.197 shows that the maximum steady state power in the load inductance is $p_{L_{max}} = 1.1557.10^7$ [W] = 11.557 [MW].

(e) The average power in the load resistance is given by $P_R = P_1 = \frac{1}{2}.e_{R_{max}}.i_{R_{max}} = \frac{1}{2}PR_{max}$. Therefore, according to item d, $P_1 = 10.219/2 = 5.1095$ [MW]. On the other hand, in a pure inductance, in a constant sinusoidal regime of angular frequency $\omega = 2\pi f$,

$$p_L(t) = E_{ef}I_{ef}\sin 2\omega t = \frac{1}{2}.e_{L_{max}}.i_{L_{max}}.\sin 2\omega t = Q_L \sin 2\omega t = p_{L_{max}} \sin 2\omega t.$$

$$(6.56)$$

Therefore, $Q_L = Q_1 = p_{L_{max}}$. So, $Q_1 = 11.557$ [MVAR]. An alternative is to use

$$Q_L = \frac{1}{2}X_L(i_{L_{max}})^2 = \frac{1}{2}.452.5781.(226.04)^2 = 11,557,207.91 \ [VAR]$$
$$= 11.5572 \ [MVAR]$$

$$(6.57)$$

So, the triangle of power of the load, before the connection of the capacitor bank, is given by the complex power

$$S_1 = P_1 + jQ_1 = 5.1095 + j11.557 = 12.6361e^{j66.1492°} \ [MVA].$$

Therefore the elements of the power triangle are:

$$P_1 = 5.1095 \ [MW]; Q_1 = 11.5570 \ [MVAR]; N_1 = 12.6361 \ [MVA]; \varphi_1$$
$$= 66.1492°.$$

The respective power triangle is recorded at the end of the present example.

2. **Switches k_1 and k_2 closed.**

(f) The system supply current, $i(t)$, both before and after the connection of the capacitor bank at time $t = 250$ [ms], is shown in Fig. 6.198, from time $t = 0$. Figure 6.199 shows a first zoom in the system supply current, applied slightly before the closing of switch k_2 at $t = 250$ [ms].
Figure 6.200 shows a new zoom in the supply current, with emphasis on the transitional phase generated by the input of the capacitive bank. With the cursor, it is possible to see, from Fig. 6.200, that $i(250.10_+^{-3}) = -206.88$ [A]; and that the new maximum asymmetric supply current of the industrial system is worth 266.61 [A] and occurs at time $t = 253.5$ [ms].
In Fig. 6.201 a new zoom is shown in Fig. 6.198, but now in the steady state phase. From it it is possible to determine that the new supply current in permanent sinusoidal regime has a maximum value of 92.136 [A].

(g) Fig. 6.202 shows the voltage, $e_C(t)$, and current curves $i_C(t)$, for energizing the capacitor bank. (Note that $e_A(t) = e_C(t)$ for $t > 250$ [ms]). And

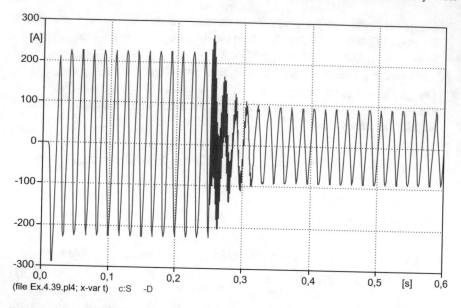

Fig. 6.198 System supply current $i(t)$ with switches k_1 and k_2 closed

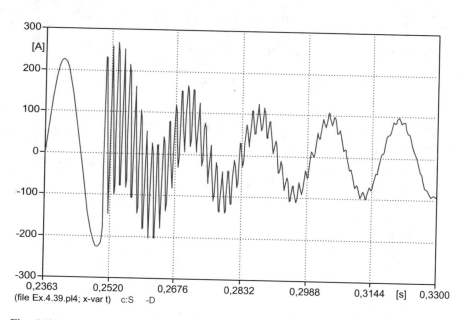

Fig. 6.199 Zoom in the supply current $i(t)$

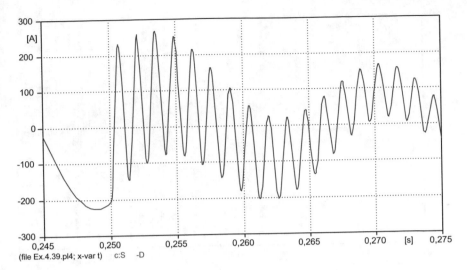

Fig. 6.200 New zoom in the supply current $i(t)$

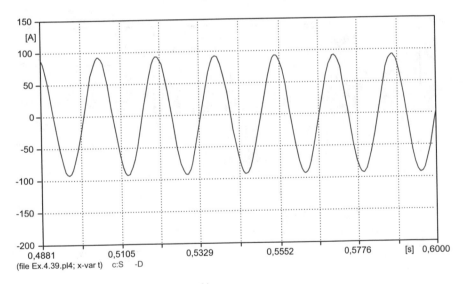

Fig. 6.201 Steady-state supply current $i(t)$

Fig. 6.203 shows a zoom in the same curves, close to the closing of switch k_2, that is, close to $t = 250$ [ms].

From Fig. 6.203, it can be extracted that the "initial value" $e_C(0.25_+) = 0.0000$ [V], which is corroborated by the fact that there is no impulse in the current $i_C(t)$.. The maximum value of the transient phase of the capacitor energization voltage is 119.64 [kV], occurring at time

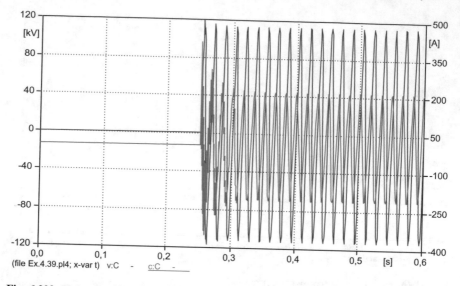

Fig. 6.202 Voltage $e_C(t)$ and current $i_C(t)$ of energization of the capacitor bank

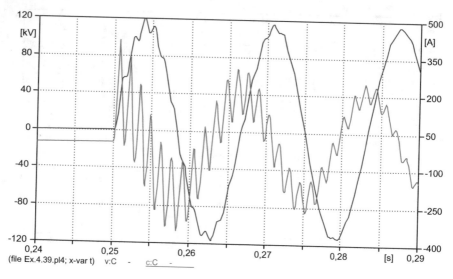

Fig. 6.203 Zoom at Fig. 6.203

$t = 253.88$ [ms]. And the maximum steady state voltage is 112.6[kV], obtained by zooming at the end of Fig. 6.202. From the same Fig. 6.203 it can be seen that the "initial" current $i_C(0.25_+) = 0.0000$ [A]; the maximum asymmetric value of $i_C(t)$ is 409.99 [A], occurring at time t = 250.7 [ms]. The corresponding steady state current has a maximum value of 212.25 [A], obtained from the same zoom applied at the end of the curve in Fig. 6.202.

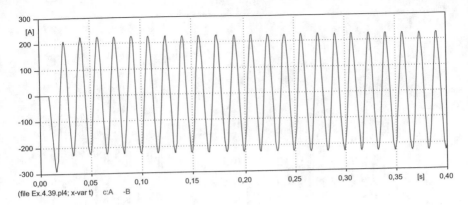

(file Ex.4.39.pl4; x-var t)   c:A   -B

Fig. 6.204 Current $i_F(t)$

(h) Fig. 6.204 shows the curve of the new current in the industry, $i_F(t)$, starting from $t = 0\,[\text{s}]$ and with $t_{max} = 0.4\,[\text{s}] = 400\,[\text{ms}]$. And Fig. 6.205 shows a zoom applied to it, in the vicinity of $t = 250\,[\text{ms}]$. The latter results in $i_F(0.25+) = -206.88\,[\text{A}]$ and that the maximum asymmetric current is $228.44\,[\text{A}]$, occurring at time $t = 257.135\,[\text{ms}]$. Zooming in at the end of Fig. 6.204 shows that the new maximum value for the permanent current of the industry current is $227.64\,[\text{A}]$. Considering that before the closing of switch k_2 it was $226.04\,[\text{A}]$, it can be seen that the energization of the capacitor bank has practically not changed the steady-state current in the industry, which, in fact, Fig. 6.204 shows.

It is worth noting that there is a discontinuity in the load voltage, $e_A(t)$, at the time of closing of switch k_2, since $e_A(250.10^{-3}-) = -136.76\,[\text{V}]$ and

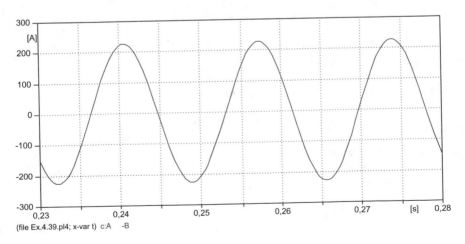

(file Ex.4.39.pl4; x-var t)   c:A   -B

Fig. 6.205 Zoom at current $i_F(t)$

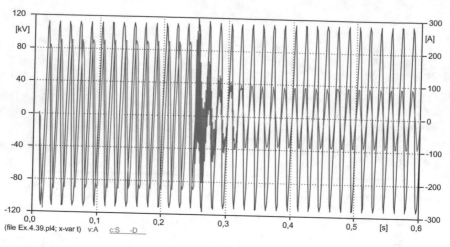

Fig. 6.206 Load voltage $e_A(t)$ and supply current $i(t)$

$e_A(250.10^{-3}+) = 0[V] = e_C(250.10^{-3}+)$. But the same does not happen with the current, because $i_F(0.25_-) = i(0.25_-) = -206.85\,[A]$ and $i_F(0.25_+) = -206.88\,[A]$.

(i) In Fig. 6.206, the load voltage curves, $e_A(t)$, and the supply current, $i(t)$, are plotted for $t > 0$.

Figure 6.207 shows a zoom, already in the steady state phase, of the curves of Fig. 6.206.

With more zoom and cursor, it can be obtained from Fig. 6.207 that:

The maximum voltage, 112.6 [kV], occurs at time $t_1 = 0.58751$ [s].

The maximum current, 92.136 [A], occurs at time $t_2 = 0.58739$ [s].

Therefore, the delay, in seconds, of the current in relation to the voltage is

$$t_2 - t_1 = -0.00012\,[\mathrm{s}]$$

Converting the lag to degrees, it comes that:

$$\begin{cases} 0.01667\,[\mathrm{s}] \rightarrow 360[°] \\ -0.00012\,[\mathrm{s}] \rightarrow \varphi_2[°] \end{cases}$$

$$\varphi_2 = -0.00012 \times 360/0.01667 = -2.5915[°].$$

And the new power factor is $fp_2 = \cos\varphi_2 = 0.9990_{CAP} \cong 1$.

An analytical solution, through the impedance of the parallel connection of the load with the capacitance, is as follows

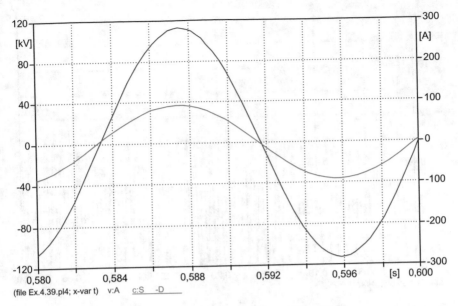

Fig. 6.207 Zoom in the steady-state regime of the curves of Fig. 6.206

$$Z_p = \frac{(200 + j452.3893)(-j530.5165)}{(200 + j452.3893) + (-j530.5165)} = \frac{239,999.9881 - j106,103.3000}{200 - j78.1272} =$$

$$= \frac{262,407.8973\angle - 23.8501°}{214.7181\angle - 21.3374°} = 1,222.1042\angle - 2.5127°$$

$$= 1,220.9292 - j53.5781 \ [\Omega]$$

Since the impedance angle is the power factor angle, then,

$$\varphi_2 = -2.5127°$$

$$fp_2 = \cos\varphi_2 = \cos(-2.5127°) = 0.9990_{CAP}$$

(j) Fig. 6.208 shows the instantaneous power curve, $p_R(t)$, in the load resistance.

Figure 6.209 shows a zoom in the permanent regime of the curve of Fig. 6.208.

From Fig. 6.209, with the aid of the cursor, and more zoom, it can be obtained that now

$$p_{R_{max}} = 10.364 \ [MW] = 2P_2. \text{ Then,}$$

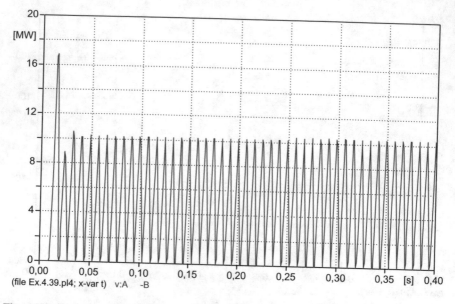

Fig. 6.208 Instantaneous power $p_R(t)$ in the load resistor

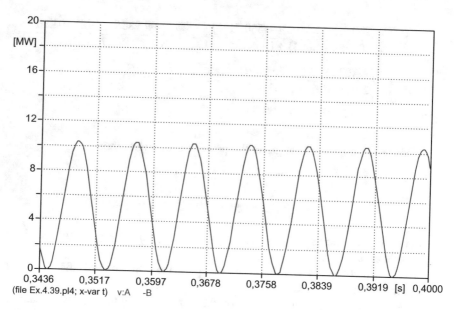

Fig. 6.209 Zoom on the load resistance power curve, $p_R(t)$

$$P_2 = 5.1820\,[MW]$$

Therefore, the other elements of the new power triangle are:

$$Q_2 = P_2.tg\varphi_2 = 5.1820.tg(-2.5915°) = -0.2345\,[MVAR]$$

$$P_2 = N_2\cos\varphi_2 \text{ and } N_2 = P_2/\cos\varphi_2 = 5.1820/0.9990 = 5.1872\,[MVA].$$

Alternatively, an analytical solution for the reactive and apparent powers is:

$$Q_2 = Q_P = \frac{1}{2}X_P(I_{max})^2 = \frac{1}{2}.(-53.5781).(92.1315)^2 = -0.2274\,[MVAR]$$

$$N_2 = \sqrt{P_2^2 + Q_2^2} = \sqrt{(5.1820)^2 + (-0.2274)^2} = 5.1870\,[MVA].$$

So, the new complex power, using data obtained from the ATPDraw program, is:

$$S_2 = P_2 + jQ_2 = 5.1820 - j0.2345 = 5.1873\epsilon^{-j2.5910°}\,[MVA].$$

The new reactive power, Q_2, can also be obtained from the instantaneous energy curves in the load inductance, $W_L(t)$, and in the capacitance of the capacitor bank, $W_C(t)$, as follows:
Figure 6.210 shows the energy curves for inductance and capacitance, from $t = 0\,[s]$ to 400 [ms]. And Fig. 6.211 shows a zoom done on them, already in the permanent regime.
From Fig. 6.211, it is possible to obtain, with another zoom and cursor, that:

$$W_{L_{max}} = 31.094\,[kJ].$$
$$W_{C_{max}} = 31.700\,[kJ].$$

But,

$$W_{L_{max}} = \frac{1}{2}L(i_{L_{max}})^2 = \frac{1}{2}L\left(\sqrt{2}I_{L_{ef}}\right)^2 = L(I_{L_{ef}})^2$$
$$W_{C_{max}} = \frac{1}{2}C(e_{C_{max}})^2 = \frac{1}{2}C\left(\sqrt{2}E_{C_{ef}}\right)^2 = C(E_{C_{ef}})^2$$

Since $Q = XI_{ef}^2 = X\left(\frac{I_{max}}{\sqrt{2}}\right)^2 = \frac{1}{2}X(I_{max})^2$, then

$$Q_L = X_L(I_{L_{ef}})^2 = \omega L(I_{L_{ef}})^2$$

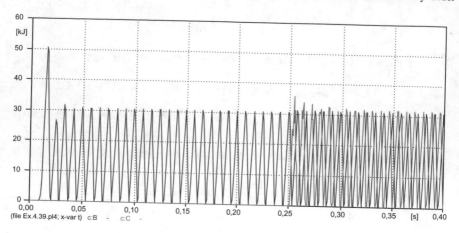

Fig. 6.210 Energy $W_L(t)$ and $W_C(t)$

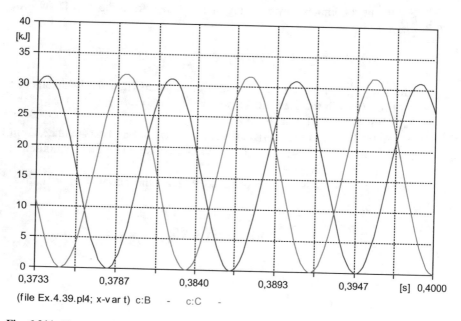

Fig. 6.211 Zoom applied at the end of Fig. 6.210

Then,

$$Q_L = \omega.W_{L_{max}} \tag{6.58}$$

$$Q_C = X_C(I_{C_{ef}})^2 = \left(-\frac{1}{\omega C}\right).(I_{C_{ef}})^2 = -\omega C(E_{C_{ef}})^2,$$

for $E_{C_{ef}} = I_{C_{ef}}/\omega C$ and $(I_{C_{ef}})^2 = (\omega C)^2.(E_{C_{ef}})^2.$
Therefore,

$$Q_C = -\omega.W_{C_{max}} \tag{6.59}$$

Then,

$$Q_L = 2\pi.60 \times 31.094 = 11,722.1618 \ [kVAR]$$
$$Q_C = -2\pi.60 \times 31.700 = -11,950.6185 \ [kVAR].$$

Considering that a conservation law applies both to the active power P and the reactive power Q, no matter how the individual loads are connected, then,

$$Q_2 = Q_L + Q_C = -228.4567[kVAR] = -0.2285 \ [MVAR].$$

The sum of the preceding reactive powers stems from the fact that you cannot add apparent powers, but you can add complex powers. Like this,

$$S_2 = S_{RL} + S_C = (P_R + jQ_L) + (0 + jQ_C) = P_R + j(Q_L + Q_C) = P_2 + jQ_2$$

Therefore,

$$P_2 = P_R,$$

$$Q_2 = Q_L + Q_C.$$

So, the elements of the new power triangle are as follows:

$$P_2 = 5.1820 \ [MW]; Q_2 = -0.2345[MVAR]; N_2 = 5.1873 \ [MVA]; \varphi_2$$
$$= -2.5915°.$$

(k) With the results obtained in items e and j, the power triangles corresponding to the pre and post-correction phases of the power factor of the industrial electrical system were constructed, shown, respectively, on the left and on the right in Fig. 6.212.

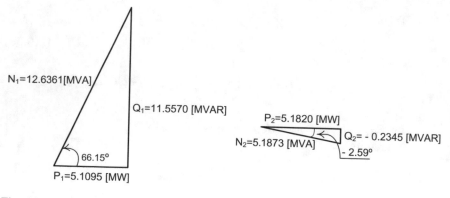

Fig. 6.212 Power triangles pre and post power factor correction

Example 6.40 Solve analytically Example 6.39.

Solution

To make it easier, the circuit in Fig. 6.175 is reproduced in Fig. 6.213.

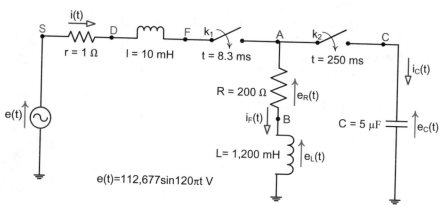

Fig. 6.213 Electric circuit of Example 6.40

1. **k_1 CLOSED and k_2 OPEN**

Considering, provisionally, the time $t = 8.3$ [ms], of the closing of the switch k_1, as new $t = 0$, then the voltage source becomes

$$e(t) = e(t + 8.3.10^{-3}) = 112,677\sin(120\pi t + 3.1290) \text{ [V]}.$$

(a) The circuit mesh equation is

$$(L+l)i'(t) + (R+r)i(t) = e(t)$$

Introducing the numerical values, comes that

$$i'(t) + 166.1157i(t) = 93,121.4876\sin(120\pi t + 3.1290)U_{-1}(t) =$$
$$93,121.4876\sin(120\pi t + 179.2785°)U_{-1}(t) = 1,172.6156U_{-1}(t) + \cdots, i(0_-) = 0.$$
$$(6.60)$$

The solution of the corresponding homogeneous equation is

$$i_H(t) = k\epsilon^{-166.1157t}$$

Therefore, the time constant is $\tau = 1/166.1157 = 6.0199$ [ms].
Using the phasor method, the particular solution comes from

$$(j120\pi + 166.1157)\dot{I}_p = 93,121.4876\angle179.2785°$$
$$\dot{I}_p = 226.0412\angle113.0585°$$

The particular solution is therefore

$$i_P(t) = 226.0412\sin(120\pi t + 113.0585°)$$
$$= 226.0412\sin(120\pi t + 1.9732\text{rad})$$

So the general solution is

$$i(t) = k\epsilon^{-166.1157t} + 226.0412\sin(120\pi t + 113.0585°)$$

From Eq. (6.60) we have that $i(0_+) = 0[\text{A}]$ and that $i'(0_+) = 1,172.6156[\text{A/s}]$. So,

$$0 = k + 207.9817 \text{ and } k = -207.9817$$

Finally,

$$i(t) = \left[-207.9817\epsilon^{-166.1157t} + 226.0412\sin(120\pi t + 113.0585°)\right]U_{-1}(t)[\text{A}]$$
$$(6.61)$$

Note, in passing, that, being $e_L(t) = Li'_F(t) = Li'(t)$, $e_L(0_+) = Li'(0_+)$ $= 1.2 \times 1,172.6156 = 1,407.1387[\text{V}]$. That is, there will be a discontinuity in the load inductance voltage $e_L(t)$, because $e_L(0_-) = 0$.

To reference $i(t)$ at time $t = 0$ of the voltage source, it is necessary to replace t with $t - 8.3.10^{-3}$. Doing this, it follows that:

$$i(t) = \left[-825.6826\epsilon^{-166.1157t} + 226.0412 \sin\left(120\pi t - 66.2225°\right)\right][A]$$
$$\textbf{\textit{for }} t \geq 0.0083\ [s]$$

(6.62)

Checking out:

$$i(0.0083_+) = 0.0015\ [A] = 1.5\ [mA].$$

$$i(0.0083 + \Delta t) = i(0.008301) = 0.0026\ [A] = 2.6\ [mA].$$

The maximum asymmetric supply current is

$$i(0.015225) = -290.0091\ [A].$$

The maximum value of the symmetric current, or steady state, is 226.0412 [A].

$$i(0.250_-) = -206.8543\ [A].$$

(b) The load resistor voltage is

$$e_R(t) = Ri_F(t) = Ri(t) = 200i(t).$$

Therefore,

$$e_R(t) = -165,136.52\epsilon^{-166.1157t} + 45,208.24 \sin\left(120\pi t - 66.2225°\right)[V]$$
$$\textbf{\textit{for}}\ t \geq 0.0083\,[s]$$

(6.63)

Checking out:

$$e_R(0.0083_+) = 0.3030\ [V] = 303\ [mV].$$

$$e_R(0.0083 + \Delta t) = e_R(0.008301) = 0.5343\ [V] = 534.3\ [mV].$$

The maximum asymmetric voltage in the load resistance is

$$e_R(0.015225) = -58,001.8191\ [V].$$

The steady state voltage in the load resistor has a maximum value of 45,208.24 [V].

$$e_R(0.250_-) = -41,370.8648 \text{ [V]}.$$

The voltage at the load inductance is

$$e_L(t) = L\frac{di_F(t)}{dt} = L\frac{di(t)}{dt} = 1.2i'(t).$$

Performing the calculations it results that

$$\boldsymbol{e_L(t) = 164,590.6117\epsilon^{-166.1157t} + 102,258.6298\sin(120\pi t + 23.7775°)\,[V]},$$
for $t > 0.0083\,[s]$.

$$(6.64)$$

Checking out:

$$e_L(0.0083_+) = 1,408.3493 \text{ [V]}.$$

$$e_L(0.0083 + \Delta t) = e_L(0.008301) = 1,365.9949 \text{ [V]}.$$

The maximum asymmetric voltage in the load inductance is

$$e_L(0.019661) = 108,502.0395 \text{ [V]}.$$

The steady state voltage at the load inductance has a maximum value of $102,258.6298$ [V].

$$e_L(0.250_-) = 41,229.6519 \text{ [V]}.$$

The voltage on the industrial load is given by

$$e_A(t) = e_R(t) + e_L(t).$$

By making substitutions and calculations, it results that

$$\boldsymbol{e_A(t) = -545.9083\epsilon^{-166.1157t} + 111,806.1373\sin(120\pi t - 0.0726°)\,[V]}$$
for $t > 0.0083\,[s]$.

$$(6.65)$$

Checking out:

$$e_A(0.0083_+) = 1,409.0971 \text{ [V]}.$$

$$e_A(0.0083 + \Delta t) = e_A(0.008301) = 1,366.9740 \, [\text{V}].$$

The maximum asymmetric voltage in the load is

$$e_A(0.012504) = -111,874.5319 \, [\text{V}].$$

The steady state voltage on the load has a maximum value of $111,806.1373 \, [\text{V}]$.

$$e_A(0.250_-) = -141.6558 \, [\text{V}].$$

(c)   The angle of the power factor in the industrial load is the difference between the angles of $e_A(t)$ and $i_F(t) = i(t)$, in steady state. That is, it is the angle of the load impedance. So,

$$\varphi_1 = \sphericalangle e_A(t) - \sphericalangle i(t) = -0.0726° - (-66.2225°) = 66.1499°$$

So, the load power factor is

$$fp_1 = \cos \varphi_1 = \cos 66.1499° = 0.4043_{IND}.$$

(d)   The instantaneous power in the load resistor is

$$p_R(t) = Ri^2(t) = 200i^2(t) =$$
$$200\left[-825.6826\epsilon^{-166.1157t} + 226.0412 \sin(120\pi t - 66.2225°)\right]^2$$

Performing the calculations, and simplifying, it turns out that

$$p_R(t) = 136,350,351.2\epsilon^{-332.2314t} - 74,655,314.28\epsilon^{-166.1157t} \sin(120\pi t - 66.2225°)$$
$$+ 5,109,462.42 - 5,109,462.42 \cos(240\pi t - 132.4450°) \, [\text{W}], \textit{for } t \geq 0.0083[\text{s}]$$

$$(6.66)$$

Checking out:

$$p_R(0.0083_+) = 0.0200 \, [\text{W}].$$

The maximum asymmetric power in the load resistance is

$$p_R(0.015225) = 16,821,054.73 \, [\text{W}] = 16.8211 \, [\text{MW}].$$

The steady state power is

$$p_R(t) = 5,109,462.42 - 5,109,462.42 \cos(240\pi t - 132.4450°) \, [\text{W}] \geq 0.$$

Since the average cosine value is zero, then the average power in the load resistance is

$$P_R = P_1 = 5,109,462.42 \, [\text{W}] = 5.1095 \, [\text{MW}].$$

The instantaneous power at the load inductance is

$$p_L(t) = e_L(t) \cdot i(t)$$
$$= \left[164,590.6117\epsilon^{-166.1157t} + 102,258.6298 \sin(120\pi t + 23.7775°) \right]$$
$$\times \left[-825.6826\epsilon^{-166.1157t} + 226.0412 \sin(120\pi t - 66.2225°) \right]$$

Performing the calculations, and simplifying, comes that

$$p_L(t) = -135,899,604.2\epsilon^{-332.2314t} + 92,266,555.9\epsilon^{-166.1157t} \sin(120\pi t - 132.4425°) \quad (6.67)$$
$$+ 11,557,331.7 \sin(240\pi t - 132.4450°) \, [\text{W}], for \, t > 0.0083 \, [\text{s}].$$

Checking out:

$$p_L(0.0083_+) = -3.5110 \, [\text{W}].$$

The maximum asymmetric power in the load inductance is

$$p_L(0.017619) = -16,531,036.9 \, [\text{W}] = -16.5310 \, [\text{MW}]$$

The steady state power is

$$p_L(t) = 11,557,331.7 \sin(240\pi t - 132.4450°) \, [\text{W}].$$

Since the average sine value is zero, $P_L = 0$, but the reactive power is

$$Q_L = Q_1 = 11,557,331.7 \, [VAR] = 11.5573 \, [MVAR].$$

(e) Being $P_1 = 5.1095 \, [\text{MW}]; \varphi_1 = 66.1499°; Q_1 = 11.5573 \, [MVAR]$,   the apparent power is

$$N_1 = \sqrt{(P_1)^2 + (Q_1)^2} = 12.6364 \, [MVA].$$

2. k_1 and k_2 CLOSED

Considering, provisionally, the time $t = 0.25 \, [\text{s}] = 250 \, [\text{ms}]$, in which the switch k_2 is closed, as new $t = 0 \, [\text{s}]$ then the external voltage source $e(t)$ must be corrected exchanging t for $t + 0.25$. Thus, the voltage source becomes

$$e(t) = 112,677 \sin[120\pi(t + 0.25)] = 112,677 \sin(120\pi t + 94.2478 \, \text{rad})$$
$$= 112,677 \sin(120\pi t + 5,400.0012°) = 112,677 \sin(120\pi t + 0.0012°).U_{-1}(t) \, [\text{V}]$$

548

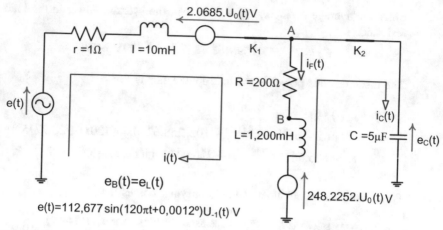

Fig. 6.214 Electric circuit of Example 6.40, when the two switches are closed

The current running through the inductances immediately before closing key k_2 was $i(0.250_-) = -206.8543[\text{A}]$. Therefore, $i(0_-) = -206.8543[\text{A}]$. So the equivalent Thévenin sources for the two inductances are:

$$li(0_-)U_0(t) = 10.10^{-3}.(-206.8543)U_0(t) = -2.0685U_0(t) \ [\text{V}].$$

$$Li(0_-)U_0(t) = 1,200.10^{-3}.(-206.8543)U_0(t) = -248.2252U_0(t) \ [\text{V}].$$

There is no Thévenin equivalent of capacitance because it is de-energized at $t = 0_-$.

The circuit for the condition in which the two switches are closed is then the one shown in Fig. 6.214.

Left mesh equation:

$$201i(t) + 1.21i'(t) - 200i_C(t) - 1.2i'_C(t)$$
$$= 112,677sin(120\pi t + 0.0012°)U_{-1}(t) - 250.2937U_0(t).$$

Right mesh equation:

$$-200i(t) - 1.2i'(t) + 2.10^5 \int i_C(t)dt + 200i_C(t) + 1.2i'_C(t) = 248.2252U_0(t).$$

Or,

$$(1.21D + 201)i - (1.2D + 200)i_C = 112,677sin(120\pi t + 0.0012°)U_{-1}(t) - 250.2937U_0(t)$$
$$- D(1.2D + 200)i + (1.2D^2 + 200D + 200,000)i_C = 248.2252U_1(t)$$

$$(6.68)$$

(f) Solving the system of differential (Eq. 6.68) for $i(t)$ results in the following differential equation:

$$\left(0.012D^3 + 3.2D^2 + 242,200D + 40,200,000\right)i(t)$$
$$= -2.4822U_2(t) - 410.8681U_1(t) + 915,598.68U_0(t)$$
$$+ 9,120,587,648\sin(120\pi t + 68.6679°)U_{-1}(t), \text{ with } i(0_-)$$
$$= i'(0_-) = i''(0_-) = 0. \tag{6.69}$$

Equation (6.69) allows us to write that

$$i(t) \doteq C_0 U_{-1}(t).$$

$$i'(t) \doteq C_0 U_0(t) + C_1 U_{-1}(t).$$

$$i''(t) \doteq C_0 U_1(t) + C_1 U_0(t) + C_2 U_{-1}(t).$$

$$i'''(t) \doteq C_0 U_2(t) + C_1 U_1(t) + C_2 U_0(t) + C_3 U_{-1}(t).$$

Replacing in Eq. (6.69) comes that

$$0.012C_0 U_2(t) + (0.012C_1 + 3.2C_0)U_1(t) + (0.012C_2 + 3.2C_1 + 242,200C_0)U_0(t) + \cdots$$
$$\doteq -2.4822U_2(t) - 410.8681U_1(t) + 915,598.68U_0(t) + \cdots$$

The identification of similar terms allows to obtain that

$$C_0 = -206.85[A] = i(0_+)$$
$$C_1 = 20,920.9917[A/s] = i'(0_+) \tag{6.70}$$
$$C_2 = 4,245,643,459[A/s^2] = i''(0_+).$$

The characteristic roots of the corresponding polynomial operator of Eq. (6.69), obtained with the aid of the Newton-Raphson method, are:

$$r_1 = -166.1160; r_{2,3} = -50.2750 \pm j4,483.0164.$$

So the homogeneous solution of differential Eq. (6.69) is

$$i_H(t) = k_1 \epsilon^{-166.116t} + \epsilon^{-50.275t}(k_2 \cos 4,483.0164t + k_3 \sin 4,483.0164t).$$

The corresponding particular solution, for $t > 0$, of the differential equation

$$\left(0.012D^3 + 3.2D^2 + 242,200D + 40,200,000\right)i_P(t)$$
$$= 9,120,587,648 \sin(120\pi t + 68.6679°)$$

is obtained using the phasor method, replacing D with $j\omega = j120\pi$ and $i_P(t)$ with its phasor $\dot{I}_P$, as follows:

$$(-j642,946.1536 - 454,791.3709 + j91,307,248.9 + +40,200,000)\dot{I}_P$$
$$= 9,120,587,648e^{j68.6679°}$$

$$\left(98,993,421e^{j66.3284°}\right)\dot{I}_P = 9,120,587,648e^{j68.6679°}$$

$$\dot{I}_P = 92.1333\angle 2.3395°$$

So the particular solution is

$$i_P(t) = 92.1333\sin(120\pi t + 2.3395°)$$

And the general solution is

$$i(t) = k_1 e^{-166.116t} + e^{-50.275t}(k_2 \cos 4,483.0164t + k_3 \sin 4,483.0164t) + 92.1333\sin(120\pi t + 2.3395°)$$

$$(6.71)$$

So,

$$i'(t) = -166.116k_1 e^{-166.116t} - 50.275e^{-50.275t}(k_2 \cos 4,483.0164t + k_3 \sin 4,483.0164t)$$
$$+ e^{-50.275t}(-4,483.0164k_2 \sin 4,483.0164t + 4,483.0164k_3 \cos 4,483.0164t)$$
$$+ 34,733.4358\cos(120\pi t + 2.3395°)$$

$$(6.72)$$

$$i''(t) = 27,594.5255k_1 e^{-166.116t}$$
$$+ 2,527.5756e^{-50.275t}(k_2 \cos 4,483.0164t + k_3 \sin 4,483.0164t)$$
$$+ e^{-50.275t}(225,383.6495k_2 \sin 4,483.0164t - 225,383.6495k_3 \cos 4,483.0164t)$$
$$- 50.275e^{-50.275t}(-4,483.0164k_2 \sin 4,483.0164t + 4,483.0164k_3 \cos 4,483.0164t)$$
$$+ e^{-50.275t}(-20,097,436.04k_2 \cos 4,483.0164t - 20,097,436.04k_3 \sin 4,483.0164t)$$
$$- 13,094,196.81\sin(120\pi t + 2.3395°)$$

$$(6.73)$$

Taking $t = 0_+$ in Eqs. (6.71), (6.72) and (6.73) and using the values of the initial conditions contained in 6.70, the following system of algebraic equations results:

$$\begin{cases} k_1 + k_2 + 0.k_3 = -210.6109 \\ -166.1160k_1 - 50.2750k_2 + 4,483.0164k_3 = -13,783.4935 \\ 27,594.5255k_1 - 20,094,908.46k_2 - 450,767.2990k_3 = 4,246,177,972. \end{cases}$$

And the solution of this system provides that:

$$k_1 = 0.5739; k_2 = -211.1848; k_3 = -5.4217.$$

Then, according to Eq. (6.71), the supply current becomes

$$i(t) = 0.5739\epsilon^{-166.116t}$$
$$+ \epsilon^{-50.275t}(-211.1848 \cos 4,483.0164t - 5.4217 \sin 4,483.0164t)$$
$$+ 92.1333 \sin(120\pi t + 2.3395°)[A], for\ t \geq 0[s].$$

$$(6.74)$$

Checking out:

$$i(0_+) = -206.85[A].$$

Example 6.39 shows that the maximum asymmetric supply current occurs at time $t = 253.5[\text{ms}] = 0.2535[\text{s}]$. In the present case, $t = 0.2535 - 0.2500 = 0.0035[\text{s}]$ and,

$$i(0.0035) = 267.4136[A].$$

The maximum symmetric current, or steady state, is $92.1333[A]$. Equation 6.74 can also be written as

$$i(t) = \mathbf{0.5739\epsilon^{-166.116t}}$$
$$+ \mathbf{211.2544\epsilon^{-50.275t} \sin(4,483.0164t - 91.4706°)} +\qquad (6.75)$$
$$+ \mathbf{92.1333 \sin(120\pi t + 2.3395°)[A],\ for\ t \geq 0\ [s].}$$

(g, h) Solving the system of differential Eq. (6.68) for $i_C(t)$ results in the following differential equation:

$$(0.012D^3 + 3.2D^2 + 242,200D + 40,200,000)i_C(t) = -162.6429U_1(t)$$
$$+ 50,974,345.88U_0(t) + 2.1011.10^{10} \sin(120\pi t + 156.1513°)U_{-1}(t)$$
$$\text{with } i_C(0_-) = i'_C(0_-) = i''_C(0_-) = 0$$

$$(6.76)$$

Examining Eq. (6.76), it can be written that:

$$i_C(t) \dot{=} 0.U_{-1}(t).$$

$$i'_C(t) \dot{=} D_0 U_{-1}(t).$$

$$i_C''(t) \doteq D_0 U_0(t) + D_1 U_{-1}(t).$$

$$i_C'''(t) \doteq D_0 U_1(t) + D_1 U_0(t) + D_2 U_{-1}(t).$$

Replacing in Eq. (6.76) comes that

$$0.012 D_0 U_1(t) + (0.012 D_1 + 3.2 D_0) U_0(t) \doteq$$
$$- 162.6429 U_1(t) + 50,974,345.88 U_0(t)$$

The identification of similar terms allows to obtain that

$$i_C(0_+) = 0[\text{A}]$$

$$i_C'(0_+) = D_0 = -13,553.5750[\text{A/s}] \tag{6.77}$$

$$i_C''(0_+) = D_1 = 4,251,476,443\,[\text{A/s}^2].$$

As the characteristic roots are the same, namely: $r_1 = -166.1160; r_{2,3} = -50.2750 \pm j4,483.0164$, then,

$$i_{C_H}(t) = k_4 \epsilon^{-166.116t} + \epsilon^{-50.275t}(k_5 \cos 4,483.0164t + k_6 \sin 4,483.0164t).$$

On the other hand,

$$\left(98,993,421 e^{j66.3284°}\right) \dot{I}_{C_P} = 2.1011.10^{10} e^{j156.1513°}$$

$$\dot{I}_{C_P} = 212.2464 \angle 89.8229°$$

Then,

$$i_{C_P}(t) = 212.2464 \sin(120\pi t + 89.8229°)$$

Therefore,

$$i_C(t) = k_4 \epsilon^{-166.116t} + k_5 \epsilon^{-50.275t} \cos 4,483.0164t + k_6 \epsilon^{-50.275t} \sin 4,483.0164t$$
$$+ 212.2464 \sin(120\pi t + 89.8229°)$$

$$\tag{6.78}$$

Then,

$$i_C'(t) = -166.116 k_4 \epsilon^{-166.116t} - 50.275 k_5 \epsilon^{-50.275t} \cos 4,483.0164t$$
$$- 4,483.0164 k_5 \epsilon^{-50.275t} \sin 4,483.0164t - 50.275 k_6 \epsilon^{-50.275t} \sin 4,483.0164t$$
$$+ 4,483.0164 k_6 \epsilon^{-50.275t} \cos 4,483.0164t + 80,015.0077 \cos(120\pi t + 89.8229°).$$

$$\tag{6.79}$$

$$i_C''(t) = 27,594.5255k_4\epsilon^{-166.116t} + 2,527.5756k_5\epsilon^{-50.275t}\cos 4,483.0164t$$
$$+ 225,383.6495k_5\epsilon^{-50.275t}\sin 4,483.0164t + 225,383.6495k_5\epsilon^{-50.275t}\sin 4,483.0164t$$
$$- 20,097,436.04k_5\epsilon^{-50.275t}\cos 4,483.0164t + 2,527.756k_6\epsilon^{-50.275t}\sin 4,483.0164t$$
$$- 225,383.6495k_6\epsilon^{-50.275t}\cos 4,483.0164t - 225,383.6495k_6\epsilon^{-50.275t}\cos 4,483.0164t$$
$$- 20,097,436.04k_6\epsilon^{-50.275t}\sin 4,483.0164t - 30,164,947.25\sin(120\pi t + 89.8229°)$$

$$(6.80)$$

Taking $t = 0_+$ in Eqs. (6.78), (6.79) and (6.80) and using the values of the initial conditions contained in 6.77, the following system of algebraic equations results:

$$\begin{cases} k_4 + k_5 = -212.2464 \\ -166.116k_4 - 50.275k_5 + 4,483.0164k_6 = -13,800.8992 \\ 27,594.5255k_4 - 20,094,908.46k_5 - 450,767.2990k_6 = 4,281,641,246. \end{cases}$$

And the solution of this system of equations provides:

$$k_4 = 0.7020; k_5 = -212.9484; k_6 = -5.4406.$$

So, the energizing current of the capacitor bank is

$$i_C(t) = \{0.7020\epsilon^{-166.116t}$$
$$+ \epsilon^{-50.275t}[-212.9484\cos 4,483.0164t - 5.4406\sin 4,483.0164t$$
$$+ 212.2464\sin(120\pi t + 89.8229°)]\}U_{-1}(t)[A].$$

$$(6.81)$$

Checking out:

$$i_C(0_+) = -0.0010[A].$$

According to Example 6.39, the maximum asymmetric energizing current of the capacitor bank occurs at time $t = 250.7[ms]$. So,

$$i_C(0.0007) = 411.2550[A].$$

The maximum symmetric current, or steady state, is $212.2464[A]$. Equation 6.81 can also be written as:

$$i_C(t) = [0.7020\epsilon^{-166.116t}$$
$$+ 213.0179\epsilon^{-50.275t}\sin(4,483.0164t - 91.4635°) \qquad (6.82)$$
$$+ 212.2464\sin(120\pi t + 89.8229°)]U_{-1}(t)[A].$$

According to Fig. 6.214, the current in the industrial load is $i_F(t) = i(t) - i_C(t)$. Then, made the substitutions and calculations, it results that:

$$i_F(t) = -0.1281\epsilon^{-166.116t}$$
$$+1.7637\epsilon^{-50.275t}\sin(4,483.0164t + 89.3860°) \quad (6.83)$$
$$+227.6397\sin(120\pi t - 66.3271°)[A], for\, t \geq 0[s].$$

Checking out:

$$i_F(0_+) = -206.8489)[A].$$

The maximum asymmetric current in the industrial load is

$$i_F(7.135.10^{-3}) = 228.4753[A].$$

And the maximum current of steady state is 227.6397[A].
The voltage in the load resistor is $e_R(t) = R.i_F(t) = 200i_F(t)$. Replacing and making the calculations, it comes that

$$e_R(t) = -25.62\epsilon^{-166.116t}$$
$$+352.74\epsilon^{-50.275t}\sin(4,483.0164t + 89.3860°) \quad (6.84)$$
$$+45,527.94\sin(120\pi t - 66.3271°), for\, t \geq 0[s].$$

Checking:

$$e_R(0_+) = -41,369.7827[V] = -41.3698[kV].$$

The maximum asymmetric voltage at the load resistance is

$$e_R(7.135.10^{-3}) = 200.i(7.135.10^{-3}) = 45,695.06[V].$$

From Eq. (6.84),

$$e_R(7.135.10^{-3}) = 45,695.0736[V].$$

And the maximum steady state voltage is 45.5279[kV].
The voltage at the load inductance is obtained from $e_L(t) = Li'_F(t) = 1.2i'_F(t)$.
From Eq. (6.83),

$$i'_F(t) = 21.2795\epsilon^{-166.116t}$$
$$+ \epsilon^{-50.275t}[7,906.696\cos(4,483.0164t + 89.3860°)$$
$$- 88.67\sin(4,483.0164t + 89.3860°)]$$
$$+ 85,818.1451\cos(120\pi t - 66.3271°)[A/s^2]$$

So, $i'_F(0_+) = 34,474.5851[A/s^2]$ and $e_L(0_+) = 1.2i'_F(0_+) = 41,369.5021$ [V]. As $e_L(0_-) = 41,229.6519[V]$, there is a discontinuity in the load inductance voltage of 139.8502[V], at the time of closing of switch k_2.

Once the replacement of $i'_F(t)$ has been made and the calculations made, then the voltage at the load inductance results in

$$e_L(t) = 25.5354\varepsilon^{-166.116t}$$
$$- 9,488.6318\varepsilon^{-50.275t}\sin(4,483.0164t + 0.0285°) \qquad (6.85)$$
$$+ 102,981.7741\sin(120\pi t + 23.6729°)[V], \textit{ for } t > 0[s].$$

Checking:

$$e_L(0_+) = 41,369.5053[V].$$

In the digital solution, ATPDraw presents the value of 108.74[kV], at time $t = 252.5[ms]$, for the maximum asymmetric voltage. So,

$$e_L(2.5.10^{-3}) = 108,804.0104[V] = 108.8040[kV].$$

And the maximum value of permanent regime is $102,981.7741[V]$. For $t > 0$, the voltage in the load is now $e_A(t) = e_C(t) = e_R(t) + e_L(t)$. Using Eqs. (6.84) and (6.85) it results that:

$$e_A(t) = -0.0846\epsilon^{-166.116t}$$
$$- 9,491.2325\epsilon^{-50.275t}\sin(4,483.0164t - 2.1012°) \qquad (6.86)$$
$$+ 112,596.7989\sin(120\pi t - 0.1772°)\ [V], \textit{ for } t > 0[s].$$

Checking:

$$e_A(0_+) = -0.3222[V].$$

From the digital solution, the maximum asymmetric value occurs at $t = 253.88[ms]$ and is equal to 119.64[kV]. So,

$$e_A(0.00388) = 119.6607[\text{kV}].$$

And the maximum value of the steady state voltage on the load is $112,596.7989[\text{V}] = 112.5968[\text{kV}]$.

(i) The lag between voltage and current in the industrial load is now given by $\varphi_2 = \angle e_A(t) - \angle i(t) = -0.1772^\circ - 2.3395^\circ = -2.5167^\circ$. Therefore, the new power factor is

$$fp_2 = \cos \varphi_2 = 0.9990_{CAP}.$$

(j) The instantaneous power in the load resistance is given by $p_R(t) = e_R(t).i_F(t)$. Replacing $e_R(t)$ and $i_F(t)$ with their analytical expressions given by Eqs. (6.84) and (6.83), it follows that:

$$p_R(t) = [-25.62\epsilon^{-166.116t}$$
$$+ 352.74\epsilon^{-50.275t} \sin(4,483.0164t + 89.3860^\circ)$$
$$+ 45,527.94 \sin(120\pi t - 66.3271^\circ)] \times [-0.1281\epsilon^{-166.116t}$$
$$+ 1.7637\epsilon^{-50.275t} \sin(4,483.0164t + 89.3860^\circ)$$
$$+ 227.6397 \sin(120\pi t - 66.3271^\circ)].$$

Performing the calculations and simplifications results in the expression

$$p_R(t) = 3.2819\epsilon^{-332.232t} - 90.3720\epsilon^{216.391t}\sin(4,483.0164t + 89.3860^\circ)$$
$$- 11,664.2582\epsilon^{-166.116t}\sin(120\pi t - 66.3271^\circ)$$
$$+ 80,297.6278\epsilon^{-50.275t}[-\sin(4,106.0253t + 65.7131^\circ) + \sin(4,860.0075t - 66.9411^\circ)]$$
$$+ 311.0638\epsilon^{-100.55t}[1 + \sin(8,966.0328t + 88.7720^\circ)]$$
$$+ 5,181,983.3[1 - \sin(240\pi t - 42.6542^\circ)][\text{W}], \textit{for } t > 0 \text{ [s]}$$

$$(6.87)$$

Checking:

$$p_R(0_+) = 8,557,294.6166[\text{W}] = 8.5573[\text{MW}].$$

$$p_R(0.007135) = 10,440,177.81[\text{W}] = 10.4402[\text{MW}].$$

And the average power in the industrial load, obtained from Eq. (6.87), is

$$P_R = P_2 = 5,181,983.3[\text{W}] = 5.1820[\text{MW}].$$

$$Q_2 = P_2.tg\varphi_2 = 5.1820.(-0.0440) = -0.2278[MVAR].$$

The new complex power is

$$S_2 = 5.1820 - j0.2278 = 5.1870\angle - 2.5171° \ [MVA]$$

Therefore, the new apparent power is worth $N_2 = 5.1870[MVA]$.
The expressions contained in Eqs. (6.75) and (6.82) to (6.87) are referred
to the instant $t = 0[s]$ taken, provisionally, on $t = 250[ms]$, corresponding
to the instant of closing of the switch k_2. To reference them at time $t = 0[s]$
of the sinusoidal voltage source that supplies the industrial system, it is
necessary to replace t with $t - 0.250$ in the aforementioned expressions.
Doing this results in the following expressions:

$$i(t) = 6.2323.10^{17}.\epsilon^{-166.116t}$$
$$- 60,721,785.41\epsilon^{-50.275t}\sin(4,483.0164t - 45.9504°) \quad (6.88)$$
$$+ 92.1333\sin(120\pi t + 2.3383°)[A], \textit{for } t \geq 0.25[s].$$

Checking:

$$i(0.25_+) = -206.8528[A].$$

$$i(253.5.10^{-3}) = 267.4289[A].$$

$$i_C(t) = [7.6235.10^{17}.\epsilon^{-166.116t}$$
$$- 61,228,676\epsilon^{-50.275t}\sin(4,483.0164t - 45.9433°) \quad (6.89)$$
$$+ 212.2464\sin(120\pi t + 89.8217°)].U_{-1}(t)[A].$$

Checking:

$$i_C(0.25_+) = -0.0066[A].$$

$$i_C(250.7.10^{-3}) = 411.2634[A].$$

$$i_F(t) = -1.3911.10^{17}.\epsilon^{-166.116t}$$
$$+ 506,948.0821\epsilon^{-50.275t}\sin(4,483.0164t - 45.0938°) \quad (6.90)$$
$$+ 227.6397\sin(120\pi t - 66.3283°)[A], \textit{for } t \geq 0.25[s].$$

Checking:

$$i_F(0.25_+) = -206.8466[A].$$

$$i_F(257.135.10^{-3}) = 228.4753[A].$$

$$e_R(t) = -2.7822.10^{19}.\epsilon^{-166.116t}$$

$$+ 101,389,616.4\epsilon^{-50.275t}\sin(4,483.0164t - 45.0938°) \quad (6.91)$$

$$+ 45,527.94\sin(120\pi t - 66.3283°)[V], for\ t \geq 0.25[s].$$

Checking:

$$e_R(0.25_+) = -41,369.3166[V].$$

$$e_R(257.135.10^{-3}) = 45,695.0701[V].$$

$$e_L(t) = 2.7731.10^{19}.\epsilon^{-166.116t}$$

$$+ 2,727,359,354\epsilon^{-50.275t}\sin(4,483.0164t + 45.5487°)$$

$$+ 102,981.7741\sin(120\pi t + 23.6729°)[V], for\ t > 0.25[s].$$

$$(6.92)$$

Checking:

$$e_L(0.25_+) = 41,369.4835[V].$$

$$e_L(252.5.10^{-3}) = 108,804.6699[V] = 108.8047[kV].$$

$$e_A(t) = -9.1873.10^{16}.\epsilon^{-166.116t}$$

$$+ 2,728,106,885\epsilon^{-50.275t}\sin(4,483.0164t + 43.4190°)$$

$$+ 112,596.7989\sin(120\pi t - 0.1772°)[V], for\ t > 0.25[s].$$

$$(6.93)$$

Checking:

$$e_A(0.25_+) = -0.8034[V].$$

$$e_A(253.88.10^{-3}) = 119,686.6655[V] = 119.6867[kV].$$

$$p_R(t) = 3.8704.10^{36}.\epsilon^{-332.232t}$$

$$- 2.8209.10^{25}.\epsilon^{-216.391t}\sin(4,483.0164t - 45.0938°)$$

$$- 1.2667.10^{22}.\epsilon^{-166.116t}\sin(120\pi t - 66.3283°)$$

$$+ 2.3080.10^{10}.\epsilon^{-50.275t}[-\sin(4,106.0253t - 68.7655°)$$

$$- \sin(4,860.0075t - 21.4221°)]$$

$$+ 2.5700.10^{13}.\epsilon^{-100.55t}[1 - \sin(8,966.0328t - 0.1876°)]$$

$$+ 35,181,983.3[1 - \sin(240\pi t - 42.6565°)][W], for\ t > 0.25[s].$$

$$(6.94)$$

Checking:

$$p_R(0.25_+) = 8,557,441.121[W] = 8.5574[MW].$$

$$p_R(0.257135) = 10,440,162.03[W] = 10.4402[MW].$$

This Example 6.40 shows how the connection of a single capacitance can transform a first order circuit directly into a third order circuit - this is because with the closing of the switch k_2, incorporating the capacitance of $5[\mu F]$, the two inductances are no longer in series connection, and can no longer be replaced by an equivalent of $210[mH]$.

With the closing of only the switch k_1 at time $t = 8.3[ms]$, energizing the industrial system, and the circuit being of the first order, its complete analytical solution, containing the transient and steady-state regimes, can be considered as a *reasonable procedure*. On the other hand, after closing switch k_2 at time $t = 250[ms]$, and the circuit becoming third order, despite the elegance of the results, as shown, the complete analytical solution completely loses its sense of reasonability. As the analytical solution carried out shows, even if it omits most of the extensive calculations, the work of complete analytical solution requires an extreme effort of concentration, and causes an indescribable physical and mental wear, besides the loss of a precious time, counted in days. However, no digital simulation of the same problem, performed in Example 6.39, required more than three seconds. And the conclusion that can be drawn is that it makes no sense to proceed to the analytical solution of a circuit whose order is higher than the second. The path is always the digital solution, and ATPDraw is there for that.

6.6 Proposed Problems

P.6.1 Find $i_1(0_+)$ and $i_2(0_+)$ in the circuit of Fig. 6.215, for $M = \sqrt{L_1L_2}$.

Answer:

$$i_1(0_+) = \frac{L_2E}{R_1L_2 + R_2L_1}$$
$$i_2(0_+) = \frac{-ME}{R_1L_2 + R_2L_1}.$$

P.6.2 In Example 6.10, Fig. 6.12, find

(a) $h(0_+)$ and $h'(0_+)$;
(b) $e_0(0_+)$; $e_0'(0_+)$ if $e_0(0_-) = e_0'(0_-) = 0$ and $e(t) = E\cos(\omega t + \theta)U_{-1}(t)$.

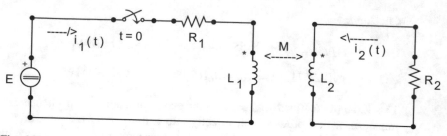

Fig. 6.215 Electric circuit of Problem P.6-1

Answer:

(a) $h(0_+) = 1/2RC$; $h'(0_+) = -(R^2C + L)/(4R^2C^2L)$.
(b) $e_0(0_+) = 0$; $e_0'(0_+) = E \cos \theta / 2RC$.

P.6.3 Considering the circuit in Fig. 6.216:

(a) Determine the initial conditions $e_0(0_+)$, $e_0'(0_+)$ and $e_0''(0_+)$ in the circuit initially de-energized in Fig. 6.216.
(b) Approach $e_0(t)$ by a finite series of singular functions, truncated in the second degree parabolic function $U_{-3}(t)$. c) Using the result of item b, calculate $e_0(0.001)$, $e_0(0.002)$ and $e_0(0.003)$.

  Data: $e(t) = E \cos(\omega t + \theta) U_{-1}(t)$; $E = 100$[V], $\omega = 400$[rad/s], $\theta = 25[°]$, $R_1 = 1[\Omega]$, $R_2 = 2[\Omega]$, $L = 1$[H], $C = 1$[F].

Answer:

(a) $e_0(0_+) = 0$[V]; $e_0'(0_+) = 30.21$[V/s]; $e_0''(0_+) = -5,665.12$[V/s²].
(b) $e_0(t) = 30.21 U_{-2}(t) - 5,665.12 U_{-3}(t) = \left(30.21t - 5,665.12 \cdot \frac{t^2}{2!}\right) U_{-1}(t)$[V].
(c) $e_0(0.001) = 0.027$[V]; $e_0(0.002) = 0.049$[V]; $e_0(0.003) = 0.065$[V].

P.6.4 Determine $e_0(0_+)$ and $e_0'(0_+)$ in the circuit of Fig. 6.217, knowing that it is de-energized at $t = 0_-$ and that $e(t) = E\epsilon^{-\alpha t} \cos \omega t . U_{-1}(t)$.

Answer: $e_0(0_+) = E$; $e_0'(0_+) = -\left(\alpha + \frac{3}{RC}\right)E$.

P.6.5 Knowing that the initial conditions at $t = 0_-$ are null in the circuit of Fig. 6.218, determine $i_2(0_+)$, $i_2'(0_+)$ and $i_2''(0_+)$ for:

(a) $e_1(t) = U_0(t)$;
(b) $e_1(t) = E_1 \cos(\omega t + \theta).U_{-1}(t)$.

Answer:

(a) $i_2(0_+) = \frac{M}{M^2 - L_1 L_2}$; $i_2'(0_+) = 0$; $i_2''(0_+) = \frac{ML_1}{C(M^2 - L_1 L_2)^2}$.
(b) $i_2(0_+) = 0$; $i_2'(0_+) = \frac{ME_1 \cos \theta}{M^2 - L_1 L_2}$; $i_2''(0_+) = \frac{-M\omega E_1 \sin \theta}{M^2 - L_1 L_2}$.

Fig. 6.216 Electric circuit of
Problem P.6-3

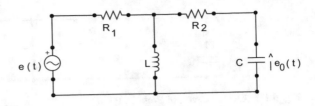

Fig. 6.217 Electric circuit of
Problem P.6-4

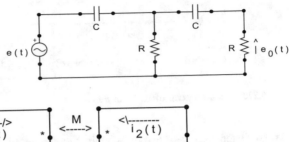

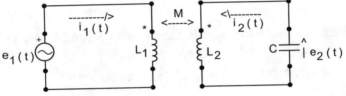

Fig. 6.218 Electric circuit of Problem P.6-5

P.6.6 Determine the expression of the voltage $e_1(t)$ in the circuit of Fig. 6.219, for
$t > 0$, knowing that $e_1(0_-) = E$ and $e_2(0_-) = 0$.

Answer: $e_1(t) = E(0.7236\epsilon^{-0.382t/RC} + 0.2764\epsilon^{-2.618t/RC})$, for $t \geq 0$.

P.6.7 Find the impulse response $h(t)$ of the circuit in Fig. 6.220.

Answer: $h(t) = U_0(t) + (16t - 8)\epsilon^{-4t}.U_{-1}(t).[V]$.

P.6.8 Find the impulse response of circuits (a), (b) and (c) of Fig. 6.221. In circuit
(a), consider that $2R = \sqrt{L/C}$.

Answer:

(a) $h(t) = \frac{2t}{L\sqrt{LC}}\epsilon^{-t/2RC}.U_{-1}(t)$.

(b) $h(t) = U_0(t) + \alpha(0.065\epsilon^{-0.382\alpha t} - 3.065\epsilon^{-2.618\alpha t}).U_{-1}(t), \alpha = R/L$.

(c) $h(t) = \frac{1}{T}(0.423\epsilon^{-0.536t/T} + 1.577\epsilon^{-7.464t/T}).U_{-1}(t), T = RC$.

P.6.9 Find the response to step $r(t)$ of the circuit in Fig. 6.222.

Answer: $r(t) = (1.261\epsilon^{-5.386t} - 0.261\epsilon^{-1.114t}).U_{-1}(t)[V]$

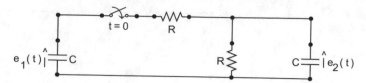

Fig. 6.219 Electric circuit of Problem P.6-6

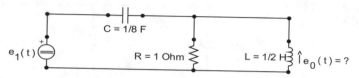

Fig. 6.220 Electric circuit of Problem P.6-7

P.6.10 Determine the current $i_L(t)$ in the circuit of Fig. 6.223, for $t > 0$, if the current source is $i(t) = 100U_{-1}(-t)[A]$.

Answer: $i_L(t) = 107.737\epsilon^{-0.268t} - 7.737\epsilon^{-3.732t}[A]$, for $t \geq 0$.

P.6.11 Find the voltage $e_C(t)$, for $t > 0$, in the circuit of Fig. 6.224, being the voltage source $e(t) = \begin{cases} 10[V], & for\ t < 0[s] \\ 5[V], & for\ t > 0[s] \end{cases}$

Answer: $e_C(t) = 1.21(\epsilon^{-3.28t} - \epsilon^{-1.22t})[V]$, *for* $t \geq 0$.

P.6.12 a) Obtain the analytical expression of the voltage $e_1(t)$, for $t > 0$, in the circuit of Fig. 6.225, knowing that $e_1(0_-) = E$ and that the other elements of the circuit are de-energized at $t = 0_-$. b) Find the numerical solution, voltage and current, in all circuit elements for $E = 100[V]$, $L = 100[mH]$ and $C = 1,000[\mu F]$.

Answer:

(a) $e_1(t) = E(0.276 \cos 0.618\omega_0 t + 0.724 \cos 1.618\omega_0 t)$, *for* $t \geq 0$, $\omega_0 = 1/\sqrt{LC}$.
(b) The circuit built for the solution by ATPDraw is shown in Fig. 6.226; and the solutions are shown in the Figs. 6.227, 6.228, 6.229 and 6.230.

P.6.13 Switch k in the circuit of Fig. 6.231 has been closed for a long time. If it is opened at time $t = 0$, find the analytical expression of the voltage $e_k(t)$ at its terminals, for $t > 0$.

Answer: $e_k(t) = 4 - 2.25\epsilon^{-2.5t}[V]$, *for* $t > 0$.

P.6.14 The circuit shown in Fig. 6.232 was operating in steady-state when the switch was closed at time $t = 0$. Find the analytical expression of the current $i_k(t)$, for $t > 0$.

Answer: $i_k(t) = \frac{E}{(2k+1)R}\left[1 - \epsilon^{-(2k+1)Rt/L}\right].U_{-1}(t)$.

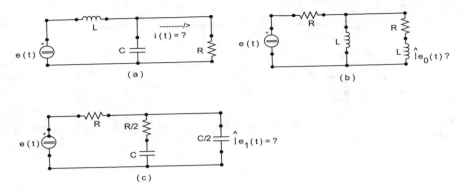

Fig. 6.221 Electric circuit of Problem P.6-8

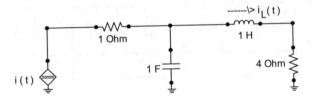

Fig. 6.222 Electric circuit of Problem P.6-9

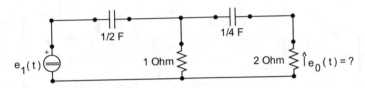

Fig. 6.223 Electric circuit of Problem P.6-10

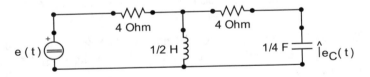

Fig. 6.224 Electric circuit of Problem P.6-11

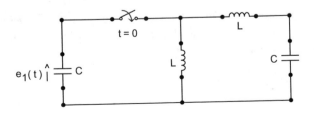

Fig. 6.225 Electric circuit of Problem P.6-12

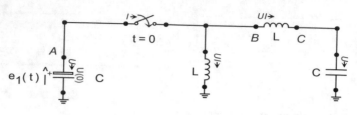

Fig. 6.226 Electric circuit of Problem P.6-12 for the numerical solution by the ATPDraw program

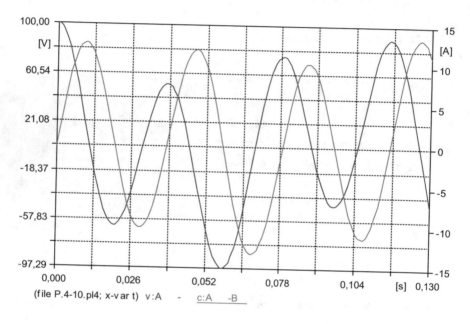

(file P.4-10.pl4; x-v ar t) v:A   -   c:A   -B

Fig. 6.227 Voltage in the left capacitor and current across the switch of Fig. 6.226

P.6.15 The circuit in Fig. 6.233 operated in a steady-state when, at time $t = 0$, switch k was opened. Establish the expression of the voltage $e_k(t)$ at the open switch terminals, for $t > 0$.

Answer: $e_k(t) = \frac{(k+1)LE}{(2k+1)R} U_0(t) + E\left[1 + \frac{k^2}{2k+1} \epsilon^{-\frac{(k+1)Rt}{L}}\right].U_{-1}(t)$.

P.6.16 The circuit shown in Fig. 6.234 was operating steadily when switch k was closed at time $t = 0$. Establish the analytical expression of the current $i_k(t)$ that will circulate through the switch for $t > 0$.

Answer: $i_k(t) = \frac{E}{3R} \epsilon^{-t/3RC}.U_{-1}(t)$.

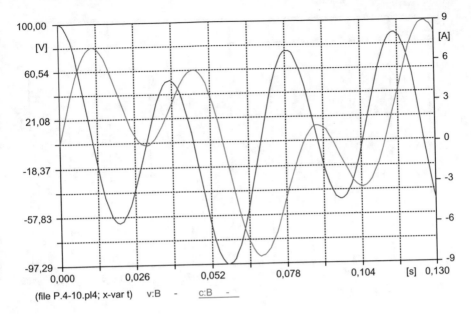

Fig. 6.228 Voltage and current in the left inductor of Fig. 6.226

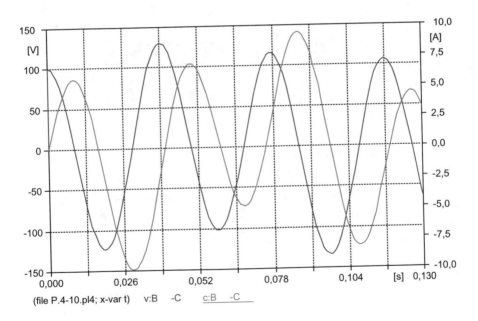

Fig. 6.229 Voltage and current in the right inductor of Fig. 6.226

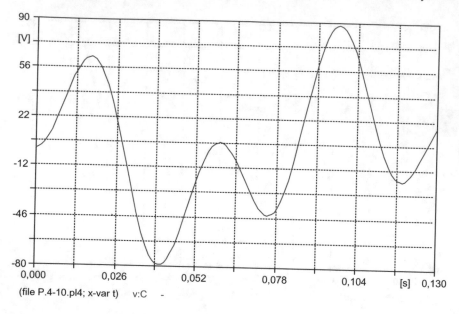

(file P.4-10.pl4; x-var t)   v:C   -

Fig. 6.230 Voltage at the right capacitor of Fig. 6.226

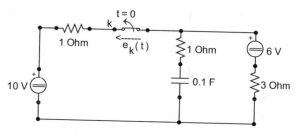

Fig. 6.231 Electric circuit of Problem 6-13

P.6.17 The circuit shown in Fig. 6.235 operated in a steady state fed by a continuous source of $E[V]$. At time $t = 0$, the source collapses, decreasing exponentially, with time constant $T[s]$. Determine the output voltage $e_0(t)$, for all t, that is, $-\infty < t < \infty$.

Answer: $e_0(t) = \begin{cases} \frac{3E}{5}, & for -\infty < t \leq 0 \\ \frac{6E}{10T-3RC} \left(T\epsilon^{-t/T} - \frac{3RC}{10}\epsilon^{-10t/3RC} \right), & for\ 0 \leq t < \infty. \end{cases}$

P.6.18 Switch k in the circuit of Fig. 6.236 is closed at time $t = 0$, the same operating in steady state for $t < 0$. Find the current expression $i_k(t)$, circling the switch, for $t > 0$.

Answer: $i_k(t) = \frac{E}{2R} \left(2 - \epsilon^{-Rt/L} \right).U_{-1}(t).$

6.6 Proposed Problems

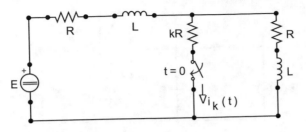

Fig. 6.232 Electric circuit of Problem 6-14

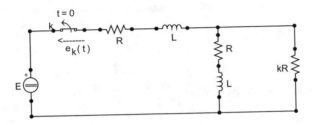

Fig. 6.233 Electric circuit of Problem 6-15

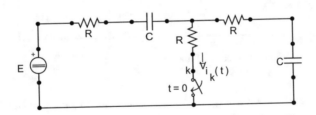

Fig. 6.234 Electric circuit of Problem 6-16

P.6.19 Switch k in the circuit of Fig. 6.237 has been closed for a long time. If it is opened at time $t = 0$, obtain the expression of the voltage at its terminals, $e_k(t)$, for $t > 0$.

Answer: $e_k(t) = \left(10 + 5\epsilon^{-16t}\right).U_{-1}(t)[V]$.

P.6.20 Switch k in the circuit of Fig. 6.238 was opened at time $t = 0$. Find the analytical expression of the voltage $e_k(t)$ at its terminals, for $t > 0$.

Answer: $e_k(t) = \left(0.284\epsilon^{-10t} - 10\right).U_{-1}(t)[V]$.

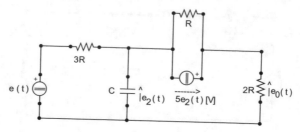

Fig. 6.235 Electric circuit of Problem 6-17

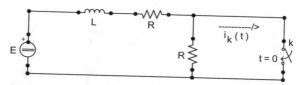

Fig. 6.236 Electric circuit of Problem 6-18

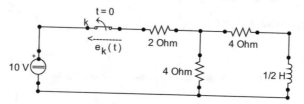

Fig. 6.237 Electric circuit of Problem 6-19

P.6.21 The circuit shown in Fig. 6.239 was operating in steady state when, at time $t = 0$, switch k was opened. Determine the voltage value $e_k(t)$, at the switch terminals, at times $t = 0$ and $t = 0_+$.

Answer:

At $t = 0$, $e_k(t) = \dfrac{LE}{2R} U_0(t)$.

At $t = 0_+$, $e_k(0_+) = E$.

P.6.22 In the circuit of Fig. 6.240:

(a) determine $e_1(t)$, $e_2(t)$, $i(t)$ and $e_R(t)$; $p_1(t)$, $p_2(t)$ and $p_R(t)$, for $t \geq 0$, considering that $e_1(0_-) = E$ and $e_2(0_-) = E/3$.
(b) Perform the numerical solution by ATPDraw for the following values: $E = 90[\text{V}]$, $R = 6[\Omega]$ and $C = 1{,}000[\mu\text{F}]$.

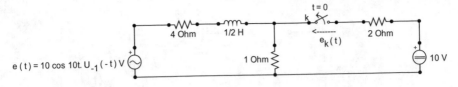

Fig. 6.238 Electric circuit of Problem 6-20

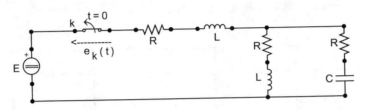

Fig. 6.239 Electric circuit of Problem 6-21

Answer:

(a) $e_1(t) = \frac{E}{9}\left(7 + 2\epsilon^{-3t/RC}\right)$, *for* $t \geq 0$.

$e_2(t) = \frac{E}{9}\left(7 - 4\epsilon^{-3t/RC}\right)$, *for* $t \geq 0$.

$$i(t) = \frac{2E}{3R}\epsilon^{-3t/RC}, \; for \; t > 0.$$

$$e_R(t) = \frac{2E}{3}\epsilon^{-3t/RC}, \; for \; t > 0.$$

$$p_1(t) = -\frac{E^2}{27R}\left(14\epsilon^{-3t/RC} + 4\epsilon^{-6t/RC}\right), \; for \; t > 0.$$

$$p_2(t) = \frac{E^2}{27R}\left(14\epsilon^{-3t/RC} - 8\epsilon^{-6t/RC}\right), \; for \; t > 0.$$

$$p_R(t) = \frac{4E^2}{9R}\epsilon^{-6t/RC}, \; for \; t > 0.$$

(b) (Figs. 6.241 and 6.242)

P.6.23 The circuit shown in Fig. 6.243 has been operating for a long time with switch k closed. If it is opened at time $t = 0$, determine the analytical expression of the current $i_2(t)$, for $t > 0$.

Answer: $i_2(t) = \frac{E}{R}\left(\frac{2}{5} - \frac{2}{9}\epsilon^{-Rt/L} - \frac{8}{45}\epsilon^{-5Rt/2L}\right).U_{-1}(t)$.

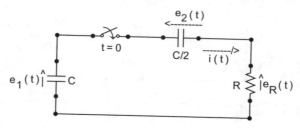

Fig. 6.240 Electric circuit of Problem 6-22

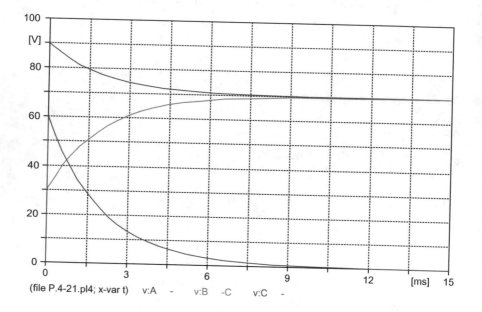

Fig. 6.241 Voltages at capacitances and resistance of Fig. 6.240

P.6.24 (a) The circuit of Fig. 6.244 was operating in steady state when switch k was opened at $t = 0$. Knowing that at the time of switching $i_k(t) = 0$, with $i'_k(t) < 0$, determine $e_L(t)$ for $t > 0$. If $E = 100[\mathrm{V}]$, $f = 60[\mathrm{Hz}]$ and $L = 100[\mathrm{mH}]$, find the numerical solution for (b) $C = 10[\mu\mathrm{F}]$, (c) $C = 40[\mu\mathrm{F}]$ and d) $C = 100[\mu\mathrm{F}]$.

Answer:

(a) $e_L(t) = \frac{\omega^2 E}{\omega_0^2 - \omega^2}(\cos \omega_0 t - \cos \omega t).U_{-1}(t)$, with $\omega_0^2 = 1/LC$.

(b) Figs. 6.245, 6.246 and 6.247

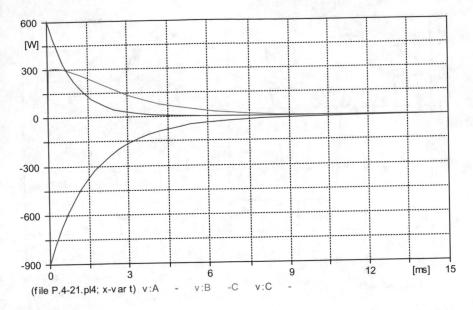

(file P.4-21.pl4; x-var t) v:A   -   v:B   -C   v:C   -

Fig. 6.242 Power at capacitances and resistance of Fig. 6.240

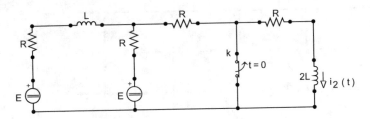

Fig. 6.243 Electric circuit of Problem 6-23

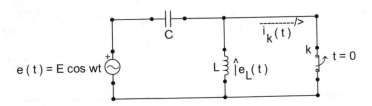

Fig. 6.244 Electric circuit of Problem 6-24

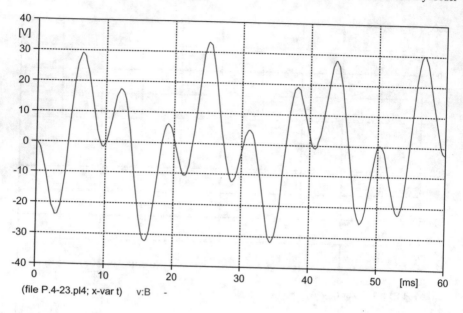

(file P.4-23.pl4; x-var t)   v:B   -

Fig. 6.245   Voltages at inductance for $C = 10[\mu F]$ of Fig. 6.244

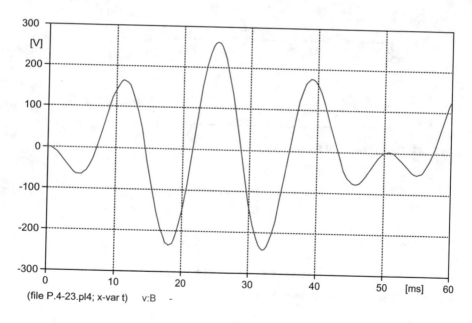

(file P.4-23.pl4; x-var t)   v:B   -

Fig. 6.246   Voltages at inductance for $C = 40[\mu F]$ of Fig. 6.244

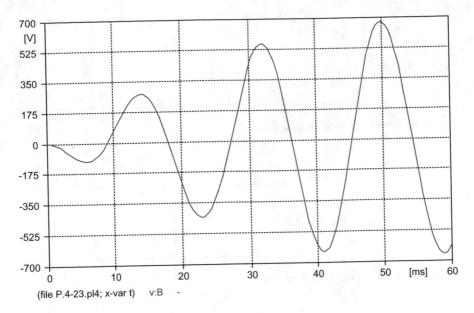

(file P.4-23.pl4; x-var t)   v:B   -

Fig. 6.247 Voltages at inductance for $C = 100[\mu F]$ of Fig. 6.244

P.6.25 With the circuit of Fig. 6.248 operating in steady-state, switch k was closed at $t = 0$, this time when $e(t) = 0$ with $e'(t) < 0$. Find the analytical expression of $e_C(t)$, for $t > 0$. Consider that $\omega_0 = 1/\sqrt{LC} = 2\omega$.

Answer: $e_C(t) = -\frac{2E}{35}\left[\sqrt{2}\sin(2\sqrt{2}\omega t) + 10\sin(\omega t)\right]$, for $t \geq 0$.

(Without origin correction, that is, $t = 0$ being the instant of closing the switch).

P.6.26 The circuit in Fig. 6.249 operated in a steady state when, at time $t = 0$, switch k was opened. Knowing that at the instant of switching the current in the inductance was minimal, determine the voltage $e_k(t)$ at the switch terminals, for $t > 0$.

Answer: $e_k(t) = \frac{\omega_0^2 E}{\omega_0^2 - \omega^2}\left(\sin\omega t - \frac{\omega_0}{\omega}\sin\omega_0 t\right).U_{-1}(t)$, where $\omega_0 = 1/\sqrt{LC}$.

(No source correction).

P.6.27 Switch k in the circuit of Fig. 6.250 was closed at time $t = 0$. Before $t = 0$, the circuit was operating in a steady state. In addition, at the moment of switching the current in the inductances was maximum. Determine the current $i_k(t)$ circulating through the switch for $t > 0$.

Answer: $i_k(t) = \frac{\omega E}{L(\omega^2 - 2\omega_0^2)}\left(\cos\omega t - \cos\sqrt{2}\omega_0 t\right).U_{-1}(t)$, where $\omega_0 = 1/\sqrt{LC}$.

(No source correction).

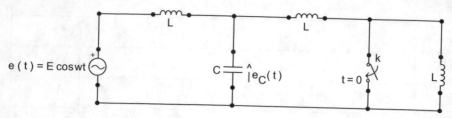

Fig. 6.248 Electric circuit of Problem 6-25

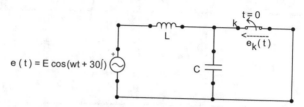

Fig. 6.249 Electric circuit of Problem 6-26

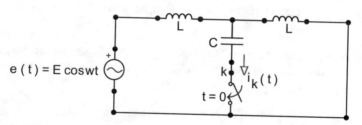

Fig. 6.250 Electric circuit of Problem 6-27

P.6.28 The circuit shown in Fig. 6.251 was operating in steady-state when switch k was opened at time $t = 0$. Determine the voltage $e_k(t)$ at the switch terminals, for $t > 0$, knowing that at the time of switching $i_k(t) = 0$ with $i'_k(t) < 0$.

Answer:

$$e_k(t) = \frac{E}{1-(\omega/\omega_0)^2}\left[-\cos\omega t + \frac{1}{2-(\omega/\omega_0)^2}\cos\omega_0 t\right].U_{-1}(t), \text{ with } \omega_0 = 1/\sqrt{LC}.$$

($t = 0$ being the instant the switch k is opened).

P.6.29 The circuit in Fig. 6.252 operated in a steady state when switch k was closed at time $t = 0$. Find the expression of the current $i_k(t)$ circulating through the switch, for $t > 0$.

Answer: $i_k(t) = \left[\frac{t}{2L} + \sqrt{\frac{C}{L}}sin\omega_0 t + \frac{\sqrt{2}}{4}\sqrt{\frac{C}{L}}sin\left(\sqrt{2}\omega_0 t\right)\right]E.U_{-1}(t); \ \omega_0 = 1/\sqrt{LC}.$

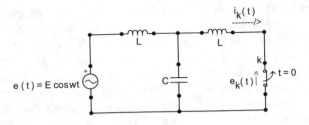

Fig. 6.251 Electric circuit of Problem 6-28

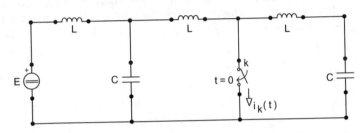

Fig. 6.252 Electric circuit of Problem 6-29

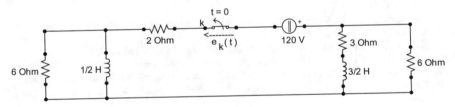

Fig. 6.253 Electric circuit of Problem 6-30

P.6.30 In the circuit of Fig. 6.253 the switch k is opened at $t = 0$, after having been closed for a long time. Determine the voltage $e_k(t)$ at its terminals, for $t > 0$.

Answer: $e_k(t) = \left(120 + 180\epsilon^{-12t} + 120\epsilon^{-6t}\right).U_{-1}(t)[V]$

6.7   Conclusions

In this chapter it was presented the analytical and digital solution for ordinary differential equations of any order, i.e. for circuits with any number of capacitances and inductances. In general, the order of the differential equation for a variable of interest in a given circuit is at most equal to the sum of the number of capacitances and inductances it contains. Logically, before calculating this sum, it is necessary to reduce the capacitances or inductances that constitute series, parallel or series-parallel connections to their equivalent capacitances or inductances. Initially, circuits powered by independent sources, starting at $t = 0$, containing capacitors and inductors without energy initially stored at $t = 0_-$, that is, circuits initially de-energized, were addressed. Then those with energy initially stored at $t = 0_-$ were studied, with or without external sources. The latter were solved, analytically, with the use of the Thévenin and Norton equivalents of capacitances and inductances initially energized. Typical examples of this second situation, that is, circuits containing capacitive elements with initial voltage at $t = 0_-$ and inductors with initial current at $t = 0_-$, the transients caused by switching. Whenever the examples are not exclusively literal, the respective numerical solutions were also presented using the ATPDraw program. Therefore, after the studying this chapter completely, a good understanding of analytical and digital solution of transients in fundamental circuits is possible.

References

1. Adkins WA, Davidson MG (2012) Ordinary differential equations, undergraduate texts in mathematics. Springer, Berlin
2. Boyce WE, Diprima RC (2002) Elementary differential equations and boundary value problems, 7th ed. LTC
3. Chicone C (2006) Ordinary differential equations with applications, 22nd. ed. Springer, Berlin
4. Close CM (1975) Linear circuits analysis, LTC
5. Doering CI, Lopes AO (2008) Ordinary differential equations, 3rd ed. IMPA
6. Irwin JD (1996) Basic engineering circuit analysis. Prentice Hall
7. Kreider DL, Kluler RG, Ostberg DR (2002) Differential equations, Edgard Blucher Ltda
8. Leithold L (1982) The calculus with analytical geometry, vol.1, 2nd. ed. Harper & How of Brasil
9. Perko L (2001) Differential equations and dynamical systems, 3rd ed. Springer, New York
10. Stewart J (2010) Calculus. vol 1, 6th. ed. Thomson
11. Zill DG, Cullen MR (2003) Differential equations, 3rd ed. Makron Books

Chapter 7
Switching Transients Using Injection of Sources

7.1 Introduction

This chapter presents another methodology for the study of transient phenomena caused by the opening or closing of one or more switches, in circuits that are or are not operating in a steady-state, before switching. It is a technique applicable, preferably, to linear circuits and invariant over time, and whose principle is to simulate the closing of a switch by injecting a voltage source into the circuit; or the simulation of opening a switch by injecting a current source. In essentially linear circuits, the procedure is based on the *substitution theorem*, also called the *compensation theorem*, and on the *superposition theorem*. In the development of the method it is assumed that the independent sources present in the circuit are sinusoidal, making it clear, at first, that it applies equally to other types of independent sources. In contrast to the preceding chapters, the various examples presented here are solved using both the classic method of ordinary differential equations and the operational method of Laplace transforms.

Considering that the superposition theorem is well known [1–6], having even been applied numerous times in the examples in previous chapters, only a brief recap of the substitution theorem is made.

Substitution Theorem

In any circuit, not necessarily linear, in which all voltages and currents of all its branches are known, if any branch k is associated with a voltage $e_k(t)$ and a current $i_k(t)$, it is possible to substitute it by another one, among the three mentioned below, without this substitution altering the general behavior of the circuit, that is, without this altering the other voltages and currents of that circuit:

(1) A voltage independent source of value $e_k(t)$, traveled by current $i_k(t)$;
(2) An independent current source of value $i_k(t)$, subject to voltage $e_k(t)$;
(3) A resistor of resistance $R_k = e_k(t)/i_k(t)$, or conductance $G_k = i_k(t)/e_k(t)$.

© The Author(s), under exclusive license to Springer Nature Switzerland AG 2021
J. C. Goulart de Siqueira and B. D. Bonatto, *Introduction to Transients in Electrical Circuits*, Power Systems, https://doi.org/10.1007/978-3-030-68249-1_7

7.2 Method of Voltage Source Injection

It is a way to *simulate the closing of a switch*, at any instant t_0, in a circuit that is or is not operating in a steady-state; and basically consists of the injection of an independent voltage source of equal value and contrary to the voltage that would continue to exist at the switch terminals, after the time t_0, if it had not been closed at that time t_0. The method is explained in the following five steps, assuming sinusoidal sources and that the circuit is operating in steady-state before time t_0.

1. Referring to Fig. 7.1, the steady-state voltage $e_k(t)$ is initially calculated for all t, between terminals a and b of switch k, and the other steady-state voltages and currents in the various circuit elements, interior to the active networks N_1 and N_2, on which there is some interest.
   Assuming that $e_k(t) = \boldsymbol{E}\cos(\omega t + \theta)$, then $i_k(t) = 0$; $p_k(t) = e_k(t).i_k(t) = 0$.
2. According to the substitution theorem, the open switch k, in Fig. 7.1, can be replaced by the ideal independent voltage source F_1, of the same value as the voltage that would continue to exist at the switch terminals if it were not closed at the instant $t = t_0$. Logically, this *substitute voltage source F_1* must be traversed by a zero current whenever acting alone between terminals a and b, as shown in Fig. 7.2.
   In Fig. 7.2, we have, then, that

$$e_1(t) = e_k(t) = \boldsymbol{E}\cos(\omega t + \theta) = \boldsymbol{E}\cos(\omega t + \theta).[U_{-1}(-t+t_0) + U_{-1}(t - t_0)];$$

$$i_k(t) = 0;\ p_{F_1}(t) = p_k(t) = 0.$$

The sum of steps $U_{-1}(-t+t_0) + U_{-1}(t - t_0)$ serves only to emphasize that $e_1(t) = E\cos(\omega t + \theta)$ for both $t < t_0$ and $t > t_0$.
3. To simulate the closing of switch k at the time $t = t_0$, a second ideal source of voltage F_2, of opposite polarity to that of F_1, is injected in series with the source F_1, from the same time $t = t_0$. Its value, therefore, is equal to the *future voltage of switch k*, that is, the voltage that would continue to exist between terminals a and b of switch k, if it had not been closed at the time $t = t_0$, nor after it. Thus, the total voltage between terminals a and b will be zero after the insertion of voltage source F_2, that is, for $t > t_0$, corresponding, therefore, to the closing of switch k, as shown in Fig. 7.3.
   In Fig. 7.3,

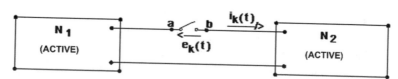

Fig. 7.1 Circuit with switch k open

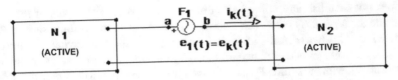

Fig. 7.2 Circuit in Fig. 7.1 with switch k open, replaced by the ideal voltage source F_1

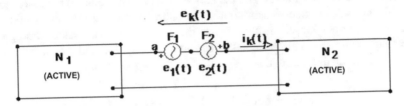

Fig. 7.3 Circuit after the closing of switch k, simulated with the injection of the voltage source F_2

$$e_1(t) = \mathbf{E}\cos(\omega t + \theta);\ e_2(t) = \mathbf{E}\cos(\omega t + \theta).U_{-1}(t - t_0);\ i_k(t) \neq 0, p_k(t) = 0;$$

$$e_k(t) = e_1(t) - e_2(t).$$

In detail, we have then:
For $t < t_0$, switch k open,

$$e_1(t) = E\cos(\omega t + \theta);\ e_2(t) = 0\ (F_2 \text{ in short-circuit}) \text{ and } e_k(t) = E\cos(\omega t + \theta).$$

For $t > t_0$, switch k closed,

$$e_1(t) = E\cos(\omega t + \theta);\ e_2(t) = E\cos(\omega t + \theta) \text{ and } e_k(t) = 0.$$

Since $e_k(t) = e_1(t) - e_2(t)$, $e_k(t).i_k(t) = e_1(t).i_k(t) - e_2(t).i_k(t)$.
So, the instantaneous power on the switch is

$$p_k(t) = p_{F_1}(t) + p_{F_2}(t) = 0,$$

showing that: **the power released by source F_2 is consumed by source F_1.**
In detail:
For $t < t_0$, switch k open, $e_k(t) = e_1(t) \neq 0$, $i_k(t) = 0$. Therefore, $p_k(t) = 0$.
For $t > t_0$, switch k closed, $e_k(t) = 0$, $i_k(t) \neq 0$. Therefore, $p_k(t) = 0$.
The current $i_k(t)$ is a consequence of the exclusive action of the source F_2, for $t > t_0$, which, due to the serial connection of the two sources, also circulates through the source F_1. Therefore, $p_k(t) = 0$.

4. The calculation of voltages and currents in all circuit elements that are of interest for $t > t_0$, that is, after the closing of switch k, is done by applying the superposition theorem. As the power at source F_1 is zero in the absence of source F_2, that is, with source F_2 at rest (short-circuited), because in this condition $i_k(t) = 0$ (Fig. 7.2), it

does not establish any voltage or current on the passive elements of the circuit inside the N_1 and N_2 networks. Therefore, **the effect of the voltage source F_1 on the circuit is always null**—after all, it only replaces the open switch! On the other hand, the voltages and currents produced by the independent sources contained in N_1 and N_2, with the source F_2 at rest, have already been previously calculated in the initial phase of the process (Fig. 7.1). Therefore, it remains to calculate only the voltages and currents produced by the voltage source F_2, acting alone, that is, with the voltage source F_1 at rest and the networks N_1 and N_2 reduced to passive form, with the voltage independent sources in short-circuit and current independent sources in open circuit. In this phase, the current $i_k(t)$ circulating through the voltage source F_2 is also calculated, as shown in Fig. 7.4, and that, in fact, is the current circulating through switch k after its closing at $t = t_0$.

In Fig. 7.4,

$$e_2(t) = \mathbf{E}\cos(\omega t + \theta).U_{-1}(t - t_0); \; i_k(t) \neq 0;$$

$$e_k(t) = -e_2(t) = -\mathbf{E}\cos(\omega t + \theta), \quad \text{for } t > t_0; \; p_{F_2}(t) = -e_2(t).i_k(t) \neq 0.$$

5. Finally, the resulting values of voltage and current in the circuit elements as a whole, after closing the switch k at the time instant $t = t_0$, are obtained by superimposing the values found with the isolated update of the source F_2 (Fig. 7.4) with those obtained before the closing of switch k (Fig. 7.1). As already mentioned in 4, the current through the voltage source F_2, when it acts in isolation, is the current itself circulating through the switch k after it is closed at $t = t_0$.

It is necessary to point out that if the closing moment of the switch k, $t = t_0$, is taken as a **new moment** $t = 0$, that is, if the time will be computed from the moment of closing of the switch k, then the analytical expressions of all voltages and currents in the circuit, both those that were given and those that were calculated before the closing of switch k (Fig. 7.1), should undergo corrections in their initial phases, to be referred to this new origin of time. And this is done simply by adding the value ωt_0 to their respective phases, that is, replacing t with $t + t_0$ in the corresponding expressions. So, for example, $e_2(t)$ would be

$$e_2(t) = E\cos[\omega(t + t_0) + \theta)]U_{-1}[(t + t_0) - t_0] = E\cos(\omega t + \theta + \omega t_0).U_{-1}(t).$$

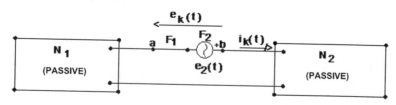

Fig. 7.4 Circuit reduced to passive for the calculation of voltages and currents produced exclusively by the source F_2, for $t > t_0$

Example 7.1 The circuit in Fig. 7.5 operated in a steady-state when switch k was closed at time $t = 0$. Find the expression of the current $i_k(t)$ flowing through the switch for $t > 0$. Also find the expression of the current $i_L(t)$ for $t > 0$.

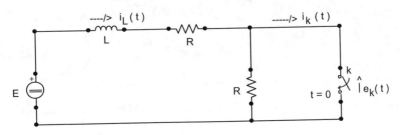

Fig. 7.5 Electric circuit for Example 7.1

Solution
In steady-state, the inductance behaves like a short-circuit, with switch k open and for $t < 0$, then,

$$i_L(t) = \frac{E}{2R}; \ e_k(t) = \frac{E}{2}.$$

Thus, for $t > 0$, the circuit that must be resolved is that of Fig. 7.6, with the inductance completely de-energized, the original voltage source at rest and the exclusive actuation of the source F_2.

The mesh equations for the circuit in Fig. 7.6 are:

$$LDi_L(t) + Ri_L(t) + R[i_L(t) - i_k(t)] = 0$$

$$R[i_k(t) - i_L(t)] = \frac{E}{2}U_{-1}(t).$$

(The voltage source has positive polarity with respect to the bottom node of the circuit).

Or,

$$(LD + 2R)i_L(t) - Ri_k(t) = 0$$

$$-2Ri_L(t) + 2Ri_k(t) = EU_{-1}(t).$$

The solution of this system of equations results in

$$i_L(t) = \frac{E}{2R}\left(1 - \epsilon^{-Rt/L}\right).U_{-1}(t)$$

$$i_k(t) = \frac{E}{2R}\left(2 - \epsilon^{-Rt/L}\right).U_{-1}(t).$$

Superimposing these results with those of steady-state, with the switch k open, it comes that:

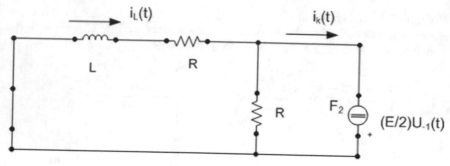

Fig. 7.6 Electric circuit of Example 7.1 with the exclusive performance of the F_2 source

$$i_L(t) = \frac{E}{2R} + \frac{E}{2R}\left(1 - \epsilon^{-Rt/L}\right).U_{-1}(t) = \frac{E}{2R}\left(2 - \epsilon^{-Rt/L}\right), \; for \; t \geq 0.$$

$$i_k(t) = 0 + \frac{E}{2R}\left(2 - \epsilon^{-Rt/L}\right).U_{-1}(t) = \frac{E}{2R}\left(2 - \epsilon^{-Rt/L}\right).U_{-1}(t).$$

As could be expected, $i_L(t) = i_k(t)$, for $t > 0$, with a new steady-state regime equal to E/R.

Example 7.2 The circuit of Fig. 7.7 has been operating with switch k open for a long time. If it is closed at time $t = 0$, find the current $i_k(t)$ circulating through it, for $t > 0$.

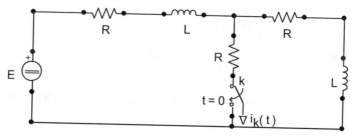

Fig. 7.7 Electric circuit for Example 7.2

Solution

The steady-state solution, with the switch k open, is shown in Fig. 7.8.

The circuit for calculating the exclusive effects of the (injected) source F_2 is shown in Fig. 7.9.

The corresponding **Transformed Circuit, in the domain of the complex frequency** $s = \sigma + j\omega$, is shown in Fig. 7.10. The Laplace transform of the injected voltage source was taken and the inductances were replaced by their operational impedances sL; and $i_k(t)$ was replaced by its Laplace transform $I_k(s)$.

Figure 7.10 shows that the current of the source is the quotient of its voltage by the operational impedance seen by it, that is,

7.2 Method of Voltage Source Injection

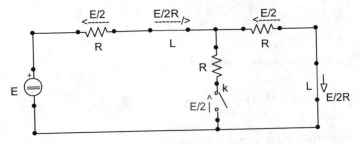

Fig. 7.8 Electric circuit of Example 7.2 with the steady-state solution

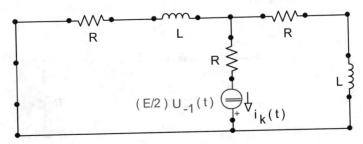

Fig. 7.9 Electric circuit of Example 7.2 for calculating the exclusive effects of the (injected) source F_2

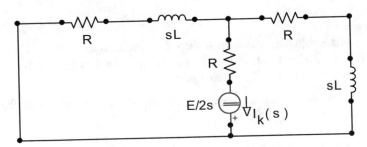

Fig. 7.10 Transformed circuit of from that of Fig. 7.9, in the domain of the complex frequency

$$I_k(s) = \frac{\frac{E}{2s}}{R + \frac{R}{2} + \frac{sL}{2}} = \frac{\frac{E}{2s}}{\frac{3}{2}R + \frac{1}{2}sL} = \frac{E}{s} \cdot \frac{1}{sL + 3R} = \frac{E}{L} \cdot \frac{1}{s(s + 3R/L)}$$
$$= \frac{E}{L} \cdot \frac{L}{3R} \left(\frac{1}{s} - \frac{1}{s + 3R/L} \right)$$

$$I_k(s) = \frac{E}{3R} \left(\frac{1}{s} - \frac{1}{s + 3R/L} \right).$$

In the expansion in partial fractions above, it was used the identity

$$\frac{k}{(s+a)(s+b)} = \frac{k}{b-a}\left(\frac{1}{s+a} - \frac{1}{s+b}\right) \tag{7.1}$$

Finally, taking the inverse transform of $I_k(s)$ results

$$i_k(t) = \frac{E}{3R}\left(1 - \epsilon^{-3Rt/L}\right).U_{-1}(t).$$

Example 7.3 The circuit in Fig. 7.11 was operating in steady-state with switch k open. If it is closed at time $t = 0$, establish the expressions of $i_k(t)$, $e_1(t)$, $e_2(t)$ and $e_3(t)$, for $t > 0$, knowing that $e_3(t) = 0$ for $t < 0$.

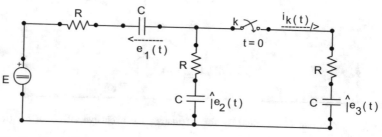

Fig. 7.11 Electric circuit for Example 7.3

Solution

The steady-state solution, with switch k open, is obtained from Fig. 7.12. In it, the capacitances on the left were replaced by open circuits, and the one on the right by a short-circuit, as it was initially de-energized.

With the switch k open, the capacitors on the left are in series and, therefore, with the same charge $q(t)$. So,

$$e_1(t) = \frac{q(t)}{C}; \ e_2(t) = \frac{q(t)}{C}; \ e_1(t) + e_2(t) = E.$$

Therefore,

$$2\frac{q(t)}{C} = E \text{ and } q(t) = \frac{CE}{2}; \ e_1(t) = e_2(t) = \frac{E}{2}; \ e_3(t) = 0.$$

$$e_k(t) = e_2(t) - e_3(t) = \frac{E}{2}; \ i_k(t) = 0.$$

For $t > 0$, that is, with switch k closed, the circuit with the injection of the source F_2 is shown in Fig. 7.13.

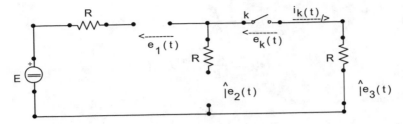

Fig. 7.12 Electric circuit of Example 7.3 for steady-state solution

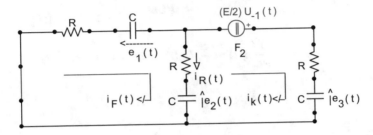

Fig. 7.13 Electric circuit of Example 7.3 with the injection of the source F_2

The mesh equations for the circuit in Fig. 7.13 are

$$\left(2R + \frac{2}{CD}\right)i_F(t) - \left(R + \frac{1}{CD}\right)i_k(t) = 0.$$

$$-\left(R + \frac{1}{CD}\right)i_F(t) + \left(2R + \frac{2}{CD}\right)i_k(t) = \frac{E}{2}U_{-1}(t).$$

or,

$$2(RCD + 1)i_F(t) - (RCD + 1)i_k(t) = 0.$$

$$-(RCD + 1)i_k(t) + 2(RCD + 1)i_k(t) = \frac{CE}{2}U_0(t).$$

So,

$$\begin{vmatrix} 2(RCD + 1) & -(RCD + 1) \\ -(RCD + 1) & 2(RCD + 1) \end{vmatrix} i_F(t) = \begin{vmatrix} 0 & -(RCD + 1) \\ \frac{CE}{2}U_0(t) & 2(RCD + 1) \end{vmatrix}.$$

$$(RCD + 1)^2 \cdot \begin{vmatrix} 2 & -1 \\ -1 & 2 \end{vmatrix} i_F(t) = (RCD + 1) \begin{vmatrix} 0 & -1 \\ \frac{CE}{2}U_0(t) & 2 \end{vmatrix}.$$

$$i_F'(t) + \frac{1}{RC}i_F(t) = \frac{E}{6R}U_0(t), \ \ with \ i_F(0_-) = 0.$$

$$i_F(t) = \frac{E}{6R}\epsilon^{-t/RC}.U_{-1}(t).$$

Similarly,

$$(RCD+1)^2.\begin{vmatrix} 2 & -1 \\ -1 & 2 \end{vmatrix}i_k(t) = \begin{vmatrix} 2(RCD+1) & 0 \\ -(RCD+1) & \frac{CE}{2}U_0(t) \end{vmatrix}$$

$$= (RCD+1).\begin{vmatrix} 2 & 0 \\ -1 & \frac{CE}{2}U_0(t) \end{vmatrix}.$$

$$i_k'(t) + \frac{1}{RC}i_k(t) = \frac{E}{3R}U_0(t), \ \ with \ i_k(0_-) = 0.$$

$$i_k(t) = \frac{E}{3R}\epsilon^{-t/RC}.U_{-1}(t).$$

Then,

$$e_1(t) = \frac{1}{C}\int_{0_+}^{t} i_F(\lambda)d\lambda = \frac{E}{6RC}\int_{0_+}^{t} \epsilon^{-\lambda/RC}d\lambda = \frac{E}{6}\left(1 - \epsilon^{-t/RC}\right).U_{-1}(t).$$

By superposition,

$$e_1(t) = \frac{E}{2} + \frac{E}{6}\left(1 - \epsilon^{-t/RC}\right) = \frac{E}{6}\left(4 - \epsilon^{-t/RC}\right), \ \ for \ t \geq 0.$$

$$i_R(t) = i_F(t) - i_k(t) = -\frac{E}{6R}\epsilon^{-t/RC}.U_{-1}(t).$$

On the other hand,

$$e_2(t) = \frac{1}{C}\int_{0_+}^{t} i_R(\lambda)d\lambda = -\frac{E}{6RC}\int_{0_+}^{t} \epsilon^{-\lambda/RC}d\lambda = \frac{E}{6}\left(\epsilon^{-t/RC} - 1\right).U_{-1}(t).$$

By superposition,

$$e_2(t) = \frac{E}{2} + \frac{E}{6}\left(\epsilon^{-t/RC} - 1\right) = \frac{E}{6}\left(2 + \epsilon^{-t/RC}\right), \ \ for \ t \geq 0.$$

Finally,

$$e_3(t) = \frac{1}{C} \int_{0_+}^{t} i_k(\lambda)d\lambda = \frac{E}{3RC} \int_{0_+}^{t} \epsilon^{-\lambda/RC}d\lambda = \frac{E}{3}\left(1 - \epsilon^{-t/RC}\right).U_{-1}(t).$$

Example 7.4   The capacitances C_1 and C_2 of the circuit of Fig. 7.14 are charged and with voltages E_1 and E_2, respectively, for $t < 0$. If the switch k is closed at time $t = 0$, find the expressions of $e_1(t)$ and $e_2(t)$, for $t > 0$.

Fig. 7.14 Electric circuit for Example 7.4

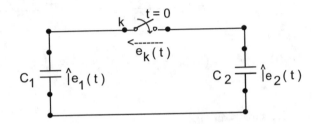

Solution

The *future voltage* at the terminals of switch k, that is, the voltage existing between its terminals if it were not closed at $t = 0$, is $e_k(t) = E_1 - E_2$. So, the circuit with the injection of the source F_2 is the one in Fig. 7.15.

The mesh equation for the circuit in Fig. 7.15 is

$$\frac{1}{C_1}\int i_k(t)dt + \frac{1}{C_2}\int i_k(t)dt = (E_1 - E_2)U_{-1}(t).$$

or,

$$\left(\frac{1}{C_1} + \frac{1}{C_2}\right)i_k(t) = (E_1 - E_2)U_0(t).$$

Therefore,

$$i_k(t) = \frac{C_1C_2(E_1 - E_2)}{C_1 + C_2}U_0(t).$$

So,

$$e_1(t) = -\frac{1}{C_1}\int_{0_-}^{t} i_k(\lambda)d\lambda = -\frac{1}{C_1}\cdot\frac{C_1C_2(E_1 - E_2)}{C_1 + C_2}\int_{0_-}^{t} U_0(\lambda)d\lambda$$

$$= \frac{-C_2(E_1 - E_2)}{C_1 + C_2}U_{-1}(t).$$

Fig. 7.15 Electric circuit of
Example 7.4 with the
injection of the source F_2

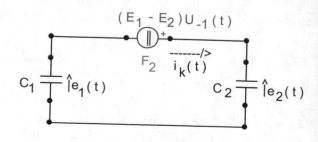

By superposition,

$$e_1(t) = E_1 + \frac{-C_2(E_1 - E_2)}{C_1 + C_2} = \frac{C_1 E_1 + C_2 E_2}{C_1 + C_2}, \quad \text{for } t > 0.$$

$$e_2(t) = \frac{1}{C_2} \int_{0_-}^{t} i_k(\lambda)d\lambda = \frac{1}{C_2} \cdot \frac{C_1 C_2(E_1 - E_2)}{C_1 + C_2} \int_{0_-}^{t} U_0(\lambda)d\lambda = \frac{C_1(E_1 - E_2)}{C_1 + C_2} U_{-1}(t).$$

By superposition,

$$e_2(t) = E_2 + \frac{C_1(E_1 - E_2)}{C_1 + C_2} = \frac{C_1 E_1 + C_2 E_2}{C_1 + C_2}, \quad \text{for } t > 0.$$

As could be expected, $e_1(t) = e_2(t)$, for $t > 0$.

Example 7.5 The circuit in Fig. 7.16 was operating in a steady-state when switch
k was closed at time $t = 0$. Find the expression of the current $i_0(t)$, for $t > 0$.

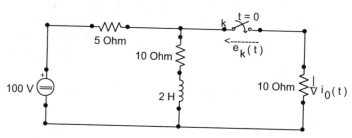

Fig. 7.16 Electric circuit for Example 7.5

Solution

In the DC steady-state, the inductance behaves like a short-circuit and the voltage at
the open switch terminals is

$$e_k(t) = \frac{10 \times 100}{10 + 5} = \frac{200}{3} \ [\text{V}].$$

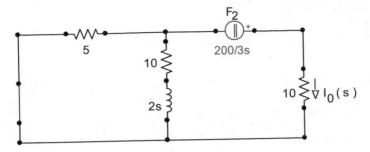

Fig. 7.17 Transformed circuit of Example 7.5, after the insertion of the F_2 source

The transformed circuit, after the insertion of the F_2 source, is shown in Fig. 7.17.

Combining the operational impedances, results the circuit in Fig. 7.18.

From Fig. 7.18,

$$I_0(s) = \frac{\frac{200}{3s}}{\frac{30s + 200}{2s + 15}} = \frac{400s + 3000}{90s(s + 20/3)} = \frac{A_1}{s} + \frac{A_2}{s + 20/3}.$$

$$A_1 = \lim_{s \to 0} \frac{400s + 3000}{90(s + 20/3)} = 5$$

$$A_2 = \lim_{s \to -20/3} \frac{400s + 3000}{90s} = -\frac{5}{9}$$

Fig. 7.18 Transformed circuit of Example 7.5, combining the operational impedances

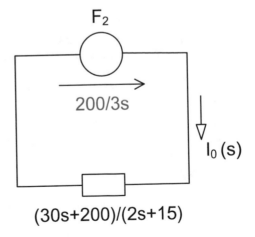

F_2

200/3s

I_0 (s)

(30s+200)/(2s+15)

Therefore,

$$I_0(s) = \frac{5}{s} + \frac{-5/9}{s + 20/3}.$$

Then,

$$i_0(t) = 5\left(1 - \frac{1}{9}\epsilon^{-20t/3}\right) \cdot U_{-1}(t) \, [A].$$

Example 7.6 The circuit shown in Fig. 7.19 operated in a steady-state when, at time $t = 0$, switch k was closed. Determine the analytical expression of the current $i_k(t)$, which will circulate through the switch for $t > 0$.

Solution
In DC steady-state regime, with a constant source, there is neither self nor mutual induction. Therefore, $e_k(t) = E$. And the circuit in the domain of the complex frequency, with the injected voltage source F_2, is shown in Fig. 7.20.

Fig. 7.19 Electric circuit for Example 7.6

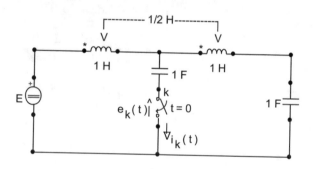

The loop equations in Fig. 7.20, in this case where the polarity is additive, as the currents enter the inductances through the marked terminals, are:

$$s(I_k + I_L) + \frac{s}{2}I_L + \frac{1}{s}I_K = \frac{E}{s}.$$

$$s(I_k + I_L) + \frac{s}{2}I_L + sI_L + \frac{s}{2}(I_k + I_L) + \frac{1}{s}I_L = 0.$$

Rearranging, it follows that:

$$(2s^2 + 2)I_k + 3s^2 I_L = 2E.$$

$$3s^2 I_k + (6s^2 + 2)I_L = 0.$$

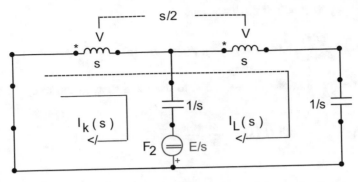

Fig. 7.20 Transformed circuit of Example 7.6, in the domain of the complex frequency with the injected voltage source F_2

So,

$$I_k(s) = \frac{(12s^2 + 4)E}{3s^4 + 16s^2 + 4} = \frac{\left(4s^2 + \frac{4}{3}\right)E}{s^4 + \frac{16}{3}s^2 + \frac{4}{3}} = \frac{\left(4s^2 + \frac{4}{3}\right)E}{(s^2 + 0.2629)(s^2 + 5.0704)}$$

$$= \frac{\left(4s^2 + \frac{4}{3}\right)E}{\left[s^2 + (0.513)^2\right]\left[s^2 + (2.252)^2\right]} \triangleq I_1(s) + I_2(s).$$

Logically, the expression of $I_k(s)$ can be expanded into four partial fractions. However, the process is extremely laborious. The alternative is the search for a shortcut, through the following:

Assuming that $F(s)$ contains a simple quadratic factor of the form $(s + \alpha)^2 + \omega^2$ in its denominator, that is, assuming that

$$F(s) = \frac{Q(s)}{(s + \alpha)^2 + \omega^2}, \tag{7.2}$$

and that $f(t)$ represents the temporal response due to the conjugated complex poles $-\alpha \pm j\omega$, then

$$f(t) = L^{-1}[F(s)] = \frac{M}{\omega} \epsilon^{-\alpha t} \sin(\omega t + \theta), \tag{7.3}$$

where

$$M\angle\theta = Q(s)|_{s = -\alpha + j\omega} = Q(-\alpha + j\omega). \tag{7.4}$$

Obviously, if $\alpha = 0$, it turns out that

$$F(s) = \frac{Q(s)}{s^2 + \omega^2}; \; M\angle\theta = Q(j\omega); f(t) = L^{-1}[F(s)] = \frac{M}{\omega}\sin(\omega t + \theta). \quad (7.5)$$

In the expression of $I_k(s)$, logically, to use the displayed shortcut,

$$I_1(s) = \frac{Q_1(s)}{s^2 + (0.513)^2}, \text{ with } Q_1(s) = \frac{\left(4s^2 + \frac{4}{3}\right)E}{s^2 + (2.252)^2} \text{ and}$$

$$I_2(s) = \frac{Q_2(s)}{s^2 + (2.252)^2}, \text{ with } Q_2(s) = \frac{\left(4s^2 + \frac{4}{3}\right)E}{s^2 + (0.513)^2}.$$

So, according to the expressions in Eq. 7.5,

$$i_1(t) = \frac{M_1}{0.513}\sin(0.513t + \theta_1)$$

$$M_1\angle\theta_1 = Q_1(j0.513) = 0.0583E\angle 0°$$

$$i_1(t) = 0.114E\sin(0.513t) \; [\text{A}].$$

Similarly,

$$i_2(t) = \frac{M_2}{2.252}\sin(2.252t + \theta_2)$$

$$M_2\angle\theta_2 = Q_2(j2.252) = 3.942E\angle 0°$$

$$i_2(t) = 1.750E\sin(2.252t) \; [\text{A}].$$

Finally,

$$i_k(t) = L^{-1}[I_k(s)] = L^{-1}[I_1(s) + I_2(s)] = L^{-1}[I_1(s)] + L^{-1}[I_2(s)] = i_1(t) + i_2(t)$$

$$i_k(t) = [0.114E\sin(0.513t) + 1.750E\sin(2.252t)].U_{-1}(t) \; [\text{A}].$$

Example 7.7 The circuit shown in Fig. 7.21 was operating in a steady-state when switch k was closed at time $t = 0$.

(a) If at the moment of switching the voltage $e_k(t)$ at the switch terminals was minimal, find the current $i_k(t)$ circulating through it, for $t > 0$. Also determine the expression of the source current, $i(t)$, for $t > 0$.
(b) Using the ATPDraw program, perform the numerical solution for the following values: $E = 1000 \; [\text{V}], f = 60 \; [\text{Hz}], L = 100 \; [\text{mH}]$ and $C = 10 \; [\mu\text{F}]$.

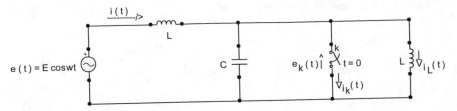

Fig. 7.21 Electric circuit for Example 7.7

Solution

(a) AC steady-state solution with switch k open, that is, for $t<0$.

The circuit in the single frequency domain is shown in Fig. 7.22.

The impedance seen by the phasor voltage source $\dot{E} = E\angle 0°$ is

$$Z(\omega) = j\omega L + \frac{j\omega L \cdot \frac{1}{j\omega C}}{j\omega L + \frac{1}{j\omega C}} = j\omega L + \frac{j\omega L}{1 - \omega^2 LC} = \cdots$$

$$= j\omega L \cdot \frac{2 - \left(\omega/\omega_0\right)^2}{1 - \left(\omega/\omega_0\right)^2}, \quad \text{where } \omega_0^2 \triangleq \frac{1}{LC}.$$

So, the phasor current $\dot{I}$ is worth:

$$\dot{I} = \frac{\dot{E}}{Z(\omega)} = \frac{E\angle 0°}{\omega L \cdot \frac{2 - \left(\omega/\omega_0\right)^2}{1 - \left(\omega/\omega_0\right)^2}\angle 90°} = \frac{\left[1 - \left(\omega/\omega_0\right)^2\right] E}{\left[2 - \left(\omega/\omega_0\right)^2\right]\omega L}\angle - 90°.$$

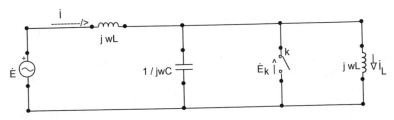

Fig. 7.22 Electric circuit of Example 7.7 for AC steady-state solution

Applying a current divider, we have that:

$$\dot{I}_L = \frac{\frac{1}{j\omega C} \cdot \dot{I}}{j\omega L + \frac{1}{j\omega C}} = \frac{\dot{I}}{1 - \omega^2 LC} = \frac{1}{1 - \left(\omega/\omega_0\right)^2} \cdot \frac{1 - \left(\omega/\omega_0\right)^2}{2 - \left(\omega/\omega_0\right)^2} \cdot \frac{E}{\omega L} \angle -90°$$

$$= \frac{E}{\omega L \left[2 - \left(\omega/\omega_0\right)^2\right]} \angle -90°.$$

Thus, the phasor voltage at the open switch terminals is worth:

$$\dot{E}_k = j\omega L \cdot \dot{I}_L = \frac{E}{2 - \left(\omega/\omega_0\right)^2} \angle 0° = E_k \angle 0°,$$

where

$$E_k \triangleq \frac{E}{2 - \left(\omega/\omega_0\right)^2}. \qquad (7.6)$$

Therefore, in a sinusoidal steady-state,

$$e(t) = E \cos \omega t.$$

$$i(t) = \frac{\left[1 - \left(\omega/\omega_0\right)^2\right] E}{\left[2 - \left(\omega/\omega_0\right)^2\right] \omega L} \cos(\omega t - 90°) = \frac{\left[1 - \left(\omega/\omega_0\right)^2\right]}{\omega L} E_k \sin \omega t.$$

$$i_L(t) = \frac{E}{\omega L \left[2 - \left(\omega/\omega_0\right)^2\right]} \cos(\omega t - 90°) = \frac{E_k}{\omega L} \sin \omega t.$$

$$e_k(t) = E_k \cos \omega t.$$

All of these expressions are referred to at time $t = 0$ of the voltage source $e(t)$.

Since switch k is closed when the voltage at its terminals is minimal, then $e_k(t_0) = -E_k$. Therefore, $-E_k = E_k \cos \omega t_0$ and $\omega t_0 = \arccos(-1) = \pi$. That is, the switch is closed, in relation to the source's $t = 0$ $e(t)$, at time $t_0 = \pi/\omega$.

Considering, however, that, according to the statement, the switch k was closed in an instant taken as **another** $t = 0$, it is necessary to correct the previous expressions, changing its phases from ωt to $\omega t + \omega t_0 = \omega t + \pi = "\omega t + 180°"$. Then,

$$e(t) = E\cos(\omega t + 180°) = -E\cos\omega t.$$

$$i(t) = \frac{\left[1 - \left(\omega/\omega_0\right)^2\right]}{\omega L} E_k \sin(\omega t + 180°) = -\frac{\left[1 - \left(\omega/\omega_0\right)^2\right]}{\omega L} E_k \sin\omega t.$$

$$i_L(t) = \frac{E_k}{\omega L}\sin(\omega t + 180°) = -\frac{E_k}{\omega L}\sin\omega t.$$

$$e_k(t) = E_k\cos(\omega t + 180°) = -E_k\cos\omega t.$$

Then, the time domain circuit, for $t > 0$, containing only the injected voltage source F_2, is the one in Fig. 7.23.

Transposing the circuit of Fig. 7.23 to the domain of the complex frequency, results the circuit of Fig. 7.24, in which

$$E_k(s) = L[E_k\cos\omega t.U_{-1}(t)] = \frac{sE_k}{s^2 + \omega^2}.$$

The operational impedance seen by source F_2 is

$$Z(s) = \frac{1}{Y(s)} = \frac{1}{\frac{1}{sL} + \frac{1}{sL} + sC} = \frac{sL}{2 + s^2LC} = \frac{sL}{LC(s^2 + 2/LC)} = \frac{s}{C\left[s^2 + \left(\sqrt{2}\omega_0\right)^2\right]}.$$

On the other hand,

$$I_k(s) = -\frac{E_k(s)}{Z(s)} = -\frac{\frac{sE_k}{s^2 + \omega^2}}{\frac{s}{C\left[s^2 + \left(\sqrt{2}\omega_0\right)^2\right]}} = -\frac{CE_k\left[s^2 + \omega^2 + \left(\sqrt{2}\omega_0\right)^2 - \omega^2\right]}{s^2 + \omega^2}$$

$$= -CE_k\left[1 + \frac{\left(\sqrt{2}\omega_0\right)^2 - \omega^2}{s^2 + \omega^2}\right] = -CE_k\left[1 + \frac{\left(\sqrt{2}\omega_0\right)^2 - \omega^2}{\omega}\cdot\frac{\omega}{s^2 + \omega^2}\right].$$

Carrying out the inverse transform, comes that

Fig. 7.23 Circuit of Example 7.7, after the insertion of the F_2 source

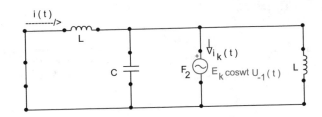

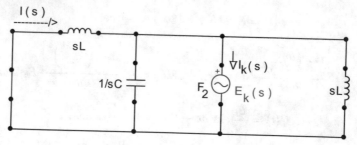

Fig. 7.24 Transformed circuit of Example 7.7, after the insertion of the F_2 source, to the domain of the complex frequency

$$i_k(t) = -CE_k\left[U_0(t) + \frac{\left(\sqrt{2}\omega_0\right)^2 - \omega^2}{\omega}\sin\omega t\right].U_{-1}(t).$$

Substituting E_k for its value, given by Eq. 7.6, we have that:

$$i_k(t) = -\frac{CE}{2 - \left(\omega/\omega_0\right)^2}U_0(t) - \frac{\left[\left(\sqrt{2}\omega_0\right)^2 - \omega^2\right]CE}{\omega\left[2 - \left(\omega/\omega_0\right)^2\right]}\sin\omega t.U_{-1}(t).$$

Simplifying the second parcel, it follows that

$$i_k(t) = -\frac{CE}{2 - \left(\omega/\omega_0\right)^2}U_0(t) - \frac{E}{\omega L}\sin\omega t.U_{-1}(t). \qquad (7.7)$$

From the circuit of Fig. 7.24, considering that the left sL impedance is in parallel with the voltage source, then,

$$I(s) = -\frac{E_k(s)}{sL} = -\frac{\frac{sE_k}{s^2 + \omega^2}}{sL} = -\frac{E_k}{L(s^2 + \omega^2)} = -\frac{\frac{E}{2 - \left(\omega/\omega_0\right)^2}}{L(s^2 + \omega^2)}$$

$$= -\frac{E}{\omega L\left[2 - \left(\omega/\omega_0\right)^2\right]} \cdot \frac{\omega}{s^2 + \omega^2}.$$

Therefore,

$$i(t) = L^{-1}[I(s)] = -\frac{E}{\omega L\left[2 - \left(\omega/\omega_0\right)^2\right]}\sin\omega t.U_{-1}(t).$$

Superimposing, for t > 0,

$$i(t) = -\frac{\left[1-\left(\omega/\omega_0\right)^2\right]}{\omega L}E_k\sin\omega t - \frac{E}{\omega L\left[2-\left(\omega/\omega_0\right)^2\right]}\sin\omega t$$

$$= -\frac{\left[1-\left(\omega/\omega_0\right)^2\right]E}{\omega L\left[2-\left(\omega/\omega_0\right)^2\right]}\sin\omega t - \frac{E}{\omega L\left[2-\left(\omega/\omega_0\right)^2\right]}\sin\omega t$$

$$i(t) = -\frac{E}{\omega L}\sin\omega t, \ para\, t \geq 0. \tag{7.8}$$

Obviously, Eqs. 7.7 and 7.8 are referred to at time $t = 0$ of the closing of switch k. To refer to the (original) time $t = 0$ of the voltage source $e(t)$, it is necessary to correct them. For this, the ωt phase must be replaced by $\omega t - \pi$, that is, t must be replaced by $t - \pi/\omega$. Thus, from Eqs. 7.7 and 7.8, it follows that

$$i_k(t) = -\frac{CE}{2-\left(\omega/\omega_0\right)^2}U_0(t-\pi/\omega) - \frac{E}{\omega L}\sin(\omega t - \pi).U_{-1}(t-\pi/\omega).$$

$$i(t) = -\frac{E}{\omega L}\sin(\omega t - \pi), \quad for\ t - \pi/\omega \geq 0.$$

$$i_k(t) = -\frac{CE}{2-\left(\omega/\omega_0\right)^2}U_0(t-\pi/\omega) + \frac{E}{\omega L}\sin\omega t.U_{-1}(t-\pi/\omega). \tag{7.9}$$

$$i(t) = \frac{E}{\omega L}\sin\omega t, \quad for\ t \geq \pi/\omega. \tag{7.10}$$

Summing up,

$$i_k(t) = \begin{cases} 0, & for\ 0 \leq t < \pi/\omega \\ -\dfrac{CE}{2-\left(\omega/\omega_0\right)^2}U_0(t-\pi/\omega), & for\ t = \pi/\omega \\ \dfrac{E}{\omega L}\sin\omega t, & for\ t > \pi/\omega. \end{cases} \tag{7.11}$$

$$i(t) = \begin{cases} \dfrac{\left[1-\left(\omega/\omega_0\right)^2\right]E}{\omega L\left[2-\left(\omega/\omega_0\right)^2\right]}\sin\omega t, & for\ 0 \leq t \leq \pi/\omega \\ \dfrac{E}{\omega L}\sin\omega t, & for\ t \geq \pi/\omega. \end{cases} \tag{7.12}$$

Needless to say, Eqs. 7.11 and 7.12 refer to the closing of switch k at time $t = \pi/\omega$, and are related to time $t = 0$ of the voltage source $e(t)$.

Entering the numerical values, Eqs. 7.11 and 7.12 become:

$$i_k(t) = \begin{cases} 0, & \textit{for } 0 \leq t < 8.33\,[\text{ms}] \\ -5.382U_0(t - 8.33)\,[\text{A}], & \textit{for } t = 8.33\,[\text{ms}] \\ 26.526\sin(376.99t)\,[\text{A}], & \textit{for } t > 8.33\,[\text{ms}]. \end{cases} \tag{7.13}$$

$$i(t) = \begin{cases} 12.248\sin(376.99t)\,[\text{A}], & \textit{for } 0 \leq t \leq 8.33\,[\text{ms}] \\ 26.526\sin(376.99t)\,[\text{A}], & \textit{for } t \geq 8.33\,[\text{ms}]. \end{cases} \tag{7.14}$$

(b) The circuit for the numerical solution by the ATPDraw program is shown in Fig. 7.25.

Figure 7.26 shows the source and switch voltages. From the source,

$$e(t) = E\cos\omega t = 1,000\cos 376.99t\,[\text{V}], \quad \textit{for } t > 0.$$

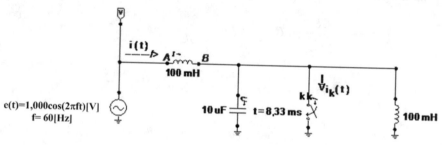

Fig. 7.25 Electric circuit of Example 7.7 for the numerical solution by the ATPDraw program

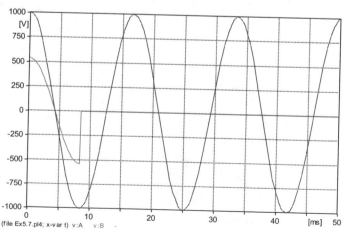

Fig. 7.26 Voltage $e(t)$ voltage $e_k(t)$

The voltage on the capacitance, which is the same as on the switch,

$$e_k(t) = \frac{E}{2 - \left(\omega/\omega_0\right)^2} \cos \omega t = 538.248 \cos(376.99t) \text{ [V]}, \quad for \ 0 < t < 8.33 \text{ [ms]}.$$

$$e_k(t) = 0, \quad for \ t > 8.33 \text{ [ms]}.$$

Figure 7.27 shows the waveform of current $i(t)$ passing through the source, which is given analytically by Eq. 7.14.

As given by the expressions in Eq. 7.13, there is an impulse in the current circulating through switch k, at the moment of its closing at $t_0 = 8.33$ [ms], which explains the numerical oscillation in the current waveform shown in Fig. 7.28.

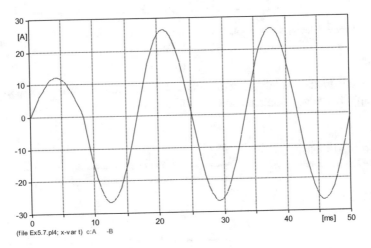

(file Ex5.7.pl4; x-var t) c:A   -B

Fig. 7.27 Current $i(t)$

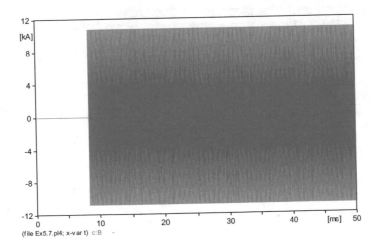

(file Ex5.7.pl4; x-var t) c:B   -

Fig. 7.28 Current $i_k(t)$

Example 7.8 The circuit in Fig. 7.29 was operating in a steady-state when switch k was closed at the instant taken as $t = 0$.

(a) Knowing that at the moment of switching $i = 0$, with $i'(t) < 0$, find the analytical expression of the current $i_k(t)$, flowing through the switch k, for $t > 0$.
(b) Using the ATPDraw program, obtain the numerical solution for $E = 100\,[\text{V}]$, $f = 60\,[\text{Hz}]$, $L = 10\,[\text{mH}]$, $K = 3$ and $C = 100\,[\mu\text{F}]$.

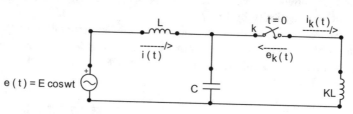

Fig. 7.29 Electric circuit for Example 7.8

Solution

(a) The steady-state solution, with switch k open, comes from the circuit in the simple frequency domain shown in Fig. 7.30.

Fig. 7.30 Electric circuit for Example 7.8, for steady-state solution

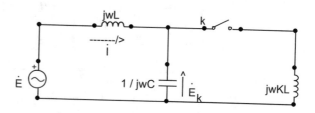

$$\dot{I} = \frac{\dot{E}}{Z} = \frac{\dot{E}}{j\omega L + 1/j\omega C} = \frac{j\omega CE\angle 0^\circ}{1 - \omega^2 LC} = \frac{\omega CE\angle 90^\circ}{1 - \left(\omega/\omega_0\right)^2}, \ where \ \omega_0 = 1/\sqrt{LC}.$$

$$\dot{E}_k = \frac{1}{j\omega C}\cdot\dot{I} = \frac{E\angle 0^\circ}{1 - \left(\omega/\omega_0\right)^2}.$$

Therefore,

$$i(t) = \frac{\omega CE}{1 - \left(\omega/\omega_0\right)^2} \cos(\omega t + 90°) = -\frac{\omega CE}{1 - \left(\omega/\omega_0\right)^2} \sin \omega t \text{ and}$$

$$i'(t) = -\frac{\omega^2 CE}{1 - \left(\omega/\omega_0\right)^2} \cos \omega t.$$

$$e_k(t) = \frac{E}{1 - \left(\omega/\omega_0\right)^2} \cos \omega t = \frac{\omega_0^2 E}{\omega_0^2 - \omega^2} \cos \omega t = E_k \cos \omega t; \ E_k = \frac{\omega_0^2 E}{\omega_0^2 - \omega^2}.$$

For $t = 0$, $i = 0$ and $i'(t) < 0$, assuming $\omega_0 > \omega$, which usually happens, meaning that the closing time of switch k coincides with the instant $t = 0$ (reference) of the voltage source $e(t)$. Thus, there is no need for any change in the expression of voltage $e_k(t)$ at the terminals of switch k, operating in open circuit, and that $e_k(t)$ is in phase with $e(t)$.

Then, the time domain circuit, for $t > 0$, containing only the injected voltage source F_2, is the one shown in Fig. 7.31.

Transposing the circuit of Fig. 7.31 to the domain of the complex frequency, through the Laplace Transform, results the equivalent circuit of Fig. 7.32.

$$E_k(s) = \frac{sE_k}{s^2 + \omega^2}.$$

$$Z(s) = sKL + \frac{sL \cdot \frac{1}{sC}}{sL + \frac{1}{sC}} = sKL + \frac{sL}{LCs^2 + 1}$$

$$= \frac{KL^2 Cs^3 + (K+1)Ls}{LCs^2 + 1} = \frac{sL\left[Ks^2/\omega_0^2 + (K+1)\right]}{s^2/\omega_0^2 + 1}.$$

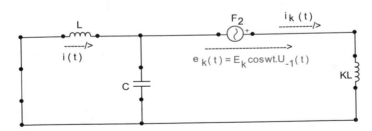

Fig. 7.31 Electric circuit of Example 7.8 for calculating the exclusive effects of the (injected) source F_2

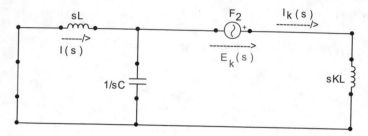

Fig. 7.32 Transformed circuit of Example 7.8, in the domain of the complex frequency

Or,

$$Z(s) = \frac{KLs\left[s^2 + \left(\frac{K+1}{K}\right)\omega_0^2\right]}{s^2 + \omega_0^2}.$$

$$I_k(s) = \frac{E_k(s)}{Z(s)} = \frac{sE_k}{s^2 + \omega^2} \cdot \frac{s^2 + \omega_0^2}{KLs\left[s^2 + \left(\frac{K+1}{K}\right)\omega_0^2\right]}$$

$$= \frac{E_k}{KL} \cdot \frac{s^2 + \omega_0^2}{(s^2 + \omega^2)\left[s^2 + \left(\sqrt{1 + \frac{1}{K}}\omega_0\right)^2\right]}.$$

On the other hand, considering that

$$\frac{u + A}{(u + B)(u + C)} = \frac{A_1}{u + B} + \frac{A_2}{u + C},$$

where $A_1 = \lim\limits_{u \to -B} \frac{u+A}{u+C} = \frac{A-B}{C-B}$

$A_2 = \lim\limits_{u \to -C} \frac{u+A}{u+B} = -\frac{A-C}{C-B},$

results the identity:

$$\frac{u + A}{(u + B)(u + C)} = \frac{1}{C - B}\left(\frac{A - B}{u + B} - \frac{A - C}{u + C}\right) \tag{7.15}$$

Considering, then, that $u = s^2$, $A = \omega_0^2$, $B = \omega^2$ and $C = (1 + 1/K)\omega_0^2$, Eq. 7.15 allows to write that

$$I_k(s) = \frac{E_k}{KL} \cdot \frac{1}{(1 + \frac{1}{K})\omega_0^2 - \omega^2}\left[\frac{\omega_0^2 - \omega^2}{s^2 + \omega^2} - \frac{\omega_0^2 - \left(1 + \frac{1}{K}\right)\omega_0^2}{s^2 + \left(\sqrt{1 + \frac{1}{K}}\omega_0\right)^2}\right].$$

Considering also that, from the steady-state solution,

$E_k = \frac{\omega_0^2 E}{\omega_0^2 - \omega^2}$, then,

$$I_k(s) = \frac{\omega_0^2 E}{KL(\omega_0^2 - \omega^2)} \cdot \frac{1}{\left(1 + \frac{1}{K}\right)\omega_0^2 - \omega^2} \cdot \left[\frac{\omega_0^2 - \omega^2}{s^2 + \omega^2} + \frac{\frac{\omega_0^2}{K}}{s^2 + \left(\sqrt{1 + \frac{1}{K}}\omega_0\right)^2} \right]$$

$$= \frac{\omega_0^2 E}{L(\omega_0^2 - \omega^2)} \cdot \frac{1}{(K+1)\omega_0^2 - K\omega^2} \cdot \left[\frac{\omega_0^2 - \omega^2}{s^2 + \omega^2} + \frac{\frac{\omega_0^2}{K}}{s^2 + \left(\sqrt{1 + \frac{1}{K}}\omega_0\right)^2} \right]$$

$$= \frac{E}{\omega L\left[(K+1) - K\frac{\omega^2}{\omega_0^2}\right]} \cdot \frac{\omega}{s^2 + \omega^2}$$

$$+ \frac{E}{\sqrt{1 + \frac{1}{K}}\omega_0 KL \cdot \left[1 - \left(\frac{\omega}{\omega_0}\right)^2\right] \cdot \left[(K+1) - K\frac{\omega^2}{\omega_0^2}\right]} \cdot \frac{\sqrt{1 + \frac{1}{K}}\omega_0}{s^2 + \left(\sqrt{1 + \frac{1}{K}}\omega_0\right)^2} \cdot$$

Finally,

$$i_k(t) = \frac{E}{\omega L\left[(K+1) - K\left(\frac{\omega}{\omega_0}\right)^2\right]} \sin(\omega t)$$

$$+ \frac{E}{\sqrt{K(K+1)}\omega_0 L\left[1 - \left(\frac{\omega}{\omega_0}\right)^2\right]\left[(K+1) - K\left(\frac{\omega}{\omega_0}\right)^2\right]}$$

$$\sin\left(\sqrt{1 + \frac{1}{K}}\omega_0 t\right),$$

for $t \geq 0$.

For the given numerical values,

$$T = \frac{1}{f} = \frac{1}{60} = 16.667 \text{ [ms]}; \quad \omega = 2\pi f = 120\pi = 376.9911 \left[\frac{\text{rad}}{\text{s}}\right].$$

$$\omega_0 = \frac{1}{\sqrt{LC}} = 1000 \left[\frac{\text{rad}}{\text{s}}\right]; \quad f_0 = \frac{\omega_0}{2\pi} = 159.1549 \left[\frac{\text{rad}}{\text{s}}\right]; \quad T_0 = \frac{1}{f_0} = 6.2832 \text{ [ms]}.$$

$$i_k(t) = 7.4226\sin(376.9911t) + 0.9416\sin(1,154.7005t), \quad t \geq 0.$$

Then,

$$i_k(0) = 0\,[\text{A}]$$

$$i_k(0.005) = 7.059 - 0.459 = 6.600\,[\text{A}]$$

$$i_k(0.010) = -4.363 - 0.802 = -5.165\,[\text{A}]$$

$$i_k(0.008333) = i(T/2) = -0.002 - 0.186 = -0.188\,[\text{A}]$$

$$i_k(0.020) = 7.059 - 0.840 = 6.219\,[\text{A}]$$

(b) The circuit for the numerical solution is shown in Fig. 7.33.

$$E_k = \frac{\omega_0^2 E}{\omega_0^2 - \omega^2} = \frac{10^6 \times 10^2}{10^6 - (120\pi)^2} = \frac{10^8}{857{,}877.6966} = 116.5667\,[\text{V}].$$

The results are shown in Fig. 7.34, with $\Delta t = 1\,[\mu s]$.
With the cursor, you can see that:

$I_k\,(0.005) = 6.598\,[\text{A}]$

$I_k\,(0.010) = -5.167\,[\text{A}]$

$I_k\,(0.008334) = -0.189\,[\text{A}]$

$I_k\,(0.020) = 6.218\,[\text{A}]$.

If the analytical solution in item *a* had not been requested, the steady-state digital solution would be necessary to determine the instant when the current satisfies condition $i(t) = 0$, with $i'(t) < 0$, as well such as the voltage curve at the switch terminals, $e_k(t)$, in order to establish the *future voltage* of the source F_2, to be injected. And the circuit for this is shown in Fig. 7.35. In Fig. 7.36, the curves of $e_k(t)$ and $i(t)$ are plotted. And it clearly shows that the condition $i(t) = 0$, with $i'(t) < 0$ occurs exactly at the original $t = 0$, therefore coinciding with the time

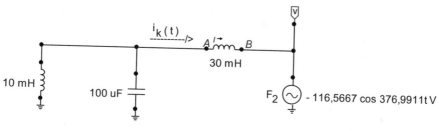

Fig. 7.33 Electric circuit of Example 7.8 for the numerical solution by the ATPDraw program

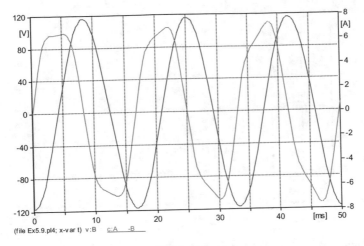

Fig. 7.34 Source voltage F_2 and current $i_k(t)$ in the switch

$t = 0$ of the closing of the switch k, no thus requiring any phase correction in the quantities of interest. From the same Fig. 7.36, the maximum values of permanent regime of 116.57 [V] for $e_k(t)$ and 4.3945 [A] for $i(t)$ are also obtained with the cursor. These being exactly the values that are provided by

$$E_k = \frac{\omega_0^2 E}{\omega_0^2 - \omega^2} \text{ and } I = \frac{\omega C E}{1 - \left(\omega/\omega_0\right)^2}.$$

Substituting the numerical values, results in $E_k = 116.5667$ [V], $I = 4.3945$ [A].

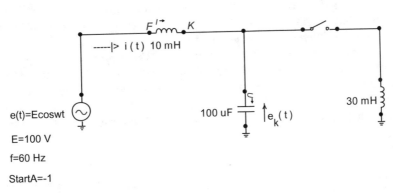

Fig. 7.35 Electric circuit for the calculation of steady-state digital solution from Example 7.8

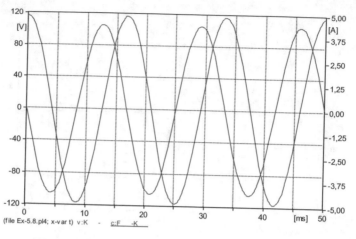

Fig. 7.36 Steady-state digital solution for switch voltage $e_k(t)$ and inductor current $i(t)$

7.3 Current Source Injection Method

It is a technique that can be applied to any linear circuit and is based on the theorems of substitution and superposition. In a way, it is the dual procedure of the voltage source injection method, and its main objective is to replace the process of opening a switch, at any time $t = t_0$, with an equivalent assembly, that is, simulating the opening of a switch at time t_0. It basically consists of the injection of a current source equal to and contrary to that which would continue to pass through the switch, after the t_0 instant, if the switch was not opened at that instant. It can be applied to circuits already operating in a steady-state, as well as to those still in transitory phase. It is described in the following five steps.

1. Referring to Fig. 7.37, the current $i_k(t)$ circulating through the switch k and the other voltages and currents in the elements inside the networks N_1 and N_2, which are of interest, are initially calculated.
   Assuming that $i_k(t) = I\cos(\omega t + \theta)$, then, $e_k(t) = 0, p_k(t) = e_k(t).i_k(t) = 0$.
2. Then, the switch k is replaced by an ideal current source F_1, whose current $i_1(t)$ has the same value as the current that circulated and that would continue to circulate through switch k, if it were not opened at time $t = t_0$. As the switch is ideal, that is, without voltage at its terminals when closed, then the substitutive

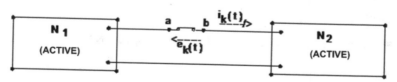

Fig. 7.37 Circuit before opening the switch at time $t = t_0$

current source F_1 always has zero voltage at its terminals, whenever acting alone between terminals a and b, as shown in Fig. 7.38.

$$i_1(t) = i_k(t) = I\cos(\omega t + \theta) = I\cos(\omega t + \theta)[U_{-1}(-t+t_0) + U_{-1}(t-t_0)];$$

$$e_k(t) = 0; p_{F_1}(t) = p_k(t) = 0.$$

3. To portray the opening of switch k at time $t = t_0$, a second ideal current source F_2, in the opposite direction, is injected in parallel with the source F_1, from the same time $t = t_0$. Its value is equal to that of the *future current through the switch*, that is, the current that would continue to circulate through the switch if it had not been opened at time $t = t_0$, nor after it. Thus, the total current $i_k(t)$, circulating through the branch containing the switch k, will be zero, after the injection of the current source F_2, that is, for $t > t_0$, which is equivalent to the opening of the switch, as illustrated in Fig. 7.39.
In Fig. 7.39,

$$i_1(t) = I\cos(\omega t + \theta); \quad i_2(t) = I\cos(\omega t + \theta).U_{-1}(t-t_0); \quad p_k(t) = 0.$$

In node b: $i_k(t) = i_1(t) - i_2(t)$.
In detail, we have, then, that:
For $t < t_0$, switch k closed,

$i_1(t) = I\cos(\omega t + \theta)$, $i_2(t) = 0$ (F_2 is an open circuit) and $i_k(t) = I\cos(\omega t + \theta)$.
For $t > t_0$, switch k open,
$i_1(t) = i_2(t) = I\cos(\omega t + \theta)$ and $i_k(t) = 0$.
Since $i_k(t) = i_1(t) - i_2(t)$, $e_k(t).i_k(t) = e_k(t).i_1(t) - e_k(t).i_2(t)$.
So, the instantaneous power on the switch is

$$p_{F_1}(t) + p_{F_2} = 0,$$

showing that the **power released by source F_2 is consumed by source F_1**.
In detail:
For $t < t_0$, switch k closed, $i_k(t) = i_1(t) \neq 0$, $e_k(t) = 0$. Therefore, $p_k(t) = 0$.
For $t > t_0$, switch k open, $i_k(t) = 0$, $e_k(t) \neq 0$.

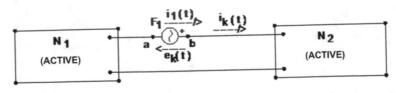

Fig. 7.38 Circuit in Fig. 7.37 with switch k closed, replaced by the current source F_1

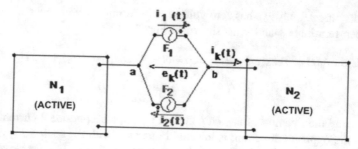

Fig. 7.39 Circuit after opening the switch k, simulated by injecting the current source F_2

The voltage $e_k(t)$ is a consequence of the exclusive action of the current source F_2, for $t > t_0$, which, due to the parallelism of the two current sources, also appears at the terminals of the source F_1. Therefore, $p_k(t) = 0$.

4. The calculation of voltages and currents over all elements of the circuit, on which there is interest for $t > t_0$, that is, with the switch k open, is done using the superposition theorem, a property of linear systems. As the power at source F_1 is zero in the absence of source F_2, that is, with source F_2 at rest—open, because in this condition $e_k(t) = 0$ (Fig. 7.38), it does not establish any voltage or current on the passive elements of the circuit inside the N_1 and N_2 networks. Therefore, **the effect of the current source F_1 on the circuit is always null.** On the other hand, the voltages and currents produced by the sources internal to the N_1 and N_2 networks, with the source F_2 at rest, have already been previously calculated in the initial phase of the process (Fig. 7.37). Thus, it remains to calculate only the voltages and currents produced by the current source F_2, acting alone for $t > t_0$, that is, with the source F_1 at rest and the networks N_1 and N_2 reduced to the passive form, that is, with all their independent voltage sources short-circuited and current sources opened. In this phase, the voltage $e_k(t)$ at the terminals of the current source F_2 is also calculated, as shown in Fig. 7.40, and that is actually the voltage itself at the terminals of switch k, after it was opened at time $t = t_0$.

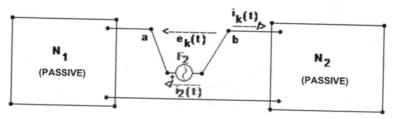

Fig. 7.40 Circuit reduced to passive for the calculation of voltages and currents produced exclusively by the current source F_2

In Fig. 7.40,

$$i_2(t) = I\cos(\omega t + \theta).U_{-1}(t - t_0); \quad e_k(t) \neq 0, \quad i_k(t) = -i_2(t) = -I\cos(\omega t + \theta).$$

For $t > t_0$, $p_{F_2} = -e_k(t).i_2(t) \neq 0$.

5. Finally, the resulting values of voltage and current in the circuit elements, after opening the switch k at time $t = t_0$, are obtained by superposition of the values found with the isolated action of the current source F_2 (Fig. 7.40) with those obtained before opening the switch k (Fig. 7.37).

It is necessary to point out that if the opening time of the switch k, $t = t_0$, is taken as a **new time** $t = 0$, that is, if the time will be computed from the time of opening of the switch k, then the analytical expressions of all voltages and currents in the circuit, both those that were given and those that were calculated before the opening of switch k (Fig. 7.37), must undergo corrections in their initial phases, to be related to this new origin of time. And this is done simply by adding the value ωt_0 to their respective phases, that is, replacing t with $t + t_0$ in the corresponding expressions. So, for example, $i_2(t)$ would be

$$i_2(t) = I\cos[\omega(t + t_0) + \theta)]U_{-1}[(t + t_0) - t_0] = I\cos(\omega t + \theta + \omega t_0).U_{-1}(t).$$

Example 7.9, although it does not refer to switching transients, because the circuit contains neither capacitance nor inductance, nor switch opening, it is quite useful for a better understanding of the current source injection method in solving circuit electric problems.

Example 7.9 Use the substitution and superposition theorems to determine the variation that current I undergoes in the circuit of Fig. 7.41, when the resistance R undergoes an increase ΔR, that is, it changes from R to $R + \Delta R$. Determine, also, the new current I and the new voltage e_r, in the resistance r.

Fig. 7.41 Electric circuit for Example 7.9

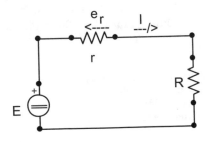

Solution
The problem can be replaced as shown in Fig. 7.42, where the resistive addition ΔR, of resistance R, is placed in parallel with a switch k, which changes from the closed to the open position, at $t = 0$.

Fig. 7.42 Adapted electric
circuit for Example 7.9

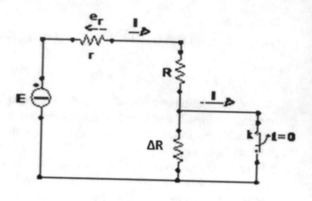

1. The current I flowing through switch k, at whose terminals the voltage is zero, is

$$I = \frac{E}{r+R}.$$

And the voltage at resistance r, by voltage divider, is

$$e_r = \frac{rE}{r+R}.$$

2. Switch k, closed, can be replaced by an ideal current source F_1, of value I and zero voltage, as shown in Fig. 7.43.
3. To simulate the opening of switch k, at time $t = 0$, an ideal current source F_2 is injected in parallel with F_1, with the same value as this one, but in the opposite direction, as shown in Fig. 7.44, valid for $t > 0$.
4. The solution of the circuit of Fig. 7.44 is made using the superposition theorem, calculating, initially, the simultaneous effects of the independent voltage source E and the current source F_1 and, later, the effects of the source F_2. It turns out that this first calculation was already done in phase 1, as **the source F_1 is nothing more than the replacement of the closed k switch** (Fig. 7.42). So, it

Fig. 7.43 Electric circuit of
Example 7.9 with switch k,
closed, replaced by an ideal
current source F_1

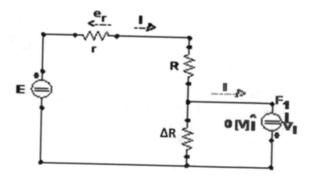

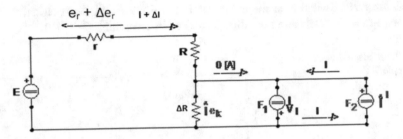

Fig. 7.44 Electric circuit of Example 7.9 with an ideal current source F_2 injected in parallel with F_1

is only necessary to calculate the exclusive effects of the F_2 source, from the circuit shown in Fig. 7.45.

From the circuit of Fig. 7.45, by current divider, we have to

$$\Delta I = -\frac{\Delta R.I}{r+R+\Delta R} = -\frac{\Delta R.\frac{E}{r+R}}{r+R+\Delta R} = -\frac{\Delta R.E}{(r+R)(r+R+\Delta R)}.$$

So,

$$\Delta e_r = r.\Delta I = -\frac{r.\Delta R.E}{(r+R)(r+R+\Delta R)}.$$

5. Thus, the new values for current I and voltage e_r, by superposition, are:

$$I = \frac{E}{r+R} - \frac{\Delta R.E}{(r+R)(r+R+\Delta R)} = \frac{E}{r+R+\Delta R}.$$

$$e_r = \frac{rE}{r+R} - \frac{r\Delta R.E}{(r+R)(r+R+\Delta R)} = \frac{rE}{r+R+\Delta R}.$$

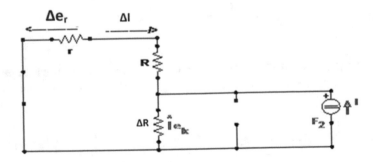

Fig. 7.45 Electric circuit of Example 7.9 for calculating the exclusive effects of the (injected) source F_2

Example 7.10 Switch k in the circuit of Fig. 7.46 has long been closed. If it is opened at $t = 0$, determine the voltage $e_k(t)$ at its terminals, for $t > 0$.

Fig. 7.46 Electric circuit for Example 7.10

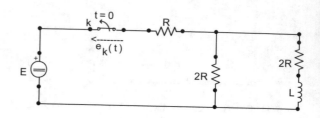

Solution

In the steady-state, with switch k closed, the circuit is the one shown in Fig. 7.47.

From Fig. 7.47,

$$i_k(t) = \frac{E}{2R}.$$

Then, the circuit for $t > 0$, containing only the injected current source, F_2, is the one shown in Fig. 7.48.

The right mesh equation is

$$L\frac{d}{dt}i_L(t) + 2Ri_L(t) + 2R\left[i_L(t) + \frac{E}{2R}U_{-1}(t)\right] = 0.$$

$$\left(D + \frac{4R}{L}\right)i_L(t) = -\frac{E}{L}U_{-1}(t), \text{ with } i_L(0_-) = 0.$$

$$i_L(t) = -\frac{E}{4R} + k_1\epsilon^{-4Rt/L}.$$

$i_L(0_+) = i_L(0_-) = 0 = -\frac{E}{4R} + k_1$. Then, $k_1 = E/4R$ and

$$i_L(t) = \frac{E}{4R}\left(\epsilon^{-4Rt/L} - 1\right).U_{-1}(t).$$

From the left mesh of Fig. 7.48, **considering the presence of the independent voltage source E, and its effects on resistors R and $2R$, and using superposition,** it comes that:

$$e_k(t) = e_R(t) + e_{2R}(t) + E = \left(-\frac{E}{2} + \frac{E}{2}\right) + \left[-\frac{E}{2} + E + \frac{E}{2}\left(\epsilon^{-4Rt/L} - 1\right)\right] + E.$$

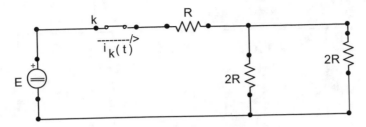

Fig. 7.47 Electric circuit of Example 7.10 for steady-state solution

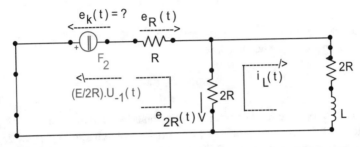

Fig. 7.48 Electric circuit of Example 7.10 for calculating the exclusive effects of the (injected) source F_2

Therefore,

$$e_k(t) = \left(E + \frac{E}{2}\epsilon^{-4Rt/L}\right).U_{-1}(t).$$

Se a fonte independente de tensão E não for considerada no cálculo da tensão $e_k(t)$, representada por um curto-circuito na Figura 7.48, então não é necessário considerar também os seus efeitos sobre as resistências R e $2R$, e nem aplicar superposição, ou seja:

If the voltage independent source E is not considered in the calculation of the voltage $e_k(t)$, represented by a short-circuit in Fig. 7.48, then it is not necessary to consider its effects on resistors R and $2R$, nor to apply superposition, that is:

$$e_k(t) = e_R(t) + e_{2R}(t) = R.\frac{E}{2R}.U_{-1}(t) + 2R.\left[\frac{E}{2R} + \frac{E}{4R}\left(\epsilon^{-4Rt/L} - 1\right)\right].U_{-1}(t)$$

$$= \left(E + \frac{E}{2}\epsilon^{-4Rt/L}\right).U_{-1}(t).$$

Example 7.11 The circuit in Fig. 7.49 has been operating in a steady-state for a long time. If the switch k is opened at $t = 0$, find the expression of the voltage $e_k(t)$ at its terminals, for $t > 0$.

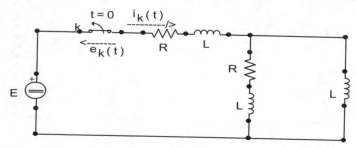

Fig. 7.49 Electric circuit for Example 7.11

Solution

As the inductances behave like short-circuits in a DC steady-state regime, then the current flowing through switch k is:

$$i_k(t) = E/R.$$

Thus, for $t > 0$, the circuit containing only the injected current source, F_2, which contributes to the simulation process of opening the switch k, is shown in Fig. 7.50.

Transposing the circuit of Fig. 7.50 to the domain of the complex frequency, it results what is shown in Fig. 7.51.

The operational impedance seen by the current source is

$$Z(s) = R + sL + \frac{sL(R + sL)}{sL + R + sL} = \frac{3L^2s^2 + 4RLs + R^2}{2Ls + R}.$$

Then,

Fig. 7.50 Electric circuit of Example 7.11 for calculating the exclusive effects of the (injected) source F_2

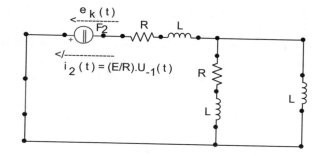

Fig. 7.51 Transformed circuit of Example 7.11, in the domain of the complex frequency

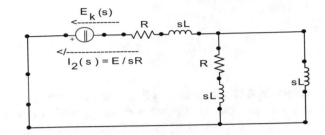

$$E_k(s) = Z(s).I_2(s) = \frac{E}{R} \cdot \frac{3L^2 s^2 + 4RLs + R^2}{2Ls^2 + Rs} = \frac{E}{R}\left(\frac{3L}{2} + \frac{\frac{5}{2}RLs + R^2}{2Ls^2 + Rs}\right),$$

where the expression in parentheses was obtained by direct division, and using the identity $N/D = Q + R/D$, that is,

$$3L^2 s^2 + 4RLs + R^2 \underline{\smash{/\,2Ls^2 + Rs}}$$

$$\underline{-3L^2 s^2 - (3/2)RLs} \qquad \frac{3L}{2}$$

$$(5/2)RLs + R^2$$

On the other hand,

$$\frac{\frac{5}{2}RLs + R^2}{2Ls^2 + Rs} = \frac{\frac{5}{2}RLs + R^2}{2Ls\left(s + \frac{R}{2L}\right)} = \frac{A_1}{s} + \frac{A_2}{s + \frac{R}{2L}}.$$

$$A_1 = \lim_{s \to 0} \frac{\frac{5}{2}RLs + R^2}{2L\left(s + \frac{R}{2L}\right)} = R.$$

$$A_2 = \lim_{s \to -\frac{R}{2L}} \frac{\frac{5}{2}RLs + R^2}{2Ls} = \frac{R}{4}.$$

Therefore,

$$E_k(s) = \frac{E}{R}\left(\frac{3L}{2} + \frac{R}{s} + \frac{R}{4} \cdot \frac{1}{s + \frac{R}{2L}}\right).$$

Finally,

$$e_k(t) = \frac{3LE}{2R} U_0(t) + E\left(1 + \frac{1}{4}\epsilon^{-Rt/2L}\right).U_{-1}(t).$$

Example 7.12 The circuit in Fig. 7.52 was operating in a steady-state when, at the moment taken as $t = 0$, switch k was opened. Knowing that at the moment of switching the current $i_k(t)$ passed through zero, with a negative derivative:

(a) find the voltage $e_k(t)$, at the terminals of switch k, for $t > 0$. Consider that $\omega_0 = 1/\sqrt{LC} > \sqrt{2}\omega$.
(b) Perform the numerical solution using the ATPDraw program for $E = 500$ [V], $f = 60$ [Hz], $L = 70.362$ [mH] and $C = 25$ [µF].

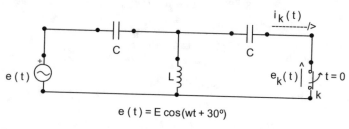

$$e(t) = E\cos(wt + 30°)$$

Fig. 7.52 Electric circuit for Example 7.12

Solution

(a) The circuit for the AC steady-state solution, using the nodal method, is shown in Fig. 7.53.

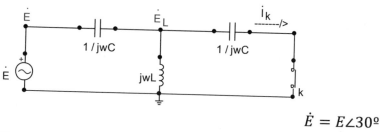

$$\dot{E} = E\angle 30°$$

Fig. 7.53 Electric circuit of Example 7.12 for AC steady-state solution

The equation of the intermediate node, of phasor tension $\dot{E}_L$, is

$$j\omega C(\dot{E}_L - \dot{E}) + \frac{\dot{E}_L}{j\omega L} + j\omega C\dot{E}_L = 0.$$

$$-\omega^2 LC(\dot{E}_L - \dot{E}) + \dot{E}_L - \omega^2 LC\dot{E}_L = 0.$$

$$\left(-\frac{2\omega^2}{\omega_0^2} + 1\right)\dot{E}_L = -\frac{\omega^2}{\omega_0^2}\dot{E}.$$

$$\dot{E}_L = \frac{\omega^2}{2\omega^2 - \omega_0^2}\dot{E} = -\frac{1}{\left(\omega_0/\omega\right)^2 - 2}\dot{E}.$$

So,

$$\dot{I}_k = j\omega C.\dot{E}_L = -\frac{\omega CE}{\left(\omega_0/\omega\right)^2 - 2}\angle 120°.$$

As $\omega_0 = \dfrac{1}{\sqrt{LC}} > \sqrt{2}\omega$, $\omega_0^2 > 2\omega^2$, $(\omega_0/\omega)^2 > 2$, and $\dfrac{\omega CE}{(\omega_0/\omega)^2 - 2} > 0$. So,

$$\dot{I}_k = \frac{\omega CE}{\left(\omega_0/\omega\right)^2 - 2}\angle - 60°.$$

Therefore,

$$i_k(t) = \frac{\omega CE}{\left(\omega_0/\omega\right)^2 - 2}\cos(\omega t - 60°).$$

$$i'_k(t) = -\frac{\omega^2 CE}{\left(\omega_0/\omega\right)^2 - 2}\sin(\omega t - 60°).$$

The current $i_k(t)$ passes through zero for $\omega t - 60° = \pi/2$ or $\omega t - 60° = 3\pi/2$, that is, $\omega t = 150°$ or $\omega t = 330°$.

For $\omega t = 150°$, $i'_k = -\dfrac{\omega^2 CE}{\left(\omega_0/\omega\right)^2 - 2} < 0$, because $\left(\omega_0/\omega\right)^2 > 2$.

For $\omega t = 330°$, $i'_k = \dfrac{\omega^2 CE}{\left(\omega_0/\omega\right)^2 - 2} > 0$.

Therefore, at the instant of opening the switch k, (**new $t = 0$**), $\omega t = 5\pi/6$, that is, 150°, in relation to the **initial $t = 0$**, that is, the one that is a reference for the voltage $e(t)$. Thus, in relation to the new $t = 0$, the current $i_k(t)$ must be corrected as follows:

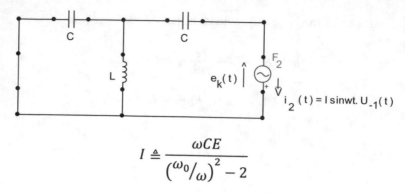

$$I \triangleq \frac{\omega CE}{\left(\omega_0/_\omega\right)^2 - 2}$$

Fig. 7.54 Electric circuit of Example 7.12 with the injection of the source F_2, for $t > 0$ (new origin)

$$i_k(t) = \frac{\omega CE}{\left(\omega_0/_\omega\right)^2 - 2} \cos(\omega t - 60° + 150°) = -\frac{\omega CE}{\left(\omega_0/_\omega\right)^2 - 2} \sin(\omega t).$$

(Sketching the $e(t)$ and $i_k(t)$ curves helps to better understand this phase correction).

Then, the circuit with the injection of the source F_2, for $t > 0$ (new origin), is what is shown in Fig. 7.54.

In the domain of complex frequency, we have to

$$I_2(s) = I\frac{\omega}{s^2 + \omega^2}.$$

The operational impedance seen by the current source is

$$Z(s) = \frac{1}{sC} + \frac{sL.\frac{1}{sC}}{sL + \frac{1}{sC}} = \frac{1}{sC} + \frac{sL}{s^2LC + 1} = \frac{1}{sC} + \frac{s}{C\left(s^2 + 1/_{LC}\right)} = \frac{1}{C} \cdot \frac{2s^2 + \omega_0^2}{s\left(s^2 + \omega_0^2\right)}.$$

As the voltage and current arrows on the F_2 source are in opposition, so:

$$-E_k(s) = Z(s).I_2(s) = \frac{I}{C} \cdot \frac{\omega\left(2s^2 + \omega_0^2\right)}{s(s^2 + \omega^2)\left(s^2 + \omega_0^2\right)}$$

$$= \frac{\omega^2 E}{\left(\omega_0/_\omega\right)^2 - 2} \cdot \frac{\left(2s^2 + \omega_0^2\right)}{s(s^2 + \omega^2)\left(s^2 + \omega_0^2\right)}. \tag{7.16}$$

Defining $\frac{\left(2s^2 + \omega_0^2\right)}{s(s^2 + \omega^2)\left(s^2 + \omega_0^2\right)} = \frac{A_1}{s} + F_1(s) + F_2(s)$, it follows that:

$$A_1 = \lim_{s \to 0} \frac{(2s^2 + \omega_0^2)}{(s^2 + \omega^2)(s^2 + \omega_0^2)} = \frac{1}{\omega^2}.$$

For the determination of $f_1(t) = L^{-1}[F_1(s)]$ and $f_2(t) = L^{-1}[F_2(s)]$ the procedure applied in Example 7.6, Eq. 7.5, is used:

$$f_1(t) = \frac{M_1}{\omega} \sin(\omega t + \theta_1); \quad M_1 \angle \theta_1 = Q_1(s)/_{s=j\omega}, \quad Q_1(s) = \frac{(2s^2 + \omega_0^2)}{s(s^2 + \omega_0^2)}.$$

$$M_1 \angle \theta_1 = \frac{-2\omega^2 + \omega_0^2}{j\omega(-\omega^2 + \omega_0^2)} = \frac{\omega_0^2 - 2\omega^2}{\omega(\omega_0^2 - \omega^2)} \angle -90°$$

$$f_1(t) = \frac{\omega_0^2 - 2\omega^2}{\omega^2(\omega_0^2 - \omega^2)} \sin(\omega t - 90°) = -\frac{\omega_0^2 - 2\omega^2}{\omega^2(\omega_0^2 - \omega^2)} \cos \omega t.$$

$$f_2(t) = \frac{M_2}{\omega_0} \sin(\omega_0 t + \theta_2), \quad M_2 \angle \theta_2 = Q_2(s)/_{s=j\omega_0}, \quad Q_2(s) = \frac{(2s^2 + \omega_0^2)}{s(s^2 + \omega^2)}.$$

$$M_2 \angle \theta_2 = \frac{-2\omega_0^2 + \omega_0^2}{j\omega_0(-\omega_0^2 + \omega^2)} = \frac{-\omega_0}{-\omega_0^2 + \omega^2} \angle -90° = \frac{\omega_0}{\omega_0^2 - \omega^2} \angle -90°$$

$$f_2(t) = \frac{1}{\omega_0^2 - \omega^2} \sin(\omega_0 t - 90°) = -\frac{1}{\omega_0^2 - \omega^2} \cos \omega_0 t.$$

From Eq. 7.16 comes that

$$-E_k(s) = \frac{\omega^2 E}{(\omega_0/\omega)^2 - 2} \cdot \left[\frac{A_1}{s} + F_1(s) + F_2(s)\right].$$

So,

$$e_k(t) = \left[-\frac{\omega^2 E}{\omega_0^2 - 2\omega^2} + \frac{\omega^2 E}{\omega_0^2 - \omega^2} \cos \omega t + \frac{\omega^4 E}{(\omega_0^2 - 2\omega^2)(\omega_0^2 - \omega^2)} \cos \omega_0 t\right].U_{-1}(t)$$

$$\tag{7.17}$$

$e_k(0_+) = 0.$

Equation 7.17 is referenced at time $t = 0$ of the opening of switch k. To reference it at time $t = 0$ of the voltage source $e(t)$, it is necessary to replace t with $t - 5\pi/6\omega$, which corresponds to replace ωt with $\omega t - 150°$, that is, $\omega t - 5\pi/6$. Then,

$$e_k(t) = E\left[-\frac{\omega^2}{\omega_0^2 - 2\omega^2} + \frac{\omega^2}{\omega_0^2 - \omega^2}\cos\omega\left(t - \frac{5\pi}{6\omega}\right)\right.$$
$$\left. + \frac{\omega^4}{(\omega_0^2 - 2\omega^2)(\omega_0^2 - \omega^2)}\cos\omega_0\left(t - \frac{5\pi}{6\omega}\right)\right] \cdot U_{-1}\left(t - \frac{5\pi}{6\omega}\right).$$

or,

$$e_k(t) = E\left[-\frac{\omega^2}{\omega_0^2 - 2\omega^2} + \frac{\omega^2}{\omega_0^2 - \omega^2}\cos\left(\omega t - \frac{5\pi}{6}\right)\right.$$
$$\left. + \frac{\omega^4}{(\omega_0^2 - 2\omega^2)(\omega_0^2 - \omega^2)}\cos\left(\omega_0 t - \frac{5\pi\omega_0}{6\omega}\right)\right], \quad t \geq \frac{5\pi}{6\omega} \tag{7.18}$$

Substituting the numerical values given in item b in Eq. 7.18, it results that

$$e_k(t) = -250 + 166.667\cos(376.991t - 2.618)$$
$$+ 83.333\cos(753.982t - 5.236)\,[\text{V}], \quad t \geq 6.944\,[\text{ms}] \tag{7.19}$$

Since $\omega_0/\omega = 753.982/376.991 = 2$, $e_k(t)$ is periodic, according to the conclusion of Example 5.22.

(b) The circuit assembled for the numerical simulation using the ATPDraw program is shown in Fig. 7.55.

And the result of the simulation is plotted in Fig. 7.56. It is interesting to note that the voltage at the open switch terminals, $e_k(t)$, is never positive.
From Eq. 7.19, we have that:

$$e_k(0.016) = -339.247\,[\text{V}]; \quad e_k(0.023) = -13.095\,[\text{V}]; \quad e_k(0.040) = -2.7346\,[\text{V}].$$

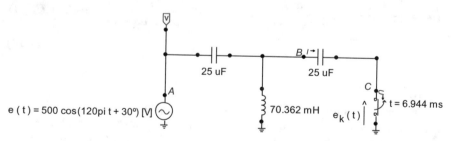

Fig. 7.55 Electric circuit of Example 7.12 for the numerical solution by the ATPDraw program

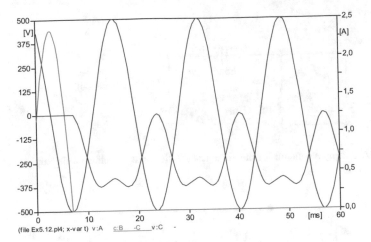

Fig. 7.56 Source voltage $e(t)$, current in switch $i_k(t)$, and voltage in switch $e_k(t)$

From Fig. 7.56,

$$e_k(0.016) = -339.250\,[\text{V}]; \quad e_k(0.023) = -13.094\,[\text{V}]; \quad e_k(0.040) = -2.7343\,[\text{V}].$$

The steady-state current flowing through the closed switch, after replacing the numerical values given in item (b), is

$$i_k(t) = 2.3562\cos(376.991t - 60°)\,[\text{A}].$$

If the interest is essentially related only to the waveform of the voltage e_k at the open switch terminals, regardless of the reference to the time $t = 0$ of the applied voltage source $e(t)$, then an alternative numerical solution can be obtained of the circuit with injection of the current source F_2, shown in Fig. 7.57, and which corresponds to the opening of switch k at time t = 0. That is, with the new instant t = 0, opening the switch k, exactly over the instant $t = 5\pi/6\omega = 6.944\,[\text{ms}]$, which was computed from the original $t = 0$ instant of the voltage source $e(t)$.

In Fig. 7.58 the waveforms of the voltage $e_k(t)$ and the current injected by the source F_2 are plotted, that is, $i_2(t)$.

Whereas $t = 6.944\,[\text{ms}]$ in Fig. 7.56 corresponds to $t = 0\,[\text{ms}]$ in Fig. 7.58, then $t = 16\,[\text{ms}]$ corresponds to $9.056\,[\text{ms}]$; $t = 23\,[\text{ms}]$ corresponds to $16.056\,[\text{ms}]$ and $t = 40\,[\text{ms}]$ corresponds to $33.056\,[\text{ms}]$. So, from Fig. 7.58,

$$e_k(9.056\,\text{ms}) = -339.250\,[\text{V}]; \quad e_k(16.056\,\text{ms}) = -13.076\,[\text{V}];$$
$$e_k(33.056\,\text{ms}) = -2.7256\,[\text{V}], \text{ values compatible with those obtained previously.}$$

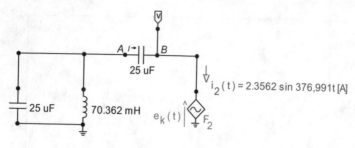

Fig. 7.57 Alternative circuit for the numerical solution with injection of the current source F_2

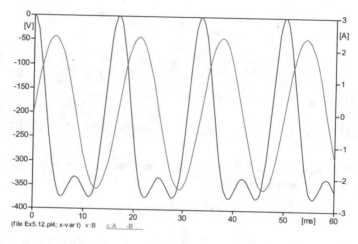

Fig. 7.58 Voltage $e_k(t)$, injected current $i_2(t)$

Example 7.13 The circuit in Fig. 7.59 was operating in a steady-state when, at an instant considered to be $t = 0$, switch k was opened. If at the time of switching the source voltage was $E/2$, with $e'(t) > 0$,

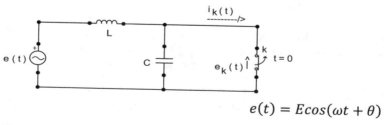

$$e(t) = E\cos(\omega t + \theta)$$

Fig. 7.59 Electric circuit for Example 7.13

(a) determine the expression of the voltage $e_k(t)$, at the terminals of the switch k, for $t > 0$;

(b) perform the digital solution, using the ATPDraw program, for the following numerical values: $E = 2000\,[\text{V}], f = 400\,[\text{Hz}], \theta = 30° \; or \; \theta = \pi/6\,[\text{rad}]$, $C = 4\,[\mu\text{F}], f_0 = 6f = 2400\,[\text{Hz}], L = 1/C\omega_0^2 = 1.0994\,[\text{mH}]$.

Solution

(a) The AC steady-state current circulating through switch k is

$$i_k(t) = \frac{E}{\omega L}\cos(\omega t + \theta - 90°) = \frac{E}{\omega L}\sin(\omega t + \theta).$$

When $e(t) = \frac{E}{2}$, then $\frac{E}{2} = E\cos(\omega t + \theta)$. Thus, $\cos(\omega t + \theta) = \frac{1}{2}$ and $\omega t + \theta = \pm 60°$.

$$e'(t) = -\omega E\sin(\omega t + \theta). \tag{7.20}$$

For $\omega t + \theta = +60°$, $e'(t) = -\omega E\sin 60° = -\frac{\sqrt{3}}{2}\omega E < 0$.

For $\omega t + \theta = -60°$, $e'(t) = -\omega E\sin(-60°) = +\frac{\sqrt{3}}{2}\omega E > 0$.

Therefore, so that $e(t) = E/2$, with $e'(t) > 0$, it would be necessary that $\omega t + \theta = -60°$. It follows that $t = -(\theta + \pi/3)/\omega\,[\text{s}] < 0$. However, in this condition, the new origin of time, the $t = 0$ of the opening of switch k, would have to be located to the left of the first origin of time, the one to which the voltage source $e(t)$ is referred, which is unacceptable, as it would imply attributing negative values for time, in relation to the initial origin. Therefore, the condition $\omega t + \theta = \pm 60°$ cannot be accepted, since the opening of the switch must occur only for positive values of t, in relation to the initial origin. In addition, the ATPDraw program does not process simulations with negative t values.

The next condition for $\cos(\omega t + \theta) = 1/2$ is that $\omega t + \theta = \mp 60° + 360°$. That is, that $\omega t + \theta = 5\pi/3\,[\text{rad}] \equiv 300°$, or that $\omega t + \theta = 7\pi/3\,[\text{rad}] \equiv 420°$, since $\cos 300° = \cos 420° = 0.5$. (Make a sketch of the $e(t)$ and $i_k(t)$ curves best illustrates this issue). From Eq. 7.20, it follows that:

for $\omega t + \theta = 300°$, $e'(t) = -\omega E\sin 300° = +\frac{\sqrt{3}}{2}\omega E > 0$.

for $\omega t + \theta = 420°$, $e'(t) = -\omega E\sin 420° = -\frac{\sqrt{3}}{2}\omega E < 0$.

So the condition for switching $e(t) = E/2$ with $e'(t) > 0$ is that $\omega t + \theta = 5\pi/3\,[\text{rad}] \; or \; \omega t + \theta = 300°$. That is, that $t = \frac{5\pi/3 - \theta}{\omega}$.

As it was stated that the switch k was opened at a new time taken as $t = 0$, it is necessary to correct the expression of the steady-state current $i_k(t)$ for the new origin of time. Therefore,

$$i_k(t) = \frac{E}{\omega L}\sin[(\omega t + \theta) + (300° - \theta)] = \frac{E}{\omega L}\sin(\omega t + 300°).$$

Therefore, the circuit with the injected source, F_2, is the one shown in Fig. 7.60. The voltage node equation $e_k(t)$ in Fig. 7.60 is

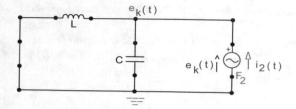

$$i_2(t) = \frac{E}{\omega L} sin(\omega t + 300^\circ). U_{-1}(t)$$

Fig. 7.60 Electric circuit of Example 7.13 for calculating the exclusive effects of the (injected) source F_2

$$C \frac{d}{dt} e_k(t) + \frac{1}{L} \int_{0_+}^{t} e_k(\tau)d\tau = \frac{E}{\omega L} sin(\omega t + 300^\circ).U_{-1}(t).$$

Or,

$$e_k''(t) + \omega_0^2 e_k(t) = -\frac{\sqrt{3}}{2} \cdot \frac{\omega_0^2 E}{\omega} U_0(t) + \omega_0^2 E \cos(\omega t + 300^\circ) U_{-1}(t)$$

$$= -\frac{\sqrt{3}}{2} \cdot \frac{\omega_0^2 E}{\omega} U_0(t) + \frac{\omega_0^2 E}{2} U_{-1}(t) + \dots, \text{ with } e_k(0_-) = e_k'(0_-) = 0.$$

For $t > 0$,

$$\left(D^2 + \omega_0^2\right)e_k(t) = \omega_0^2 E \cos(\omega t + 300^\circ).$$

$$e_{k_H}(t) = k_1 \cos \omega_0 t + k_2 \sin(\omega_0 t)$$

$$\left(-\omega^2 + \omega_0^2\right)\dot{E}_{k_P} = \omega_0^2 E \angle 300^\circ$$

$$\dot{E}_{k_P} = \frac{\omega_0^2 E}{\omega_0^2 - \omega^2} \angle 300^\circ \text{ and } e_{k_P}(t) = \frac{\omega_0^2 E}{\omega_0^2 - \omega^2} \cos(\omega t + 300^\circ)$$

$$e_k(t) = \frac{\omega_0^2 E}{\omega_0^2 - \omega^2} \cos(\omega t + 300^\circ) + k_1 \cos(\omega_0 t) + k_2 \sin(\omega_0 t).$$

$$e_k'(t) = -\frac{\omega \omega_0^2 E}{\omega_0^2 - \omega^2} \sin(\omega t + 300^\circ) - \omega_0 k_1 \sin(\omega_0 t) + \omega_0 k_2 \cos(\omega_0 t).$$

The last member of the differential equation of $e_k(t)$ allows you to write that

$$e_k(t) \doteq 0 . U_{-1}(t).$$

$$e'_k(t) \doteq C_0 U_{-1}(t).$$

$$e''_k(t) \doteq C_0 U_0(t) + C_1 U_{-1}(t).$$

Substituting in the differential equation of $e_k(t)$, then

$$C_0 U_0(t) + C_1 U_{-1}(t) \doteq -\frac{\sqrt{3}}{2} \cdot \frac{\omega_0^2 E}{\omega} U_0(t) + \frac{\omega_0^2 E}{2} U_{-1}(t).$$

$$C_0 = -\frac{\sqrt{3}}{2} \cdot \frac{\omega_0^2 E}{\omega}.$$

$$C_1 = \frac{\omega_0^2 E}{2}.$$

So,

$$e_k(0_+) = 0 = \frac{\omega_0^2 E}{\omega_0^2 - \omega^2} \cos(300°) + k_1.$$

$$e'_k(0_+) = C_0 = -\frac{\sqrt{3}}{2} \cdot \frac{\omega_0^2 E}{\omega} = -\frac{\omega \omega_0^2 E}{\omega_0^2 - \omega^2} \sin(300°) + \omega_0 k_2.$$

Therefore,

$$k_1 = -\frac{1}{2} \cdot \frac{\omega_0^2 E}{\omega_0^2 - \omega^2}.$$

$$k_2 = -\frac{\sqrt{3}}{2} \cdot \frac{\omega_0 E}{\omega} - \frac{\sqrt{3}}{2} \cdot \frac{\omega \omega_0 E}{\omega_0^2 - \omega^2}$$

$$= -\frac{\sqrt{3} \omega_0 E}{2} \left(\frac{1}{\omega} + \frac{\omega}{\omega_0^2 - \omega^2} \right) = -\frac{\sqrt{3} \omega_0}{2\omega} \cdot \frac{\omega_0^2 E}{\omega_0^2 - \omega^2}.$$

Finally,

$$e_k(t) = \frac{\omega_0^2 E}{\omega_0^2 - \omega^2} \left[\cos(\omega t + 300°) - \frac{1}{2} \cos(\omega_0 t) - \frac{\sqrt{3}}{2} \cdot \frac{\omega_0}{\omega} \cdot \sin(\omega_0 t) \right] . U_{-1}(t).$$

$$(7.21)$$

Equation 7.21 is referenced at time $t = 0$ of the opening of switch k. To reference it at time $t = 0$ of the voltage source $e(t)$, it is necessary to replace t with $t - (300° - \theta)/\omega$, which corresponds to replace ωt with $\omega t - 300° + \theta$.

So,

$$e_k(t) = \frac{\omega_0^2 E}{\omega_0^2 - \omega^2} \left\{ \cos(\omega t + \theta) - \frac{1}{2} \cdot \cos\left[\omega_0\left(t - (300° - \theta)/\omega\right)\right] \right. $$
$$\left. - \frac{\sqrt{3}}{2} \cdot \frac{\omega_0}{\omega} \cdot \sin\left[\omega_0\left(t - (300° - \theta)/\omega\right)\right] \right\} U_{-1}\left[t - (300° - \theta)/\omega\right].$$

Or,

$$e_k(t) = \frac{E}{1 - \left(\frac{\omega}{\omega_0}\right)^2} \left\{ \cos(\omega t + \theta) - \frac{1}{2}\cos\left[\omega_0 t - \frac{\omega_0}{\omega}\left(\frac{5\pi}{3} - \theta\right)\right] \right.$$
$$\left. - \frac{\sqrt{3}}{2} \cdot \frac{\omega_0}{\omega} \cdot \sin\left[\omega_0 t - \frac{\omega_0}{\omega}\left(\frac{5\pi}{3} - \theta\right)\right] \right\}, \quad for\, t \geq \frac{\frac{5\pi}{3} - \theta}{\omega}. \tag{7.22}$$

Substituting the given numerical values, Eq. 7.22 becomes

$$e_k(t) = 2057,1429 \cos\left(800\pi t + \frac{\pi}{6}\right) - 1028,5714 \cos(4,800\pi t - 9\pi)$$
$$- 10,689.2281 \sin(4,800\pi t - 9\pi), \quad for\, t \geq 1.8750\,[\text{ms}].$$

Or,

$$e_k(t) = 2057,1429 \cos(2513,2741 t + 0.5236)$$
$$+ 10,738.6012 \cos(15,079.6447 t - 26.6076), \quad t \geq 1.8750\,[\text{ms}]. \tag{7.23}$$

The periods are:

$$T = 1/f = 1/400 = 2.5\,[\text{ms}] \text{ and } T_0 = 1/f_0 = 1/2.400 = 0.4167\,[\text{ms}].$$

(b) The circuit built for the numerical solution is shown in Fig. 7.61.

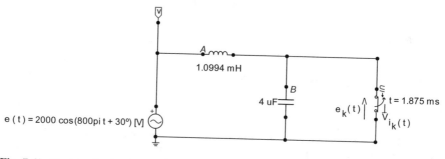

Fig. 7.61 Electric circuit of Example 7.13 for the numerical solution by the ATPDraw program

7.3   Current Source Injection Method

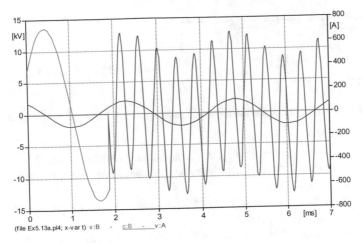

Fig. 7.62   Voltage $e_k(t)$, switch current $i_k(t)$, source voltage $e(t)$

As at the moment of opening the switch k the current is worth $i_k = \frac{E}{\omega L}\sin(\omega t_{open} + \theta) = 727.1333\sin(800\pi.1.875 \times 10^{-3} + \pi/6) = -629.72\,[A]$, it was established that $Imar = 630\,[A]$. The answers are plotted in Fig. 7.62, with $\Delta t = 10^{-8}\,[s]$.

It is interesting to note that because $\omega_0 = 6\omega$, the voltage $e_k(t)$ is periodic for period $T = 2.5\,[ms]$, and that this can be seen in Fig. 7.62 between $t = 1.875\,[ms]$ and $t = 1.875 + 2.5 = 4.375\,[ms]$.

From Eq. 7.23,

$$e_k(t_{open}) = e_k(1.875\,\text{ms}) = e_k(0.001875) = -0.0811\,[V].$$

$$e_k(2\,\text{ms}) = e_k(0.002) = -8319,4239\,[V].$$

$$e_k(4\,\text{ms}) = e_k(0.004) = -7951,8777\,[V].$$

From the curve plotted in Fig. 7.62,

$$e_k(t_{open}) = e_k(1.875\,\text{ms}) = 0.0000\,[V].$$

$$e_k(0.002) = -8320,2\,[V].$$

$$e_k(0.004) = -7951,7\,[V].$$

If what is desired is only the waveform of the voltage $e_k(t)$ at the open switch terminals, it does not matter to reference it at time $t = 0$ of the applied voltage source $e(t)$, that is, regardless of the that happened before $t = 1.875\,[ms]$, then an alternative numerical solution can be obtained from the circuit with injection of the

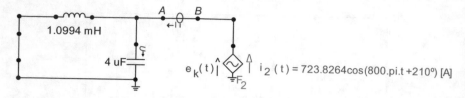

Fig. 7.63 Alternative circuit for the numerical solution with injection of the current source F_2

current source F_2, shown in Fig. 7.63, and that corresponds to the opening of switch k at time $t = 0$. That is, the time $t = 0$ of the opening of the switch is located on the time $t = 1.875$ [ms].

In Fig. 7.64 the curves of $e_k(t)$ and $i_2(t)$ are plotted, the current that was injected by the current source F_2.

Considering that $t = 1.875$ [ms] in Fig. 7.62 corresponds to $t = 0$ in Fig. 7.64, then $t = 2$ [ms] corresponds to $2-1.875 = 0.125$ [ms] and $t = 4$ [ms] corresponds to $4-1.875 = 2.125$ [ms]. Thus, from Fig. 7.64, we have that:

$$e_k(0.125) = -8319,7 \,[\text{V}].$$

$$e_k(2.125) = -7951,8 \,[\text{V}].$$

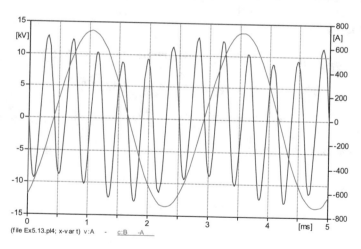

(file Ex5.13.pl4; x-var t) v:A - c:B -A

Fig. 7.64 Voltage $e_k(t)$, injected current $i_2(t)$

7.4 Displacement of Current Source Method

The method called displacement of current source (Displacement I) is useful for transforming a circuit containing an ideal current source $i(t)$, directly connected between nodes A and B of the same, in another equivalent circuit, in which a current source $i(t)$ it is placed in parallel with each branch, or network, which forms a loop with the original current source. It is useful, for example, to generate Norton equivalents, in situations where the current source is not in parallel with any element of the circuit. Figures 7.65 and 7.66 illustrate the method.

In Fig. 7.65, the source current, $i(t)$, leaves node B and enters node A. In Fig. 7.66, current $i(t)$ continues to exit node B and enter node A. On nodes C and D,

Fig. 7.65 Illustration of the displacement of current source method—original electric circuit

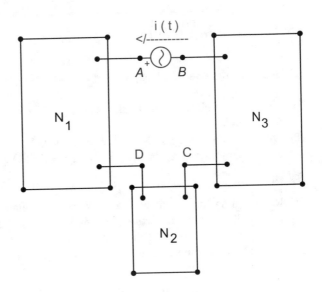

Fig. 7.66 Illustration of the displacement of current source method—equivalent electric circuit

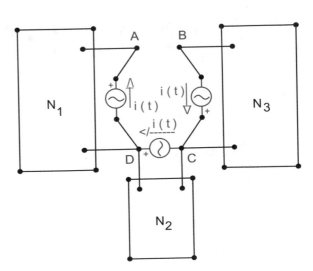

however, current $i(t)$ does not act, because it enters one branch and leaves another, at the same time. That is, the equation obtained with the Kirchhoff currents law is not changed in nodes C and D by the presence of the current source $i(t)$.

Example 7.14 The circuit in Fig. 7.67 is operating in steady-state with switch k closed. If it is opened at any time $\tau > 0$,

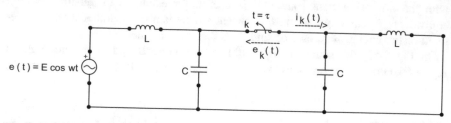

Fig. 7.67 Electric circuit of Example 7.14

(a) establish the analytical expression of the voltage $e_k(t)$, at its terminals, for $t > \tau$;
(b) study the following particular cases: (1) $\tau = 0$ [s], (2) $\tau = \pi/2\omega$ [s], (3) $\tau = \pi/\omega$ [s];
(c) with $E = 1000$ [V], $f = 60$ [Hz], $L = 6.25$ [mH] and $C = 70.36$ [µF], simulate all the particular cases of item b, using the ATPDraw program.

Solution

(a) Representing the circuit of Fig. 7.67 in the domain of simple frequency and then replacing the left part of the switch with its Thévenin equivalent and the right part with its equivalent impedance, the circuit shown in Fig. 7.68 results from the which the steady-state current $i_k(t)$ can be determined, circulating through the switch k.

So, the phasor current flowing through the switch is

$$\dot{I}_k = \frac{\dot{E}_{TH}}{j\frac{2\omega L}{1-\omega^2 LC}} = \frac{\frac{\dot{E}}{1-\omega^2 LC}}{j\frac{2\omega L}{1-\omega^2 LC}} = \frac{\dot{E}}{j2\omega L} = \frac{E}{2\omega L}\angle -90°.$$

Therefore,

$$i_k(t) = \frac{E}{2\omega L}\cos(\omega t - 90°) = \frac{E}{2\omega L}\sin \omega t.$$

To simulate the opening of switch k at time $t = \tau$ it is necessary to inject the current source F_2, of value $i_2(t) = (E/2\omega L)\sin(\omega t).U_{-1}(t - \tau)$, in the opposite

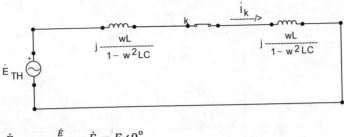

$$\dot{E}_{TH} = \frac{\dot{E}}{1-\omega^2 LC}\,;\, \dot{E} = E\angle 0°.$$

Fig. 7.68 Electric circuit of Example 7.14 for AC steady-state solution

direction to of $i_k(t)$, and short-circuit the external voltage source $e(t)$, as shown in Fig. 7.69.

Using displacement I, the circuit of Fig. 7.69 can be converted to the equivalent circuit shown in Fig. 7.70.

The Kirchhoff voltage law applied to the ABB'A'A closed path allows us to conclude that:

$$e_k(t) = e_1(t) + e_2(t).$$

Due to the symmetry that the circuit of Fig. 7.68 presents, it is easily verified that $e_2(t) = e_1(t)$. So,

$$e_k(t) = 2e_1(t). \tag{7.24}$$

The equation for node A is

$$C\tfrac{d}{dt}e_1(t) + \tfrac{1}{L}\int_{\tau_+}^{t} e_1(\lambda)d\lambda = i_2(t), \text{ and,}$$

$$e_1''(t) + \omega_0^2 e_1(t) = \tfrac{1}{C}i_2'(t), \ \omega_0^2 \triangleq 1/LC.$$

Since $i_2(t) = \tfrac{E}{2\omega L}\sin\omega t.U_{-1}(t - \tau)$, then

$$i_2'(t) = \tfrac{E}{2\omega L}\sin\omega\tau.U_0(t - \tau) + \tfrac{E}{2L}\cos\omega t.U_{-1}(t - \tau).$$

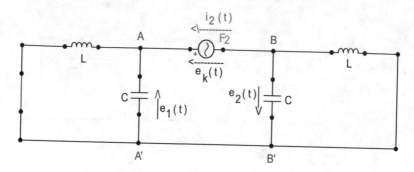

$$i_2(t) = \frac{E}{2\omega L}\sin(\omega t).U_{-1}(t-\tau)$$

Fig. 7.69 Electric circuit of Example 7.14 for calculating the exclusive effects of the (injected) source F_2

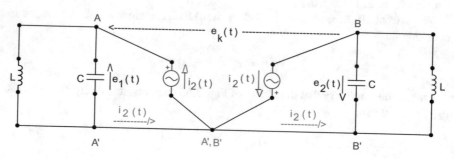

Fig. 7.70 Equivalent electric circuit of Example 7.14 with displacement of the (injected) current source F_2

Therefore,

$$e_1''(t) + \omega_0^2 e_1(t) = \frac{\omega_0^2 E}{2\omega}\sin(\omega\tau).U_0(t-\tau) + \frac{\omega_0^2 E}{2}\cos(\omega t).U_{-1}(t-\tau)$$

$$= \frac{\omega_0^2 E}{2\omega}\sin(\omega\tau).U_0(t-\tau) + \frac{\omega_0^2 E}{2}\cos(\omega\tau).U_{-1}(t-\tau) \quad (7.25)$$

$$+ \ldots; \; e_1(\tau_-) = e_1'(\tau_-) = 0$$

For $t > \tau$,

$$(D^2 + \omega_0^2)e_1(t) = \frac{\omega_0^2 E}{2}\cos\omega t$$

$$e_{1_H}(t) = k_1\cos(\omega_0 t) + k_2\sin(\omega_0 t)$$

$$\left(-\omega^2 + \omega_0^2\right)\dot{E}_{1_P} = \frac{\omega_0^2 E}{2}\angle 0°; \dot{E}_{1_P} = \frac{\omega_0^2 E}{2\left(\omega_0^2 - \omega^2\right)}\angle 0°.$$

$$e_{1_P} = \frac{\omega_0^2 E}{2\left(\omega_0^2 - \omega^2\right)}\cos(\omega t).$$

So,

$$e_1(t) = \frac{\omega_0^2 E}{2\left(\omega_0^2 - \omega^2\right)}\cos(\omega t) + k_1\cos(\omega_0 t) + k_2\sin(\omega_0 t). \qquad (7.26)$$

$$e_1'(t) = -\frac{\omega\omega_0^2 E}{2\left(\omega_0^2 - \omega^2\right)}\sin(\omega t) - \omega_0 k_1\sin(\omega_0 t) + \omega_0 k_2\cos(\omega_0 t).$$

From Eq. 7.25, it is easy to conclude that

$$e_1(\tau_+) = 0.$$

$$e_1'(\tau_+) = \frac{\omega_0^2 E}{2\omega}\sin(\omega\tau).$$

So,

$$k_1\cos(\omega_0\tau) + k_2\sin(\omega_0\tau) = -\frac{\omega_0^2 E\cos(\omega\tau)}{2\left(\omega_0^2 - \omega^2\right)}. \qquad (7.27)$$

$$-\omega_0 k_1\sin(\omega_0\tau) + \omega_0 k_2\cos(\omega_0\tau) = \frac{\omega_0^2 E}{2\omega}\sin(\omega\tau)$$
$$+ \frac{\omega\omega_0^2 E}{2\left(\omega_0^2 - \omega^2\right)}\sin(\omega\tau) = \frac{\omega_0^4 E\sin(\omega\tau)}{2\omega\left(\omega_0^2 - \omega^2\right)} \qquad (7.28)$$

The system solution formed by Eqs. 7.27 and 7.28, for k_1 and k_2, provides that:

$$k_1 = \frac{-\omega_0^2 E}{2\left(\omega_0^2 - \omega^2\right)}\left[\cos(\omega\tau).\cos(\omega_0\tau) + \frac{\omega_0}{\omega}\sin(\omega\tau).\sin(\omega_0\tau)\right]. \qquad (7.29)$$

$$k_2 = \frac{-\omega_0^2 E}{2\left(\omega_0^2 - \omega^2\right)}\left[\cos(\omega\tau).\sin(\omega_0\tau) - \frac{\omega_0}{\omega}\sin(\omega\tau).\cos(\omega_0\tau)\right]. \qquad (7.30)$$

With the constants k_1 and k_2 calculated, the voltage $e_1(t)$, expressed by Eq. 7.26, is completely established. Replacing, then, $e_1(t)$ in Eq. 7.24, it finally results that:

$$e_k(t) = \frac{\omega_0^2 E}{\omega_0^2 - \omega^2} (\cos\omega t + A\cos\omega_0 t + B\sin\omega_0 t).U_{-1}(t - \tau), \qquad (7.31)$$

where

$$A = -\left(\cos\omega\tau.\cos\omega_0\tau + \frac{\omega_0}{\omega}.sen\omega\tau.\sin\omega_0\tau\right) \qquad (7.32)$$

$$B = -\left(\cos\omega\tau.\sin\omega_0\tau - \frac{\omega_0}{\omega}.\sin\omega\tau.\cos\omega_0\tau\right) \qquad (7.33)$$

$$\omega_0 = 1/\sqrt{LC}.$$

(b) Particular Cases

1. $\tau = 0$

In this case the switch k is opened at the instant when the voltage source $e(t)$ is at its maximum value E, and the current $i_k(t)$, through the switch, is null, with a positive derivative.

From Eqs. 7.32 and 7.33, for $\tau = 0, A = -1$ and $B = 0$. Therefore, from Eq. 7.31,

$$e_k(t) = \frac{E}{1 - (\omega/\omega_0)^2} (\cos\omega t - \cos\omega_0 t).U_{-1}(t). \qquad (7.34)$$

2. $\tau = \pi/2\omega$.

In this case, where $\omega\tau = \pi/2$, the current $i_k(t)$ through the switch is passing its maximum value, $E/2\omega L$, at the moment of switch opening.

From Eqs. 7.32 and 7.33,

$$A = -\frac{\omega_0}{\omega}\sin(\omega_0\pi/2\omega)$$

$$B = \frac{\omega_0}{\omega}\cos(\omega_0\pi/2\omega)$$

So, from Eq. 7.31,

$$e_k(t) = \frac{E}{1 - (\omega/\omega_0)^2}\left[\cos\omega t - \frac{\omega_0}{\omega}\sin(\omega_0\pi/2\omega)\cos\omega_0 t\right.$$
$$\left. + \frac{\omega_0}{\omega}\cos(\omega_0\pi/2\omega)\sin\omega_0 t\right]U_{-1}(t - \pi/2\omega)$$

$$e_k(t) = \frac{E}{1 - (\omega/\omega_0)^2}\left[\cos\omega t + \frac{\omega_0}{\omega}\sin(\omega_0 t - \omega_0\pi/2\omega)\right]U_{-1}(t - \pi/2\omega).$$

Or,

$$e_k(t) = \frac{E}{1 - (\omega/\omega_0)^2} \left\{ \cos\omega t + \frac{\omega_0}{\omega} \sin[\omega_0(t - \pi/2\omega)] \right\}.U_{-1}(t - \pi/2\omega). \quad (7.35)$$

To obtain Eq. 7.35, the trigonometric identity was used:
sina cosb − sinb cosa = sin(a − b).

3. $\tau = \pi/\omega$.

Now the switch is opened at the instant the source $e(t)$ is at its minimum value, $-E$, and the current passing through the switch k is null, with a negative derivative.
From Eqs. 7.32 and 7.33,

$$A = \cos(\omega_0\pi/\omega)$$

$$B = \sin(\omega_0\pi/\omega).$$

From Eq. 7.31,

$$e_k(t) = \frac{E}{1 - (\omega/\omega_0)^2} [\cos\omega t + \cos(\omega_0\pi/\omega)\cos\omega_0 t + \sin(\omega_0\pi/\omega)\sin\omega_0 t]$$
$$U_{-1}(t - \pi/\omega)$$

Or,

$$e_k(t) = \frac{E}{1 - (\omega/\omega_0)^2} \left\{ \cos\omega t + \cos[\omega_0(t - \pi/\omega)] \right\}.U_{-1}(t - \pi/\omega). \quad (7.36)$$

(c)

1. Substituting the numerical values in Eq. 7.34, (for $\tau = 0$), comes that

$$e_k(t) = 1066,6647(\cos376.9911t - \cos1507,9852t).U_{-1}(t) \text{ [V]}. \quad (7.37)$$

The circuit for this case is shown in Fig. 7.71, and the result of the simulation, in Fig. 7.72.
From Eq. 7.37,

$$e_k(2\,\text{ms}) = 1835,8244 \text{ [V]}; \ e_k(8\,\text{ms}) = -1993,0643 \text{ [V]}; \ e_k(10\,\text{ms}) = 0.1298 \text{ [V]}$$

From Fig. 7.72,

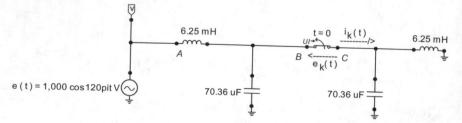

Fig. 7.71 Electric circuit of Example 7.14 for the numerical solution by the ATPDraw program

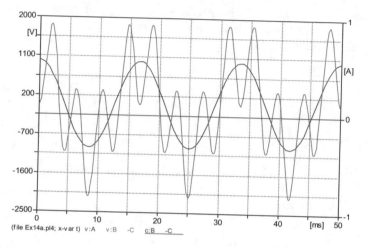

Fig. 7.72 Voltages $e(t)$, $e_k(t)$ and current $i_k(t)$ accross the switch for this case

$$e_k(2\,\text{ms}) = 1835, 8\,[\text{V}]; \; e_k(8\,\text{ms}) = -1993, 1\,[\text{V}]; \; e_k(10\,\text{ms}) = 0.1299\,[\text{V}].$$

2. Introducing the numerical values in Eq. 7.35, (for $\tau = \pi/2\omega$), it turns out that

$$e_k(t) = 1066, 6647\cos 376.9911t + 4266, 7655\sin(1507, 9852t - 6.2833)\,[\text{V}],$$
$$for\ t \geq 4.1667\,[\text{ms}].$$

$$(7.38)$$

The ATPDraw circuit is the same as in Fig. 7.71, except that the switch is now open at time $t = \tau = \pi/2\omega = 4.1667$ [ms]. Figure 7.73 shows the results obtained. From Eq. 7.38,

$$e_k(5\,\text{ms}) = 3728, 3031\,[\text{V}]; \; e_k(7\,\text{ms}) = -4795, 4654\,[\text{V}]; \; e_k(15\,\text{ms})$$
$$= -1645, 6694\,[\text{V}].$$

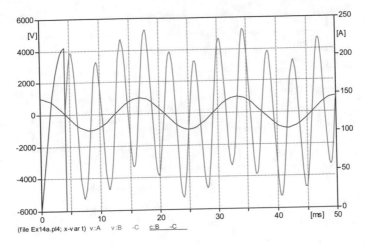

Fig. 7.73 Voltages $e(t)$, $e_k(t)$ and current $i_k(t)$ accross the switch for this case

From Fig. 7.73,

$$e_k(5\,\text{ms}) = 3726,7\,[\text{V}]; \; e_k(7\,\text{ms}) = -4793,3\,[\text{V}]; \; e_k(15\,\text{ms}) = -1641,7\,[\text{V}].$$

3. With the numerical values, Eq. 7.36, (for $\tau = \pi/\omega$), becomes

$$e_k(t) = 1066,6647\{\cos 376.9911t + \cos[1507,9852(t - 0.0083333)]\}\,[\text{V}],$$
$$\text{for } t \geq 8.3333\,[\text{ms}].$$

$$(7.39)$$

The ATPDraw circuit is the same as in Fig. 7.71, now with the switch k opening at time $t = \tau = \pi/\omega = 8.3333$ [ms]. And the results of the simulation are shown in Fig. 7.74.

From Fig. 7.74,

$$e_k(10\,\text{ms}) = -1725,9\,[\text{V}]$$

$$e_k(13\,\text{ms}) = 977.37\,[\text{V}]$$

$$e_k(14\,\text{ms}) = -108.47\,[\text{V}]$$

$$e_k(16.7\,\text{ms}) = 2131,9\,[\text{V}].$$

From Eq. 7.39,

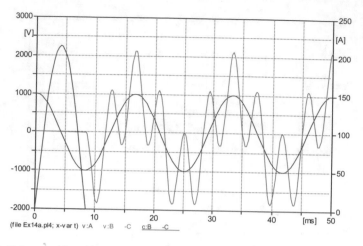

Fig. 7.74 Voltages $e(t)$, $e_k(t)$ and current $i_k(t)$ accross the switch for this case

$$e_k(10\,\text{ms}) = -1725,9530\,[\text{V}]$$

$$e_k(13\,\text{ms}) = 977.3305\,[\text{V}]$$

$$e_k(14\,\text{ms}) = -108.5082\,[\text{V}]$$

$$e_k(16.7\,\text{ms}) = 2131,8859\,[\text{V}].$$

7.5   Proposed Problems

P.7-1 Determine the expression of the voltage $e_1(t)$, for $t > 0$, in the circuit of Fig. 7.75, knowing that $e_1(0_-) = E$ and that $e_2(0_-) = E/k$.

Answer: $e_1(t) = \frac{E}{k+1}\left[2 + (k-1)\epsilon^{-t/kRC}\right]$,   *for* $t \geq 0$.

P.7-2 The circuit shown in Fig. 7.76 was operating steadily when switch k was closed at time $t = 0$. Establish the expression of the current $i_k(t)$ that will circulate through the switch, for $t > 0$.

Answer: $i_k(t) = \frac{E}{3R}\epsilon^{-t/3RC}.U_{-1}(t)$.

P.7-3 The circuit in Fig. 7.77 operated in a steady-state when switch k was closed at the instant taken as $t = 0$. Determine the current $i_k(t)$, for $t > 0$.

Answer: $i_k(t) = \frac{E}{(2k+1)R}\left[1 - \epsilon^{-(2k+1)Rt/L}\right].U_{-1}(t)$.

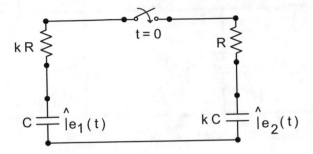

Fig. 7.75 Electric circuit for Problem P.7-1

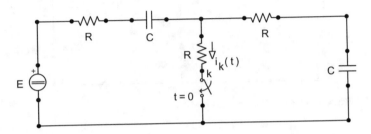

Fig. 7.76 Electric circuit for Problem P.7-2

Fig. 7.77 Electric circuit for
Problem P.7-3

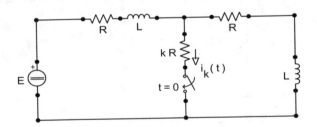

P.7-4 The circuit in Fig. 7.78 operated in a steady-state when switch k was closed at time $t = 0$. Find the expression of the current $i_k(t)$ circulating through the switch, for $t > 0$.

Answer: $i_k(t) = \left[\frac{t}{2L} + \sqrt{\frac{C}{L}} \sin \omega_0 t + \frac{\sqrt{2}}{4} \sqrt{\frac{C}{L}} \sin(\sqrt{2}\omega_0 t) \right] E.U_{-1}(t);\ \omega_0 = 1/\sqrt{LC}.$

P.7-5 Switch k in the circuit of Fig. 7.79 is closed at the moment taken as $t = 0$, after the steady-state has been reached. Find the voltage $e_o(t)$, for $t > 0$.

Answer: $e_o(t) = 12(1 - \epsilon^{-2t}).U_{-1}(t)\,[\mathrm{V}].$

P.7-6 The circuit contained in Fig. 7.80 was operating in a steady-state when, at time $t = 0$, switch k was closed. Find the expression of the voltage $e_0(t)$, for $t > 0$. Data: $E = 100\,[\mathrm{V}]$, $R = 5\,[\Omega]$, $L_1 = M = 1\,[\mathrm{H}]$ and $L_2 = 2\,[\mathrm{H}]$.

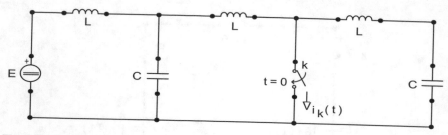

Fig. 7.78 Electric circuit for Problem P.7-4

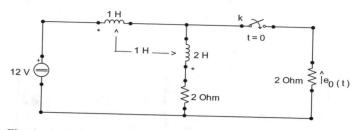

Fig. 7.79 Electric circuit for Problem P.7-5

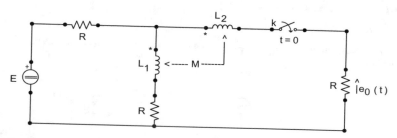

Fig. 7.80 Electric circuit for Problem P.7-6

Answer: $e_0(t) = \frac{25}{3}\left(4 - 3\epsilon^{-5t} - \epsilon^{-15t}\right).U_{-1}(t)\,[\mathrm{V}]$.

P.7-7 The permanent regime was established in the circuit of Fig. 7.81 with the k switch open. At time $t = 0$ it is closed. Determine $i_C(t)$, for $t > 0$, with $E = 125\,[\mathrm{V}]$, $R_1 = 50\,[\Omega]$, $R_2 = 250\,[\Omega]$, $R_3 = 200\,[\Omega]$, $L = 1\,[\mathrm{H}]$, $C = 0.2\,[\mathrm{F}]$.

Answer: $i_C(t) = 100.25\left(\epsilon^{-0.125t} - \epsilon^{-50t}\right).U_{-1}(t)\,[\mathrm{A}]$.

P.7-8 The circuit shown in Fig. 7.82 was operating in a steady-state when switch k was closed at the time considered as $t = 0$. Find the voltage $e_C(t)$, for $t > 0$, knowing that at the time of switching $e(t) = 0$ with $e'(t) < 0$. Consider that $\omega_0 = 1/\sqrt{LC} = 2\omega$.

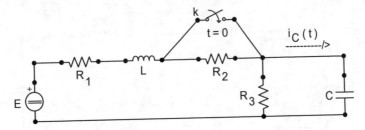

Fig. 7.81 Electric circuit for Problem P.7-7

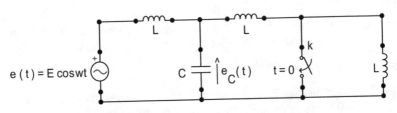

Fig. 7.82 Electric circuit for Problem P.7-8

Answer: $e_C(t) = -\frac{2E}{35}\left[\sqrt{2}\sin(2\sqrt{2}\omega t) + 10\sin\omega t\right]$ [V],   *for* $t > 0$.
  (Without origin correction, that is, $t = 0$ is the closing time of switch k).

P.7-9 With the circuit of Fig. 7.83 operating in steady-state, switch k was closed at time $t = 0$. At the time of switching, the current in the inductances was maximum. Determine the current $i(t)$, for $t > 0$.

Answer: $i(t) = \frac{E}{2L(\omega^2 - 2\omega_0^2)}\left(\frac{-2\omega_0^2}{\omega}\cos\omega t + \omega\cos\sqrt{2}\omega_0 t\right)$ [A], *for* $t \geq 0$.
  (No correction of time origin).

P.7-10 The circuit shown in Fig. 7.84 was operating in a steady-state when, at time $t = 0$, switch k was closed. Determine the current $i_1(t)$, for $t > 0$, knowing that at the instant of switching the voltage of the sinusoidal source was maximum.

Data: $e(t) = 100\cos 2t$ [V], $L = 1$ [H], $M = 0.5$ [H] and $C = 1/16$ [F].

Answer: $i_1(t) = 46.156\sin(2t) - 10.361\sin(4.62t)$ [A],   *for* $t \geq 0$.

Fig. 7.83 Electric circuit for
Problem P.7-9

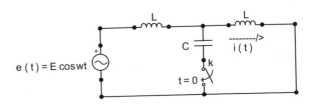

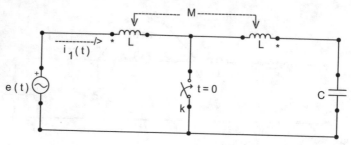

Fig. 7.84 Electric circuit for Problem P.7-10

P.7-11 The circuit shown in Fig. 7.85 operated in a steady-state when, at time $t = 0$, switch k was opened. Determine the voltage $e_k(t)$ at its terminals for $t > 0$.

Answer: $e_k(t) = E\left(1 - \frac{1}{4}\epsilon^{-t/2RC}\right).U_{-1}(t)$.

P.7-12 The circuit in Fig. 7.86 was operating in a steady-state when switch k was opened at $t = 0$. Determine the voltage $e_k(t)$ and the current $i(t)$, for $t > 0$.

Answer: $e_k(t) = 2U_0(t) + \left(6 + \frac{8}{3}\epsilon^{-2t/3}\right).U_{-1}(t)$ [V].

$i(t) = 3 + 2\epsilon^{-2t/3}$ [A],   *for* $t > 0$.

P.7-13 After being closed for a long time, switch k in the circuit of Fig. 7.87 was opened at time $t = 0$. Calculate the voltage $e_k(t)$ at its terminals, for $t > 0$.

Answer: $e_k(t) = 20U_0(t) + \left(75 + 5\epsilon^{-8t} - 40\epsilon^{-t/4}\right)\dot{U}_{-1}(t)$ [V].

P.7-14 The circuit in Fig. 7.88 operated in a steady-state when switch k was opened at time $t = 0$. Determine the expression of the voltage $e_k(t)$ at the switch terminals, for $t > 0$.

Answer: $e_k(t) = (1 + 2\epsilon^{-2t})E.U_{-1}(t)$ [V].

Fig. 7.85 Electric circuit for
Problem P.7-11

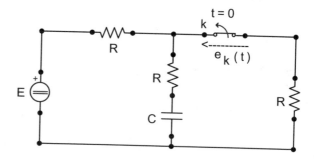

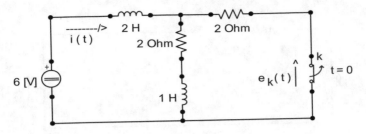

Fig. 7.86 Electric circuit for Problem P.7-12

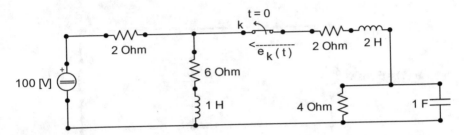

Fig. 7.87 Electric circuit for Problem P.7-13

P.7-15 The linear and invariant circuit shown in Fig. 7.89 is operating in a steady-state with switch k closed. If it is opened at $t = 0$, find $e_1(t)$, for $t \geq 0$, where $R_1 = R_2 = 1\ [\Omega]$, $L = 1\ [H]$, $C = 1\ [F]$ and $e(t) = \cos t\ [V]$.

Answer: $e_1(t) = 0.612\epsilon^{-t/2} \cos\left(\frac{\sqrt{3}}{2}t - 11.3°\right)\ [V]$,   *for* $t \geq 0$.

P.7-16 The circuit shown in Fig. 7.90 was operating steadily when switch k was opened at $t = 0$. Knowing that at the time of switching the voltage of the source $e(t)$ was at its minimum value (maximum negative), determine the voltage $e_2(t)$ for $t > 0$.

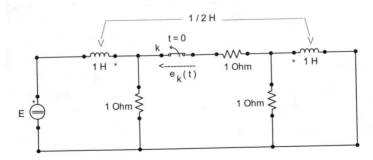

Fig. 7.88 Electric circuit for Problem P.7-14

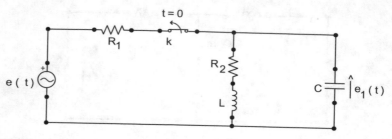

Fig. 7.89 Electric circuit for Problem P.7-15

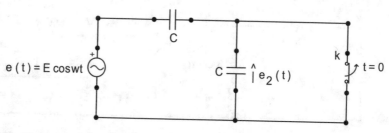

Fig. 7.90 Electric circuit for Problem P.7-16

Answer: $e_2(t) = \frac{E}{2}(1 - \cos\omega t).U_{-1}(t)$.

(Without origin correction, that is, t = 0 is the opening time of switch k).

P.7-17 The circuit shown in Fig. 7.91 has long been operating with switch k closed. However, at the moment considered as $t = 0$ it was opened. Knowing that at the moment of opening the current $i_k(t)$, through it, was zero, with a positive derivative, determine the analytical expression of the voltage $e_k(t)$ at its terminals, for $t > 0$.

Answer: $e_k(t) = [-0.32\epsilon^{-37.33t} + 120\sin(40t + 34.5°)].U_{-1}(t)$ [V].

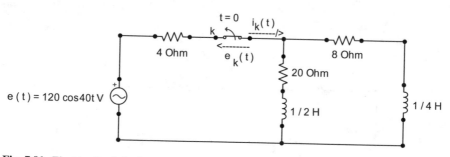

Fig. 7.91 Electric circuit for Problem P.7-17

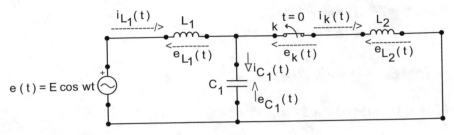

Fig. 7.92 Electric circuit for Problem P.7-18

P.7-18 The circuit in Fig. 7.92 was operating in a steady-state when, at time $t = 0$, switch k was opened. At the instant the switch was opened, the current through it was $i_k = 0$, with $i'_k > 0$.

(a) Find the analytical, literal, steady-state expressions of the currents $i_k(t)$, $i_{L_1}(t)$ and $i_{C_1}(t)$, and the voltages $e_{L_1}(t)$, $e_{C_1}(t)$ and $e_{L_2}(t)$.

(b) Establish the expressions of currents and voltages in the preceding item to the following values: $E = 2,000$ [V], $f = 60$ [Hz], $L_1 = 10$ [mH], $L_2 = 30$ [mH] and $C = 40$ [μF].

(c) Using the ATPDraw program, find the steady-state solution for all currents and voltages in the circuit.

(d) Using the current source injection method, find the analytical, literal expressions of the voltages $e_{L_2}(t)$, $e_k(t)$ and $e_{C_1}(t)$, for $t > 0$, that is, for the open switch k.

(e) Obtain the corresponding expressions of the voltages in item (d) for the numerical values given in item (b).

(f) Using the ATPDraw program, establish the numerical solution of $e_{L_2}(t)$, $e_k(t)$ and $e_{C_1}(t)$, for $t > 0$.

(g) Justify the numerical oscillation in $e_{L_2}(t)$ and $e_k(t)$.

(h) Repeat the previous item for $k_p = 10$, in the specification of L_2.

(i) Repeat the simulation for *Imar* $= 0$, with $k_p = 0$ and then $k_p = 10$.

Answer:

(a) $i_k(t) = I_k \sin(\omega t)$; $I_k = \dfrac{E}{\omega \left[(L_1 + L_2) - L_2 \left(\omega/\omega_1 \right)^2 \right]}$; $\omega_1^2 = \dfrac{1}{L_1 C_1}$.

$i_{L_1}(t) = (1 - \omega^2 L_2 C_1) I_k \sin(\omega t)$.

$i_{C_1}(t) = -\omega^2 L_2 C_1 I_k \sin(\omega t)$.

$e_{L_1}(t) = \omega \left[L_1 - L_2 \left(\omega/\omega_1 \right)^2 \right] I_k \cos(\omega t)$.

$e_{C_1}(t) = e_{L_2}(t) = \omega L_2 I_k \cos(\omega t).$

(b) $I_k = 138.5358\,[\text{A}]$; $\omega_1 = 1581,1388\,[\text{rad/s}].$

$i_k(t) = 138.5358 \sin(376.9911t)\,[\text{A}].$

$i_{L_1}(t) = 114.9154 \sin(376.9911t)\,[\text{A}].$

$i_{C_1}(t) = -23.6268 \sin(376.9911t)\,[\text{A}].$

$e_{L_1}(t) = 433.2014 \cos(376.9911t)\,[\text{V}].$

$e_{C_1}(t) = e_{L_2}(t) = 1566,8029 \cos(376.9911t)\,[\text{V}].$

(c) See Figs. 7.93 and 7.94.

(d) $e_k(t) = e_{C_1}(t) = \dfrac{\omega_1^2 E}{C_1\left(\omega_1^2 - \omega^2\right)\left[(L_1 + L_2)\omega_1^2 - L_2\omega^2\right]}$

$$\left\{\left[1 + L_2 C_1\left(\omega_1^2 - \omega^2\right)\right]\cos(\omega t) - \cos(\omega_1 t)\right\}, \quad t > 0.$$

$e_{L_2}(t) = 0, \ t > 0.$

(e) $e_k(t) = e_{C_1}(t) = 2120,5523 \cos(376.9911t)$

$$-553.7481 \cos(1581,1388t)\,[\text{V}], \quad t > 0.$$

$e_{L_2}(t) = 0, \ t > 0.$

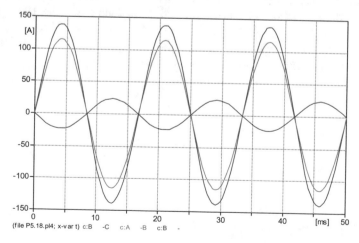

Fig. 7.93 $i_k(t), i_{L_1}(t), i_{C_1}(t)$

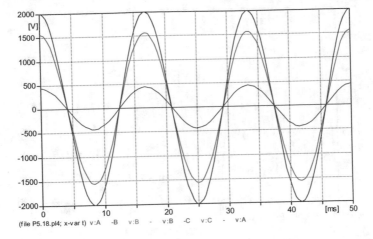

Fig. 7.94 $e_{L_1}(t), e_{C_1}(t) = e_{L_2}(t), e_k(t), e(t)$

(f) See Fig. 7.95.

(g) The ATPDraw program actually opens the switch k at time $t = \Delta t = 1\,[\mu s]$, when the current $i_k = i_{L_2} = 138.5358 \sin(376.9911 \times 10^{-6}) = 0.052227\,[A]$, and not exactly at $t = 0$, when $i_k = i_{L_2} = 0$. This discontinuity in the current of inductance L_2 causes the appearance of impulsive voltage at the terminals of the same and, consequently, at the terminals of switch k, causing the numerical oscillations seen in the plot of item (f). As the current through the switch is not zero at the time of opening, it is necessary to take Im$ar > 0.052227\,[A]$, so that switch k

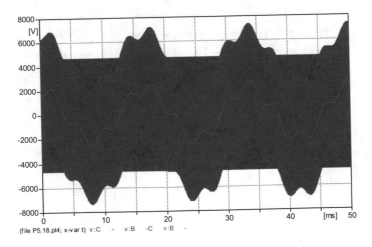

Fig. 7.95 $e_{L_2}(t), e_k(t), e_{C_1}(t)$

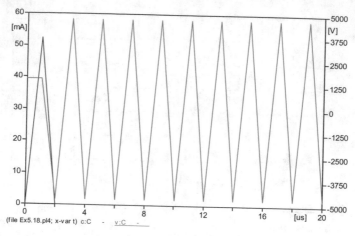

Fig. 7.96 Zoom in $i_{L_2}(t)$ and $e_{L_2}(t)$

can be opened as requested in (f). This explanation is corroborated by the zoom shown in Fig. 7.96. Also, $e_{L_2}(\Delta t) = 1566, 8\,[\text{V}]$.

(h) See Fig. 7.97.

(i) With *Imar* = 0 the switch *k* is opened at time $t = 8.3340\,[\text{ms}] \cong T/2$ (Figs. 7.98 and 7.99).

P.7-19 The circuit shown in Fig. 7.100 operated in steady-state when switch *k* was opened at time $t = 0$. If at the moment of switching the current $i_k(t)$, passing through the switch, was maximum,

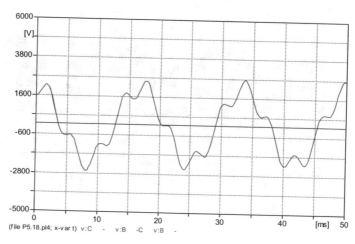

Fig. 7.97 $e_{L_2}(t), e_k(t) = e_{C_1}(t)$

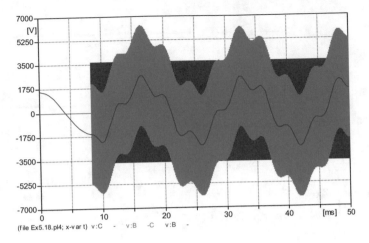

Fig. 7.98 $e_{L_2}(t), e_k(t), e_{C_1}(t)$, for $k_p = 0$

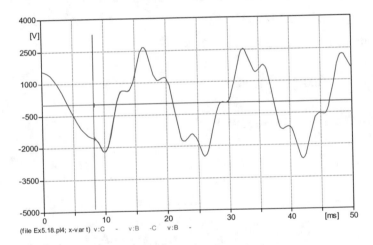

Fig. 7.99 $e_{L_2}(t), e_k(t), e_{C_1}(t)$, for $k_p = 10$

(a) determine the analytical expression of the voltage $e_k(t)$, for $t > 0$, that is, at the open switch terminals. Consider, when calculating the permanent sinusoidal regime, that "$1/j\omega C_1 \cong \infty$". After injecting the current source F_2, to simulate the opening of switch k, consider that $\omega_1^2 - \omega^2 \cong \omega_1^2$, and that $\omega^2/\omega_1^2 \cong 0$, where $\omega_1 = 1/\sqrt{L_1 C_1}$.

(b) Establish the result of item (a) for the numerical values: $E = 1,000\,[\mathrm{V}]$, $f = 60\,[\mathrm{Hz}]$, $L_1 = 10\,[\mathrm{mH}]$, $L_2 = 30\,[\mathrm{mH}]$ and $C_1 = 25\,[\mu\mathrm{F}]$.

(c) Using the ATPDraw program, find the "exact" numerical solution for $e_k(t)$.

(d) Compare the results of items (b) and (c), at times $t = 10\,[\mathrm{ms}]$, $t = 17.5\,[\mathrm{ms}]$ and $t = 20\,[\mathrm{ms}]$, establishing the percentage error in each case (Fig. 7.100).

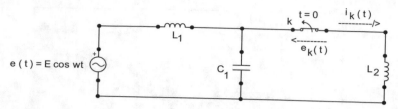

Fig. 7.100 Electric circuit for Problem P.7-19

Answer:

(a) $e_k(t) \cong \begin{cases} 0, & \text{for } t < \pi/2\omega \\ \frac{L_2 E}{\omega(L_1 + L_2)} U_0(t - \pi/2\omega), & \text{for } t = \pi/2\omega \\ E\left[\cos \omega t + \frac{\omega_1 L_1}{\omega(L_1 + L_2)} \sin(\omega_1 t - \pi\omega_1/2\omega)\right], & \text{for } t > \pi/2\omega. \end{cases}$

(b) $e_k(t) \cong \begin{cases} 0, & \text{for } t < 4.1667 \text{ [ms]} \\ 1.9894 U_0(t - 4.1667 \times 10^{-3}), & \text{for } t = 4.1667 \text{ [ms]} \\ 1,000 \cos 376.9911t + 1326, 2913 \sin(2,000t - 8.3333), & \text{for } t > 4.1667 \text{ [ms]}. \end{cases}$

(c) See Figs. 7.101, 7.102 and 7.103.

(d) From the Figs. 7.101, 7.102 and 7.103,

$$e_k(10 \text{ ms}) = e_k(0.01) = -1946, 7 \text{ [V]}$$

$$e_k(17.5 \text{ ms}) = e_k(0.0175) = 2397, 8 \text{ [V]}$$

$$e_k(20 \text{ ms}) = e_k(0.020) = 668.73 \text{ [V]}.$$

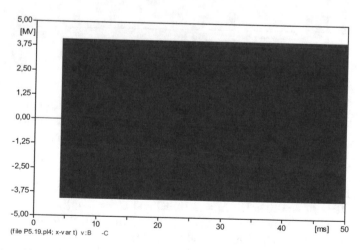

Fig. 7.101 $e_k(t)$, for $k_p = 0$ in L_2

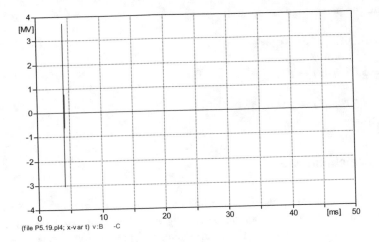

Fig. 7.102 $e_k(t)$, for $k_p = 10$ in L_2

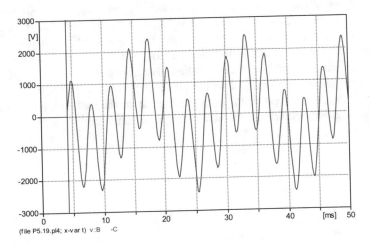

Fig. 7.103 $e_k(t)$, for $k_p = 10$ in L_2, after changing the scale

From item (b),

$$e_k(10\,\text{ms}) = e_k(0.01) \cong -1847,67\,[\text{V}]$$

$$e_k(17.5\,\text{ms}) = e_k(0.0175) \cong 2276,45\,[\text{V}]$$

$$e_k(20\,\text{ms}) = e_k(0.020) \cong 638.14\,[\text{V}].$$

And the percentage errors are:

$\epsilon = 5.09\%$ at $t = 10\,[\text{ms}]$.

$\epsilon = 5.06\%$ at $t = 17.5\,[\text{ms}]$.

$\epsilon = 4.57\%$ at $t = 20\,[\text{ms}]$.

7.6 Conclusions

In this chapter it was presented another methodology for the study of transient phenomena caused by the opening or closing of one or more switches, in circuits that are or are not operating in a steady-state, before switching. It is a technique applicable, preferably, to linear circuits and invariant over time, and whose principle is to simulate the closing of a switch by injecting a voltage source into the circuit; or the simulation of opening a switch by injecting a current source. In essentially linear circuits, the procedure is based on the *substitution theorem*, also called the *compensation theorem*, and on the *superposition theorem*. In the development of the method it is assumed that the independent sources present in the circuit are sinusoidal, making it clear, at first, that it applies equally to other types of independent sources. In contrast to the preceding chapters, the various examples presented here are solved using both the classic method of ordinary differential equations and the operational method of Laplace transforms.

References

1. Alexander CK, Sadiku MNO (2008) Fundamentals of electrical circuits, 3rd edn. McGraw-Hill of Brazil Ltda, São Paulo
2. Boylestad RL (2012) Introduction to circuit analysis, 12th edn. Pearson Education of Brazil, São Paulo
3. Close CM (1966) The analysis of linear circuits. Harcourt, Brace & World Inc., New York
4. Mckelvey JP, Grotch H (1979) Physics, vols 1 and 3. Harper & Row do Brazil Ltda, São Paulo
5. Nilson JW, Riedel SA (2009) Electrical circuits, 8th edn. Pearson Education of Brazil, São Paulo
6. O'Malley JR (1992) Circuit analysis. Schaum's outline, 2nd edn. McGraw-Hill, New York

Appendix A
Processing in the ATP

In this appendix, a brief description is given on how to use the **ATP—Alternative Transient Program** in the numerical solution of electrical circuits. It is an alternative version, for microcomputers, of the famous **EMTP—Electromagnetic Transient Program**, which was developed in the second half of the 1960s by Dr. Hermann W. Dommel. It is hoped that the material exposed here about the version 5.6 used here, will be sufficient for the reader to understand the multiple examples presented and to solve the numerous problems proposed in this book. Of course, more information about it can be found in its online help. Obviously, it is assumed that the reader is already familiar with the use of the Microsoft Windows operating system and that the **ATPDraw** program is already installed on his microcomputer—certainly in a more recent version, and appearing as an icon on the respective desktop. In order to better understand the step by step of this short tutorial, it will be taken as a parallel goal to solve Example 6.12, Fig. 6.19. It should be noted that the best way to learn how to use ATP is to try to use it, after having acquired reasonable knowledge about its fundamentals. Furthermore, trial and error are certainly part of the learning process, in addition to the old and good intuition.

Processing in ATP basically consists of three main steps, namely:

A. **ATPDraw**—It consists of drawing the circuit and creating the input file.
B. **Run ATP**—It consists of running the simulation algorithm and generating the output files. This second stage is the solution of the circuit itself.
C. **PlotXY**—In this last step, the output files are read and the graphics of the requested output variables are generated.

A. To **design the circuit and create the input file**, the procedure is as follows:

1. Double click on the **ATPDraw** icon, with the left mouse button (**Lmb**), on the desktop. A table will appear with its header shown in the background as in Fig. A.1.

J. C. Goulart de Siqueira and B. D. Bonatto, *Introduction to Transients in Electrical Circuits*, Power Systems, https://doi.org/10.1007/978-3-030-68249-1

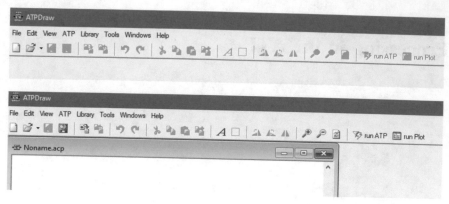

Fig. A.1 ATPDraw workspace window

2. Open a new file by clicking on **File**, in the upper left part of the open frame and, immediately, **New**. A second frame will appear, superimposed on the first one, entitled **Noname.acp**, also shown in Fig. A.1.
3. Again, in File, click on **Save As** and assign a name to the circuit, **without spacing in the name typing**. In this case, **Ex6.12**. The second frame is then called **Ex6.12.acp**, as shown in Fig. A.2.
4. Click once with the right mouse button (**Rmb**) on the second screen to open the **Component Menu**. A list of Vertical titles will then appear, containing: **Probes and 3-phase; Branch Linear; Branch Nonlinear; Line/Cables; Switches; Sources;** ..., also shown in Fig. A.2.

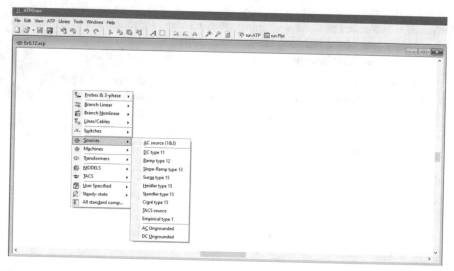

Fig. A.2 ATPDraw menu window

Then, placing the cursor over **Sources**, for example, a second vertical list appears: **AC source (1 and 3)**; **DC type 11**; **Ramp type 12**; ..., also shown in Fig. A.2.

When placing the cursor on **Branch Linear**, the vertical list appears: **Resistor; Capacitor; Inductor; RLC; ...; C:U(0); L:I(0).**

If the cursor is placed on **Switches**, the list consisting of: **Switch time controlled; Switch time 3-ph**; ...

When executing, for example, the sequence: **Rmb, cursor in Linear and Lmb in Resistor**, a green resistor will appear in the window that was previously opened; and it can be moved to the desired position in the window, clicking on it with the **Lmb** and dragging, or rotating 90° to the right, clicking with the **Rmb** on it and rotating one or more times. Any green component can be eliminated using the Delete key on the keyboard.

The selection sequence: (1) **Sources > DC type 11**, (2) **Switches > switch time controlled**, (3) **Linear > Resistor**, (4) **Linear > Inductor** and (5) **Linear > Capacitor** produce the result shown in Fig. A.3.

By clicking on each element with the **Lmb**, so that it changes from red to green again, and using the procedures described in the penultimate paragraph, the elements of the circuit under construction can be repositioned over the open window, as shown in Fig. A.4.

5. To make the connection between the various elements of the circuit, it is necessary to click on a node of a component with the **Lmb**, transforming the mouse arrow (cursor) into a "little hand", drag it to the terminal node of the other component and finish the operation with a simple click, thus connecting the two components in question.

6. To ground any node in the circuit under construction, click on it with the **Rmb**, marking **Ground** and then **OK**. This was done at the bottom terminal of the capacitor. After the execution of items 5 and 6, results the circuit of Fig. A.5.

When clicking on a voltage source, continuous (DC) or alternating (AC), it will always appear grounded (with one of its two nodes as a reference node). Remember that the ATPDraw program equates the circuit using the Nodal Method.

Fig. A.3 Placing circuit components in **ATPDraw** window

Fig. A.4 Moving circuit components in **ATPDraw** window

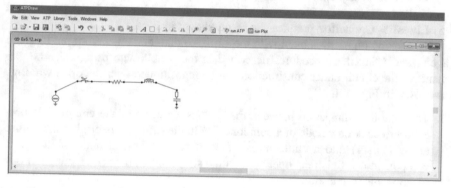

Fig. A.5 Connecting circuit elements in **ATPDraw** window

7. You can give names to the circuit nodes, with alphanumeric characters of up to 8 digits, by clicking with the **Rmb** on one of the nodes of each component. In the **Node Data** box, type the node name in **To/From**, check the **Name on Screen** box and click **OK**. In the circuit under construction, nodes A, B, C and D were designated.

8. The assignment of numerical values to the circuit elements is done with **two clicks** on each element with the **Lmb** and the assignment of the corresponding numerical value in the then opened frame, called **Component**: for the circuit under construction, in the voltage source, **Component: DC1PH**, it should be marked in **Value**: 10 for Volt. Keeping −1 for **T start** means that the voltage source has been operating for a long time, that is, well before t = 0 [s]. Keeping 1,000 [s] for **T stop** means that the source will act "indefinitely". It is necessary to close the box by clicking **OK**. For the key, **Component: TSWITCH**, it should be marked in **Value**: 0 [s] for **T-cl** and keep 1,000 [s] for **T-op**, to indicate that the key "will never" be opened, and then click **OK**.

For the resistor, **Component: RESISTOR**, mark in **Value** 10 [Ω] and close with **OK**.

For the inductor, **Component: IND_RP**, in **Value**, mark 100 [mH] and click **OK**.

For the capacitor, **Component: CAP_RS**, in **Value** adopt 1,000 [μF] and click **OK**.

(Note that the inductance and capacitance units are standardized in millihenrys and microfarads. Furthermore, the decimal notation is American).

9. The definition of the unknowns, that is, the output variables in the circuit elements (voltage, current, voltage and current, power, energy, power and energy) is usually done together with item 8, and it must be carefully observed, **the direction indicated for the current**, that is, the direction of the arrow that the program places next to each component, considering that **associated directions are used**, that is: that the voltage and current arrows on an element connected between the nodes k and m, $e_{km}(t)$ and $i_{km}(t)$ have **opposite directions**.

After performing the procedures described in items 7, 8 and 9, the circuit under construction is the one shown in Fig. A.6.

Note that voltage and current in the resistor (UI), voltage (U) in the inductance and voltage (U) in the capacitance were requested. And that a **Probe Volt** was also included, connected to node A, to detect the voltage between node A and the ground (reference), that is, the voltage of the source. The **Probe Volt** was inserted in the same way that the other components of the circuit were inserted.

10. To label the circuit, click with the **Lmb** on the blue letter **A** in the upper bar and then next to the component to be subtitled. Then, a small rectangle will appear, within which the desired typing is made. If it is necessary to move, vertically or horizontally, what was typed, to improve aesthetics, just click on what was typed and drag. After this procedure, the circuit under construction is the one shown in Fig. A.7. Since **ATPDraw** is not a text editor, Greek letters widely used in circuits are not available, such as: $\theta, \mu, \pi, \varphi, \omega, \Omega$.

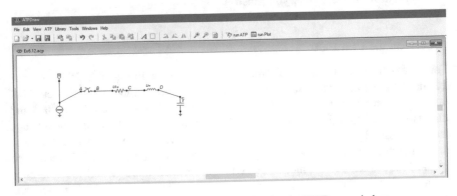

Fig. A.6 Circuit components and node name identification in **ATPDraw** window

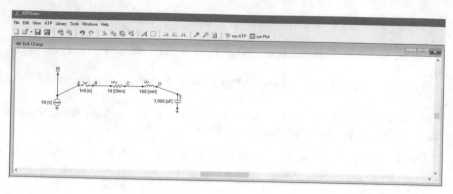

Fig. A.7 Label identification in **ATPDraw** window

11. **Save the file** by clicking on **File** and immediately **Save**. After this operation, it is interesting to perform the following test: close the window containing the circuit and then reopen, clicking on the icon below **Edit**, to see if the title has changed **to the proper directory**. If this has not happened, it is a sign that a mistake has been made. It is then necessary to review the entire procedure completed. Sometimes it is better to reset everything and start over.

12. Define the integration time step size Δt and the maximum processing time (simulation) t_{max}. For this, click on **ATP** and then on **Settings** and establish the desired values, paying attention to the appropriate choice of values according to the circuit time constants and frequency range of the simulated phenomena (Nyquist). The **ATPDraw** program uses **Delta T: 1E-6** as standard, which means that $\Delta t = 1 \, [\mu s]$. **Delta T: 1E-6** and **Tmax : $2E - 1$** were adopted for the circuit under consideration, that is $t_{max} = 0.2 \, [s]$.

B. For this step, which is the solution of the circuit, just click on **Run ATP** with the **Lmb**. Then, a very fast movement of numbers on the computer screen is noticed.

C. To perform this last step, you must:

1. click on the **Run Plot**.

For the example in question, the result is the table shown in Fig. A.8. In it, you can see the name of the file (**File Name; Ex6-12.pl4**), the number of variables (**# of vars; 6**), the number of points that have been calculated (**# of Points; 200,001**) and maximum simulation time (**Tmax; 0.2**). In **Variables, v: B -C** means the voltage between nodes B and C of the circuit, that is, the voltage in the resistance; **v: A** means the voltage of node A, that is, the voltage of the source, which is measured by the Probe Volt; **v: D** - means the voltage between node D and the reference (ground), that is, the voltage in the capacitance. Obviously, **c: B -C** means the current flowing from node B to node C, which is the current circulating in the circuit of Fig. A.7.

Fig. A.8 Picture resulting
from the operation **Run Plot**

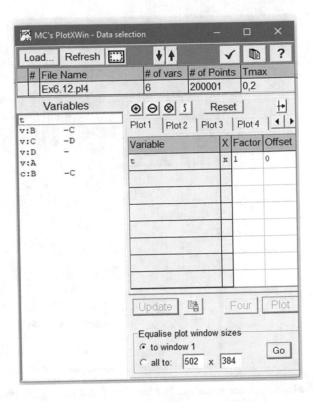

2. Choose the variables to be plotted, that is, represented graphically.

By clicking on the four items below the variable t (time) with the **Lmb** the voltages
in the three passive elements are selected, as well as the voltage of the source. By
clicking with the **Rmb** on the last item, the current circulating in the circuit is
selected. After selecting the variables, the box remains with the configuration
shown in Fig. A.9.

Clicking with the left mouse button, the chosen variable will have its scale
registered in the vertical axis on the left of the graph. By clicking with the right
button, the selected quantity will have its scale registered in the axis on the right.
This option is useful so that you can plot quantities on different graphs with different units (see Fig. A.9).

3. Click on **Plot** in the table registered in Fig. A.9.

After this operation, for the circuit in Fig. A.7, the result is that shown in Fig. A.10.

The first icon at the bottom of the graph in Fig. A.10 provides the possibility to
label it, opening a rectangle at its top. The second opens a small frame through
which it is possible to grid the graph. The third allows changes to be made to the
vertical scales of the graph, in case any adjustment is necessary. The fourth icon
allows the insertion of a cursor (a full vertical line), through which it is possible to
measure the values of the variables (electrical quantities) at different times, moving
it horizontally. Such values are shown in a small separate box.

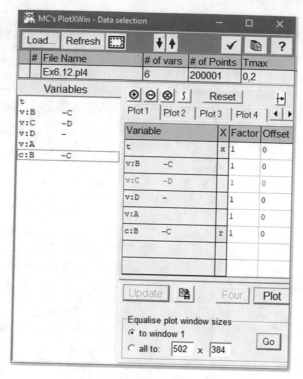

Fig. A.9 Selecting the variables for plotting and running the **Plot** program

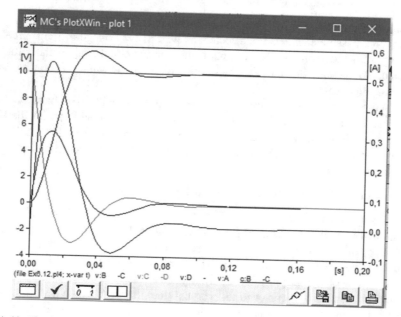

Fig. A.10 Plotting results from the transient simulation with **ATPDraw**

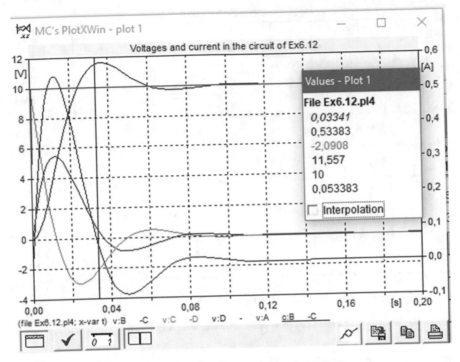

Fig. A.11 Detailed results from the transient simulation with **ATPDraw**

After using these resources, the result is that shown in Fig. A.11.

In the position where the cursor is in Fig. A.11, we have, then, that for the instant $t = 0.03341$ [s], the voltage in the resistance (v: B -C) is 0.53383 [V]; the voltage at the inductance (v: C -D) is -2.09080 [V]; the voltage at the capacitance (v: D -) is 11.55700 [V]; the voltage at the source (v: A) is 10 [V]. And the current circulating in the circuit (c: B -C) is worth 0.053383 [A].

The penultimate icon at the bottom of Fig. A.11 allows you to copy the curves plotted by the **PlotXY** program. By clicking on it (**Copy to Clipboard**) and then OK, the image is available to be pasted into the editor's file, using the command Ctrl + V. Performing this operation in the graph of Fig. A.11, results in Fig. A.12.

It is also possible to zoom in on a selected part of an obtained graph, aiming, for example, a more accurate view of the plotted curves. To do this, just click on a point on the graph and drag until you reach the section of interest and release the **Lmb**. Figure A.13 is the result of one of them. Obviously, you can zoom over zoom.

It is possible to move the cursor over a graph, in a kind of fine adjustment, obtaining more accurate readings, using the horizontal scroll arrows on the keyboard.

It is possible to plot one variable against another. To do this, just select the variable that will be the abscissa (horizontal axis), and then remove the time variable (t). For example, $v(t)$ as a function of $i(t)$.

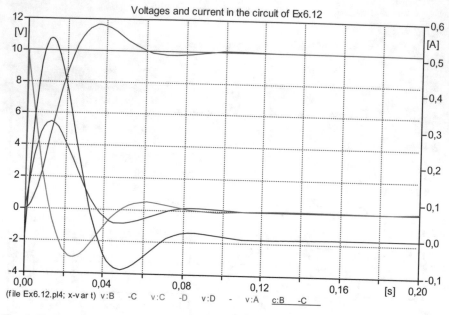

Fig. A.12 Plotted curves from the transient simulation with **ATPDraw**

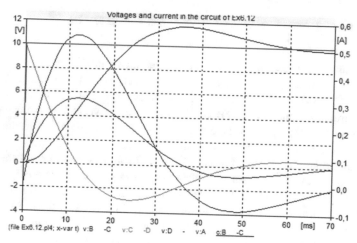

Fig. A.13 Zoom on plotted curves from the transient simulation with **ATPDraw**

You can copy the circuit design created in **ATPDraw** and paste it in a text editor, such as MSWord. To do this, select the drawing (Ctrl + A, or Edit -> Select -> All menu), which will be all green and surrounded by an equally green rectangle. To copy the image, select from the Edit -> Copy Graphics menu, or type Ctrl + W. The image will be available to be pasted into the editor's file, using the command Ctrl + V.

Appendix B
Main Relations Involving Singular Functions

1. See Fig. B.1.

2. $U_{k-1}(t) = \int U_k(t)dt = \int\limits_{-\infty}^{t} U_k(\lambda)d\lambda$

   $U_k(t) = \frac{d}{dt}U_{k-1}(t),\ k = 0,\ \pm 1, \pm 2, \pm \ldots$

   (a) $U_{-1}(t) = \int\limits_{-\infty}^{t} U_0(\lambda)d\lambda$

   (b) $U_{-2}(t) = \int\limits_{-\infty}^{t} U_{-1}(\lambda)d\lambda$

   (c) $U_0(t) = \int\limits_{-\infty}^{t} U_1(\lambda)d\lambda$

   (d) $U_0(t) = \frac{d}{dt}U_{-1}(t)$

   (e) $U_{-1}(t) = \frac{d}{dt}U_{-2}(t)$

   (f) $U_1(t) = \frac{d}{dt}U_0(t)$

3. $U_{-(m+1)}(t) = \frac{t^m}{m!}U_{-1}(t),\ m = 1, 2, 3, \ldots$

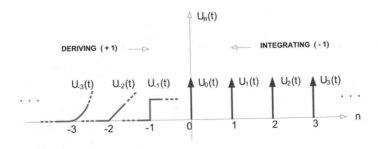

Fig. B.1 Ilustrative representation of singular functions

(a) $U_{-2}(t) = tU_{-1}(t)$

(b) $U_{-3}(t) = \frac{t^2}{2}U_{-1}(t)$

4. $\int_0^t f(\lambda) \cdot U_{-1}(\lambda - a)d\lambda = \left[\int_a^t f(\lambda)d\lambda\right] \cdot U_{-1}(t - a)$

5. $\int_{-\infty}^{+\infty} U_n(t - a)f(t)dt = (-1)^n f^{(n)}(a), \ n = 0, 1, 2, \ldots$

(a) $\int_{-\infty}^{+\infty} U_0(t)f(t)dt = f(0)$

(b) $\int_{-\infty}^{+\infty} U_1(t)f(t)dt = -f'(0)$

6. $\int_{-\infty}^{+\infty} U_0(t - a)f(t)dt = \int_{-\infty}^{+\infty} U_0(t)f(t + a)dt = f(a)$

7. $\int_{-\infty}^{+\infty} U_0(at)f(t)dt = \frac{1}{|a|} \int_{-\infty}^{+\infty} U_0(t)f(\frac{t}{a})dt = \frac{1}{|a|}f(0), \ a \neq 0$

8. $\int_{-\infty}^{+\infty} U_1(at)f(t)dt = \frac{1}{a|a|} \int_{-\infty}^{+\infty} U_1(t)f(t)dt = \frac{-1}{a|a|}f'(0), \ a \neq 0$

9. $f(t)U_n(t - a) = \sum_{k=0}^{n}(-1)^k \frac{n!}{k!(n-k)!}f^{(k)}(a)U_{n-k}(t - a), \ n = 1, 2, 3, \ldots$

(a) $f(t)U_1(t) = f(0)U_1(t) - f'(0)U_0(t)$

(b) $f(t)U_2(t) = f(0)U_2(t) - 2f'(0)U_1(t) + f''(0)U_0(t)$

10. $U_n(-t) = (-1)^n U_n(t), \ n = 0, 1, 2, \ldots$

(a) $U_0(-t) = U_0(t)$

(b) $U_1(-t) = -U_1(t)$

11. $U_{-1}(t)U_0(t) = \frac{1}{2}U_0(t)$

12. $\Phi(t) \cdot U_{-1}(t) = \sum_{n=0}^{+\infty} \Phi^{(n)}(0_+)U_{-(n+1)}(t)$

(a) $\sin t \cdot U_{-1}(t) = U_{-2}(t) - U_{-4}(t) + U_{-6}(t) - U_{-8}(t) + \cdots$

(b) $\cos t \cdot U_{-1}(t) = U_{-1}(t) - U_{-3}(t) + U_{-5}(t) - U_{-7}(t) + \cdots$

(c) $\varepsilon^{-\alpha t} \cdot U_{-1}(t) = U_{-1}(t) - \alpha U_{-2}(t) + \alpha^2 U_{-3}(t) - \alpha^3 U_{-4}(t) + \cdots$

(d) $\sin(\omega t + \theta) \cdot U_{-1}(t) = \sin\theta \cdot U_{-1}(t) + \omega \cos\theta \cdot U_{-2}(t) - \omega^2\sin\theta \cdot U_{-3}(t)$

$- \omega^3 \cos\theta \cdot U_{-4}(t) + \omega^4\sin\theta \cdot U_{-5}(t) + \cdots$

(e) $\cos(\omega t + \theta) \cdot U_{-1}(t) = \cos\theta \cdot U_{-1}(t) - \omega\sin\theta \cdot U_{-2}(t) - \omega^2 \cos\theta \cdot U_{-3}(t)$

$+ \omega^3\sin\theta \cdot U_{-4}(t) + \omega^4 \cos\theta \cdot U_{-5}(t) - \cdots$

13. $f(t)U_0(t - a) = \frac{f(a_+)+f(a_-)}{2}U_0(t - a)$

14. $f(t) = f(t_0)U_{-1}(t - t_0) + \int_{t_0}^{+\infty} f'(\lambda)U_{-1}(t - \lambda)d\lambda, \ \text{for } t \geq t_0$

15. $f(t) * U_0(t) = \int\limits_{-\infty}^{+\infty} f(\lambda)U_0(t - \lambda)d\lambda = f(t)$

16. $f(t) * U_1(t) = \int\limits_{-\infty}^{+\infty} f(\lambda)U_1(t - \lambda)d\lambda = f'(t)$

17. $f(t) * U_{-1}(t) = \int\limits_{-\infty}^{t} f(\lambda)U_{-1}(t - \lambda)d\lambda = \int\limits_{-\infty}^{t} f(\lambda)d\lambda = \int f(t)dt$

   (a) $f(t) * U_{-1}(t - a) = \int\limits_{-\infty}^{t-a} f(\lambda)d\lambda$

   (b) $f(t) * [U_{-1}(t - a) - U_{-1}(t - b)] = \int\limits_{t-b}^{t-a} f(\lambda)d\lambda, \ a < b$

   (c) $f(t) * \dfrac{1}{b - a}[U_{-1}(t - a) - U_{-1}(t - b)] = \dfrac{1}{b - a}\int\limits_{t-b}^{t-a} f(\lambda)d\lambda$

                 $= $ average of $f(t)$ over the range $[t - b, t - a], a < b$, for all t

18. $AU_m(t - a) * BU_n(t - b) = ABU_{m+n}(t - a - b)$.

Appendix C
Laplace Transform Properties

$f(t) = \mathcal{L}^{-1}[F(s)]$	$F(s) = \mathcal{L}[f(t)]$
1. $k_1 f_1(t) + k_2 f_2(t)$	$k_1 F_1(s) + k_2 F_2(s)$
2. $f(at)$	$\frac{1}{a}F\left(\frac{s}{a}\right)$
3. $\varepsilon^{\pm at}f(t)$	$F(s \mp a)$
4. $f(t-a)U_{-1}(t-a)$	$\varepsilon^{-as}F(s)$
5. $f'(t)$	$sF(s) - f(0_-)$
6. $f''(t)$	$s^2 F(s) - sf(0_-) - f'(0_-)$
7. $f^{(n)}(t) = \frac{d^n}{dt^n}f(t)$	$s^n F(s) - s^{n-1}f(0_-) - s^{n-2}f'(0_-) - \ldots - f^{(n-1)}(0_-)$
8. $-tf(t)$	$F'(s)$
9. $t^2 f(t)$	$F''(s)$
10. $(-1)^n t^n f(t)$	$F^{(n)}(s) = \frac{d^n}{ds^n}F(s)$
11. $\int\limits_0^t f(\lambda)d\lambda$	$\frac{1}{s}F(s)$
12. $f_1(t) * f_2(t)$	$F_1(s) \cdot F_2(s)$
13. $\frac{f(t)}{t}$	$\int\limits_s^{+\infty} F(\tau)d\tau$
14. $f(t+T) = f(t)$	$\frac{1}{1-\varepsilon^{-sT}}\int\limits_0^T \varepsilon^{-s\tau}F(\tau)d\tau$
15. *Initial value theorem*	15. $f(0_+) = \lim\limits_{s\to\infty} sF(s)$
16. *Final value theorem*	16. $f(\infty) = \lim\limits_{s\to 0} sF(s)$

© The Editor(s) (if applicable) and The Author(s), under exclusive license to
Springer Nature Switzerland AG 2021
J. C. Goulart de Siqueira and B. D. Bonatto, *Introduction to Transients
in Electrical Circuits*, Power Systems, https://doi.org/10.1007/978-3-030-68249-1

$$F(s) = \mathcal{L}[f(t)] = \int\limits_{0_-}^{+\infty} f(t)\varepsilon^{-st}dt, \; s = \sigma + j\omega$$

$$f(t) = \mathcal{L}^{-1}[F(s)] = \frac{1}{2\pi j} \int\limits_{c-j\infty}^{c+j\infty} F(s)\varepsilon^{st}ds$$

c is a damping factor that ensures the convergence of the integral, which is carried out on the complex plane along the line $s = c$, that is, from $c - j\infty$ to $c + j\infty$.

Appendix D
Laplace Transform Pairs

$f(t)$	$F(s)$
1. $U_0(t)$	1
2. $1, U_{-1}(t)$	$\frac{1}{s}$
3. $t, U_{-2}(t)$	$\frac{1}{s^2}$
4. $U_1(t)$	s
5. $U_2(t)$	s^2
6. $U_n(t), \ n = 0, \pm 1, \pm 2, \pm \ldots$	s^n
7. $\frac{t^{n-1}}{(n-1)!}, \ n = 1, 2, 3, \ldots, 0! = 1$	$\frac{1}{s^n}$
8. $\varepsilon^{\pm at}$	$\frac{1}{s \mp a}$
9. $\frac{t^{n-1}\varepsilon^{\pm at}}{(n-1)!}, \ n = 1, 2, 3, \ldots$	$\frac{1}{(s \mp a)^n}$
10. $\sin \omega t$	$\frac{\omega}{s^2 + \omega^2}$
11. $\cos \omega t$	$\frac{s}{s^2 + \omega^2}$
12. $\varepsilon^{-at} \sin \omega t$	$\frac{\omega}{(s+\alpha)^2 + \omega^2}$
13. $\varepsilon^{-at} \cos \omega t$	$\frac{s+\alpha}{(s+\alpha)^2 + \omega^2}$
14. $\sin(\omega t + \theta)$	$\frac{\omega \cos\theta + s\sin\theta}{s^2 + \omega^2}$
15. $\cos(\omega t + \theta)$	$\frac{s \cos\theta - \omega\sin\theta}{s^2 + \omega^2}$
16. $\varepsilon^{-at} \sin(\omega t + \theta)$	$\frac{\omega \cos\theta + (s+\alpha)\sin\theta}{(s+\alpha)^2 + \omega^2}$
17. $\varepsilon^{-at} \cos(\omega t + \theta)$	$\frac{(s+\alpha) \cos\theta - \omega\sin\theta}{(s+\alpha)^2 + \omega^2}$
18. $\sinh \beta t$	$\frac{\beta}{s^2 - \beta^2}$
19. $\cosh \beta t$	$\frac{s}{s^2 - \beta^2}$
20. $\frac{\varepsilon^{-at} - \varepsilon^{-\beta t}}{\beta - \alpha}, \ \alpha \neq \beta$	$\frac{1}{(s+\alpha)(s+\beta)}$
21. $\varepsilon^{-at} + \varepsilon^{-\beta t}$	$\frac{2s + (\alpha+\beta)}{(s+\alpha)(s+\beta)}$

(continued)

© The Editor(s) (if applicable) and The Author(s), under exclusive license to
Springer Nature Switzerland AG 2021
J. C. Goulart de Siqueira and B. D. Bonatto, *Introduction to Transients in Electrical Circuits*, Power Systems, https://doi.org/10.1007/978-3-030-68249-1

(continued)

$f(t)$	$F(s)$
22. $\frac{\varepsilon^{-\alpha t}}{(\alpha-\beta)(\alpha-\gamma)} + \frac{\varepsilon^{-\beta t}}{(\beta-\alpha)(\beta-\gamma)} + \frac{\varepsilon^{-\gamma t}}{(\gamma-\alpha)(\gamma-\beta)}$	$\frac{1}{(s+\alpha)(s+\beta)(s+\gamma)}$, $\alpha \neq \beta \neq \gamma$
23. $\frac{\alpha\varepsilon^{-\alpha t} - \beta\varepsilon^{-\beta t}}{\alpha-\beta}$	$\frac{s}{(s+\alpha)(s+\beta)}$
24. $t\sin\omega t / 2\omega$	$\frac{s}{(s^2+\omega^2)^2}$
25. $(\sin\omega t + \omega t\cos\omega t)/2\omega$	$\frac{s^2}{(s^2+\omega^2)^2}$
26. $(\sin\omega t - \omega t\cos\omega t)/2\omega^3$	$\frac{1}{(s^2+\omega^2)^2}$
27. $\cos\omega t - \frac{1}{2}\omega t\sin\omega t$	$\frac{s^3}{(s^2+\omega^2)^2}$
28. $t\cos\omega t$	$\frac{s^2-\omega^2}{(s^2+\omega^2)^2}$
29. $U_0(t) - \alpha\varepsilon^{-\alpha t}$	$\frac{s}{s+\alpha}$
30. $U_0(t) + \frac{\beta^2\varepsilon^{-\beta t} - \alpha^2\varepsilon^{-\alpha t}}{\alpha-\beta}$	$\frac{s^2}{(s+\alpha)(s+\beta)}$
31. $\frac{1}{\omega^2}(1-\cos\omega t)$	$\frac{1}{s(s^2+\omega^2)}$
32. $\frac{1}{\alpha^2+\omega^2}\left[1 - \sqrt{1+\left(\frac{\alpha}{\omega}\right)^2}\varepsilon^{-\alpha t}\sin\left(\omega t + tg^{-1}\frac{\omega}{\alpha}\right)\right]$	$\frac{1}{s\left[(s+\alpha)^2+\omega^2\right]}$
33. $\frac{1}{\omega^2}\left(t - \frac{1}{\omega}\sin\omega t\right)$	$\frac{1}{s^2(s^2+\omega^2)}$
34. $\frac{1}{\omega^2}\left[A + \sqrt{A^2+\omega^2}\sin\left(\omega t - tg^{-1}\frac{A}{\omega}\right)\right]$	$\frac{s+A}{s(s^2+\omega^2)}$
35. $\frac{1}{\omega_0^2-\omega^2}\left(\frac{1}{\omega}\sin\omega t - \frac{1}{\omega_0}\sin\omega_0 t\right)$	$\frac{1}{(s^2+\omega^2)(s^2+\omega_0^2)}$
36. $\frac{1}{\omega_0^2-\omega^2}(\cos\omega t - \cos\omega_0 t)$	$\frac{s}{(s^2+\omega^2)(s^2+\omega_0^2)}$
37. $\frac{1}{\omega_0^2-\omega^2}(\omega_0\sin\omega_0 t - \omega\sin\omega t)$	$\frac{s^2}{(s^2+\omega^2)(s^2+\omega_0^2)}$
38. $\frac{M}{\omega}\varepsilon^{-\alpha t}\sin(\omega t + \theta)M\angle\theta = Q(-\alpha+j\omega)$	$\frac{Q(s)}{(s+\alpha)^2+\omega^2}$

Appendix E
Heaviside Expansion Theorem

In the calculation of transient phenomena of electrical circuits with concentrated or lumped parameters, using the operational method of Laplace transforms, it is very common for transforms to have the general form of a rational fraction of the type:

$$F(s) = \frac{F_1(s)}{F_2(s)} = \frac{a_0 s^m + a_1 s^{m-1} + a_2 s^{m-2} + \cdots + a_m}{b_0 s^n + b_1 s^{n-1} + b_2 s^{n-2} + \cdots + b_n}, \quad m < n, \qquad (E.1)$$

where the coefficients a_k and b_k are real numbers and the polynomials $F_1(s)$ and $F_2(s)$ have no common roots. And the problem, then, is to find the inverse transform of $F(s)$, that is, $f(t)$. One of the ways to do this is to use the tables of transform pairs, like the one in Appendix D. However, it often happens that the table does not include the $F(s)$ transform that needs to be inverted to obtain $f(t)$. And in order not to have to perform the Laplace inverse transform using the inversion formula— which requires a non-trivial integration in the complex plane, numerical methods are used. This is what the Heaviside expansion theorem is all about.

To find $f(t) = \mathcal{L}^{-1}[F(s)]$ it is necessary, initially, to expand $F(s)$ by adding simpler fractions. And, depending on the nature of the roots of $F_2(s) = 0$, three main cases can be presented.

1st Case: Distinct Roots

If the n roots of $F_2(s) = 0$, that is, if $s_1, s_2, \ldots, s_k, \ldots, s_n$, called poles of $F(s)$, are distinct, then $F_2(s)$ can be written in factored form, as in Eq. (E.2).

$$F_2(s) = b_0(s - s_1)(s - s_2)\ldots(s - s_k)\ldots(s - s_n) \qquad (E.2)$$

Thus, the expansion of $F(s)$ in partial fractions is

J. C. Goulart de Siqueira and B. D. Bonatto, *Introduction to Transients in Electrical Circuits*, Power Systems, https://doi.org/10.1007/978-3-030-68249-1

$$\frac{F_1(s)}{F_2(s)} = \frac{A_1}{s - s_1} + \frac{A_2}{s - s_2} + \cdots + \frac{A_k}{s - s_k} + \cdots + \frac{A_n}{s - s_n} = \sum_{k=1}^{n} \frac{A_k}{s - s_k} \qquad \text{(E.3)}$$

So,

$$(s - s_k) \sum_{k=1}^{n} \frac{A_k}{s - s_k} = \frac{(s - s_k)F_1(s)}{F_2(s)} \qquad \text{(E.4)}$$

Making s tend to s_k, it comes that

$$A_k = \lim_{s \to s_k} \frac{(s - s_k)F_1(s)}{F_2(s)} \qquad \text{(E.5)}$$

As the factor $(s - s_k)$ is also in the denominator of Eq. (E.5), as indicated in Eq. (E.2), after simplification, Eq. (E.5) will always provide the value of the coefficient A_k. But Eq. (E.5) can be presented in another form, if the mentioned simplification is not carried out.

Considering that s_k is not the root of $F_1(s) = 0$, Eq. (E.5) can be written as

$$A_k = F_1(s_k) \cdot \lim_{s \to s_k} \frac{s - s_k}{F_2(s)} \qquad \text{(E.6)}$$

But the expression $(s - s_k)/F_2(s)$, for $s \to s_k$, represents an indeterminacy of type $0/0$. Therefore, by the L'Hospital rule, deriving numerator and denominator, we have:

$$A_k = F_1(s_k) \cdot \lim_{s \to s_k} \frac{1 - 0}{F_2'(s)} = \frac{F_1(s_k)}{F_2'(s_k)} \qquad \text{(E.7)}$$

Thus, substituting Eq. (E.7) in Eq. (E.3), it follows that

$$F(s) = \frac{F_1(s)}{F_2(s)} = \sum_{k=1}^{n} \frac{F_1(s_k)}{F_2'(s_k)} \cdot \frac{1}{s - s_k} \qquad \text{(E.8)}$$

And Laplace's inverse transform is:

$$f(t) = \mathcal{L}^{-1}\left[\frac{F_1(s)}{F_2(s)}\right] = \sum_{k=1}^{n} \frac{F_1(s_k)}{F_2'(s_k)} \varepsilon^{s_k t}, \textbf{ for } t > 0 \qquad \text{(E.9)}$$

This is one of the forms of the Heaviside expansion theorem.

Example E.1 Find the antitransform of the function $F(s) = \frac{s}{(s^2 + \omega^2)(s^2 + \omega_0^2)}$.

Solution

$F_1(s) = s$

$$F_2(s) = (s^2 + \omega^2)(s^2 + \omega_0^2) = s^4 + (\omega^2 + \omega_0^2)s^2 + \omega^2\omega_0^2$$

$$F_2'(s) = 4s^3 + 2(\omega^2 + \omega_0^2)s$$

The poles are: $s_1 = j\omega$; $s_2 = -j\omega$; $s_3 = j\omega_0$; $s_4 = -j\omega_0$

$F_1(s_1) = j\omega$; $F_1(s_2) = -j\omega$; $F_1(s_3) = j\omega_0$; $F_1(s_4) = -j\omega_0$

$$F_2'(s_1) = 4(j\omega)^3 + 2(\omega^2 + \omega_0^2)(j\omega) = j2\omega(\omega_0^2 - \omega^2)$$

$$F_2'(s_2) = 4(-j\omega)^3 + 2(\omega^2 + \omega_0^2)(-j\omega) = -j2\omega(\omega_0^2 - \omega^2)$$

$$F_2'(s_3) = 4(j\omega_0)^3 + 2(\omega^2 + \omega_0^2)(j\omega_0) = -j2\omega_0(\omega_0^2 - \omega^2)$$

$$F_2'(s_4) = 4(-j\omega_0)^3 + 2(\omega^2 + \omega_0^2)(-j\omega_0) = j2\omega_0(\omega_0^2 - \omega^2)$$

According to Eq. (E.9),

$$f(t) = \frac{j\omega}{j2\omega(\omega_0^2 - \omega^2)} \cdot \varepsilon^{j\omega t} + \frac{-j\omega}{-j2\omega(\omega_0^2 - \omega^2)} \cdot \varepsilon^{-j\omega t}$$

$$+ \frac{j\omega_0}{-j2\omega_0(\omega_0^2 - \omega^2)} \cdot \varepsilon^{j\omega_0 t} + \frac{-j\omega_0}{j2\omega_0(\omega_0^2 - \omega^2)} \cdot \varepsilon^{-j\omega_0 t}$$

$$= \frac{1}{\omega_0^2 - \omega^2}\left(\frac{\varepsilon^{j\omega t} + \varepsilon^{-j\omega t}}{2}\right) - \frac{1}{\omega_0^2 - \omega^2}\left(\frac{\varepsilon^{j\omega_0 t} + \varepsilon^{-j\omega_0 t}}{2}\right)$$

Therefore,

$$f(t) = \frac{1}{\omega_0^2 - \omega^2}(\cos \omega t - \cos \omega_0 t),$$

confirming the pair of transforms of number 36 in Appendix D.

2nd Case: A Null Root

Heaviside established his expansion theorem primarily for determining the response of an electrical circuit to a unitary step. In such cases the denominator of Eq. (E.1) has a factor s, that is, there is a zero root as pole of $F(s)$. This situation is depicted by Eq. (E.10).

$$F(s) = \frac{F_1(s)}{sF_2(s)} \tag{E.10}$$

Supposing that $F_2(s) = 0$ has n distinct and non-zero roots, then, according to Eq. (E.9), it comes that

$$f(t) = \mathcal{L}^{-1}\left[\frac{F_1(s)}{sF_2(s)}\right] = \frac{F_1(0)}{\left\{\frac{d}{ds}[sF_2(s)]\right\}_{s=0}} + \sum_{k=1}^{n}\frac{F_1(s_k)\cdot\varepsilon^{s_k t}}{\left\{\frac{d}{ds}[sF_2(s)]\right\}_{s=s_k}}$$

$$= \frac{F_1(0)}{[F_2(s)+sF_2'(s)]_{s=0}} + \sum_{k=1}^{n}\frac{F_1(s_k)\cdot\varepsilon^{s_k t}}{[F_2(s)+sF_2'(s)]_{s=s_k}}$$

Since $F_2(s_k) = 0$, because s_k is the root of $F_2(s) = 0$, it finally results that

$$f(t) = \mathcal{L}^{-1}\left[\frac{F_1(s)}{sF_2(s)}\right] = \frac{F_1(0)}{F_2(0)} + \sum_{k=1}^{n}\frac{F_1(s_k)}{s_k \cdot F_2'(s_k)}\cdot\varepsilon^{s_k t}, \textbf{ for } t > 0 \qquad (E.11)$$

This is the original form of the Heaviside expansion theorem. It is applicable even when there is an infinite number of roots at $F_2(s) = 0$, as in the study of transient phenomena in transmission lines with distributed parameters.

Example E.2 Find the inverse Laplace transform of the function

$$F(s) = \frac{s+A}{s(s^2+\omega^2)}$$

Solution

$$F_1(s) = s+A; F_2(s) = s^2 + \omega^2; F_2'(s) = 2s; s_1 = j\omega; s_2(s) = -j\omega$$

From Eq. (E.11),

$$f(t) = \frac{F_1(0)}{F_2(0)} + \frac{F_1(s_1)}{s_1 F_2'(s_1)}\cdot\varepsilon^{s_1 t} + \frac{F_1(s_2)}{s_2 F_2'(s_2)}\cdot\varepsilon^{s_2 t}$$

So,

$$f(t) = \frac{A}{\omega^2} + \frac{(j\omega+A)\cdot\varepsilon^{j\omega t}}{(j\omega)(j2\omega)} + \frac{(-j\omega+A)\cdot\varepsilon^{-j\omega t}}{(-j\omega)(-j2\omega)} = \frac{A}{\omega^2} - \frac{A}{\omega^2}\cos\omega t + \frac{1}{\omega}\sin\omega t$$

Considering that $y(t) = -\frac{A}{\omega^2}\cos\omega t + \frac{1}{\omega}\sin\omega t = -\frac{A}{\omega^2}\sin(\omega t + 90°) + \frac{1}{\omega}\sin\omega t$, then the phasor of $y(t)$ is

$$\dot{Y} = -\frac{A}{\omega^2}\varepsilon^{j90°} + \frac{1}{\omega}\varepsilon^{j0°} = \frac{1}{\omega} - j\frac{A}{\omega^2} = \sqrt{\frac{1}{\omega^2}+\frac{A^2}{\omega^4}}\varepsilon^{-j arctg(A/\omega)}$$

Returning to the time domain, we have to

$$y(t) = \frac{\sqrt{A^2 + \omega^2}}{\omega^2} \sin(\omega t - arctgA/\omega)$$

So, finally,

$$f(t) = \frac{1}{\omega^2}\left[A + \sqrt{A^2 + \omega^2}\sin\left(\omega t - tg^{-1}\frac{A}{\omega}\right)\right],$$

confirming the pair of transforms number 34 in Appendix D.

3rd Case: Multiple Roots

If $F_2(s) = 0$ has n distinct roots, $s_k, k = 1, 2, \ldots, n$, and an α root of multiplicity m, then the Heaviside expansion theorem takes the form

$$f(t) = \mathcal{L}^{-1}\left[\frac{F_1(s)}{F_2(s)}\right] = \sum_{k=1}^{n}\frac{F_1(s_k)}{F_2'(s_k)}\varepsilon^{s_k t} + \varepsilon^{\alpha t}\sum_{k=1}^{m}\frac{B_k}{(m-k)!}t^{m-k} \qquad \text{(E.12)}$$

$$B_k = \frac{1}{(k-1)!}\lim_{s\to\alpha}\frac{d^{k-1}}{ds^{k-1}}\left[(s-\alpha)^m\frac{F_1(s)}{F_2(s)}\right] \qquad \text{(E.13)}$$

Example E.3 Find the Laplace antitransform of the function

$$F(s) = \frac{2s+3}{s(s+1)(s+2)^2}$$

Solution
$F(s)$ is as contained in Eq. (E.10), where

$$F_1(s) = 2s+3; F_2(s) = (s+1)(s+2)^2 = s^3 + 5s^2 + 8s + 4;$$
$$F_2'(s) = 3s^2 + 10s + 8$$

Therefore, from Eqs. (E.11) and (E.12), it follows that:

$$f(t) = \frac{F_1(0)}{F_2(0)} + \frac{F_1(s_1)}{s_1 F_2'(s_1)} \cdot \varepsilon^{s_1 t} + (B_1 t + B_2)\varepsilon^{\alpha t}$$

$$F_2(s) = 0 \to s_1 = -1, \alpha = -2, m = 2$$

$$F_1(0) = 3; F_2(0) = 4; F_1(s_1) = F_1(-1) = 1; F_2'(s_1) = F_2'(-1) = 1$$

From Eq. (E.13),

$$B_1 = \lim_{s \to -2} \frac{2s+3}{s(s+1)} = -\frac{1}{2}$$

$$B_2 = \lim_{s \to -2} \frac{d}{ds}\left[\frac{2s+3}{s^2+s}\right] = \lim_{s \to -2} \frac{(s^2+s)(2) - (2s+3)(2s+1)}{(s^2+s)^2} = \frac{1}{4}$$

So,

$$f(t) = \frac{3}{4} - \varepsilon^{-t} + \left(-\frac{t}{2} + \frac{1}{4}\right)\varepsilon^{-2t}, \; for \; t > 0$$

Example E.4 Redo Example E.1 using the formula in item 38 of Appendix D.
Solution

$$F(s) = \frac{s}{(s^2+\omega^2)(s^2+\omega_0^2)} \to f(t) = f_1(t) + f_2(t)$$

$$F(s) = \frac{\left[\frac{s}{s^2+\omega^2}\right]}{s^2+\omega_0^2} \to f_1(t) = \frac{M_1}{\omega_0}\sin(\omega_0 t + \theta_1);$$

$$M_1 \varepsilon^{j\theta_1} = Q_1(s)\big|_{s=j\omega_0} = \frac{s}{s^2+\omega^2}\bigg|_{s=j\omega_0} = \frac{j\omega_0}{-\omega_0^2+\omega^2} = -\frac{\omega_0}{\omega_0^2-\omega^2}\varepsilon^{j90°}$$

$$\therefore M_1 = -\frac{\omega_0}{\omega_0^2 - \omega^2}, \; \theta_1 = 90°$$

$$f_1(t) = -\frac{1}{\omega_0^2 - \omega^2}\sin(\omega_0 t + 90°) = -\frac{1}{\omega_0^2 - \omega^2}\cos\omega_0 t$$

$$F(s) = \frac{\left[\frac{s}{s^2+\omega_0^2}\right]}{s^2+\omega^2} \to f_2(t) = \frac{M_2}{\omega}\sin(\omega t + \theta_2);$$

$$M_2 \varepsilon^{j\theta_2} = Q_2(s)\big|_{s=j\omega} = \frac{s}{s^2+\omega_0^2}\bigg|_{s=j\omega} = \frac{j\omega}{-\omega^2+\omega_0^2} = \frac{\omega}{\omega_0^2-\omega^2}\varepsilon^{j90°}$$

$$\therefore M_2 = \frac{\omega}{\omega_0^2 - \omega^2}, \; \theta_2 = 90°$$

$$f_2(t) = \frac{1}{\omega_0^2 - \omega^2}\sin(\omega_0 t + 90°) = \frac{1}{\omega_0^2 - \omega^2}\cos\omega t$$

Finally,

$$f(t) = \frac{1}{\omega_0^2 - \omega^2}(\cos\omega t - \cos\omega_0 t)$$

Printed in the United States
by Baker & Taylor Publisher Services